IMMS' GENERAL TEXTBOOK OF ENTOMOLOGY

OF ENTOMOLOGY

Volume I

IMMS' GENERAL TEXTBOOK OF ENTOMOLOGY

TENTH EDITION

Volume 1: Structure, Physiology and Development

O. W. RICHARDS

M.A., D.Sc., F.R.S.

Emeritus Professor of Zoology
and Applied Entomology,
Imperial College, University of London

and

R. G. DAVIES

M.Sc.

Reader in Entomology, Imperial College,
University of London

LONDON

CHAPMAN AND HALL

A Halsted Press Book
John Wiley & Sons, New York

First published 1925
by Methuen and Co., Ltd.
Second edition, revised, 1930
Third edition, revised and enlarged, 1934
Fourth edition, 1938
Fifth edition, 1942
Sixth edition, 1947
Seventh edition, 1948
Eighth edition, 1951
Ninth edition, revised by
O. W. Richards and R. G. Davies, 1957
Tenth edition published in two volumes, 1977
by Chapman and Hall Ltd.,
11 New Fetter Lane, London EC4P 4EE
Reprinted 1979

© *1977 O. W. Richards and R. G. Davies*

Filmset in 'Monophoto' Ehrhardt 11 on 12 pt.
and printed in Great Britain by
Richard Clay (The Chaucer Press), Ltd.,
Bungay, Suffolk

Volume 1 ISBN 0 412 15210 X

CONTENTS

VOLUME II

PART III. The Orders of Insects

PREFACE TO THE
TENTH EDITION

In the twenty years that have elapsed since our last complete revision of this textbook, entomology has developed greatly, both in extent and depth. There are now over 8000 publications on the subject each year (excluding the applied literature) and the difficulty of incorporating even a fraction of the more important new results has occupied us considerably. We have nevertheless retained the original plan of the book, especially as it has the merit of familiarity for many readers, but we have made a number of appreciable changes in the text as well as innumerable smaller alterations. We have decided, with some reluctance, to dispense with the keys to families that were formerly given for most of the orders of insects. These are increasingly difficult to construct because specialists tend to recognize ever larger numbers of families, often based on regional revisions and therefore applicable with difficulty, if at all, to the world fauna. Our revision of the text has also entailed extensive changes in the bibliographies, which have been brought more or less up to date. In doing this we have had to be rigorously selective and we have tended to give some emphasis to review articles or recent papers at the expense of older works. We recognize that this has sometimes done less than justice to the contributions of earlier authorities, but the immense volume of literature left little alternative and we apologize to those who feel our choice of references has sometimes been almost arbitrary.

Every chapter has been revised in detail, many of them include new sections, and some have been extensively rewritten. In a few groups such as the Plecoptera and Heteroptera the higher classification has been recast; more often we have made smaller amendments in the number and arrangement of families so as to bring the scheme into broad but conservative agreement with modern views. The general chapters now include some information on ultrastructure and we have retained and tried to modernize the physiological sections; as non-specialists in this field we owe a great debt to the textbooks of Wigglesworth and of Rockstein. Inevitably the book has grown in size with the development of the subject. It may, indeed, be argued that the day of the general textbook has passed and that it must be replaced by a series of special monographs. We believe, however, that there are some advantages in a more unified viewpoint and it is our hope that the new balance we have reached between the various aspects of entomology will seem as appropriate now as the original balance was when Dr A. D. Imms' textbook was first published over fifty years ago.

There are 35 new figures, all based on published illustrations, the sources of which are acknowledged in the captions. We are grateful to the authors concerned and also to Miss K. Priest of Messrs Chapman & Hall, who saved us from many errors and omissions, and to Mrs R. G. Davies for substantial help in preparing the bibliographies and checking references.

London   O.W.R.
May 1976   R.G.D.

Part I
ANATOMY AND PHYSIOLOGY

Chapter 1

INTRODUCTION

Definition of the Insecta (Hexapoda)

The insects are tracheate arthropods in which the body is divided into head, thorax and abdomen. A single pair of antennae (homologous with the antennules of the Crustacea) is present and the head also bears a pair of mandibles and two pairs of maxillae, the second pair fused medially to form the labium. The thorax carries three pairs of legs and usually one or two pairs of wings. The abdomen is devoid of ambulatory appendages, and the genital opening is situated near the posterior end of the body. Postembryonic development is rarely direct and a metamorphosis usually occurs.

Relationships with Other Arthropods

The arthropods (Snodgrass, 1952; Clark, 1972) include animals differing widely in structure but agreeing in certain fundamental characters, some of which probably evolved convergently. The body is segmented and invested with a chitinous exoskeleton. A variable number of the segments carry paired jointed appendages exhibiting functional modifications in different regions of the body. The heart is dorsal and is provided with paired ostia, a pericardium is present and the body-cavity is a haemocoele. The central nervous system consists of a supra-oesophageal centre or brain connected with a ganglionated ventral nerve-cord. The muscles are composed almost entirely of striated fibres and there is a general absence of ciliated epithelium. No animals other than arthropods exhibit the above combination of characters. Apart from the Insecta, the various major divisions are as follows.

The **Trilobita** (trilobites) are an extinct group of Palaeozoic marine forms with the body moulded longitudinally into three lobes. They possess a single pair of antennae followed by a variable number of pairs of biramous limbs little differentiated among themselves. Four pairs of these appendages belong to the head and the remainder to the trunk region.

The **Chelicerata** include three classes, the Merostomata (king crabs), the Pycnogonida (sea-spiders) and the numerous Arachnida (spiders, scorpions, mites, ticks, etc.). The body is usually divided into cephalothorax and abdomen; the legs consist of four pairs and there are no antennae. The primitive forms respire by means of branchiae which, in the higher forms, are replaced by lung-books or tracheae. Spiracles when present are generally abdominal and consist at most of four pairs. The gonads open near the base of the abdomen and the excretory organs are usually Malpighian tubules. The presence of chelicerae, in place of sensory antennae, and the general characters of the remaining appendages mark off the Chelicerata very definitely from all other arthropods.

The Crustacea (lobsters, shrimps, crabs, barnacles, etc.) are characterized by the possession of two pairs of antennae and at least five pairs of legs. In the higher forms the body segments are definite in number and arranged into two regions – the cephalothorax and abdomen. Respiration almost always takes place by means of gills, and the excretory organs are, at least in part, modified coelomoducts usually represented by green glands or shell glands. The genital apertures are situated anteriorly, i.e. on the 9th postoral segment in some cases, up to the 14th in others.

The Onychophora (*Peripatus* and its allies) are in some respects annectent between the annelida and arthropods, and the reasons for their inclusion in the latter are not evident from superficial examination. They are probably to be derived from primitive Annelid ancestors which had forsaken a marine habitat and become terrestrial. The appendages are lobe-like structures (lobopodia) that have become modified for locomotion on land without having acquired the jointed arthropod character. The integument is soft, though it contains chitin, and the excretory organs take the form of metamerically repeated coelomoducts. Arthropodan features are exhibited in the possession of tracheae, salivary glands, and the terminal claws to the appendages. The presence of jaws of an appendicular nature, the paired ostia to the heart, the pericardium, the haemocoelic body-cavity and the reduced coelom are further important characters allying them with other arthropod groups.

The Myriapoda comprise four classes, whose members are characterized by the presence of a single pair of antennae and the absence of any differentiation of the trunk into thorax and abdomen; each segment usually bears appendages. The Diplopoda (millipedes) have the greater number of the body segments so grouped that each apparent somite carries two pairs of legs and two pairs of spiracles. The gonads open behind the 2nd pair of legs. The Pauropoda are characterized by the legs being arranged in single pairs although the terga are mostly fused in couples. The antennae are biramous and there are only twelve postcephalic segments, nine of which bear legs. The gonads open on the 3rd segment. The Symphyla have long antennae and most of the body segments bear a single pair of legs. The gonads open on the 4th postcephalic segment and there is a single pair of spiracles which are situated on the head. The Chilopoda (centipedes) are usually provided with a single pair of appendages and a pair of spiracles to each of the postcephalic segments. The first pair of legs is modified to form poison claws and the gonads open on the penultimate segment of the abdomen.

Two further small groups of animals, the **Tardigrada** and the **Pentastomida**, are placed in or near the arthropods. The Tardigrada (bear animalcules) are minute animals with a cuticle that is moulted and with four pairs of unjointed legs but devoid of antennae, mouth-appendages or respiratory organs. The gonads open into the intestine. The parasitic Pentastomida are worm-like and devoid of appendages except two pairs of hooks near the mouth. Their arthropodan affinities are mainly suggested by the larvae, which possess two pairs of clawed, leg-like processes.

The phylogenetic relationships of the various arthropod groups have been much discussed and widely different theories of insect origins have been proposed. The very incomplete fossil record makes it impossible to substantiate these theories palaeontologically and they are based mainly on inferences from the morphology of recent forms. In an authoritative review, Tiegs and Manton (1958) gave reasons for believing that arthropod features evolved independently in more than one phyletic line. This interpretation is now supported by a large array of data from comparative and functional morphology (Manton, 1964, 1966, 1972, 1973, etc.) and from embryology (Anderson, 1973). One group which they regard as monophyletic, however, consists of the Onychophora, Myriapoda and insects (Hexapoda). These animals have in common primitively uniramous appendages, mandibles formed from the whole limb and biting with the tips, a long mid gut either without or with only a few simple diverticula, as well as distinctive embryonic features. The group – which Manton (1972) treats as an arthropod phylum Uniramia – may be contrasted with the Chelicerata, Trilobita and Crustacea, in which there are gnathobasic mandibles, primitively biramous appendages of various kinds, and a mid gut that is usually shorter and with elaborate diverticula. The ancestral stock in the Uniramian line probably comprised soft-bodied animals with many pairs of unsegmented, lobopodial appendages. The Onychophora retained these features and evolved simple sclerotized jaws on the second cephalic segment though they did not acquire a sclerotized head capsule. Insects and Myriapoda, on the other hand, seem to have arisen from ancestral forms in which the mandibles evolved on the fourth segment and which thereafter separated into two lines, each acquiring independently a sclerotized head capsule, trunk and jointed limbs. In the Myriapodan lines of evolution the mandibles are jointed structures, biting in the transverse plane, and the many pairs of limbs are associated with the trunk by coxo-sternal articulations. In the insect lineages, the mandibles are unjointed structures in which the primitive movement is a rolling action, and there has been independent evolution of the hexapod condition (with accompanying elaboration of thoracic segments), the legs being attached and moved in various ways.

The above views imply not only the parallel development of arthropod features in the Uniramian line and among the Chelicerata, Trilobita and Crustacea, but also involve the independent separation at a relatively early stage in evolution of five major groups of insects – the primitively wingless Thysanura, Diplura, Collembola and Protura as well as the winged Pterygota, all now deserving the status of separate classes. It follows also that none of these have a particularly close relationship with any of the recent Myriapodan classes, so that theories deriving the insects from a Symphylan-like stock (Imms, 1936; Tiegs, 1945) must be given up in favour of more remote connections through unknown, extinct groups. A further corollary is that the independent evolution of the hexapod condition nullifies many attempts to trace common patterns in the thoracic structures of the five insect groups listed above or in the highly modified entognathous mouthparts of the Diplura, Collembola and Protura.

The degree of convergence involved in this view of arthropod evolution is greater than can be accepted by some zoologists, e.g. Lauterbach (1973), Hennig (1969) and Siewing (1960), who find it impossible to regard the compound eye, the jointed cuticular exoskeleton, the extensive cephalization and the specialized anterior ganglia of the central nervous system as having evolved convergently. Less radical theories therefore tend to assume a mono-phyletic Arthropoda, within which the Myriapoda and insects are allied with the Crustacea to form a subphylum Mandibulata, which may be contrasted with the Chelicerata and Trilobita (Snodgrass, 1938–58). The insects moreover are still widely – if incorrectly – regarded as a natural, monophyletic group. Some of the problems involved will be discussed in connection with the apterygote groups (pp. 433 ff.) but it must be stressed that the detailed morphological and embryological evidence now available – mainly through Manton's work – renders some of the other phylogenetic and morphological discussions (including even relatively recent ones such as Sharov, 1966) subject to serious qualification and correction.

General Organization of a Winged Insect

Examination of the structure and development of the more primitive groups enables one to reconstruct a generalized winged insect. The characters of this hypothetical form show various secondary modifications in different orders of the Pterygota but its more important features are as follows.

The head is formed by the fusion of six embryonic segments of which the 2nd and 4th to 6th carry appendages in the adult. These appendages are the antennae, mandibles, maxillae and labium (2nd maxillae). The head also carries a pair of compound eyes and three dorsal ocelli.

The thorax consists of three segments each of which bears a pair of legs, while the 2nd and 3rd segments carry a pair of dorsolateral membranous outgrowths or wings. The two pairs of wings are similar, and each is supported by a system of longitudinal cuticular ribs or veins which are formed around the pre-existing tracheae of the externally developing wing-pad. There are no true cross-veins but only an irregular network (archedictyon) formed by thickenings of the wing-membrane.

The abdomen consists of eleven segments together with a terminal region or telson: the 11th segment carries a pair of segmented appendages, the cerci.

The digestive system is divisible into the fore gut or stomodaeum, a simple sac-like mid gut or mesenteron and the hind gut or proctodaeum. A pair of salivary glands lies along the sides of the fore gut and their ducts pass forward to unite and form the main salivary duct which opens behind the hypopharynx. Six Malpighian tubules are present and arise at the junction of the mid and hind gut.

The central nervous system consists of two principal cephalic centres united with a ventral ganglionated nerve-cord. The supra-oesophageal centre or brain is formed by the fusion of the three preoral cephalic ganglia. It is

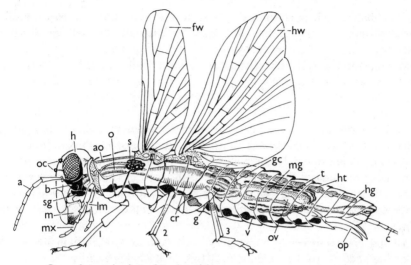

FIG. 1 General organization of a winged insect

a, antenna; *ao*, aorta; *b*, brain; *c*, cercus; *cr*, crop; *fw*, fore wing; *g*, gizzard; *gc*, gastric caeca; *h*, head; *hg*, hind gut; *ht*, heart; *hw*, hind wing; *lm*, labium; *m*, mandible; *mg*, mid gut; *mx*, maxilla; *o*, oesophagus; *oc*, ocelli; *op*, ovipositor; *ov*, ovary; *s*, salivary gland; *sg*, suboesophageal ganglion; *t*, Malpighian tubule; *v*, ventral nerve cord; *1*, *2*, *3*, fore, mid and hind legs

joined by a pair of para-oesophageal connectives with the sub-oesophageal centre. The latter is formed by the fusion of the three postoral cephalic ganglia. The ventral nerve-cord consists of three thoracic and nine abdominal ganglia united by paired longitudinal connectives. There is consequently one ganglion to each of the first twelve postcephalic segments.

The dorsal vessel consists of an abdominal portion or heart and a thoracic portion or aorta. The heart is metamerically divided into chambers, each of which is provided with paired lateral ostia. Beneath the heart is a transverse septum or pericardial diaphragm. The aorta is a narrow tubular extension arising from the first chamber of the heart and extending forwards through the thorax into the head, where it ends just behind the brain.

The respiratory system consists of segmentally repeated groups of tracheae linked by longitudinal and transverse trunks and communicating with the exterior by ten pairs of spiracles. These are situated on each of the two hinder thoracic and the first eight abdominal segments respectively.

The reproductive organs of the two sexes are very similar. In the male each testis consists of a small number of lobe-like follicles whose cavities communicate with the vas deferens. The vasa deferentia unite posteriorly and continue as a common ejaculatory duct which opens on the aedeagus. Vesiculae seminales are present as simple dilations of the vasa deferentia and paired accessory glands open into the proximal portion of the latter. In the female each ovary consists of several panoistic ovarioles comparable to the lobes of the testis. The lateral oviducts combine posteriorly to form a com-

mon oviduct which is continued posteriorly as the vagina. A median spermatheca opens on the dorsal wall of the latter and paired colleterial or
accessory glands are also present.

Metamorphosis is of the gradual or hemimetabolous type.

Number and Size of Insects

Insects comprise about 70 per cent of the known species of all kinds of
animals. Over 800 000 species of insects have been described (Sabrosky,
1952), but it is doubtful whether this number represents even one-fifth of
those existing today. The Coleoptera, with over 330 000 species, form the
largest order and among them at least 60 000 species are included in the
single family Curculionidae, while the Chrysomelidae are only slightly less
numerous.

Among living insects, the greatest size is found in individuals of the
following species. In the Coleoptera, *Megasoma elephas* attains a length up to
120 mm and *Macrodontia cervicornis* (including the mandibles) ranges up to
150 mm. Among Phasmida, *Pharnacia serratipes* may exceed 260 mm long
and the Hemipteran *Belostoma grande* attains a length of 115 mm. For the
Lepidoptera their size may, perhaps, be best gauged by the wing-span. The
latter reaches its maximum in *Erebus agrippina*, whose outspread wings
measure up to 280 mm from tip to tip and in large examples of *Attacus atlas*
they measure 240 mm. None of these compare with the extinct dragonfly
Meganeura monyi, which had a wing-span of up to 700 mm. With regard to
the smallest insects certain Coleoptera (family Ptiliidae) do not exceed a
length of 0·25 mm while egg-parasites belonging to the Hymenopteran
family Mymaridae are, in some cases, even more minute. As Folsom has
observed, some insects are smaller than the largest Protozoa and others are
larger than the smallest vertebrates.

Literature

Insects and other Arthropods

ANDERSON, D. T. (1973), *Embryology and Phylogeny in Annelids and Arthropods*,
    Pergamon Press, Oxford, 492 pp.
CLARK, K. U. (1972), *The Biology of the Arthropoda*, Arnold, London, 270 pp.
HENNIG, W. (1969), *Die Stammesgeschichte der Insekten*, Kramer, Frankfurt a. M.,
    436 pp.
IMMS, A. D. (1936), The ancestry of insects, *Trans. Soc. Br. Ent.*, 3, 1–32.
LAUTERBACH, K. E. (1973), Schlüsselereignisse in der Evolution der Stammgruppe
    der Euarthropoda, *Zool. Beitr.* (N.F.), 19, 251–299.
MANTON, S. M. (1964), Mandibular mechanisms and the evolution of arthropods,
    Phil. Trans. R. Soc., *Ser. B*, 427, 1–183.
—— (1966), The evolution of arthropodan locomotory mechanisms, Part 9:
    Functional requirements and body design in Symphyla and Pauropoda and the

relations between Myriapoda and pterygote insects, *J. Linn. Soc. (Zool.)*, **46**, 103–141.

—— (1970), Arthropods: Introduction, In: Florkin, M. and Scheer, B. T. (eds) (1970), *Chemical Zoology*, **5A**, 1–34.

—— (1972), The evolution of arthropodan locomotory mechanisms. Part 10: Locomotory habits, morphology and evolution of the hexapod classes, *J. Linn. Soc. (Zool.)*, **51**, 203–400.

—— (1973), Arthropod phylogeny – a modern synthesis, *J. Zool., Lond.*, **171**, 111–130.

SHAROV, A. G. (1966), *Basic Arthropodan Stock*, Pergamon Press, London and Oxford, 271 pp.

SIEWING, R. (1960), Zum Problem der Polyphylie der Arthropoden, *Z. wiss. Zool.*, **164**, 238–270.

SNODGRASS, R. E. (1938), Evolution of the Annelida, Onychophora and Arthropoda, *Smithson. misc. Collns*, **97** (6): 1–159, 54 figs.

—— (1950), Comparative studies on the jaws of mandibulate Arthropods, *Smithson. misc. Collns*, **116**, 85 pp.

—— (1951), *Comparative Studies on the Head of Mandibulate Arthropods*, Comstock Publishing Co., Ithaca, N.Y., 116 pp.

—— (1952), *A Textbook of Arthropod Anatomy*, Comstock Publishing Associates, Ithaca, N.Y., 363 pp.

—— (1958), Evolution of Arthropod mechanisms, *Smithson. misc. Collns*, **138**, 77 pp.

TIEGS, O. W. (1945), The postembryonic development of *Hanseniella agilis* (Symphyla), *Q. Jl microsc. Sci.*, **85**, 191–328.

TIEGS, O. W. AND MANTON, S. M. (1958), The evolution of the Arthropoda, *Biol. Rev.*, **33**, 255–337.

General Works on Insecta

BEIER, M. (ed.) (1968), Arthropoda: Insecta, In: Kükenthal's *Handbuch der Zoologie*, 2. Auflage (J.-G. Helmcke, D. Starck and H. Wermuth, eds), Bd. IV, 2. Hälfte.

BORROR, D. J. AND DELONG, D. M. (1970), *Introduction to the Study of Insects*, 3rd edn, Holt, Rinehart & Winston, New York, 591 pp.

BRUES, C. T., MELANDER, A. L. AND CARPENTER, F. M. (1954), *Classification of Insects*, 2nd edn, Harvard Univ. Press, Cambridge, Mass, 917 pp.

CHAPMAN, R. F. (1969), *The Insects: Structure and Function*, English Univ. Press, London, 818 pp.

DERKSEN, W. AND SCHEIDING-GÖLLNER, U. (1963–72), *Index Litteraturae Entomologicae*. Serie II: *Die Welt-Literatur über die gesamte Entomologie von 1864 bis 1900*.

EIDMAN, H. AND KÜHLHORN, F. (1970), *Lehrbuch der Entomologie*, Fischer, Hamburg and Berlin, 2. Auflage, 633 pp.

ESSIG, E. O. (1942), *College Entomology*, Macmillan Co., New York, 900 pp.

—— (1958), *Insects and Mites of Western North America*, Macmillan Co., New York, 2nd edn, 1056 pp.

FLORKIN, M. AND SCHEER, B. T. (eds) (1971), *Chemical Zoology (Arthropoda)*, **5**, 478 pp., **6**, 492 pp.

FROST, S. W. (1959), *Insect Life and Natural History*, Dover Publ., New York, 526 pp.

GILMOUR, D. (1960), *The Biochemistry of Insects*, Academic Press, New York and London, 343 pp.

—— (1965), *The Metabolism of Insects*, Edinburgh and London, 195 pp.

GRANDI, G. (1951), *Introduzione allo Studio dell'Entomologia*, Calderini, Bologna, 1, 950 pp.; 2, 1332 pp.

GRASSÉ, P. P. (ed.) (1949, 1951, 1974, 1975), *Traité de Zoologie: Insectes*, 8 (Fasc. I), 797 pp. (1974); 8 (Fasc. III), 910 pp. (1975); 9, 1117 pp. (1949); 10, 1948 pp. (1951), Masson et Cie., Paris.

HORN, W. AND SCHENKLING, S. (1928–29), *Index Litteraturae Entomologicae. Die Welt-Literatur über die gesamte Entomologie bis inklusive 1863*, 4 vols., 1426 pp.

KÉLER, S. VON (1963), *Entomologisches Wörterbuch mit besonderer Berücksichtigung der morphologischen Terminologie*, 4th edn, Akademie-Verlag, Berlin, 790 pp.

LAUGÉ, M. G., BERGERARD, M. J. AND LE BERRE, J. R. (1973), *Cours d'Entomologie, I: Morphologie, Anatomie, Classification. II: Physiologie des Insectes*.

MACKERRAS, I. M. (ed.) (1970), *The Insects of Australia*, Melbourne Univ. Press, Melbourne, 1029 pp. (Supplement, 1974: 146 pp.)

NEVILLE, A. C. (ed.) (1970), *Insect Ultrastructure (5th Symp., R. ent. Soc. Lond.)*, Blackwell, Oxford and Edinburgh, 196 pp.

OBENBERGER, J. (1952–64), *Entomologie*. 5 vols., Czech Acad. Sci., Prague. (In Czech.)

OLDROYD, H. (1970), *Collecting, Preserving and Studying Insects*, 2nd edn, Hutchinson, London, 336 pp.

ROCKSTEIN, M. (ed.) (1973–74), *The Physiology of Insecta*, 2nd edn, Academic Press, New York, Vols. 1–6.

ROMOSER, W. S. (1973), *The Science of Entomology*, Macmillan Co., New York and London, 544 pp.

SCHRÖDER, C. (ed.) (1925–29), *Handbuch der Entomologie*, Fischer, Jena, 3 vols.

SÉGUY, S. (1967), Dictionnaire des termes techniques d'entomologie élémentaire, *Encycl. Ent.*, 41, 1–465.

SHVANVICH, B. N. (1949), *A Course of General Entomology*, Acad. Sci. U.S.S.R., Leningrad, 900 pp. (In Russian.)

SMITH, D. S. (1968), *Insect Cells: Their Structure and Function*, Oliver & Boyd, Edinburgh, 372 pp.

SNODGRASS, R. E. (1935), *Principles of Insect Morphology*, McGraw-Hill, New York, 667 pp.

TORRE-BUENO, J. R. DE LA (1937), *A Glossary of Entomology*, Brooklyn Ent. Soc., Brooklyn, N.Y., 336 pp.

WEBER, H. (1933), *Lehrbuch der Entomologie*, Fischer, Jena, 726 pp. (Reprinted Koenigstein, 1966.)

—— (1974), *Grundriss der Insektenkunde*, Fischer, Stuttgart, 5th edn, 640 pp. (Revised by H. Weidner.)

WIGGLESWORTH, V. B. (1972), *The Principles of Insect Physiology*, Chapman & Hall, London, 7th edn, 827 pp.

Chapter 2

THE INTEGUMENT

Reviews of various general aspects of the insect integument have been given by Beament (1961, 1964), Hackman (1971, 1974), Hinton (1970), Lawrence (1967), Locke (1964, 1967), Neville (1967, 1970, 1975), Weis-Fogh (1970) and Wigglesworth (1959).

Structure, Composition and Functions

The integument consists of the following layers: (i) the cuticle, (ii) the epidermis, and (iii) the basement membrane (Fig. 2).

(1) **The Cuticle** is a complex, non-cellular layer secreted largely by the epidermis and though commonly considered non-living is actually the seat of complex biochemical changes, some at least under enzymatic control. It forms the outermost investment of the body and its appendages but is invaginated locally to form endoskeletal structures (p. 81) and also provides

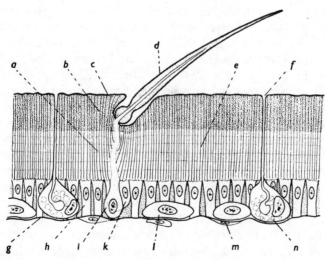

FIG. 2   Section of typical insect cuticle (*after* Wigglesworth)

a, laminated endocuticle; *b* exocuticle; *c*, epicuticle; *d*, bristle; *e*, pore-canals; *f*, duct of dermal gland; *g*, basement membrane; *h*, epidermal cell; *i*, trichogen cell; *k*, tormogen cell; *l*, oenocyte; *m*, haemocyte adherent to basement membrane; *n*, dermal gland. *From* Wigglesworth, *Principles of Insect Physiology*, 7th edn, p. 27, Fig. 19.

the lining of the tracheal system, some glands and parts of the alimentary canal and reproductive tract. When newly formed it is flexible and elastic and in many larvae it remains so over much of the body. In most insects, however, the greater part of the cuticle undergoes a process of sclerotization whereby it becomes hardened and darkened to form more or less tough, rigid sclerites separated from each other by membranous zones of unchanged soft cuticle. Such an arrangement combines rigidity with flexibility and in addition to its protective function the cuticle determines the form of the insect, its relative impermeability to water reduces desiccation and it provides a firm basis for the attachment of muscles. Two main layers may be discerned in the cuticle, an outer epicuticle and an inner procuticle, both generally compound structures. The epicuticle is a very thin layer devoid of chitin and of complex chemical composition and ultrastructure. In the terminology of Locke (1964, 1966) and Filshie (1970) there is a dense outer cuticulin layer about 10 nm thick which is very resistant to acids and organic solvents and probably contains a highly polymerized lipid. It is the first layer to be secreted in the cycle of cuticle formation, arising on top of the epidermal microvilli as small plaques which subsequently fuse to cover virtually the whole body. Beneath it lies the main epicuticular layer, the so-called protein epicuticle, also containing bound lipid. This is a homogeneous, dense, refractile layer about 1 μm thick, formed from granules that originate in the Golgi complex of the epidermal cells. It is the material of this layer to which the term 'cuticulin' was originally applied by Wigglesworth (1947). In addition the epicuticle may include a superficial lipid layer (Lockey, 1960) covered by a protective cement layer of unknown composition; the cement layer is secreted by epidermal glands and poured out over the surface. Before ecdysis the epicuticle is folded but its shape determines the overall shape of the insect (Bennet-Clark, 1963) and it bears a more or less elaborate surface pattern (Locke, 1967). Primitively this pattern comprises a network of polygons, each the product of a single epidermal cell, though such a simple arrangement is often replaced by other patterns of special functional significance (Hinton, 1970).

The procuticle, which may be absent from the tracheoles, is secreted by the epidermal cells and makes up the bulk of the integument. The mechanical properties of insect cuticle are well adapted to its skeletal function and are determined largely by the various procuticular components, mainly chitin and various proteins (Rudall and Kenchington, 1973). Together they show a high resistance to impact, high tensile, flexural and compressive strengths and a high strength–weight ratio, the latter associated with the lack of mineral salts. Typically the procuticle comprises an outer exocuticle, composed of a homogeneous electron-dense matrix, hardened through sclerotization (see below), and shed entirely when the insect moults. Within this is a softer endocuticle, capable of increasing considerably through postecdysial deposition, and showing an elaborate ultrastructure in which variously oriented protein and chitin fibres form repeated layers 0·1 to 1 μm thick.

Especially characteristic of some endocuticles is the progressive 'helicoidal' change of direction of these fibres in successive lamellae, which produces parabolic figures in sectioned material (Neville, 1970). Deposition of the endocuticle often occurs in daily growth-layers (Neville, 1963). These are visible by phase-contrast and polarization microscopy and occur in many Pterygote insects; such layers are not found in Apterygotes and are laid down with an irregular rhythm in the Coleoptera. Two further procuticular layers have been recognized: a fuchsinophil mesocuticle (Schatz, 1952) lying between endo- and exocuticle, and a basal subcuticle (Schmidt, 1956) whose granular ultrastructure suggests that the newly secreted cuticular microfibrils have not undergone orientation.

The procuticle is usually pierced by very numerous pore-canals which run perpendicularly to its surface and are initially occupied by cytoplasmic filaments from the epidermis, though they may later become filled with cuticular material. They are less than 1 μm in diameter and have a flat or twisted ribbon-like shape, through which the axial filaments run a straight course (Neville, Thomas and Zelazny, 1969; Neville and Luke, 1969a). The procuticle is probably secreted around them and they may also transport material to the outer procuticle and epicuticle. Pore-canals are absent from the rubber-like cuticle found in some insects. Sclerotization, whereby the exocuticle acquires its hard, tough, inelastic character, is due to a process of tanning, in which adjacent protein chains are cross-linked by o-quinones. The events leading up to this have been much studied in the development of the Dipteran puparium, in which the quinones apparently arise by oxidation of a phenolic substrate N-acetyldopamine, itself derived from tyrosine (Karlson and Sekeris, 1966). Darkening of the hardened regions is a separate but related process; both are involved in the repair of cuticular injuries.

The two major components of insect cuticle are the carbohydrate chitin, which accounts for 25–60 per cent of the dry weight of various cuticles, and a number of proteins. On hydrolysis, chitin yields acetic acid and glucosamine and is a high molecular weight polymer consisting mainly of anhydro-N-acetylglucosamine residues joined by β-1,4-linkages, though up to 10 per cent of the residues may be deacylated. The chains are unbranched and consist of several hundred units with a repeating structure $3 \cdot 1$ nm long. There are three different crystallographic forms of chitin – α-, β- and γ-chitin – which differ in the arrangement of the chains and in the presence of bound water; α-chitin is the most stable and the only one found in arthropod cuticle. The metabolic pathways through which chitin is formed are not fully established but enzymatic synthesis of UDP-N-acetylglucosamine from glucose has been demonstrated by Candy and Kilby (1962) and there is evidence for polymerization under the control of a chitin synthetase. In the cuticle, chitin chains are apparently joined to proteins by covalent linkages involving aspartic acid and histidine (Rudall, 1963). The chitin–protein complex is in fact a polydisperse glyco-protein in which rod-like chitin fibres $2 \cdot 5$ to $6 \cdot 5$ nm in diameter are embedded in a protein matrix (Neville and Luke, 1969b).

Chitin is insoluble in water, alkalis, dilute acids and organic solvents but dissolves with decomposition in concentrated mineral acids and sodium hypochlorite. It has a specific gravity of about 1·4, a refractive index of about 1·55 and is best detected by the van Wisselingh test – treatment with concentrated potassium hydroxide at 160° C for 20 minutes converts it to chitosan which gives a rose-violet colour with 0·2 per cent iodine in 1 per cent sulphuric acid (Campbell, 1929). A more specific histological test depends on the use of conjugates of a chitinase with fluorescent dyestuffs (Benjaminson, 1969). The main structural proteins of the insect cuticle are sometimes known collectively as arthropodin though they are a complex mixture, more than a dozen components having been identified in some insects by amino-acid analyses, electrophoresis and serological techniques (Fox and Mills, 1969; Hackman and Goldberg, 1971). The proteins differ in amino-acid composition from one species

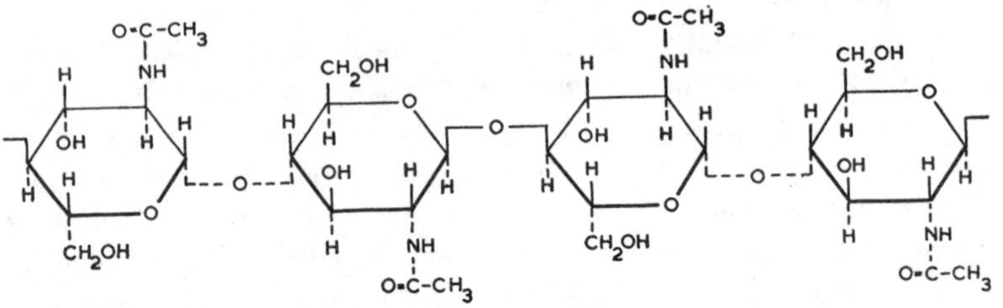

FIG. 3. Structural formula of chitin

to another but those of related species show some resemblances. Protein deposition in the cuticle is related to changes in the amino acids of the blood though the mechanisms of synthesis are not known. Of special interest is the distinctive protein resilin (Andersen and Weis-Fogh, 1964; Andersen, 1970). This resembles the elastin of vertebrate connective tissue, being an isotropic three-dimensional network of polypeptide chains held together by stable covalent cross-links and showing remarkable rubber-like properties. It is structureless and metabolically inert but forms mechanical 'springs' which are readily deformed and show perfect elastic recovery. It occurs in localized zones where these properties are functionally important, for example near the wing articulation of many insects and in the metathorax of fleas, where it forms part of the jumping mechanism (p. 945).

Other components of the cuticle account for only a very small part of its weight but are of great physiological significance. Phenolic precursors of the quinones that link amino-acid chains to form sclerotin are widely distributed, though some Apterygotes lack them and harden their cuticles by disulphide linkages (Krishnan, 1969). There are also various waxes and other lipids secreted by the oenocytes (Locke, 1969a) and responsible for waterproofing the cuticle (Beament, 1964; Gilby, 1965; Gilbert, 1967). These form a layer at or near the surface or are incorporated into inner layers of the cuticle. They are a complex mixture of up to 80 per cent hydrocarbons with smaller amounts of fatty acids, alkyl esters and other constituents. Chemically the waxes found in exuviae of Bombyx mori comprise a mixture of odd-numbered C_{27}–C_{37} paraffins with esters of even-numbered C_{26}–C_{30} fatty acids. Though the epicuticular lipids play a major role in reducing water loss through the integument, it now seems unlikely that this depends on an oriented layer of wax molecules subject to disruption at a critical transition temperature (Hackman, 1971). The integument also includes enzyme systems responsible for the synthesis and degradation of the major cuticular materials and those involved in melanization and sclerotization; many of these systems require fuller investigation. Inorganic cuticular constituents are rare but calcareous nodules may develop on the outside of Stratiomyid and Psychodid larvae while *Rhagoletis cerasi* larvae show intra-cuticular calcification (Wiesmann, 1938). Pigments occurring in the cuticle are discussed below.

(2) **The Epidermis** forms a continuous single layer of cells, the plasma membranes of which are joined by numerous septate desmosomes. The ultrastructure of the cells varies with their cycles of secretory activity (Locke, 1969*b*) but they have a microvillate surface and contain numerous mitochondria, Golgi vesicles and cisternae of smooth-surfaced endoplasmic reticulum as well as cytoskeletal structures in the form of oriented microfibres and microtubules. Rough endoplasmic reticulum is present mainly in cells secreting resilin. Scattered among the normal epidermal cells are specialized gland cells and those concerned in the formation of cuticular sensilla (see p. 124). Muscle attachments penetrate the epidermis, the myofibrillae usually being associated with tonofibrillae that run through the epidermis and into the procuticle, while the oenocytes (p. 256) which originate from epidermal cells sometimes remain closely associated with this layer. Not only does the epidermis secrete the greater part of the cuticle but it also produces the moulting fluid (Bade and Wyatt, 1962; Jeuniaux, 1963), which dissolves the old endocuticle before the immature insect moults (p. 361), it absorbs the digestion products of the old cuticle, repairs wounds and differentiates in such a way as to determine the surface patterns of the insect (Wigglesworth, 1959; Lawrence, 1967).

(3) **The Basement Membrane** is a continuous layer about 0·5 μm thick with an amorphous ultrastructure. It contains neutral mucopolysaccharides and is secreted by haemocytes (Wigglesworth, 1956). Chordotonal organs (p. 127), tracheoles, and nerves run to it or through it.

Cuticular Appendages

These structures include all outgrowths of the cuticle that are connected with it by means of a membranous joint. They arise from modified epidermal cells and may be classified into setae and spurs.

Setae or **Macrotrichia** (Fig. 4) are commonly known as hairs and each arises from a cup-like pit or *alveolus*. At its base the seta is attached by a ring of articular membrane. Setae are hollow structures developed as extensions of the exocuticle and each is produced by a single, usually enlarged, *trichogen cell*. The articular membrane is usually produced by a separate *tormogen cell*. The arrangement of the more constantly located setae (chaetotaxy) is important in the systematics of some insect groups, e.g. the Diplura, Thysanoptera, Cyclorrhaphan Diptera and larval Lepidoptera. The main kinds of setae and their modifications are listed below; the structure of some is now known in great detail through the use of specialized optical techniques and of the scanning and transmission electron microscopes (e.g. Baker and McCrae, 1966–67; Evans, 1967; Hale and Smith, 1966; Lippert and Gentil,

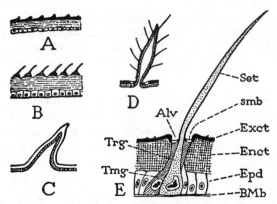

FIG. 4   External processes of the body-wall, diagrammatic. A, B, non-cellular cuticular processes. C, D, multi-cellular processes. E, a typical unicellular process or seta

Alv, şetal socket or alveolus; *Set*, seta; *smb*, setal membrane; *Tmg*, tormogen or socket-forming cell; *Trg*, trichogen or seta-forming cell. Snodgrass, *Principles of Insect Morphology*, McGraw-Hill, 1935, Fig. 28.

1959; Lukoschus, 1962; Picken, 1949). (1) *Clothing hairs* – These invest the general surface of the body or its appendages and frequently exhibit various degrees of specialization. When furnished with thread-like branches as in the Apoidea they are termed *plumose hairs*. Setae which are particularly stout and rigid are known as *bristles* and are well exhibited for example in the Tachinidae. (2) *Scales* – These are highly modified clothing hairs and are characteristic of all Lepidoptera and many Collembola: they are also present in some Thysanura, Coleoptera, Diptera and Hymenoptera. Transitional forms between ordinary clothing hairs and scales are frequent. (3) *Glandular setae* – Grouped under this heading are those setae which function as the outlet for the secretion of epidermal glands (see p. 263). If they are especially stout and rigid they are then termed *glandular bristles* as in the urticating hairs of certain lepidopterous larvae. (4) *Sensory Setae* – Very frequently the setae of certain parts of the body, particularly the appendages, are modified in special ways and become sensory in function. Sensory setae (see p. 125) are in all cases connected with the nervous system.

Spurs occur on the legs of many insects and differ from setae in being of multicellular origin.

Cuticular Processes

The external surface of the cuticle, in addition to being sculptured in various ways, bears a great variety of outgrowths which are integral parts of its substance. They are rigidly connected with the cuticle, having no membranous articulation and are therefore readily separable from cuticular appendages. The principal types of cuticular processes are as follows.

Microtrichia (fixed hairs or aculei) – These are minute hair-like structures found, for example, on the wings of the Mecoptera and certain

Diptera. They resemble very small covering hairs, but the absence of the basal articulation is their distinguishing feature (Figs. 4 and 26).

Spines – This expression has been used by various writers with considerable latitude but is here confined to outgrowths of the cuticle which are more or less thorn-like in form. Spines differ from spine-like setae in being produced by undifferentiated epidermal cells and are usually, if not always, of multicellular origin.

In addition to the above there are many other cuticular processes which either take the form of more or less conical *nodules* and *tubercles* of different shapes, or of larger projections known as *horns* which are characteristic of the males of certain Coleoptera.

Coloration

The colours of adult and immature insects may be grouped into three classes: (1) pigmentary or chemical colours, (2) structural or physical colours, and (3) combination or chemico-physical colours.

1. Pigmentary Colours – These are due to chemical substances that absorb some wavelengths of the incident light and reflect others. They may be present in the cuticle, epidermis or subepidermal tissues (usually fat-body or blood). A colour pattern often consists of an epidermal or subepidermal ground colour (which may fade rapidly after death) and overlying areas of more permanent cuticular pigmentation. Eye pigments occurring in the ommatidial cells of some insects may be of special interest since studies of their metabolism reveal something of the mode of action of genes controlling eye colour.

Biochemical aspects of insect pigmentation have been reviewed recently by Cromartie (1959), Thomson (1962), Kilby (1963), Goodwin (1971), Fuzeau-Braesch (1972) and others. The substances concerned may be classified as follows.

(a) *Melanins*. These are amorphous, highly stable, dark brown or black cuticular pigments which are generally non-granular and are insoluble in the usual solvents though they are rapidly decolorized by oxidizing agents. Their chemical nature has not been satisfactorily elucidated and it is, in fact, difficult to distinguish between cuticular darkening due to sclerotization and that caused by the presence of true melanins (Richards, A. G., 1967). The mode of formation of melanins is uncertain (Lerner and Fitzpatrick, 1950) but they probably arise through polymerization of indole compounds which are derived initially from the amino-acid tyrosine by ring-closure and oxidation under the influence of the enzyme tyrosinase. Thus, *Aedes* larvae reared on a diet deficient in tyrosine or the related phenylalanine together with the resulting pupae are unpigmented (Golberg and De Meillon, 1948). Where melanic pigmentation is discontinuous (e.g. dark spots, etc.) this is apparently the result of a localized distribution of the substrate tyrosine, the tyrosinase occurring also in areas which remain unpigmented.

(b) *Carotenoids*. These are polyene pigments usually containing 40 carbon atoms in the molecule; they are readily soluble in fat-solvents and are characteristically synthesized by plants (Lederer, 1938; Feltwell and Rothschild, 1974). When ingested by animals they accumulate in the blood and tissues unchanged or after minor oxidative alterations and in some cases they form the prosthetic group of a chromoprotein. β-carotene, derived from the tissues of the potato plant, occurs in the blood of the Colorado Potato-beetle *Leptinotarsa decemlineata* and is responsible for the red and yellow coloration of the Pentatomid *Perillus bioculatus* which preys on larvae of the beetle (Palmer and Knight, 1924). Again, in *Coccinella*, the red colour is due to the presence of the plant carotenoids lycopene and α- and β-carotene (Lederer, 1938). Astaxanthin (3:3-dihydroxy-4:4-diketo-β-carotene) and β-carotene both occur as chromoproteins in the integument of locusts and the green chromoprotein pigment of many insects (known as insectoverdin) is a complex, the yellow-orange component of which may have as its prosthetic group β-carotene (*Carausius*), lutein (*Sphinx*, *Tettigonia*) or astaxanthin.

(c) *Pteridines.* These pigments are derived from a heterocyclic pyrimidine-pyrazine ring structure and are found more often in insects than in other organisms (Ziegler and Harmsen, 1969). The majority belong to the class of pterines (2-amino-4-hydroxy-pteridines) and the natural pigment is often a mixture of two or more of these. Thus the conspicuous colours of the Heteroptera *Dysdercus*, *Pyrrhocoris*, *Oncopeltus* and *Phonoctonus* are due to the red and yellow pterines erythropterin and xanthopterin, along with the white leucopterin. Xanthopterin and leucopterin are also found in the integument of the Vespidae, while the body- and wing-scales of many Pierid butterflies contain leucopterin, xanthopterin, isoxanthopterin and erythropterin. Some pterines occur characteristically as eye pigments, e.g. a group of so-called drosopterins contribute to the red eye colour of the wild-type *Drosophila melanogaster*. Integumentary pterines lie in the epidermis and interlamellar spaces of scales and hairs as ellipsoidal granules about 0·5 μm long. In the ommatidia they occur as proteinaceous granules in the pigment cells surrounding the crystalline cone and retinulae; they are distinct from and sometimes masked by the ommochrome pigments that may also be present. In addition to their role as pigments some hydrogenated pterines can act metabolically as co-factors in hydroxylation reactions; in the Pieridae they are not only pigments but also represent the insoluble end-products that accumulate in 'storage excretion' of nitrogenous metabolites.

(d) *Ommochromes* (Linzen, 1974). These are best known as eye pigments and comprise two groups of substances, the sulphur-free ommatins of low molecular weight and the sulphur-containing ommins of high molecular weight. The latter are the main pigments occurring in the eye of insects from many orders. Xanthommatin is an eye pigment in *Drosophila*; its biosynthesis from tryptophane via kynurenine and 3-hydroxykynurenine is under genetic control. Like other ommochromes it is a redox pigment, yellow when oxidized and red when reduced. Integumentary ommochromes occur in the wings of Nymphalid butterflies and in *Schistocerca* and they are also involved in the colour changes of the stick insect *Carausius morosus*. In the dragonfly *Sympetrum striolatum* the scarlet male contains ommatins and the yellow female their reduced equivalents.

(e) *Anthraquinones.* These are confined to the Coccoidea, where the best known is carminic acid, the colouring principle of cochineal, found in the eggs and fat body of the female *Dactylopius coccus* and accounting for half its body weight. Kermesic acid is the pigment in the dyestuff kermes, from females of *Kermococcus ilicis*, and laccaic acid is the water-soluble red pigment of lac (p. 266).

(f) *Aphins.* These are polycyclic quinones which decompose immediately after the death of the insect containing them, giving rise to the erythroaphin reported from *Aphis fabae*, *Eriosoma lanigerum* and other aphids.

(g) *Miscellaneous pigments*. Many other substances play a minor role as insect pigments, occurring either in small amounts or only in a few species. Among such are haemoglobin, derivatives of chlorophyll, anthoxanthins, anthocyanins, riboflavin (which accumulates as the greenish-yellow 'entomo-urochrome' of the Malpighian tubules) and purines. The bile pigment mesobiliverdin is perhaps more important since it is the prosthetic group of the blue component of the insectoverdins (complex green pigments) mentioned above in connection with the carotenoids.

2. Structural Colours – Structural colours differ from those due to pigments in that they are changed or destroyed by physical changes in the cuticle such as result from shrinkage, swelling, distortion or permeation with liquids of the same refractive index as the cuticle. They may also be duplicated by physical models, are not destroyed by bleaching and all the component wavelengths of the incident light are to be found in either the reflected, scattered or transmitted fractions. Of the many papers on this topic, see especially Mason (1926–27) and Anderson and Richards (1942). Four types of structural coloration may be distinguished: (i) *Structural white* is caused by the scattering, reflection and refraction of light by microscopic particles large in comparison with the wavelength of light and which, in themselves, are usually transparent. Probably most insect whites have a structural basis. (ii) *Tyndall blue*, though uncommon, occurs in some Odonata and is due to the scattering of the shorter wavelengths by particles with dimensions of about the same size as the wavelengths of light. (iii) *Interference colours* are produced by optical interference between reflections from a series of superimposed laminae or ribs. This is one of the commonest types of physical coloration and the iridescent appearance is well seen, for example, in the wing-scales of *Morpho* butterflies, in the Diamond beetles *Entimus* and *Cyphus* and in the Chrysididae. (iv) *Diffraction colours*, resulting from the presence of closely spaced striae, 0·5 to 3 μm apart, occur in the Mutillidae and various beetles such as *Serica*, several Carabidae and Gyrinidae, *Nicrophorus* and others (Hinton and Gibbs, 1969, 1971). It is possible that variations in the colour and intensity of light reflected from these structures help to confuse predators as to the size and distance of the insect.

The beetle *Dynastes hercules* can change colour quite quickly from yellow to black and vice versa. This process has a structural basis elucidated by Hinton and Jarman (1973). Below the transparent epicuticle of the elytra is a very thin spongy layer; when its interstices are filled with air this appears yellow, but when they fill with water the black colour of the underlying cuticle becomes apparent.

3. Combination Colours – These are produced by a structural modification in conjunction with a layer of pigment and are much commoner than purely structural colours. In the butterfly *Teracolus phlegyas* a red pigment in the scale wall (but not in the striae) combines with a structural violet to produce magenta: in *Ornithoptera poseidon* the emerald green is due to a structural blue combined with a yellow pigment in the walls and striae of the scales. In a number of cases (e.g. Lycaenids) there is no indication of the cause of colour. The golden iridescence of *Cassida* and its allies is produced by a film of moisture beneath the surface cuticle. These insects rapidly lose their colour when dried, but it returns after soaking in water provided the drying has not been too prolonged.

Literature on the Body-wall and Coloration

ANDERSEN, S. O. (1970), Resilin, In: Florkin, M. and Stotz, E. H. (eds), *Comprehensive Biochemistry*, **26C**.

ANDERSEN, S. O. AND WEIS-FOGH, T. (1964), Resilin. A rubberlike protein in Arthropod cuticle, *Adv. Insect Physiol.*, **2**, 1–65.

ANDERSON, T. F. AND RICHARDS, A. G. (1942), An electron-microscope study of some structural colours in insects, *J. appl. Phys.*, **13**, 748–758.

BADE, M. L. AND WYATT, G. R. (1962), Metabolic conversions during pupation of the cecropia silkworm. 1. Deposition and utilization of nutrient reserves, *Biochem. J.*, **83**, 470–478.

BAKER, J. R. AND MCCRAE, J. M. (1966), The ultrastructure of the 'Podura' scale (*Lepidocyrtus curvicollis*, Collembola), as revealed by whole mounts, *J. Ultrastruct. Res.*, **15**, 516–521.

—— (1967), A further study of the *Lepidocyrtus* ('Podura') scale (Insecta: Collembola), *J. Ultrastruct. Res.*, **19**, 611–615.

BEAMENT, J. W. L. (1961), The water relations of insect cuticle, *Biol. Rev.*, **36**, 281–320.

—— (1964), The active transport and passive movement of water in insects, *Adv. Insect Physiol.*, **2**, 67–129.

BENJAMINSON, M. A. (1969), Conjugates of chitinase with fluorescein isothiocyanate or lissamine rhodamine as specific stains for chitin *in situ*, *Stain Technol.*, **44**, 27–31.

BENNET-CLARK, H. C. (1963), The relation between epicuticular folding and the subsequent size of an insect, *J. Insect Physiol.*, **9**, 43–46.

CAMPBELL, F. L. (1929), The detection and estimation of insect chitin; and the interrelation of chitinization to hardness and pigmentation of the cuticle of the American cockroach, *Periplaneta americana* L., *Ann. ent. Soc. Am.*, **22**, 401–426.

CANDY, D. J. AND KILBY, B. A. (1962), Studies on chitin synthesis in the desert locust, *J. exp. Biol.*, **39**, 129–140.

CROMARTIE, R. I. (1959), Insect pigments, *A. Rev. Ent.*, **4**, 59–76.

EVANS, J. J. T. (1967), The integument of the Queensland fruit fly, *Dacus tryoni* (Frogg.), 1. The tergal glands, *Z. Zellforsch. mikrosk. Anat.*, **81**, 49–61. 2. Development and ultrastructure of the abdominal integument and bristles, *Ibid.*, **81**, 34–48.

FELTWELL, J. AND ROTHSCHILD, M. (1974), Carotenoids in thirty-eight species of Lepidoptera. *J. Zool.*, **174**, 441–465, 1 pl.

FILSHIE, B. K. (1970), The fine structure and deposition of the larval cuticle of the sheep blowfly (*Lucilia cuprina*), *Tissue and Cell*, **2**, 479–498.

FOX, F. R. AND MILLS, R. R. (1969), Changes in haemolymph and cuticle proteins during the moulting process in the American cockroach, *Comp. Biochem. Physiol.*, **29**, 1187–1195.

FUZEAU-BRAESCH, S. (1972), Pigments and colour changes, *A. Rev. Ent.*, **17**, 403–424.

GILBERT, L. I. (1967), Lipid metabolism and function in insects, *Adv. Insect Physiol.*, **4**, 70–211.

GILBY, A. R. (1965), Lipids and their metabolism in insects, *A. Rev. Ent.*, **10**, 141–160.

GOLBERG, L. AND DE MEILLON, B. (1948), The nutrition of the larva of *Aedes aegypti* Linnaeus, 4. Protein and amino-acid requirements, *Biochem. J.*, **43**, 379–387.

GOODWIN, T. W. (1971), Pigments–Arthropoda, In: Florkin, M. and Scheer, B. T. (eds), *Chemical Zoology*, **6B**, 279–306.

HACKMAN, R. H. (1971), The integument of Arthropoda, In: Florkin, M. and Scheer, B. T. (eds), *Chemical Zoology*, **6B**, 1–62.

—— (1974), Chemistry of the insect cuticle, In: Rockstein, M. (ed.) *The Physiology of Insecta*, Academic Press, New York, 2nd edn, 6, 215–270.

HACKMAN, R. H. AND GOLDBERG, M. (1971), Studies on the hardening and darkening of insect cuticles, *J. Insect Physiol.*, **17**, 335–347.

HALE, W. G. AND SMITH, A. L. (1966), Scanning electron microscope studies of cuticular structures in the genus *Onychiurus* (Collembola), *Revue Ecol. Biol. Sol.*, **3**, 343–354.

HINTON, H. E. (1970), Some little known surface structures, In: Neville, A. C. (ed.) Insect Ultrastructure, *Symp. R. ent. Soc. Lond.*, **5**, 41–58.

HINTON, H. E. AND GIBBS, D. F. (1969), An electron microscope study of the diffraction gratings of some Carabid beetles, *J. Insect Physiol.*, **15**, 959–962.

—— (1971), Diffraction gratings in Gyrinid beetles, *J. Insect Physiol.*, **17**, 1023–1035.

HINTON, H. E. AND JARMAN, G. M. (1973), Physiological colour change in the elytra of the Hercules beetle, *Dynastes hercules, J. Insect Physiol.*, **19**, 533–549.

JEUNIAUX, C. (1963), *Chitine et Chitinolyse, un chapitre de la biologie moleculaire*, Masson et Cie, Paris, 181 pp.

KARLSON, P. AND SEKERIS, C. E. (1966), Ecdysone, an insect steroid hormone, and its mode of action, *Recent Progr. Hormone Res.*, **22**, 473–502.

KILBY, B. A. (1963), The biochemistry of the insect fat body, *Adv. Insect Physiol.*, **1**, 111–174.

KRISHNAN, G. (1969), Chemical components and mode of hardening of the cuticle of Collembola,*Acta histochemica*, **34**, 212–228.

LAWRENCE, P. A. (1967), The insect epidermal cell – a simple model of the embryo. In: Beament, J. W. L. and Treherne, J. E. (eds), *Insects and Physiology*, 53–68.

LEDERER, E. (1938), Recherches sur les caroténoïdes des invertebrés. *Bull. Soc. Chim. biol.*, **20**, 567–610.

LERNER, A. B. AND FITZPATRICK, T. B. (1950), Biochemistry of melanin formation, *Physiol. Rev.*, **30**, 91–126.

LINZEN, B. (1974), The tryptophan → ommochrome pathway in insects. *Adv. Insect Physiol.*, **10**, 117–246.

LIPPERT, W. AND GENTIL, K. (1959), Über lamellare Feinstrukturen bei den Schillerschuppen der Schmetterlinge vom *Urania*- und *Morpho*-Typ, *Z. Morph. Ökol. Tiere*, **48**, 115–122.

LOCKE, M. (1964), The structure and formation of the insect cuticle, In: *The Physiology of Insecta*, III, 379–470.

—— (1966), The structure and formation of the cuticulin layer in the epicuticle of an insect, *Calpodes ethlius* (Lepidoptera, Hesperiidae), *J. Morph.*, 118, 461–494.

—— (1967), The development of patterns in the integument of insects, *Adv. Morphogen.*, 6, 33–88.

—— (1969a), The ultrastructure of the oenocytes in the moult/intermoult cycle of an insect, *Tissue and Cell*, 1, 103–154.

—— (1969b), The structure of an epidermal cell during the development of the protein epicuticle and the uptake of molting fluid in an insect, *J. Morph.*, 127, 7–39.

LOCKEY, K. H. (1960), The thickness of some insect epicuticular wax layers, *J. exp. Biol.*, 37, 316–329.

LUKOSCHUS, F. (1962), Ausbildungsformen von Kutikula und kutikularen Kleinorganen bei der Honigbiene (*Apis mellifica* L.), *Z. Morph. Ökol. Tiere*, 51, 547–574.

MASON, C. W. (1926–27), Structural colours in insects: 1–3, *J. phys. Chem.*, 30, 383–395; 31, 321–354; 1856–1872.

NEVILLE, A. C. (1963), Daily growth layers in locust rubber-like cuticle influenced by an external rhythm, *J. Insect Physiol.*, 9, 177–186.

—— (1967), Chitin orientation in cuticle and its control, *Adv. Insect Physiol.*, 4, 213–286.

—— (1970), Cuticle ultrastructure in relation to the whole insect, In: Neville, A. C. (ed.) Insect Ultrastructure, *Symp. R. ent. Soc. Lond.*, 5, 17–39.

—— (1975), *Biology of the Arthropod Cuticle*, Springer, Berlin, 448 pp.

NEVILLE, A. C. AND LUKE, B. M. (1969a), Molecular architecture of adult locust cuticle at the electron microscope level, *Tissue and Cell*, 1, 355–366.

—— (1969b), A two-system model for chitin-protein complexes in insect cuticles, *Tissue and Cell*, 1, 689–707.

NEVILLE, A. C., THOMAS, M. G. AND ZELAZNY, B. (1969), Pore canal shape related to molecular architecutre of arthropod cuticle, *Tissue and Cell*, 1, 183–200.

PALMER, L. S. AND KNIGHT, H. H. (1924), Carotin – the principal cause of the red and yellow colours in *Perillus bioculatus* (Fab.), and its biological origin from the haemolymph of *Leptinotarsa decemlineata* Say., *J. biol. Chem.*, 59, 443–449.

PICKEN, L. E. R. (1949), Shape and molecular orientation in Lepidopteran scales, *Phil. Trans. R. Soc.*, 234 B: 1–208.

RICHARDS, A. G. (1967), Sclerotization and the localization of black and brown colors in insects, *Zool. Jb. (Anat.)*, 84, 25–62.

RUDALL, K. M. (1963), The chitin/protein complexes of insect cuticles, *Adv. Insect Physiol.*, 1, 257–313.

RUDALL, K. M. AND KENCHINGTON, W. (1973), The chitin system, *Biol. Rev.*, 49, 597–636.

SCHATZ, L. (1952), The development and differentiation of arthropod procuticle: staining, *Ann. ent. Soc. Am.*, 45, 678–686.

SCHMIDT, E. L. (1956), Observations on the subcuticular layer in the insect integument, *J. Morph.*, 99, 211–231.

THOMSON, R. H. (1962), Insect pigments, *Verhandl. XI. int. Kongr. Ent.*, 3, 21–43.

WEIS-FOGH, T. (1970), Structure and formation of insect cuticle, In: Neville, A. C. (ed.), Insect Ultrastructure, *Symp. R. ent. Soc. Lond.*, 5, 165–185.

WIESMANN, R. (1938), Untersuchungen über die Struktur der Kutikula des Puppentönnchens der Kirschfliege, *Rhagoletis cerasi* L., *Vjschr. naturf. Ges. Zürich*, **83**, 127–136.

WIGGLESWORTH, V. B. (1947), The epicuticle in an insect, *Rhodnius prolixus*, *Proc. R. Soc. Ser. B*, **134**, 163–181.

—— (1956), The haemocytes and connective tissue formation in an insect, *Rhodnius prolixus* (Hemiptera), *Q. Jl microsc. Sci.*, **97**, 89–98.

—— (1959), *The Control of Growth and Form: a Study of the Epidermal Cell in an Insect*, Cornell Univ. Press, Ithaca, N.Y., 136 pp.

ZIEGLER, I. AND HARMSEN, R. (1969), The biology of pteridines in insects, *Adv. Insect Physiol.*, **6**, 139–203.

Chapter 3

SEGMENTATION AND THE DIVISIONS OF THE BODY

The cuticle of an insect forms a more or less hardened exoskeleton and, although perfectly continuous over the whole body, it remains flexible along certain definite, usually transverse, lines. Here the cuticle becomes infolded and is membranous in character. The body of an insect therefore presents a jointed structure which is an example of *segmentation*, and is divided into a series of successive rings variously known as *segments*, *somites* or *metameres*. In many cases the definitive segment incorporates part of what was primitively an intersegmental sclerite. The flexible infolded portion of the cuticle between adjacent definitive segments is the so-called *intersegmental membrane* whose function is to allow free movement of the body.

Segmentation is not only shown in the external differentiation of the body but also involves many of the internal organs. In the Annelida and the Onychophora the internal structure of each segment is very similar to that of the segment preceding or following it. In such highly evolved animals as insects the primitive segmentation, in so far as it affects the internal anatomy, has undergone profound modifications; the segmental repetition of parts is nevertheless retained to some extent in the central nervous system, the heart, tracheal system and in the body musculature.

The cuticle also exhibits localized areas of hardening which are sometimes delimited by sutures. The latter name has been given to several somewhat different structures. It may denote (i) the external groove or sulcus corresponding to an internal ridge-like inflection of the cuticle which provides mechanical rigidity, or (ii) a line of thinner, weaker cuticle along which rupture or bending of the integument can occur at ecdysis, or (iii) a narrow, flexible, membranous zone of unsclerotized cuticle, or (iv) a linear impression without any obvious mechanical significance. In certain regions the sclerites do not come into apposition by sutures and are thus, as it were, islands of cuticle surrounded by membrane. Complete fusion of adjacent sclerites is common, particularly among the higher orders of insects, all traces of sutures being lost

The Divisions of a Body Segment

In most adult insects, and in many of their larvae, the body-wall of a typical segment is divisible into four definite sclerotized regions: a dorsal region or *tergum*, a ventral region or *sternum*, and a lateral region or *pleuron* on each side of the body. Each of these regions may be differentiated into separate sclerites. In this case the sclerites composing the tergum are known as *tergites*, those of the sternum as *sternites*, and those constituting each pleuron as *pleurites*. Between adjacent segments there may be present small detached plates or *intersegmentalia* and such sclerites belong partly to the segment in front and partly to the segment behind them. According to their position they are termed *intertergites*, *interpleurites* and *intersternites*.

The Segmental Appendages

In the embryo, each body segment may bear a pair of outgrowths or appendages which may or may not be retained in postembryonic life. Among adult insects, an appendage is normally attached to its segment between the pleuron of its side and the sternum. Typical appendages are segmented tubes invested with a dense cuticle. Between adjacent segments, the cuticle remains membranous and becomes infolded to form the articular membrane. On account of its jointed structure, the whole or part of an appendage is movable by means of its muscles. An insect appendage consists typically of a limb base and a shaft. There is no evidence of a biramous condition among the appendages in any insects.

Processes of the Body-Wall

In addition to true segmental appendages numerous other outgrowths of the body-wall are found in various insects. Unlike true appendages, processes of the body-wall are by no means invariably represented by embryonic counterparts; they may or may not be segmentally arranged, they may be originally paired or unpaired, and more than a single pair is sometimes borne on a segment. They differ from cuticular processes in containing a definite extension of the body cavity and in some cases they are freely movable. It is sometimes difficult to distinguish between such processes and true appendages but the principal types of organs which have been included under this category are: (1) *Pseudopods*, which are characteristic of many dipterous larvae. (2) *Scoli*, or thorny processes, characteristic of Nymphalid and Saturniid larvae: the anal horn of Sphingid larvae is also of a very similar nature. (3) *Branchiae* or gills which are found in most aquatic insect larvae (see p. 226). (4) *Wings* (see p. 50), which are always confined to the meso- and metathorax and attain their full development in adult insects.

The Regions of the Body

The body segments of an insect are grouped together to form three usually well-defined regions or tagmata – the *head, thorax* and *abdomen* (Fig. 1). In each tagma certain of the primary functions of the organism are concentrated. The head carries the mouthparts, which are concerned with feeding, and the organs of special sense. The thorax bears the locomotor organs, i.e. legs and wings. The abdomen is concerned with reproduction and may carry genital appendages; it is also the seat of many metabolic processes.

In most orders an intersegmental region, the neck or *cervix*, connects the head with the thorax and possibly also includes the posterior unsclerotized part of the labial segment and the anterior part of the prothorax.

References on Segmentation and the Divisions of the Body

SNODGRASS, R. E. (1935), *Principles of Insect Morphology*, McGraw-Hill, New York, 667 pp., 319 figs.
—— (1958), Evolution of arthropod mechanisms. *Smithson. misc. Collns*, **138** (2): 1–77, 24 figs.
—— (1963), A contribution toward an encyclopedia of insect anatomy. *Smithson. misc. Collns*, **146** (2): 1–48.

Chapter 4

THE HEAD AND CERVIX

The insect head presents many morphological problems but it is convenient here to give a descriptive account of the head-capsule and cephalic appendages before discussing the more important theories of the segmental composition of this region.

The Head-Capsule

The exoskeleton of the head is composed of several sclerites more or less intimately associated to form a hard, compact case or *head-capsule*, general works on which include those of Duporte (1957), Gouin (1968), Manton (1964), Matsuda (1965), Snodgrass (1947, 1960) and Strenger (1952). In the more generalized insects the wall of the head-capsule is reinforced by a number of inflected ridges or *sulci* which occupy relatively constant positions and delimit to some extent the morphological areas into which the wall of the head can be divided (Figs. 5, 6). In addition, the dorsal and facial surface of the head of immature insects is crossed by a so-called *ecdysial cleavage line*, often having the form of an inverted Y (Fig. 5). Here the sclerotized exocuticle is absent and the head-capsule splits open along the cleavage line when the insect moults. The course of the split varies from one species to another; in some its lateral arms run dorsal to the antennal insertions, in others it runs to them or below them. In some insects the ecdysial cleavage line lies concealed along the infolded portion of a sulcus, thus obscuring its real nature. The cleavage line is only rarely retained in adult insects but some species have a similarly situated sulcus of different morphological and functional significance. In older accounts the term 'epicranial suture' sometimes referred to the ecdysial cleavage line, sometimes to comparably placed sulci; it is therefore best avoided.

The *frons* is the unpaired upper facial part of the head which often lies between the lateral arms of the ecdysial cleavage line when this is present. It has been defined more precisely as the region of the head-capsule on which arise the pharyngeal dilator muscles, but even this criterion is not entirely satisfactory. It usually bears the median ocellus and is often delimited laterally by the *frontogenal sulci* and distally by the *epistomal* or *frontoclypeal sulcus* (Fig. 5), associated with which are the anterior tentorial invaginations.

The *clypeus*, on which arise the cibarial dilator muscles, lies immediately

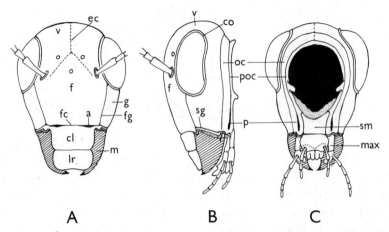

A   B   C

FIG. 5   General diagram of insect head capsule (based on Snodgrass, 1960). A,
anterior view; B, lateral view; C, posterior view

a, anterior tentorial pit; cl, clypeus; co, circumocular sulcus; ec, ecdysial cleavage
line; f, frons; fc, fronto-clypeal (= epistomal) sulcus; fg, fronto-genal sulcus; g,
gena; lr, labrum; m, mandible; max, maxilla; oc, occiput; p, posterior tentorial
pit; poc, post-occiput; sg, subgenal sulcus; sm, submentum; v, vertex.

anterior to the frons and the two sclerites are often fused owing to the
obliteration of the *frontoclypeal sulcus*. In other insects the clypeus is partially or completely divided by a transverse suture into two sclerites – the
postclypeus and the *anteclypeus* (Fig. 304). The former sclerite carries on
either side a convex process serving for articulation with the ginglymus of
the mandible. Laterally the clypeus may be delimited from the sides of the
head by the *clypeogenal sulci*.

The *labrum* is an unpaired sclerite usually movably articulated with the
clypeus by means of the *clypeolabral* suture. On its pharyngeal surface it
bears lateral sclerotized pieces known as *tormae*.

The *epicranium* is the upper region of the head from the frons to the neck
and may be divided longitudinally into two *epicranial plates* by the ecdysial
cleavage line or median sulcus. That portion of the epicranium which lies
immediately behind the frons and between the compound eyes is termed the
vertex. It sometimes carries the paired ocelli, but is not differentiated as a
separate sclerite. The *occiput* is the hinder part of the epicranium between the
vertex and the neck; it is rarely present as a distinct sclerite.

The *gena* (Fig. 5) forms the whole of the lateral area below and posterior
to the eyes on each side; near its junction with the clypeus is a facet for
articulation with the ginglymus of the mandible and proximally it bears a
cavity which receives the mandibular condyle. Crossing the hind part of the
cranium there is in some insects an *occipital suture*. When fully developed it
extends on either side to end between the two points of articulation of the
mandible. The areas posterior to this suture are the *occiput* dorsally and the
postgenae laterally. The postgenae bear the condylar articulations for the

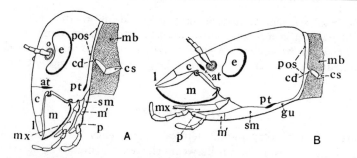

FIG. 6   A. Hypognathous head. B. Prognathous head

p, prementum; *m'*, mentum; *sm*, submentum; *gu*, gula. Other lettering as in Fig. 5. Adapted from Snodgrass.

maxillæ. The dorsal and lateral margins of the occipital foramen are commonly bordered by a narrow rim or *postocciput* with which the neck membrane is directly continuous. This rim is marked off from the rest of the cranium by a groove or *postoccipital sulcus* which ends at the posterior tentorial pit on either side and along which are inserted the dorsal prothoracic muscles moving the head.

The heads of insects are broadly divisible into three types (Fig. 6) depending on the inclination of the long axis and the position of the mouthparts. In the *hypognathous* head the long axis is vertical and the mouthparts ventral; the occipital foramen lies in or near the transverse plane. In the *prognathous* head the long axis is horizontal, or slightly inclined ventrally, while the mouthparts are anterior in position. The prognathous condition often involves an inclination of the occipital foramen or the latter may retain its transverse position owing to an elongation of the ventral region of the head. This may be achieved, as in the soldier caste of Isoptera, by a backward extension of the postmentum and genae. Or, as among Coleoptera, a median ventral sclerite or *gula* (not necessarily homologous in all the orders in which it occurs) extends from the occipital foramen to the base of the submentum (DuPorte, 1962). It occupies the area between the postoccipital sulci which, along with the posterior tentorial pits, have extended forwards on the head capsule. In many Endopterygote larvae, however, the gula is bounded by variously formed ecdysial lines along which the ventral side of the head-capsule bends or splits when the insect moults (Hinton, 1963). These have no consistent relationships with the posterior tentorial pits and the term 'gular suture' is confusing since, like the 'epicranial suture', it has been used to denote a strengthening sulcus or a weak ecdysial line. The gula and submentum are often fused into a single sclerite, the *gulamentum* (Fig. 6B). The third or opisthognathous condition occurs in the Homoptera, where the head is directed backwards so that the specialized mouthparts arise between the anterior legs.

In addition to the foregoing there are other sclerites of less general occurrence such as the following. (1) The *antennal sclerites* (Fig. 4). Each is a ring of cuticle into which the basal segment of the antennae of its side is inserted. (2) The *ocular sclerites*. These are also annular in form and each surrounds the compound eye of its side. (3) *The mandibular sclerite*. A small sclerite close to the base of the mandible and separated by a transverse suture from the gena is found in many Orthoptera.

The Antennae

These are a pair of very mobile jointed appendages which are articulated with the head in front of or between the eyes. Imms (1939, 1940) has distinguished two main types of antennae. In the Collembola and Diplura all the antennal segments except the last contain intrinsic muscles, the antenna grows postembryonically by division of the terminal segment and Johnston's organ (p. 129) is absent. In the remaining insects only the basal segment contains muscles, Johnston's organ is present and increase in the number of segments occurs through division of the 3rd segment or sometimes also of some or all of the more distal segments. In the more generalized insects the antennae are filiform and many-segmented, the segments being equal or sub-equal in size. They vary greatly in form in the higher orders, however, and some segments are frequently differentiated from their fellows. In the more specialized insects the antenna is divisible into scape, pedicel and flagellum (Fig. 7).

The *scape* is the first or basal segment of the antenna and is often conspicuously longer than any of the succeeding segments. The *pedicel* is the segment which immediately follows the scape. In geniculate antennae it forms the pivot between the scape and flagellum. The *flagellum* forms the remainder of the antenna. It varies greatly in form among different families in adaptation to the environment and habits of the species concerned. In some insects, e.g. Chalcidoidea, the flagellum is divisible into the ring-segments, the funicle and the club. The ring-segments are the basal segment or segments of the flagellum, are of much smaller calibre than the segments that follow, and are ring-like in form. The *club* is formed by the swollen or enlarged distal segments of the antenna.The *funicle* comprises those segments which intervene between the ring-segments and the club, or between the latter and the pedicel in cases when the ring-segments are not differentiated.

The antennae afford important secondary sexual characters which are particularly well shown in the pectinate or bipectinate organs of certain male Lepidoptera, and in the densely plumose antennae of male Culicidae and Chironomidae. Functionally the antennae are organs of special sense (Schneider, 1964; see also p. 139) but in a few cases they are modified for other uses. Thus in the larvae of *Chaoborus* and its allies they are adapted for seizing the prey, while those of the male of *Meloe* and several other insects are used for holding the females. In larvae of the Hymenoptera Apocrita and the higher Diptera the antennae are often reduced to minute tubercles or are atrophied. They are wanting in all Protura.

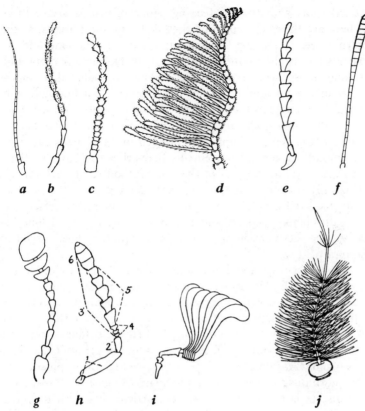

FIG. 7 Types of antennae

a, setaceous (*Blatta*); b, filiform (*Carabus*); c, moniliform (*Kalotermes*); d, pectinate (Tenthredinid, after Enslin); e, serrate (Elaterid); f, clavate (Lepidopteron); g, capitate (*Necrobia*); h, geniculate (Chalcid); i, lamellate (*Melolontha*, after Newport); j, plumose (male of *Culex*); 1, scape; 2, pedicel; 3, flagellum; 4, ring-segments; 5, funicle; 6, club.

The Mouthparts

Essentially, these organs comprise three pairs of appendicular jaws, the anterior *mandibles* followed in turn by the *maxillae* and a second pair of maxilla-like structures that fuse medially during embryonic development to form the *labium* or lower lip. Closely associated with them are two unpaired, non-appendicular structures, the *labrum* or upper lip and the median, tongue-like *hypopharynx*. The mouthparts vary in form to a greater extent than almost any other organs, the variation being correlated with the method of feeding and other uses to which they may be subjected. An examination of the structure of the mouthparts will therefore give a clue to the feeding mechanism and frequently to the nature of the food of an insect. The various modifications which these organs undergo are of considerable taxonomic importance and are dealt with in the chapters devoted to the different orders of insects. Broadly speaking the feeding habits of insects are of three main kinds. There are those with mandibulate or biting mouthparts, such as the Orthopteroid orders and the Coleoptera, those with piercing and suctorial mouthparts such as the Hemiptera, Siphunculata, Siphonaptera and some Diptera, and those with more or less elongate, haustellate mouthparts adapted for taking up liquids without piercing (e.g. Lepidoptera and some Diptera and Hymenoptera). In the Ephemeroptera and certain Lepidoptera and Diptera the mouthparts are greatly reduced or non-functional.

The *labrum* (Fig. 5) is a simple plate hinged to the clypeus and capable of a limited amount of vertical movement. It overlies the bases of the mandibles and forms part of the roof of the preoral food cavity (Moulins, 1971). Morphologically it represents the most anterior region of the head and has secondarily acquired a basal hinged attachment. Its inner surface is usually provided with chemoreceptors and is produced into a small lobe-like epipharynx in the Hymenoptera and a long epipharyngeal stylet in the Siphonaptera.

The *mandibles* of insects and Myriapods differ from those of Crustacea in representing entire limbs, biting with their tips, rather than the proximal gnathobases (Manton, 1964). They are usually adapted for cutting or crushing the food and frequently also for defence; more rarely they are modified into sickle-like or stylet-like piercing organs. In the soldiers of the Isoptera they assume grotesque and inexplicable forms and in certain Coleoptera such as *Lucanus* and *Chiasognathus* they exhibit sexual dimorphism, attaining relatively enormous proportions in the male. Typically, the mandible of Pterygote insects is a solid compact piece articulating with the head by a *ginglymus* and *condyle*. The former is a groove or cavity which articulates with a convex process of the clypeus and the condyle is a rounded head adapted to fit into a socket placed at the lower end of the gena or postgena. Each jaw is moved in the transverse plane by powerful adductor and abductor muscles. In phytophagous insects the mandibles are bluntly toothed and often bear a molar or crushing surface near the base of the biting margin. In carnivorous forms the teeth are sharply pointed, being adapted for seizing and cutting, and the molar surface is wanting. In some insects a flexible plate or *prostheca*, fringed with hairs, is present on the inner border of the mandible. Mandibles are wanting in many adult Trichoptera and most Diptera, and are absent or vestigial in almost all Lepidoptera.

In the Apterygote insects several different mandibular mechanisms occur (Manton, 1964). The mandible of *Petrobius* and other Machilidae (Fig. 9) has a single posterior articulation and is rotated by promotor–remotor movements that bring the molar regions and the apices of opposite mandibles into close apposition. The Lepismatidae have acquired an additional anterior articulation which, with other changes, enables the mandible to move by adduction and abduction in the transverse plane rather as it does in most mandibulate Pterygotes. The mandibles of Ephemeropteran larvae are in some respects intermediate between the Lepismatidae and the higher

Pterygotes (Snodgrass, 1950; Brown, 1961). In the Collembola, Diplura and Protura the mandibles are protrusible, rotatory structures enclosed with the maxillae in a gnathal pouch that is sunk into the head. Despite views to the contrary (Tuxen, 1959), this entognathous condition has probably evolved independently in the three groups concerned (Manton, 1964).

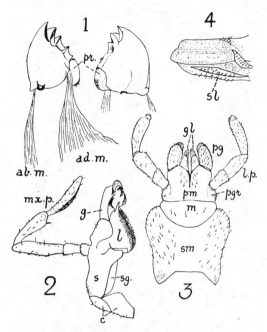

F I G. 8   Mouthparts of *Blatta*

1, Mandibles – *ab.m*, abductor muscle; *ad.m*, adductor muscle; *pr*, prostheca; 2, Maxilla – *c*, cardo; *g*, galea; *l*, lacinia; *mx.p*, maxillary palp; *s*, stipes; *sg*, subgalea; 3, Labium – *gl*, glossa; *l.p*, labial palp; *m*, mentum; *pg*, paraglossa; *pgr*, palpiger; *pm*, prementum; *sm*, submentum; 4, Hypopharynx – *sl*, part of suspensory apparatus.

The *maxillae* (Figs. 8–10) are composed of the following sclerites. The *cardo* is the first or proximal piece and, in many insects, is the only portion directly attached to the head. The *stipes* articulates with the distal border of the cardo and is sometimes divided into a *basistipes* and a *dististipes*. It bears a lateral *palpifer* and sometimes an inner sclerite, the *subgalea* (or *parastipes*). The palpifer carries the *maxillary palp*, the most conspicuous appendage of the maxilla. It is one- to seven-segmented and sensory in function. In many insects the subgalea is not a separate sclerite, being fused with the lacinia or merged into the stipes. Distally the maxilla consists of two lobes: an outer *galea* and an inner *lacinia*. The galea is often two-segmented and frequently overlaps the lacinia like a hood. The lacinia is commonly spined or toothed on its inner border and when fused with the subgalea it has the appearance of carrying the galea. A characteristic muscle, the cranial flexor of the lacinia, runs from the lacinia to the cranial wall and has been used to identify the former in specialized mouthparts (Das, 1937; Imms, 1944). In many Coleopteran larvae each maxilla carries a single lobe or *mala* which in some cases represents the galea and in others the lacinia (Das, 1937). Functionally

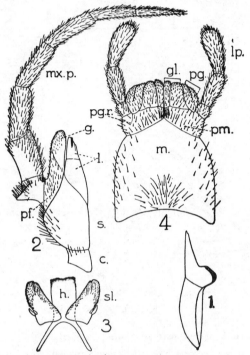

FIG. 9   Mouthparts of *Petrobius maritimus*

1, Mandible. 2, Maxilla. *pf*, palpifer. 3, Hypopharynx (*h*) and superlinguae (*sl*). 4, Labium. *m*, postmentum. Other lettering as in Fig. 8.

the maxillae are a pair of accessory jaws, their laciniae aiding the mandibles in holding the food when the latter are extended, as well as assisting in mastication. In many higher insects the maxillae are so greatly modified that they no longer retain any evidences of their primitive structure. In piercing insects they are styliform and their palps atrophied. The insect maxilla is to be regarded as a highly modified walking limb, whose main shaft is represented by the palp and base by the cardo and stipes. The palpifer is a secondarily demarcated portion of the stipes and of little morphological importance while the galea and lacinia are endites of the stipes.

The *labium* (Figs. 8, 9, 11, 12, 13) is formed by the fusion of a pair of appendages serially homologous with the maxillae. The completeness of the fusion that has taken place varies greatly in different orders of insects, and indications of the original paired condition are clearly seen among the lower

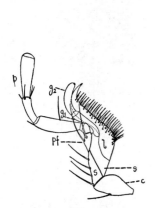

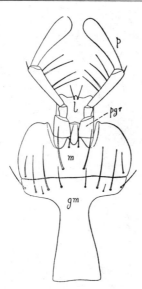

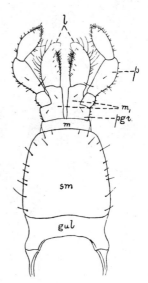

FIG. 10

Right maxilla (ventral aspect) of a beetle, *Nebria brevicollis*

c, cardo; *g₁*, *g₂*, proximal and distal points of galea; *l*, lacinia; *p*, palp; *pf*, palpifer; *s*, *s*, stipes.

FIG. 11

Labium (ventral aspect) of *Nebria brevicollis*

gm, gulamentum; *l*, ligula; *m*, mentum; *p*, palp; *pgr*, palpiger.

FIG. 12

Labium of *Forficula* (ventral aspect)

l, ligula; *gul*, gula; *m*, mentum; *p*, palp; *pgr*, palpiger; *m₁*, prementum; *sm*, submentum.

orders. The labium is divided into two primary regions – a proximal *postmentum* and a distal *prementum*, the line of division between the two being the *labial suture*. The muscles of the palps and the terminal lobes originate within the body of the prementum and consequently lie anterior to the labial suture. The median retractor muscles of the prementum, on the other hand, arise in the postmentum and have their insertions on the proximal margin of the prementum (Fig. 13). The relationships of these muscles, therefore, help in determining the homologies of the main parts of the labium. The postmentum remains undivided in the Thysanura, Isoptera and some higher orders but in many Orthopteroid insects it is divided transversely into a distal *mentum* and a proximal *submentum*. The mentum is often ill-defined and has few muscle attachments. Near the base of the prementum, on either side, is the *palpiger* which carries the *labial palp* and often resembles a basal segment of the latter. The labial palps are composed of one to four segments and function as sensory organs. Arising from the distal margin of the prementum are two pairs of lobes which collectively form the *ligula*; there is an outer pair or *paraglossae*, and an inner pair or *glossae*. More usually, the latter are fused to form a median *glossa* or the prementum may bear a single median lobe to which the general term ligula is applied.

In Fig. 14 the homologies of the sclerites of the labium with those of the maxillae are indicated. The glossae and paraglossae are the counterparts of the laciniae and

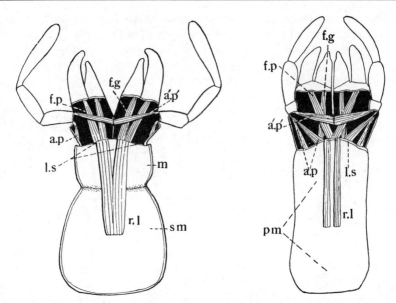

FIG. 13   Right, labium of *Grylloblatta* (adapted from Walker). Left, labium of *Mastotermes* (original). In both figures the wall of the prementum has been removed to show the musculature

a.p, abductor of palp; *a′.p′*, adductor of palp; *f.g*, flexor of glossa; *f.p*, flexor of paraglossa; *l.s*, labial suture; *m*, mentum; *p.m*, postmentum; *r.l*, median retractors of prementum; *sm*, submentum.

galeae respectively, while the labial palps are homologous with those of the maxillae. The two lobes of the primitive divided prementum are clearly traceable as the representatives of the stipites which, in most insects, undergo fusion. The only part comparable to the united cardines is the postmentum, but it is possible that the median part of the sternum of the labial segment is incorporated in this sclerite. The true morphological interpretation of the postmentum is unsolved and further data are needed (Matsuda, 1965).

Situated behind the mouth lies the median hypopharynx. This is usually a tongue-like structure and at its base on the lower side there opens the salivary duct. In the Diplura, Collembola, Machilidae and Ephemeropteran

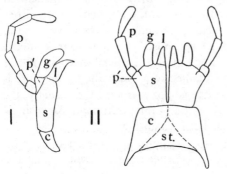

FIG. 14
Diagram showing homologous parts in maxilla (I) and labium (II)

c, cardo; *s*, stipes; *l*, lacinia; *g*, galea; *p*, palp; *p′*, palpifer (or palpiger in II); *st*, sternum.

nymphs the hypopharynx is 3-lobed, the median lingua bearing a pair of lateral superlinguae. In the Lepismatidae and Pterygota the hypopharynx is typically a simple lobe with a number of suspensory sclerites but in most Diptera it is a stylet-like structure pierced by the salivary canal and in some cases is used as a piercing organ. The embryonic development of the hypopharynx shows it to be a composite structure derived from the sternal regions of the premandibular and gnathal segments of the head (e.g. Scholl, 1969).

Segmentation of the Insect Head

After an insect has emerged from the egg the completed head shows few indications of a segmental origin apart from the fact that it carries paired appendages. There is, however, every reason to believe that the insect head arose by the coalescence of a number of body segments and a non-segmental anteriorly-placed acron homologous with the Annelid prostomium. Opinions differ on the number and characteristics of the segments involved and most of the hypotheses have been reviewed by Matsuda (1965) and Gouin (1968). The orthodox theory is based largely on embryological evidence summarized recently by Jura (1972) and Anderson (1972). It asserts that the insect head consists of the acron plus six segments. In favourable cases the latter may each be recognized embryologically by the presence of (i) a pair of hollow mesodermal somites, (ii) paired neuromeres (embryonic ganglia), and (iii) paired appendages. The composition of the insect head according to this theory may be tabulated as follows:

Segment	Neuromere		Coelom sacs	Appendages
Acron	} Protocerebrum	}		
1. Preantennal		} Brain	Sometimes present	Absent
2. Antennal	Deutocerebrum		Usually present	Antennae
3. Premandibular (= intercalary)	Tritocerebrum		Usually absent	Embryonic
4. Mandibular	Mandibular ganglion	}	Usually present	Mandibles
5. Maxillary	Maxillary ganglion	} Suboesophageal ganglion	Usually present	Maxillae
6. Labial	Labial ganglion		Usually present	Labium

A few comments on this table are necessary, especially in the light of more recent embryological work (e.g. Bruckmoser, 1965; Wada, 1966; Ullmann, 1964, 1967; Malzacher, 1968; Rohrschneider, 1968; Rempel and Church, 1969; Rempel, 1975; Scholl, 1969 and Larink, 1970). The preantennal segment is often poorly defined or unrecognizable and the protocerebrum is derived largely from the ectoderm of the lateral cephalic lobes. Such a situation has led Matsuda (1965) to claim that in most insects the head comprises an acron and only five cephalic segments. On the other hand, earlier claims that an additional seventh or labral segment is present in front of the preantennal segment now seem unlikely. The origin of the labrum from paired rudiments, interpreted as limb buds, is very uncommon (occurring only in *Locusta, Carausius* and *Rhodnius*) while the allegedly labral coelom sacs of *Carausius* have been shown on reinvestigation to belong to the preantennal segment (Scholl, 1969). Other theories of the segmental composition of the insect head are based largely on the comparative anatomy of adult forms, especially of the central nervous system. They cannot be reconciled easily with the embryological evidence and are not widely accepted. Four of these theories may be mentioned briefly:

1. Snodgrass (1963) agrees with the earlier views of Hanström (1928) and concludes that the protocerebrum, deutocerebrum and antennae are derivatives of an anterior, non-segmental blastocephalon, followed by four of the segments (premandibular to labial) recognized above.

2. Henry (1947–48) and Ferris (1947) consider that the head comprises six segments, of which the first three give rise to the labrum, the clypeus plus hypopharynx, and an oculo-antennal region broadly equivalent to the frons. The remaining three segments bear the mandibles, maxillae and labium.

3. Chaudonneret (1950) recognizes the six segments of the orthodox theory, together with a so-called superlingual segment, lying between the premandibular and mandibular segments. His views are based on a detailed anatomical study of *Thermobia domestica*, but are not supported by the embryology of the head in the closely related *Lepisma saccharina* (Larink, 1970).

4. Butt (1960) agrees with Snodgrass in denying the existence of the preantennal and antennal segments, but regards the premandibular segment as bearing appendages that fuse to form the labrum. His theory has been criticized by Manton (1960).

Finally, it is worth bearing in mind that if, as now seems likely, the process of cephalization occurred independently in several arthropodan lineages (see p. 5), interpretations of the head which rely on comparisons between the insects and other arthropod groups no longer have the validity ascribed to them in the past.

The Cervix or Neck Region

The cervix (Matsuda, 1970) is the flexible region between the head and prothorax (Fig. 1). In its membrane are embedded a variable number of *cervical sclerites* (Fig. 19). These occur in nearly all orders of insects but are best developed in the more primitive groups (Orthoptera, Dermaptera, Isoptera, Odonata); in the higher orders they are more or less reduced. In their least modified form the cervical sclerites consist of paired dorsal, lateral and ventral plates of which the lateral pair is of special importance. The lateral sclerites usually comprise two plates on either side, closely hinged together so as to form a fulcrum between the head and prothorax. The distal plate articulates with the occipital condyle of the head, while the proximal plate is hinged to the prothoracic episternum (Fig. 15). Levator muscles arising from the postoccipital rim and the prothoracic tergum are attached to the lateral cervical sclerites of their side. The contraction of these muscles widens the angle between the two plates of a pair and, in this way, causes the protraction of the head.

The morphological nature of the neck is highly problematical and the available evidence suggests that something more than an enlarged intersegmental region may be involved. The innervation of the muscles inserted on

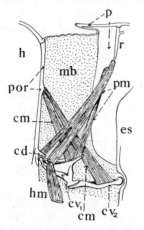

FIG. 15
Neck and cervical sclerites of a grasshopper (*Dissosteira*)

cd, occipital condyle; *es*, episternum; cephalic (*cm*), protergal (*pm*) and prosternal (*hm*) muscles of cervical sclerites (*cv₁*, *cv₂*); *h*, head; *mb*, cervical membrane; *p*, protergum; *por*, postoccipital rim; *r*, ridge of protergum. *From* Snodgrass

the cervical sclerites suggests that they are derived partly from the prothorax and partly from the labial segment of the head (Henry, 1958; Schmitt, 1959; Shepheard, 1973). An embryological study of *Carausius* by Scholl (1969) confirms this, suggesting that the neck includes a dorsal labial component and a ventral prothoracic part, though Wada (1966) regards the neck of *Tachycines* as entirely labial in origin.

Literature on the Head, Mouthparts and Neck

ANDERSON, D. T. (1972), In: Counce, S. J. and Waddington, C. H. (eds), *Developmental Systems: Insects*. Vol. 1, pp. 95–242.

BROWN, D. S. (1961), The morphology and functioning of the mouthparts of *Chloëon dipterum* L. and *Baetis rhodani* Pictet (Insecta, Ephemeroptera), *Proc. zool. Soc. Lond.*, **136**, 147–176.

BRUCKMOSER, P. (1965), Embryologische Untersuchungen über den Kopfbau der Collembole *Orchesella villosa* L., *Zool. Jb.* (*Anat.*), **82**, 299–364.

BUTT, F. H. (1960), Head development in the Arthropods, *Biol. Rev.*, **35**, 43–91.

CHAUDONNERET, J. (1950), La morphologie cephalique de *Thermobia domestica* (Packard) (Insecte Apterygote Thysanoure), *Ann. Sci. Nat.* (*Zool.*), 11. Sér., **12**, 145–302.

DAS, G. M. (1937), The musculature of the mouthparts of insect larvae, *Q. Jl microsc. Sci.*, **80**, 39–80.

DUPORTE, E. M. (1957), The comparative morphology of the insect head, *A. Rev. Ent.*, **2**, 55–70.

—— (1962), Origin of the gula in insects, *Can. J. Zool.*, **40**, 381–384.

FERRIS, G. F. (1947), The contradictions of the insect head, *Microentomology*, **12**, 59–64.

GOUIN, F. (1968), Morphologie, Histologie und Entwicklungsgeschichte der Insekten und der Myriapoden. IV. Die Strukturen des Kopfes, *Fortschr. Zool.*, **19**, 194–282.

HANSTRÖM, B. (1928), *Vergleichende Anatomie des Nervensystems der wirbellosen Tiere unter Berücksichtigung seiner Funktion*, Springer, Berlin, 628 pp.

HENRY, L. M. (1947–48), The nervous system and segmentation of the head in the Annulata, *Microentomology*, **12**, 65–110; **13**, 1–48.

—— (1958), Musculature of the cervical region in insects, *Microentomology*, **23**, 95–105.

HINTON, H. E. (1963), The ventral ecdysial lines of the head of endopterygote larvae, *Trans. R. ent. Soc. Lond.*, **115**, 39–61.

IMMS, A. D. (1939), On the antennal musculature in insects and other arthropods. *Q. Jl microsc. Sci.*, **81**, 273–320.

—— (1940), On growth processes in the antennae of insects, *Q. Jl microsc. Sci.*, **81**, 585–593.

—— (1944), On the constitution of the maxillae and labium in Mecoptera and Diptera, *Q. Jl microsc. Sci.*, **85**, 73–96.

JURA, C. (1972), In: Counce, S. J. and Waddington, C. H. (eds), *Developmental Systems: Insects*. Vol. 1, pp. 49–94.

LARINK, O. (1970), Die Kopfbildung von *Lepisma saccharina* L. (Insecta, Thysanura), *Z. Morph. Tiere*, **67**, 1–15.

MALZACHER, P. (1968), Die Embryogenese des Gehirns paurometaboler Insekten. Untersuchungen an *Carausius morosus* und *Periplaneta americana*, *Z. Morph. Ökol. Tiere*, **62**, 103–161.

MANTON, S. M. (1960), Concerning head development in the arthropods, *Biol. Rev.*, **35**, 265–282.

—— (1964), Mandibular mechanisms and the evolution of Arthropods, *Phil. Trans. R. Soc. Ser. B*, **247**, 1–183.

MATSUDA, R. (1965), Morphology and evolution of the insect head, *Mem. Am. ent. Inst.*, **4**, 334 pp.

—— (1970), Morphology and evolution of the insect thorax, *Mem. ent. Soc. Canada*, **76**, 431 pp.

MOULINS, M. (1971), La cavité préorale de *Blabera craniifer* Burm. (Insecte, Dictyoptère) et son innervation: Étude anatomo-histologique de l'épipharynx et l'hypopharynx, *Zool. Jb. (Anat.)*, **88**, 527–586.

REMPEL, J. G. (1975), The evolution of the insect head: the endless dispute, *Quaestiones entomologicae*, **11**, 9–25.

REMPEL, J. G. AND CHURCH, N. S. (1969), The embryology of *Lytta viridana* Le Conte (Coleoptera: Meloidae). V. The blastoderm, germ layers and body segments, *Can. J. Zool.*, **47**, 1157–1171.

ROHRSCHNEIDER, I. (1968), Beiträge zur Entwicklung des Vorderkopfes und der Mundregion von *Periplaneta americana*, *Zool. Jb. (Anat.)*, **85**, 537–578.

SCHMITT, J. B. (1959), The cervicothoracic nervous system of a grasshopper, In: Clarke *et al.*, Studies in invertebrate morphology, *Smithson. misc. Collns*, **137**, 307–329.

SCHNEIDER, D. (1964), Insect antennae, *A. Rev. Ent.*, **9**, 103–122.

SCHOLL, G. (1969), Die Embryonalentwicklung des Kopfes und Prothorax von *Carausius morosus* Br. (Insecta, Phasmida), *Z. Morph. Tiere*, **65**, 1–142.

SHEPHEARD, P. (1973), Musculature and innervation of the neck of the desert locust, *Schistocerca gregaria* (Forskål), *J. Morph.*, **139**, 439–464.

SNODGRASS, R. E. (1947), The insect cranium and the epicranial suture, *Smithson. misc. Collns*, **107**, 52 pp.

—— (1950), Comparative studies on the jaws of mandibulate arthropods, *Smithson. misc. Collns*, **116**, 85 pp.

—— (1960), Facts and theories concerning the insect head, *Smithson. misc. Collns*, **142**, 61 pp.

—— (1963), A contribution toward an encyclopedia of insect anatomy, *Smithson. misc. Collns*, **146**, 48 pp.

STRENGER, A. (1952), Die funktionelle und morphologische Bedeutung der Nähte am Insektenkopf, *Zool. Jb.* (*Anat.*), **72**, 467–521.

TUXEN, S. L. (1959), The phylogenetic significance of entognathy in entognathous Apterygotes, In: Clarke *et al.*, Studies in invertebrate morphology. *Smithson. misc. Collns*, **137**, 379–416.

ULLMANN, S. L. (1964), The origin and structure of the mesoderm and the formation of the coelomic sacs in *Tenebrio molitor* L. (Insecta, Coleoptera), *Phil. Trans. R. Soc. Ser. B*, **248**, 245–277.

—— (1967), The development of the nervous system and other ectodermal derivatives in *Tenebrio molitor* L. (Insecta, Coleoptera), *Phil. Trans. R. Soc. Ser. B*, **252**, 1–25.

WADA, S. (1966), Analyse der Kopf-Hals-Region von *Tachycines* (Saltatoria) in morphogenetische Einheiten. I, II, *Zool. Jb.* (*Anat.*), **83**, 185–234; 235–326.

Chapter 5

THE THORAX

Segmentation of the Thorax

The thorax is composed of three segments, the *pro-*, *meso-* and *metathorax*.In almost all insects each segment bears a pair of legs and in most adults both the meso- and metathorax carry a pair of wings. Where the legs are wanting, their absence is secondary. This apodous condition is extremely rare among the imagines but it is the rule among the larvae of the Diptera and certain families of Coleoptera. All Hymenopteran larvae, excepting the vast majority of the suborder Symphyta, are similarly devoid of legs. The absence of wings, on the other hand, may be a primitive character as in the Apterygota, but among the Pterygota it is always a secondary feature due to the loss of pre-existing organs. The thorax is exhibited in a simple form in the Thysanura, in certain more generalized Pterygota and in the larvae of many orders. In these instances the segments differ little in size and proportions, but with the acquisition of wings a correlated specialization of the thorax usually results. The meso- and metathorax become more or less intimately associated to form a pterothorax and the union is often so close that the limits of those regions can only be ascertained with difficulty. In orders where the wings are of about equal area these two thoracic segments are of equal size e.g. Isoptera, Embioptera, Odonata. Where the fore wings are markedly larger than the hind pair there is a correspondingly greater development of the mesothorax (Hymenoptera, and also Diptera where the hind wings are reduced in size and not used for flying). Where the fore wings are small or not used in flight there is a correlated reduction of the mesothorax (Coleoptera). The prothorax never bears wings and varies in size. Its dorsal region may be enlarged to form a shield as in the Orthoptera, Dictyoptera, Coleoptera and Heteroptera; in most other orders it is a narrow annular segment. For general accounts of thoracic morphology see Snodgrass (1927, 1929), Gouin (1959) and Matsuda (1970). Ferris (1940 and other papers) has an unorthodox theory of thoracic structure while Manton (1972) has brought forward many new functional and evolutionary interpretations.

The Sclerites of a Thoracic Segment

When describing the sclerites and regions of the thorax the prefixes *pro*, *meso* and *meta* are used according to the segment to which the reference applies.

Thus the expression protergum refers to the tergum of the prothorax and mesepimeron to the epimeron of the mesothorax. The prefixes *pre* and *post* are also used to designate certain sclerites of any one of the segments and in such cases the prefixes *pro*, *meso* and *meta* are usually not applied. For example the prescutum may be present on each thoracic segment in front of the scutum.

The Tergites – Despite arguments to the contrary (Duporte, 1965; Matsuda, 1970), it may be supposed that the thoracic terga consisted primitively of three simple segmental plates (the *nota*) between which lay small intersegmental sclerites (Snodgrass, 1927). In all known insects, however, a secondary segmentation has become established whereby the intersegmental sclerites have become closely associated with the notum in front or with the one behind. In the latter case the intersegmental sclerite forms a narrow band at the front of the notum and is known as the *acrotergite*. In the former case, the intersegmental sclerite becomes known as the *postnotum* and is sometimes a conspicuous plate. These transpositions of the intersegmental sclerites of the pterothorax differ in different insects. Thus, in many Apterygota, the Blattids, the Isoptera and many immature forms the meso- and metathorax each possesses an acrotergite but no postnotum. In most other orders a postnotum is present in both meso- and metathorax, but in the Orthoptera and Coleoptera the metathorax has acquired two intersegmental sclerites, having both acrotergite and postnotum while the mesothorax has accordingly no postnotum. The notum is typically divided into three sclerites, the prescutum, the scutum and the scutellum (Fig. 16). At the sides of the pronotum in many Lepidoptera are lobe-like structures known as patagia.

The Pleurites (Figs. 17–19) – It is sometimes considered that the pleural sclerites were derived from a primitive subcoxal segment of the leg which became flattened and incorporated into the pleural region of the body-wall. The dorsal elements of the subcoxa thus came to form the definitive thoracic pleuron, while a ventral element probably fused with the primitive sternum to form the definitive sternal area. Support for these views has been sought in the small, variable, pleural sclerites of the Apterygote insects, though recent work on these by Manton (1972) has cast serious doubt on the subcoxal theory. Whatever its origins, the pleuron of the Pterygotes (Fig. 17) is a relatively uniform structure consisting of an anterior sclerite or *episternum* and a posterior sclerite or *epimeron*, the two being separated by the *pleural sulcus* (also known as the pleural suture). Many insects show deviations from this simple condition because of the subdivision of the pleurites into secondary plates, or their fusion with other regions of their segment. The anterior part of the episternum is frequently marked off as a separate plate, the *pre-episternum*, while in many insects (e.g. *Chrysopa*, *Corydalis*, *Tipula*, *Tabanus*) the episternum is divided into an upper and lower sclerite, the *anepisternum* and *katepisternum* respectively. Not infrequently the lower portion of the episternum is fused with the sternum, as in Diptera, and the compound plate thus formed is the *sternopleurite* (Crampton) or *sternopleura* (Osten-Sacken). The epimeron is also sometimes divided into two sclerites by a transverse suture, giving an *anepimeron* and a *katepimeron*. When the pleuron as a whole is fused with the sternum the combined sclerite is known as the *pectus*. In many of the higher insects the pleuron is usually connected and fused with the tergum by downward prolongations of the prescutum and postnotum.

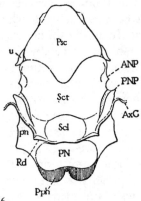

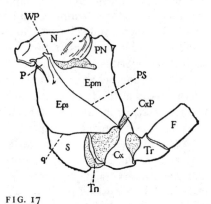

FIG. 16

Mesotergum of a cranefly showing division of notum into three sclerites (*Psc*, *Sct* and *Scl*) behind which is the postnotum (*PN*)

AxC, axillary cord; *ANP*, anterior notal wing process; *PN*, *pn*, postnotum; *PNP*, posterior notal wing process; *Pph*, postphragma; *Psc*, prescutum; *Rd*, posterior reduplication of notum; *Scl*, scutellum; *Sct*, scutum; *u*, lobe of prescutum before base of wing. *After* Snodgrass, *Proc. U.S. Nat. Mus.* 39.

FIG. 17

Metathorax of a stonefly, left side

Cx, coxa; *CxP*, pleural coxal process; *Epm*, epimeron; *Eps*, episternum; *F*, base of femur; *N*, notum; *P*, episternal parapterum; *PN*, postnotum; *PS*, pleural sulcus; *q*, sternopleural suture; *S*, sternum; *Tn*, trochantin; *Tr*, trochanter; *WP*, pleural wing process. *After* Snodgrass, *loc. cit.*

The Sternum – As in the terga, the sternal region of each segment consists typically of a segmental plate and an intersegmental sclerite, the latter associated with the segmental region in front of it. The segmental plate or eusternum is subdivided into three sclerites, the *presternum, basisternum* and *sternellum* and in generalized forms the basisternum is separated from the sternellum by a transverse suture connecting the apophyseal pits – points of cuticular invagination from which arises a pair of furcal arms forming part of the thoracic endoskeleton (p. 81). The intersegmental sclerite is known as the *spinasternum* (or *poststernellum*) and is produced internally into a peg-like apodeme which, with the furca, provides areas for the attachment of the ventral longitudinal muscles. Separate laterosternites are sometimes found at the sides of the eusternum and fusion of the sternal and pleural regions may result in the formation of precoxal and postcoxal bridges. While broad sternal plates subdivided as above may be found in many Orthopteroid insects (Fig. 19) there is often considerable specialization of the ventral region. Thus, the sterna may be narrow and extensively desclerotized (as in the cockroaches), the sutures between the various sclerites may be obliterated and the spinasternum may be lost or its spina consolidated with the eusternum. The apophyseal pits may also become closely approximated,

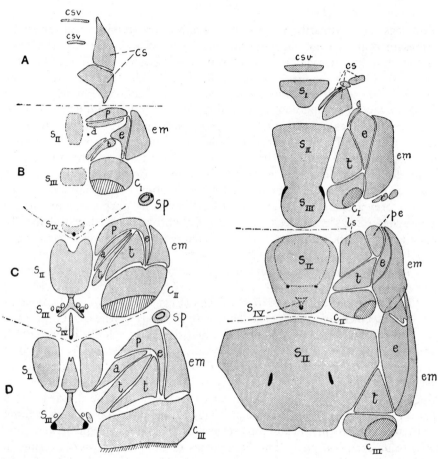

FIG. 18
Sternal and pleural sclerites of cervix and thorax of *Blatta*

A, Cervix. B, Prothorax. C, Mesothorax. D, Metathorax. *a*, antecoxal piece; *cI–cIII*, coxae; *cs*, lateral cervical sclerites; *csv*, ventral ditto; *e*, episternum; *em*, epimeron; *p*, precoxal bridge; *sp*, spiracle; *sII*, eusternum; *sIII*, sternellum; *sIV*, poststernellum; *t*, *t*, trochantin.

FIG. 19
Sternal and pleural sclerites of *Forficula*.

pe, pre-episternum; *ls*, laterosternite; *s₁*, presternum.
Other lettering as in Fig. 18.

the boundaries between pleuron and sternum may be lost by fusion and a cuticular inflection may be developed along the mid-line of the ventral surface. In an attempt to account for some of the peculiarities of the sterno-pleural region, Ferris (1940, etc.) has argued that the greater part of the ventral side of the thorax of some insects is derived from pleural structures, but further critical study of his theory is necessary.

The Legs

The legs are primarily organs for running or walking and are well represented in their normal condition in a cockroach or Carabid beetle. They exhibit, however, a wide range of adaptive modifications in different families (Fig. 20). Thus in the Gryllotalpidae, the Scarabaeidae and some others the fore legs are modified for burrowing, and in the Mantidae, Phymatidae and Mantispidae for seizing and holding the prey. In certain families of butterflies the fore legs are so much reduced that there are only two pairs of functional legs. In the saltatorial Orthoptera, and in beetles of the subfamilies Halticinae and Sagrinae, the hind femora are greatly enlarged in order to accommodate the powerful extensor muscles which are used in leaping. Among the Odonata all the legs are adapted for seizing and retaining the prey and are hardly ever used for locomotory purposes, while in the Bombyliidae the slender legs are used for alighting rather than walking. In aquatic insects they are often specially adapted as swimming organs. Each leg (Fig. 21) consists of the following parts – *coxa*, *trochanter*, *femur*, *tibia* and *tarsus* together with certain basal or *articular sclerites* and a terminal pretarsus.

The Basal Articulations of the Legs (Figs. 17 and 19) – The coxa or proximal segment of the leg articulates with the body by the coxal process of the pleuron and with the trochantin when the latter sclerite is present. A ventral, sternal articulation may also occur. The *coxal process* is situated at the ventral extremity of the pleural sulcus. The *trochantin* is the articular sclerite situated at the base of the coxa in the more primitive orders. It frequently unites with neighbouring sclerites, or it may be divided into a pair of plates. Between the single or divided trochantin and the episternum, or between the trochantin and the precoxal bridge, there is frequently an inner sclerite or *antecoxal piece*. The homologies of these small basal sclerites in different insects have been much discussed and it is possible that they are derived from an original subcoxa.

The Coxa is the functional base of the leg. It is often divided into two lobes by an inflexion of its wall where it articulates with the pleuron. The posterior lobe thus delimited is the *meron* (Larsén, 1945) which is usually the larger part of the coxa. A meron is well developed in *Periplaneta*, the Isoptera, Neuroptera and Lepidoptera.

The Trochanter is the second segment of the leg; it articulates with the coxa but is usually rigidly fixed to the femur. In the Odonata it is divided into two subsegments and among the parasitic Hymenoptera a second apparent trochanter, derived from the base of the femur, is present (see p. 1185).

The Femur – The femur usually forms the largest region of the leg and is especially conspicuous in many jumping insects.

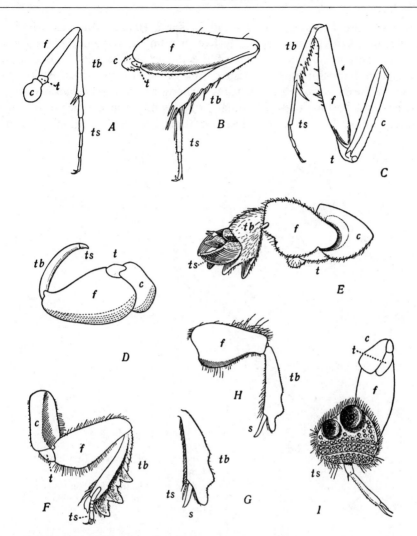

FIG. 20 Adaptive modifications of the legs

A, *Cicindela sexguttata*; B, *Nemobius vittatus*, hind leg; C, *Stagmomantis carolina*, left fore leg; D, *Pelocoris femoratus*, right fore leg; E, *Gryllotalpa borealis*, left fore leg; F, *Canthon laevis*, right fore leg; G, *Phanaeus carnifex*, fore tibia and tarsus of female; H, *P. carnifex*, fore tibia of male; I, *Dytiscus fasciventris*, right fore leg of male; *c*, coxa; *f*, femur; *s*, spur; *t*, trochanter; *tb*, tibia; *ts*, tarsus. *After Folsom, 1923.*

The Tibia – The fourth division of the leg is the tibia; it is usually slender and often equals or exceeds the femur in length. Near its distal extremity it carries one or more *tibial spurs*. In the Apocritan Hymenoptera the enlarged apical spur of the anterior tibia fits against a pectinated semicircular pit in the first tarsal segment, and the antennae are cleaned by being passed between these two organs.

The Tarsus consists primitively of a single segment, a feature which is

present in the Protura, Diplura and in some larvae. More usually it is
divided into subsegments, typically five in number, but none of these has
acquired muscles and movement of the tarsus as a whole is effected by
levator and depressor muscles arising from the apex of the tibia. At its apex
the tarsus bears a group of structures forming the *pretarsus* (Holway, 1935;
Sarkaria and Patton, 1949; Dashman, 1953) which represents the terminal
segment of the leg. In its simplest condition, seen in Collembola, Protura

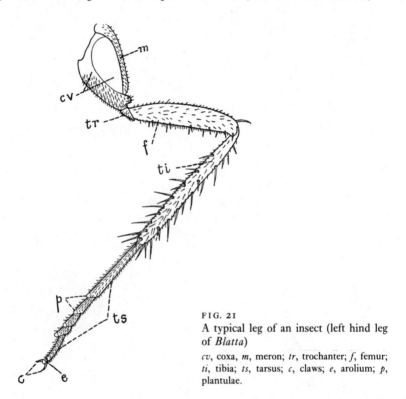

FIG. 21

A typical leg of an insect (left hind leg
of *Blatta*)

cv, coxa; *m*, meron; *tr*, trochanter; *f*, femur;
ti, tibia; *ts*, tarsus; *c*, claws; *e*, arolium; *p*,
plantulae.

and many larvae, the pretarsus is prolonged into a single claw. In most insects
the claws are paired and between them, on the ventral side, the pretarsus is
supported by a median *unguitractor plate* to which the apodeme of the flexor
muscle of the claws is attached. In front of and above this plate the pretarsus
expands into a median lobe or *arolium* (Fig. 22). Among Diptera there
are two lobes or *pulvilli* lying below the claws, often with an arolium between
them or, in place of an arolium, the plantar sclerite distal to the unguitractor
plate is prolonged into a median bristle or *empodium* (Fig. 22). On the
underside of the tarsal segments there are frequently pulvillus-like organs or
plantulae (Fig. 21). The arolium and pulvilli are pad-like organs enabling
their possessors to climb smooth or steep surfaces: the plantulae also have a
similar function. Such organs are outgrowths of the parts from which they
arise and their cavities contain blood. Various explanations have been offered

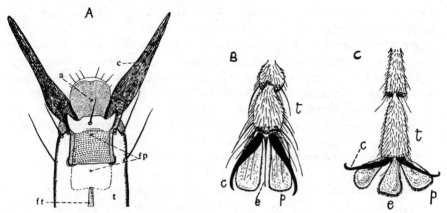

FIG. 22   Pretarsus of A, an Orthopteran, ventral view; B, male *Asilus crabroniformis*; C, male *Rhagio notata*.

a, arolium; c, claw; e, empodium; fp, unguitractor plate; ft, apodeme of flexor muscle; p, pulvillus; t, last tarsal segment. B and C after Verrall.

as to how they function but attachment probably depends on the fact that the structures concerned are covered with tubular tenent hairs, the apices of which are moistened by a glandular secretion (Fig. 23). The hairs can be applied very closely to a smooth surface and adhesion occurs, the insect

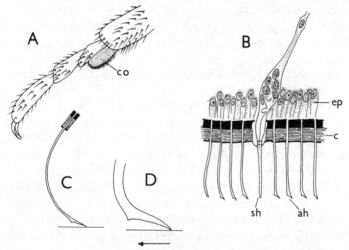

FIG. 23   Climbing organ of *Rhodnius prolixus* (*after* Gillett and Wigglesworth, 1932). A, apical part of fore tibia and tarsus to show position of climbing organ; B, section through cuticle and adhesive hairs; C, a single adhesive hair; D, apex of adhesive hair. Each hair is an obliquely truncate cylinder; in life a wedge-shaped drop of oily secretion fills the space between the substrate and the 'sole' of the hair, permitting displacement in the direction of the arrow but resisting it in the opposite direction

ah, adhesive hair; c, cuticle; co, climbing organ; ep, epidermis; sh, sensory hair.

being held by surface molecular forces (Arnhart, 1923; Gillett and Wigglesworth, 1932).

Leg Movements and Locomòtion – The legs of insects are used in three forms of locomotion – walking (or running), jumping and swimming. Walking is the most widespread form of terrestrial locomotion, during which the fore legs serve mainly as organs of traction and the mid legs as supporting structures while the hind legs exert a propulsive force. The classical view was that progression occurred in a zigzag fashion by movements of the legs in two alternating groups, the fore and hind legs of one side and the mid leg of the other side being moved forward simultaneously while the insect is balanced on a 'tripod of support' formed by the remaining three legs. In fact, cinematographic analysis shows that the legs move in various sequences and attempts have been made to reduce them to variations on a single basic pattern (Hughes, 1965; Wilson, 1966; see also Manton, 1972). As Fig. 24 illustrates, the basic pattern can give rise to others by changes in the phase difference between the stepping sequences of right and left sides. As a result, some legs may step in pairs or there may be three legs in motion at once – two on one side of the body and one on the opposite side. The last situation is the system of alternating tripods of support, which is thus seen to be only one of a larger number of possible gaits.

Jumping mechanisms vary considerably from one group of insects to another. They include the unique jumping gait of Machilidae (Manton, 1972) and the leaping of saltatorial Orthoptera (Alexander, 1968), flea-beetles (Barth, 1954), the Auchenorrhynchan Homoptera (e.g. Sander, 1956) and the fleas (see p. 945).

Swimming legs occur in several groups of aquatic insects, especially among the Heteroptera and Coleoptera (Nachtigall, 1965; Schenke, 1965; Larsén, 1966). Typically the hind legs or the mid and hind legs are flattened or clothed with lateral hair-fringes which increase the effective surface area to several times that of the leg proper. Propulsion occurs on the back-stroke (when the hairs are erect) and far less energy is expended in moving the leg forward on the recovery stroke (when the hairs collapse against the leg and so reduce the resistance to movement). In the whirligig beetle *Gyrinus natator* the elaborately adapted swimming legs have a remarkably high mechanical efficiency of about 84 per cent (Nachtigall, 1962).

In soft-bodied insect larvae, where the appendages are reduced or absent, locomotion occurs through quite different physical mechanisms. The musculature of the body-wall plays an important role and the haemocoele acts as a hydrostatic skeleton. In Lepidopteran larvae such mechanisms are supplemented by the action of the thoracic legs and the abdominal prolegs (Barth, 1937).

The Wings

The presence of wings is one of the most characteristic features of insects, and their dominance may be attributed to the possession of these organs.

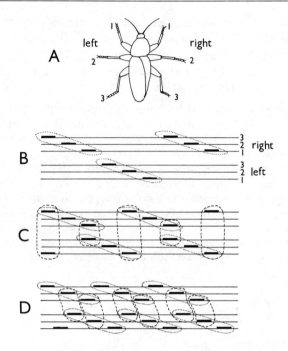

FIG. 24 Diagram illustrating how various hexapod gaits may be derived from a basic pattern of leg movements (simplified, *after* Walker, 1966). The legs are numbered as in A, the horizontal axes denote time and the heavy bars indicate the period during which the leg is off the ground in forward movement. Dotted lines encircle the three successive steps made by the legs of one side. B shows the basic sequence, in which each leg steps by itself; C shows a pattern in which the sequences of right and left sides overlap, so that some legs step in pairs (enclosed by broken lines); D shows further overlap of right and left sequences to give a tripod gait

Owing to their wide range of differentiation, wings provide some of the most useful taxonomic characters. The wing of an insect presents three margins (Fig. 25): the *anterior margin* or *costa* (*a–b*); the *outer* or *apical margin* (*b–c*) and the *inner* or *anal margin* (*c–d*). Three well-defined angles are also recognizable, viz. the *humeral angle* (*a*) at the base of the costa; the *apex* (*b*) or angle between the costal and outer margins and the *anal angle* or *tornus* (*c*) between the outer and inner margins. Although, in most insects, the wings

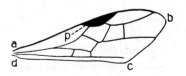

FIG. 25
Wing of a Hymenopteran (explanation given in the text)

p, pterostigma.

appear to be naked, microscopical examination often reveals the presence of fine hairs. On the other hand, in certain groups the wings are obviously clothed. In the Trichoptera and the Dipterous family Psychodidae, for example, they are closely covered with hairs, while in the Lepidoptera they are invested with overlapping scales.

Tillyard (1918*b*) has studied the hairs occurring on the wings of the most primitive groups of Holometabola. *Microtrichia* are found indiscriminately on the wing-membrane and veins alike. *Macrotrichia* or true setae, with annular bases of insertion, are found on the main veins and their branches, on the archedictyon (p. 62), less frequently on the wing-membrane and very rarely on the cross-veins. On the disappearance of the archedictyon, or of an individual vein, the macrotrichia may persist on the wing-membrane in their original positions; their presence there is regarded by Tillyard as evidence of descent from more densely veined ancestors. By plotting the positions of the macrotrichia present on the wing-membrane in such primitive forms as *Archichauliodes*, *Rhyacophila* and *Anisopus*, and joining them into a polygonal meshwork, the lost archedictyon can often, to some extent, be reconstructed (Fig. 26).

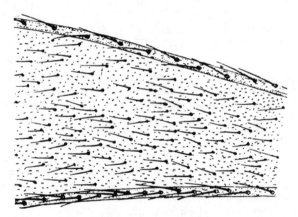

FIG. 26   Portion of a wing of *Anisopus brevis* showing
macrotrichia and microtrichia. *After* Tillyard,
Proc. Linn. Soc. N.S.W. 43

A conspicuous opaque spot is found near the costal margin of the wing in many insects, and is termed the *stigma* or *pterostigma* (Fig. 25). It is present, for example, in the fore wings of the Psocoptera and most Hymenoptera, and in both pairs of wings of the Odonata. Within the stigma is a more or less clearly defined blood sinus through which haemocytes move from the lacuna of the costa into other veins (Arnold, 1963). It seems likely that the stigma can improve the aerodynamic efficiency of the wing by acting as a passive, inertial regulator of its pitch angle, thus raising the speed limits operative both in active flight and in gliding (Norberg, 1972).

The Basal Attachment and Articular Sclerites of the Wings – Each wing is hinged to two processes of the notum of its segment, the *anterior* *notal process* and the *posterior notal process* (Fig. 27, A). The wing also articulates below with the *pleural wing process*. The posterior margin of the membrane at the base of the wing is frequently strengthened to form a cord-like structure known as the *axillary cord*. The latter arises, on either side, from the posterior lateral angle of the notum (Fig. 27, A).

Situated around the base of each wing is a variable number of *articular sclerites* which consist of the tegulae, the humeral plate and the axillaries (Fig. 27). The *tegulae* (paraptera of some authors) are a pair of small scale-like sclerites carried at the extreme base of the costa of each fore wing: they are rarely present in connection with the hind wings. Tegulae are best developed in the Lepidoptera, Hymenoptera and Diptera, being especially large in the first mentioned order. The *axillaries* (pteralia) participate in the formation of the complex joint by which the wing is articulated to the thorax. According to Snodgrass they occur in all winged insects but are much modified in the Ephemeroptera and Odonata, presumably because these insects do not flex the wings over the abdomen at rest. As a rule, three of these sclerites are present, but a fourth occurs in the Orthoptera and Hymenoptera. The *first axillary* articulates with the anterior notal process and is associated with the base of the subcostal vein. The *second axillary* articulates partly with the preceding sclerite and, as a rule, partly with the base of the radius (see p. 60). The *third axillary* usually articulates with the posterior notal process and with a group of anal veins. When a *fourth axillary* occurs it has a double articulation, i.e. with the posterior notal process proximally and with the third axillary distally. For a more detailed treatment of these sclerites see Snodgrass (1929), La Greca (1947), Tannert (1958) and Sharplin (1963–64).

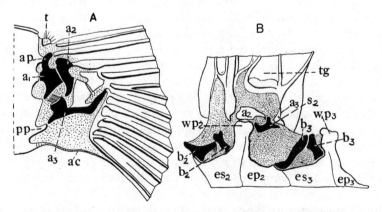

FIG. 27   Basal articulations of the wings. A. Wing base and tergal articulation. B. Upper part of pleuron of a grasshopper with base of left tegmen (*tg*) upraised

a_1–a_3, axillary sclerites; *ac*, axillary cord; *ap*, anterior notal wing process; b_2, basalar sclerites of mesothorax; b_3, basalar sclerites of metathorax; ep_2, epimeron of mesothorax; ep_3, epimeron of metathorax; es_2, episternum of mesothorax; es_3, episternum of metathorax; *pp*, posterior notal wing process; s_2, subalar sclerite of mesothorax; *t*, tegula; wp_2, pleural wing process of mesothorax; wp_3, pleural wing process of metathorax. Adapted from Snodgrass.

In addition to the foregoing, there are present in many insects small epipleural sclerites which are located below the insertions of the wings (Fig. 27). Although they are regarded as parts of the pleura they may be conveniently referred to here on account of their close association with the wing attachments. These sclerites are separated into two series by the pleural wing process. The anterior or *basalar sclerites* are never more than two in number, and lie just above the episternum, while the posterior or *subalar sclerite* is almost always single and lies behind the pleural wing process and above the epimeron.

Modifications of Wings – Although wings are usually present in adult insects many species are apterous. This condition is a constant feature of the Apterygota, where it is a primitive character, but in the Pterygota the loss of wings is secondary (La Greca, 1954). The parasitic orders Mallophaga, Siphunculata and Siphonaptera are exclusively apterous, and the same applies to the sterile castes of the Isoptera and Formicidae, and to the females of the Coccoidea, Strepsiptera and Embioptera. Among other Pterygota, apterous forms are of more casual occurrence, and often confined to a single sex or species. Thus, in a few moths (*Erannis defoliaria*, etc.), the females alone are apterous, while in the Chalcid genus *Blastophaga* it is the male which has lost the wings. Transitional forms between the apterous and the fully winged condition are found. In the moth *Diurnea fagella*, for example, the wings of the female are lanceolate appendages, little more than half the length of those of the male and useless for flight. In the winter moths (*Operophtera*), and in the fly *Clunio marinus*, they are reduced in the female to the condition of small flap-like vestiges. Reduction or loss of the wings is accompanied by various changes in the sclerites and musculature of the thorax; much information on this is available for the Heteroptera (Larsén, 1950), Coleoptera (Smith, 1964) and Hymenoptera (Reid, 1941).

Throughout the Diptera, and in the males of the Coccoidea, the hind wings are represented only by a pair of slender processes termed *halteres*. Among the Coleoptera, the fore wings are much hardened to form horny sheaths or *elytra*, which protect the hind wings when these are in repose. In *Atractocerus*, and the males of the Strepsiptera, the elytra are reduced to small scale-like appendages. On the other hand, in certain Carabidae and Curculionidae, the hind wings are atrophied and the function of flight is lost. In the Heteroptera the fore wings are thickened at their bases like elytra and are therefore termed *hemelytra*. Among the Orthoptera, Dictyoptera and Phasmida, the fore wings are hardened and leathery, being known as *tegmina*.

The Wing-coupling Apparatus – There seems little doubt that in the primitive Pterygota the fore and hind pairs of wings moved independently of each other (as in the Isoptera and Odonata), and that coincidence of motion was a later acquisition associated with the development of a wing-coupling apparatus (Fig. 28). Among the Panorpoid orders (Tillyard, 1918*a*) the coupling depends on modifications of the bases of the wings, the fore wing possessing on its posterior margin a small jugal lobe while the anterior margin of the hind wing is produced into a small humeral lobe. In the Mecoptera, both lobes bear a few relatively long bristles, the jugal ones lying on top of the hind wing in flight while the humeral bristles form the frenulum which presses against the under side of the fore wing. In the Trichoptera and some Monotrysian Lepidoptera (Philpott, 1924; 1925), the

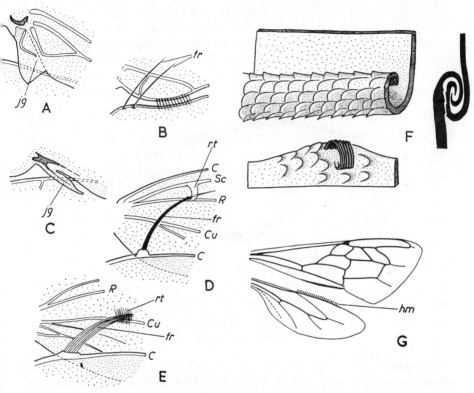

FIG. 28   Methods of wing-coupling in insects. A. *Rhyacophila* (based on Comstock, 1918). B. *Panorpa*. C. A Hepialid moth (based on Comstock, *l.c.*). D. Male frenate Lepidopteran. E. Female frenate Lepidopteran. F. An aphid, *Drepanosiphon* (*after* Weber, 1930). G. *Apis*.

fr, frenulum; *hm*, hamuli; *jg*, jugum; *rt*, retinaculum.

jugal area is produced into a lobe-like fibula or more elongate jugum which lies on top of the hind wing during flight but may be folded beneath the fore wing at rest. Frenular bristles are absent or small in these insects but there is sometimes a more distally placed series of costal spines (the pseudofrenulum) on the hind wing which functions independently of the jugum by pressing against the anal area of the fore wing (e.g. *Sabatinca*), or a series of interlocking hairs on the basal half of the hind margin of the fore wing and the fore margin of the hind wing (e.g. *Mnesarchaea*). In most higher

Lepidoptera, a pseudofrenulum occurs only in some lower families (Braun, 1919), the jugum is almost invariably lost and the frenulum usually assumes great importance (Braun, 1924). In females it generally consists of a group of stout bristles which lies beneath the extended fore wing and engages there in a retinaculum formed from a patch of hairs near the cubitus. In males, the frenular bristles are fused into a single stout structure which is normally held by a curved process from the subcostal vein of the fore wing as well. Finally, in the Papilionoidea and many Bombycoidea, the wings are coupled simply by overlapping basally – the so-called amplexiform method. Among other orders, the Hymenoptera have a hamulate type of wing-coupling in which a row of small hooks (hamuli) on part of the costal margin of the hind wing catch in a sclerotized fold along the hind margin of the fore wing. In many Hemiptera the wings are held together in flight by various small hooks or

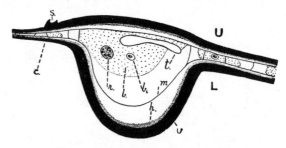

FIG. 29   Transverse section of a vein and adjacent portion of the wing-membrane of a moth, *Notodonta camelina*

U, upper surface; L, lower surface; *v*, vein; *h*, remains of epidermis; *t*, trachea; *c*, cuticle; b_1, blood corpuscle and *b*, plasma; *m*, basement membrane; *r*, 'Semper's rib'; *s*, scale socket.

folds along the wing-margins (Weber, 1930) while in the Psocoptera the costa of the hind wing is held by a spiny or hooked process of the node where the second cubital vein of the fore wing reaches the margin.

The **Structure and Development of Wings** – Wings are thin plate-like expansions of the integument which are strengthened by a framework of hollow sclerotized tubes known as *veins*. A wing is composed of upper and lower layers which may readily be separated in an insect which has just emerged from the pupa. The veins are much more strongly sclerotized than the wing-membrane and each usually encloses a small central trachea. A fine nerve fibre accompanies the larger veins of many insects (Fudalewicz-Niemczyk, 1963) and a degenerate trachea known as 'Semper's rib' is present in Lepidoptera alongside the ordinary trachea within the vein cavity (Fig. 29). When an adult insect emerges the veins contain blood which has been observed to circulate through them, and even in the fully formed wings the circulation is often still maintained (Yeager and Hendrickson, 1934; Clare and Tauber, 1940).

Detailed accounts of the development of the wings are available for the Exopterygote *Pteronarcys* (Holdsworth, 1940; 1942) and for several Endopterygotes (Köhler, 1932; Behrends, 1935; Hundertmark, 1935; Kuntze, 1935; Waddington, 1941; see also Clever, 1959). In insects with an incomplete metamorphosis, the wings develop externally and appear in the early instars along a line where the suture between tergum and pleuron later develops. They are usually so directly continuous with the tergum that they are regarded as postero-lateral outgrowths of that region. The external changes during growth are comparatively slight and consist mainly of an increase in size at each moult. In the Odonata and Orthoptera, however, the wing-pads in the later immature instars have twisted about their points of attachment so that the costal margins lie dorsally and the hind wings cover the fore wings. The wings then resume their normal position when the adult emerges. Internally, the developing wing-pad undergoes many histological changes. At first there is merely a thickening of the epidermis overlain by cuticle. Later, as the sac-like evagination develops, the bases of the epithelial cells become drawn out into

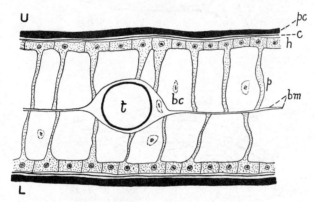

FIG. 30   Transverse section of a portion of the wing of a pupal insect

U, upper surface; L, lower surface; *pc*, pupal cuticle; *c*, cuticle of developing wing; *h*, epidermis of wing; *p*, process of epidermal cell; *bm*, basement membrane; *t*, trachea in cavity of developing vein; *bc*, blood corpuscle.

long processes, so imparting a spongy texture to the tissues of the wing-pad, and the basement membranes of the lower and upper epithelia become apposed for most of their area so as to form the so-called middle membrane (Fig. 30). Blood-filled spaces or *lacunae* soon develop in the wing-pad (appearing in the 2nd instar of *Pteronarcys*, for example) and alter as growth proceeds so that in the later instars the lacunae correspond in arrangement to the veins of the adult wings; the veins, in fact, arise by differential sclerotization of the integument adjacent to the lacunae. While each of the principal lacunae is developing, a tracheal branch and a nerve grow into it from the base of the wing, the lacunae apparently offering the paths of least resistance. Though the association between tracheae and lacunae – and therefore between tracheae and subsequent venation – is not always exact, the pattern of wing-pad tracheation is often of great value in deciding the homologies of the veins. A generalized arrangement of tracheae is depicted in Fig. 31. Two distinct groups of tracheae enter the wing – a costo-radial group and a cubito-anal group, and while in some forms (Blattidae, Plecoptera and Homoptera) the two groups remain separate, it is more usual for them to be united by a fine transverse basal trachea which is part of the basic tracheal framework (Whitten, 1962). It may be noted that tracheae do not precede the cross-veins of the adult wing and that the tracheal and venational patterns sometimes differ appreciably. Ontogenetic and phylogenetic changes occur in the tracheal arrangement (e.g. Leston, 1962; Whitten, 1962) without necessarily affecting the venational pattern. In some cases these changes may be due to the tendency of tracheoles to migrate actively towards areas of high oxygen demand,

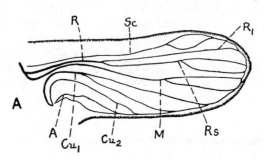

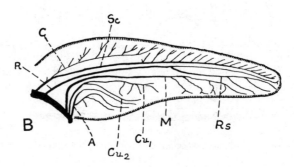

FIG. 31   Tracheation of developing fore wings of A. *Nemoura* (Plecoptera) and B. *Conocephalus* (Tettigoniidae). Adapted from Comstock. (For explanation of lettering, see p. 60)

drawing the tracheae after them (Wigglesworth, 1954; Smart, 1956). In the final stages of wing-development the epidermis secretes the wing-membrane and the thickened walls of the veins and the folded adult wings take shape within the cuticle of the last pre-imaginal instar. On the emergence of the adult the wing is inflated to its full size by blood-pressure and the cuticle hardens. The epidermis degenerates and little trace of its cells remains in the fully hardened wings.

In insects with a complete metamorphosis the wings arise from imaginal buds or thickenings of the epidermis, usually in the neighbourhood of one of the larger tracheae, and are evident in the very young larva or even the embryo. These buds become enlarged and folded or invaginated in various ways, sometimes forming pocket-like sacs, or *peripodial cavities* (Fig. 32), from the bottom of which the thickened portion of the bud ultimately becomes evaginated. At the same time, the walls of the pocket become extremely thin but retain their connection with the

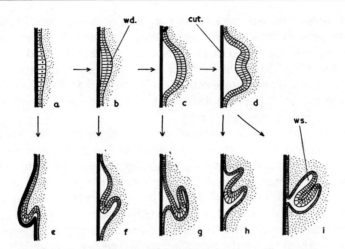

FIG. 32   Various modes of wing development (*after* Weber, 1933)

a–d denote various early stages, e–i denote the corresponding later stages. e, free external wing-bud (most Exopterygotes); f, free internal wing-bud (many Coleoptera); g, reversed wing-bud (Neuroptera, Lepidoptera); h, simple sunken wing-bud (Nematocera, some Coleoptera); i, stalked wing-bud (Cyclorrhapha); *cut*, cuticle; *wd*, wing-disc; *ws*, wing-sac.

general epidermis. At a later stage, the evaginated portion elongates and comes to hang downwards; it is this evaginated portion which eventually forms the wing. Internal histological changes, comparable to those occurring in exopterygotes, take place during the prepupal period and the wing-rudiment becomes pushed out of its pocket and comes to lie just beneath the cuticle. On the assumption of the pupal stage, the wing-rudiments become evident externally along the sides of the body. When the imago emerges, the wings appear as small wrinkled sacs which gradually become distended by blood-pressure, and attain their full development usually several hours afterwards. During their later stages of development the wing-buds become supplied with tracheoles. In *Pieris*, for example, during the 4th larval instar a series of tracheoles arise as proliferations of the epithelium of the large tracheae associated with the wing-bud. These tracheoles may be termed the larval or provisional tracheoles, and they extend in bundles into the developing lacunae. A little later, the true wing tracheae develop as tubular outgrowths of the large tracheae, and extend into the vein cavities along with the larval tracheoles, which they supplant. During the early pupal stage the latter degenerate and disappear. Although the tracheation of the pupal wings has yielded important data for ascertaining the homologies of the wing-veins of the adults, there is in some orders (e.g. Trichoptera) a wide divergence between the two systems. In such cases, comparisons among the more generalized types and palaeontological evidence may aid in settling the identity of the principal veins.

Venation – The complete system of veins of a wing is termed its *venation*. It presents characters of great systematic importance, but unfortunately the various systems of nomenclature in use are confusing both to the student and the specialist. The older systems were established by entomologists whose work was uninfluenced by ideas of evolution. As a result the terminology of an author was usually only applicable within the limits of the particular order of insects which he studied. Attempts to introduce a common terminology for the venation achieved little success until the work of Comstock and Needham (1898). By an extensive study of the tracheae which precede, and in a general sense coincide with the positions of the veins, these writers constructed a hypothetical system of venation from which all others might be derived (Comstock, 1918).

While the researches of Comstock and Needham form the basis for the interpretation of venation, their original conceptions have been modified by the later work of Lameere (1922), Hamilton (1972) and many papers by Martynov and Tillyard. A less exclusive emphasis is now placed on tracheational studies and, in spite of the opposition of Needham (1935), there is a tendency to emphasize the fact that among the lower, generalized orders the wings are longitudinally plicated after the manner of a partially opened fan. Those veins which follow the ridges are termed *convex veins* and those which follow the furrows *concave veins*. These features are well exhibited not only in the early fossil orders but also in the Ephemeroptera and Odonata where the alternation of convex and concave veins is probably a mechanical adaptation to provide increased rigidity of the thin wing-membrane (Edmunds and Traver, 1954). In the higher orders the fluting tends to become obscured by flattening of the wing-membrane or the development of secondary curvatures imposed by mechanical considerations. The fact that in the lower orders the convex or concave condition is constant for individual veins helps in determining their homologies.

The hypothetical primitive venational pattern recognized here is shown in Fig. 33 which also indicates the nomenclature and abbreviations in common use. The *costa* (C) is unbranched and convex while the *subcosta* (Sc) is rarely branched and concave. The *radius* (R) is typically 5-branched: its main stem is convex and divides into two, of which the first branch (R_1) passes directly to the wing-margin: the second branch or *radial sector* (Rs) is concave and divides into four veins (R_2 to R_5). The *media* (M) divides into an *anterior media* (MA), which is convex and 2-branched (MA_1, MA_2), and a concave *posterior media* (MP), which is 4-branched (MP_1) to MP_4). The *cubitus* (Cu) divides into two main branches, the first cubitus (Cu_1) being convex and the second cubitus (Cu_2) concave; the first cubitus may subdivide into anterior (Cu_{1a}) and posterior (Cu_{1b}) veins. There follow three *anal veins* ($1A$ to $3A$) which are usually convex, or $2A$ may be concave.

In many insects there are two strongly convex veins, R_1 and Cu_1, which are easily noted and therefore facilitate the identification of the other veins. A complete media is found in many Palaeozoic fossil insects and in the Ephemeroptera among recent forms. For the most part MA atrophies and consequently the media in recent insects is generally MP, although it is usually designated by the symbol M. The Odonata and Plecoptera, however, seem to be unusual in retaining MA and not MP, while further research is needed into the constitution of the media in other Orthopteroid insects. The

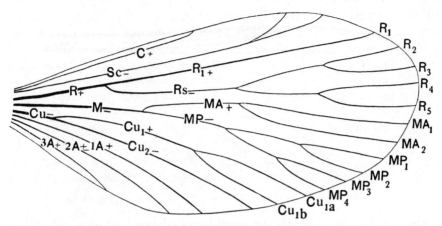

FIG. 33 Hypothetical primitive venation (for abbreviations see p. 60) (convex veins +, concave veins −)

anal veins are extremely variable and in wings with a reduced anal area one or more are atrophied. On the other hand, in insects with a well-developed anal lobe the anal veins may be freely branched, probably through sub-division of $2A$. It may be noted that some modern authorities denote by MA and MP the two main branches of the posterior media, while Cu_1 and Cu_2 are sometimes known as CuA and CuP respectively. Forbes (1933) and Snodgrass (1935) have introduced changes in the nomenclature of the cubito-anal veins which do not seem to be widely accepted while Vignon (1929) has an unorthodox modification of the Comstock–Needham system.

Deviation from the primitive venational type has occurred in two ways, by reduction and addition. In many insects the number of veins is less than in the hypothetical type, and the reduction has been brought about by the degeneration or complete atrophy of a vein, or of one or more of its branches, or by the coalescence of adjacent veins. Atrophy explains the presence of only a single well-developed anal vein in *Anisopus* (Fig. 34) and other Diptera, while the occurrence in this genus of a single vein R_{2+3}, in place of the separate veins R_2 and R_3, is due to coalescence. Similarly R_4 and R_5 have coalesced, and the single vein thus formed is referred to as R_{4+5}. Coalescence takes place in two ways. The point at which two veins diverge may become gradually pushed towards the margin of the wing until the latter is reached, and only a single vein remains evident. In the second

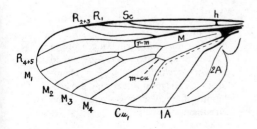

FIG. 34
Wing of *Anisopus punctatus*. (For ex-planation of lettering, see p. 60)

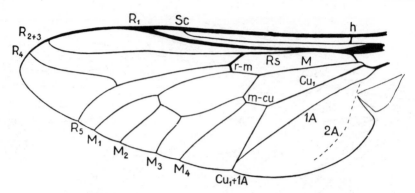

FIG. 35   Wing of *Tabanus*. (For explanation of lettering, see p. 60)

method, the apices of the two veins may approximate, and ultimately fuse at a point on the wing-margin: coalescence of this type takes place inwardly towards the base of the wing. The first type is well shown in the radial veins of *Anisopus*, and the second method in the apical fusion of $1A$ and Cu_I in *Tabanus* (Fig. 35). The homology of a particular vein is often difficult to determine and resort has to be made to comparison with allied forms (including fossils), which exhibit transitional stages in reduction, or to a study of the preceding tracheation. An increase in the number of veins may be due either to an increase in the number of branches of a principal vein, or to the development of secondary longitudinal veins, between pre-existing veins. In no instance is there any increase in the number of principal veins present. For a more detailed acquaintance with the various modifications of the wing-veins the works by Comstock (1918) and Séguy (1959) should be consulted.

In the wings of some more generalized insects, such as the fossil Palaeodictyoptera, an irregular network of veins is found between the principal longitudinal veins, but no definite cross-veins are present (Fig. 36). To this primitive meshwork Tillyard (1918*b*) gave the name *archedictyon*. It

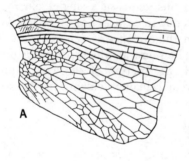

FIG. 36
A. Portion of a wing of a Carboniferous insect (*Hypermegethes*) showing archedictyon. *After* Handlirsch. B. Diagram illustrating the evolution of regular cross-veins. *After* Needham

appears to have undergone reduction in the Endopterygota, though it is probably represented by the dense reticulation present in certain orders of Exopterygota such as the Odonata. Needham (1903) from his studies of the wings of the latter order has discussed the transformation of such an irregular network into regular transverse veins (Fig. 36). Transitional stages in the evolution of definite cross-veins may also be observed in wings of the more specialized Palaeodictyoptera and among living Orthoptera, where both irregular and definite cross-veins occur in the same wing. According to Tillyard, however, true cross-veins are later developments; they are never preceded by tracheae and are almost always devoid of macrotrichia. Veinlets, on the other hand, are primitive and constitute the finer twigs of a principal vein; they are preceded by tracheae and carry macrotrichia (Fig. 37). It is

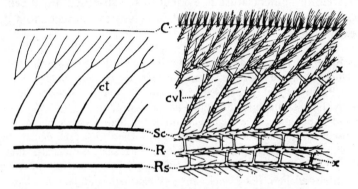

FIG. 37 Portion of costal area of fore wing of *Psychopsis elegans* (Neuroptera) with the corresponding tracheation (to the left) of the pupal wing

C, costa; *ct*, tracheae preceding the costal veinlets; *cvl*, costal veinlets; *x*, cross-veins; *R*, radius; *Rs*, radial sector; *Sc*, subcosta. *After* Tillyard, *Proc. Linn. Soc. N.S.W.* 43.

probable that homologous cross-veins do not exist in many orders but their positions in some cases are so constant that analogies, if not homologies, can be traced and similar names are applicable. The following cross-veins are the most important and their symbols are given in brackets.

The *humeral cross-vein* (*h*) extending from the subcosta to the costa, near the humeral angle of the wing.

The *radial cross-vein* (*r*) extending from R_I to the radial sector (R_s).

The *sectorial cross-vein* (*s*) extending from the stem R_{2+3} to R_{4+5} or from R_3 to R_4.

The *radio-medial cross-vein* (*r-m*) extending from the radius to the media, usually near the middle of the wing.

The *medial cross-vein* (*m*) extending between M_2 and M_3.

The *medio-cubital cross-vein* (*m-cu*) extending from the media to the cubitus.

The veins divide the wings into spaces or *cells*. In the Comstock–Needham system the terminology of the cells is derived from the veins which form their anterior margins. The cells fall into two groups, i.e. basal cells and distal cells. The former are bounded by the main stems of the principal veins, and the latter by the branches of the forked veins. Thus the cell situated behind the main stem of the radius, near the base of the wing, is cell R, while the cell behind the first branch of the radius is cell R_1. When two veins coalesce the cell that was between them becomes obliterated. Thus when veins R_2 and R_3 fuse, as in *Anisopus* (Fig. 34), the cell situated behind the vein R_{2+3} is referred to as cell R_3, and not cell R_{2+3}, cell R_2 having disappeared. Not infrequently two or more adjacent cells may become confluent owing to the atrophy of the vein or veins separating them. The compound cell is then designated by a combination of the abbreviations applied to the original separate cells. Thus, a cell resulting from the fusion of cells R and M is referred to as cell $R + M$. The advantage of this relatively simple system of nomenclature is evident in the so-called discal cell, a term used in at least four separate orders of insects with reference to a different cell in each case.

Insect Flight – The aerodynamic theory of insect flight is complicated and not fully developed. For general reviews see Hocking (1953), Johnson (1969), Nachtigall (1974) and Rainey (1976). Expressed in the simplest terms, however, normal flight depends on the creation by a propellor-like action of the wings of a zone of low pressure in front of and above the insect and one of high pressure behind and below it, the consequent movement being a resultant of the thrust provided by the insect and forces due to gravity and air-resistance. The skeleto-muscular mechanisms involved in the movements of the wings are outlined on p. 92 but other aspects of flight are discussed here (Chadwick, 1953; Weis-Fogh and Jensen, 1956; Pringle, 1957, 1965, 1968). Before flight can occur the thoracic flight-muscles must attain a sufficiently high temperature and for this reason some insects carry out preliminary vibrations of the wings before flight, thereby raising the temperature of the muscles to over 30° C, for example, in *Bombus*, and some Lepidoptera (Krogh and Zeuthen, 1941; Dorsett, 1962; Kammer, 1968).

Detailed experimental analyses of wing-movements during flight are available for only a few species and probably differ in different insects (Weis-Fogh, 1956a, 1956b; Jensen, 1956; Jensen and Weis-Fogh, 1962; Nachtigall, 1966; Neville, 1960; Vogel, 1966–67). Generally speaking, however, in normal flight the wing (or coupled wings) moves in a path such that a point on its surface describes an irregular loop or an elongate figure-of-eight with respect to the wing-base, the long axis of the plane of vibration being inclined at an angle to the long axis of the insect (Fig. 38) while the angle at which the surface of the wing is held changes throughout the cycle. The greater part of the propulsive force is generated on the downbeat and the relative sizes of the vertical and horizontal components of this force vary with the angle of the plane of vibration to the horizontal. Thus, in insects which are hovering (Weis-Fogh, 1972), the plane of vibration tends to be more horizontal, while the maximum amount of forward movement would be achieved as the plane of vibration becomes vertical. Another mechanism, producing movement in the vertical plane, is indicated by Hollick's (1940) finding that in *Muscina* a reduction in the amplitude of the wing-beat causes the thrust vector to intersect the long axis of the insect's body behind the centre of gravity, so altering the cephalo-caudal couple around this centre. Differences in amplitude between the wing-beats of one side and those of the other cause a lateral turning movement away from the side with the greater amplitude (Stellwaag, 1916; Govind, 1972). Finally, some insects are also capable of backward flight by shifting the plane of vibration, sometimes so far that the wings actually move upwards on the 'down-beat' (Fig. 38). There are considerable differences in the speed of different species in flight. *Panorpa communis*, for example, has been recorded to fly at only 0·5 metres per second whereas *Aeshna mixta* can move at 7 metres per second and it is unlikely that any insect exceeds about twice this velocity.

The great metabolic activity of insects during flight is indicated by their greatly increased oxygen consumption under these conditions (e.g. Davis and Fraenkel, 1940) and as flight continues uninterruptedly for many hours in some species, a reserve of oxidizable material is required (Sacktor, 1970). In Diptera and Hymenoptera this consists mainly of carbohydrates such as glycogen and blood

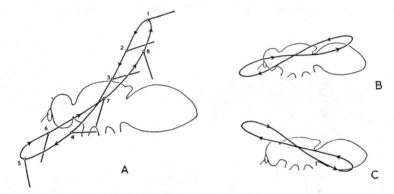

FIG. 38   Wing movements in insect flight (*partly after* Chadwick, 1953). A, forward flight; 1–8 successive positions of wing. B, hovering. C, backward flight

sugars. In Lepidoptera and Orthoptera it is usually fat, though some insects use both carbohydrate and fat and *Glossina* employs the amino-acid proline as an energy-furnishing reserve. The complicated movements required in flight are co-ordinated by a number of reflex mechanisms, some of which have now been studied in detail (Pringle, 1968; Wilson 1968). Fraenkel (1932) investigated the reflex stimulation to flight that occurs when the tarsi are deprived of contact with a substratum together with other tactile stimuli to flight perceived by different parts of the body. Continuous flight may also require the stimulus of a moving current of air which in *Schistocerca* is perceived by setae on the head (Weis-Fogh, 1950; Guthrie, 1966) and in *Muscina* by the antennae (Hollick, 1940). Reflexes involving campaniform sensilla on the costa and subcosta control various features of wing-movement in *Schistocerca* (Gettrup, 1966). Other mechanisms ensure the appropriate orientation of the insect during flight. Visual stimuli sometimes play a role here and Mittelstaedt (1950) found that Anisopteran dragonflies always fly with the dorsal surface orientated towards the light, while optomotor reactions regulate flight in relation to movements of the visual field in many insects. Reflexes mediated by the antennae were also discovered by Hollick (1940) to modify the path of the wing, and therefore the flight-characteristics, of *Muscina*, but the role of the halteres of Diptera and male Strepsiptera in controlling equilibrium during flight is perhaps the best known of these mechanisms (Pringle, 1948; Schneider, 1953). Removal of the halteres in *Calliphora* has little or no effect on the beat-frequency, amplitude and duration of flight but markedly affects the stability of the flying insect. The halteres vibrate during flight with the same frequency as the wings but in opposite phase and, by virtue of the relatively heavy terminal knob, they function as gyroscopic organs. Groups of sensilla at the base of the halteres are stimulated by deformations of the integument when the halteres vibrate and the resulting pattern of nervous impulses is modified by the addition of torques due to the turning movements of the fly. The system is apparently most sensitive to movements in the 'yawing' (horizontal) plane, and deviations from the flight-path in this plane can readily be corrected by the fly. There is also some evidence that the halteres act as stimulatory organs, increasing the sensitivity with which the fly reacts to other stimuli during flight (von Buddenbrock, 1919) and in some species removal of the halteres interferes with walking.

For certain physiological properties of the flight muscles, see p. 95.

FIG. 39
A Carboniferous insect (*Stenodictya lobata*) showing prothoracic wing-like expansions. From Carpenter, *after* Handlirsch

Origin of Wings and Flight – It is now generally accepted that wings arose, perhaps in the early Devonian, as lateral expansions of the thoracic terga (Hamilton, 1971). Such expansions, often known as paranota, are found on the prothorax of the Palaeodictyoptera and other fossil insects (Fig. 39), in the early stages of *Kalotermes*, and in various Mantids, Lepismatids and Hemiptera. There is, indeed, an inherent tendency towards the development of paranota in various Arthropods (e.g. the so-called pteromorphs of Oribatid mites discussed by Woodring, 1962). Further, in a few cases such as *Lepisma* (Šulc, 1927) and *Hemidoecus* (Evans, 1939) the paranotal expansions are supplied with tracheae arranged in a pattern somewhat similar to that in developing wing-pads (Fig. 40). It has been suggested that the paranotal lobes of insects were at first connected with epigamic behaviour (Alexander and Brown, 1963) but it is perhaps more likely that they were associated with locomotion from the beginning. They may, for example, have assisted take-off into the winds or convection currents that disperse small insects in the 'aerial plankton' (Wigglesworth, 1963). Quite small paranota could also control the attitude of falling insects and might subsequently have increased in size through an additional selective process to improve the angle at which a gliding flight occurred (Hinton, 1963; Flower,

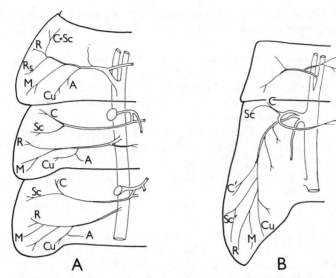

FIG. 40   Tracheation of thoracic paranota of *Lepisma saccharina* (A), and of fore wing pad in *Cloeon dipterum* (B) (*after* Šulc, 1927). The terminology of the paranotal tracheae follows the homologies suggested by Šulc

A, anal; C, costa; Cu, cubitus; M, media; R, radius; Rs, radial sector; Sc, subcostal. In *Cloeon* the tracheae C and Sc are eliminated during the later developmental stages and their places are taken by the secondary tracheae denoted as C' and Sc'

1964). Sustained flight could then arise at a still later stage through the evolution of a basal articulation and the development of flapping movements using tergocoxal muscles that were concerned primitively with leg movements (Tiegs, 1955). The remains of the earliest winged insect from the Upper Carboniferous give no indications of how and where the Pterygota arose. The fact that in many insect groups the newly emerged adults show a slow gliding type of flight linked to dispersal suggests, however, that the earliest winged insects may have evolved in temporary habitats of small, erect plants in which dense insect populations sometimes developed (Johnson, 1963; Leston, 1963). For a modern restatement of the theory that wings arose from thoracic gills see Wigglesworth (1973).

Literature on the Thorax and its Appendages

ALEXANDER, R. D. AND BROWN, W. L. (1963), Mating behaviour and the origin of insect wings, *Occ. Paper. Mus. Zool. Univ. Mich.*, **628**, 1–19.

ALEXANDER, R. M. (1968), *Animal Mechanics*, Sidgwick & Jackson, London, 346 pp.

ARNHART, L. (1923), Das Krallenglied der Honigbiene, *Arch. Bienenkunde*, **5**, 37–86.

ARNOLD, J. W. (1963), A note on the pterostigma in insects, *Can. Ent.*, **95**, 13–16.

BARTH, R. (1937), Muskulatur und Bewegungsart der Raupen, zugleich ein Beitrag zur Spannbewegung und Schreckstellung der Spannerraupen, *Zool. Jb.* (*Anat.*), **62**, 507–566.

—— (1954), O aparelho saltatorio do Halticineo *Homophoeta sexnotata* Ha. (Coleoptera), *Mem. Inst. Osw. Cruz*, **52**, 365–376.

BEHRENDS, J. (1935), Ueber die Entwicklung des Lakunen- Ader- und Tracheensystems während Puppenruhe im Flügel der Mehlmotte *Ephestia kühniella* Zeller, *Z. Morph. Ökol. Tiere*, **30**, 573–596.

BRAUN, A. F. (1919), Wing structure of Lepidoptera and the phylogenetic and taxonomic value of certain persistent Trichopterous characters, *Ann. ent. Soc. Am.*, **12**, 349–366.

—— (1924), The frenulum and its retinaculum in the Lepidoptera, *Ann. ent. Soc. Am.*, **17**, 234–256.

BUDDENBROCK, W. VON (1919), Die vermutliche Lösung der Halterenfrage, *Pflügers Arch. ges. Physiol.*, **175**, 125–164.

CHADWICK, L. E. (1953), In: Roeder, K. D., *Insect Physiology*, Wiley, New York, pp. 577–655.

CLARE, S. AND TAUBER, O. E. (1940), Circulation of haemolymph in the wings of the cockroach *Blattella germanica* L. I. In normal wings, *Iowa St. J. Sci.*, **14**, 107–127.

CLEVER, U. (1959), Über experimentelle Modifikationen des Geäders und die Beziehungen zwischen den Versorgungssystemen im Schmetterlingsflügel. Untersuchungen an *Galleria mellonella*, *Arch. EntwMech. Org.*, **151**, 242–279.

COMSTOCK, J. H. (1918), *The Wings of Insects*, Comstock Publ. Co., New York, 430 pp.

COMSTOCK, J. H. AND NEEDHAM, J. G. (1898–99), The wings of insects, *Am. Nat.*, **32**, 43–48, 81–89, 231–257, 413–422, 560–565, 769–777, 903–911; **33**, 117–126, 573–582, 845–860.

DASHMAN, T. (1953), Terminology of the pretarsus, *Ann. ent. Soc. Am.*, **46**, 56–62.

DAVIS, R. A. AND FRAENKEL, G. (1940), The oxygen consumption of flies during flight, *J. exp. Biol.*, **17**, 402–407.

DORSETT, D. A. (1962), Preparation for flight by hawkmoths, *J. exp. Biol.*, **39**, 579–588.

DUPORTE, E. M. (1965), The lateral and ventral sclerites of the insect thorax, *Can. J. Zool.*, **43**, 141–154.

EDMUNDS, G. F. AND TRAVER, J. R. (1954), The flight mechanics and evolution of the wings of Ephemeroptera, with notes on the archetype insect wing, *J. Wash. Acad. Sci.*, **44**, 390–400.

EVANS, J. W. (1939), The morphology of the thorax of the Peloridiidae, *Proc. R. ent. Soc. Lond.*, (A), **14**, 143–150.

FERRIS, G. F. (1940), The myth of the thoracic sternites of insects, *Microentomology*, **5**, 87–90.

FLOWER, J. W. (1964), On the origin of flight in insects, *J. Insect Physiol.*, **10**, 81–88.

FORBES, W. T. M. (1933), The axillary venation of the insects, *Proc. 5th int. Congr. Ent.*, **2**, 277–284.

FRAENKEL, G. (1932), Untersuchungen über die Koordination von Reflexen und automatisch-nervösen Rhythmen bei Insekten. I–IV. *Z. vergl. Physiol.*, **16**, 371–462.

FUDALEWICZ-NIEMCZYK, W. (1963), L'innervation et les organes sensoriels des

Diptères et comparaison avec l'innervation des ailes d'insectes d'autres ordres, *Acta Zool. cracov.*, **8**, 351–462.

GETTRUP, E. (1966), Sensory regulation of wing twisting in locusts, *J. exp. Biol.*, **44**, 1–16.

GILLETT, J. D. AND WIGGLESWORTH, V. B. (1932), The climbing organ of an insect, *Rhodnius prolixus* (Hemiptera, Reduviidae), *Proc. R. Soc.* (B), **111**, 364–376.

GOUIN, F. J. (1959), Le thorax imaginal des insectes à la lumière des travaux récents, *Année biol.*, **35**, 269–303.

GOVIND, C. K. (1972), Differential activity in the coxo-subalar muscle during directional flight in the milkweed bug *Oncopeltus*, *Can. J. Zool.*, **50**, 901–905.

GUTHRIE, D. M. (1966), The function and fine structure of the cephalic airflow receptor in *Schistocerca gregaria*, *J. Cell Sci.*, **1**, 463–470.

HAMILTON, K. G. A. (1971), The insect wing. Part I. Origin and development of wings from notal lobes, *J. Kansas ent. Soc.*, **44**, 421–433.

—— (1972), The insect wing. Part II. Vein homology and the archetypal insect wing, *J. Kansas ent. Soc.*, **45**, 54–58.

HINTON, H. E. (1963), The origin of flight in insects, *Proc. R. ent. Soc. Lond.* (C), **28**, 24–25.

HOCKING, B. (1953), The intrinsic range and speed of flight of insects, *Trans. R. ent. Soc. Lond.*, **104**, 223–345.

HOLDSWORTH, R. P. (1940), The histology of the wing-pads of the early instars of *Pteronarcys proteus* Newport (Plecoptera), *Psyche*, **47**, 112–119; 714–715.

—— (1942), The wing development of *Pteronarcys proteus* (Pteronarcidae: Plecoptera), *J. Morph.*, **70**, 431–462.

HOLLICK, F. S. J. (1940), The flight of the dipterous fly *Muscina stabulans* Fallén, *Phil. Trans. R. Soc.*, *Ser. B*, **230**, 357–390.

HOLWAY, R. T. (1935), Preliminary note on the structure of the pretarsus and its possible phylogenetic significance, *Psyche*, **42**, 1–24.

HUGHES, G. M. (1965), Locomotion: terrestrial, In: Rockstein, M. (ed.), *The Physiology of Insecta*, Academic Press, New York, **2**, 227–254.

HUNDERTMARK, A. (1935), Die Entwicklung der Flügel des Mehlkäfers *Tenebrio molitor*, mit besonderer Berücksichtigung der Häutungsvorgänge, *Z. Morph. Ökol. Tiere*, **30**, 506–543.

JENSEN, M. (1956), Biology and physics of locust flight. III. The aerodynamics of locust flight, *Phil. Trans. R. Soc.*, *Ser. B*, **239**, 511–552.

JENSEN, M. AND WEIS-FOGH, T. (1962), Biology and physics of locust flight. V. Strength and elasticity of locust cuticle, *Phil. Trans. R. Soc.*, *Ser. B*, **245**, 137–169.

JOHNSON, C. G. (1963), The origin of flight in insects, *Proc. R. ent. Soc. Lond.* (C), **28**, 26–27.

—— (1969), *Migration and Dispersal of Insects by Flight*, Chapman & Hall, London, 763 pp.

KAMMER, A. E. (1968), Motor patterns during flight and warm-up in Lepidoptera, *J. exp. Biol.*, **48**, 89–109.

KÖHLER, W. (1932), Die Entwicklung der Flügel bei der Mehlmotte *Ephestia kühniella* Zeller mit besonderer Berücksichtigung des Zeichnungsmusters, *Z. Morph. Ökol. Tiere*, **24**, 582–681.

KROGH, A. AND ZEUTHEN, E. (1941), The mechanism of flight preparation in some insects, *J. exp. Biol.*, **18**, 1–10.

KUNTZE, H. (1935), Die Flügelentwicklung bei *Philosamia cynthia* Drury, mit besonderer Berücksichtigung des Geäders, der Lakunen und der Tracheensysteme, *Z. Morph. Ökol. Tiere*, **30**, 544–572.

LA GRECA, M. (1947), Morphologia funzionale dell'articolazione alare degli Ortotteri, *Archo zool. ital.*, **32**, 271–327.

—— (1954), Riduzione e scomparsa delle ali negli insetti Pterigoti, *Archo zool. ital.*, **39**, 361–440.

LAMEERE, A. (1922), Sur la nervation alaire des insectes, *Bull. Acad. roy. Bruxelles*, **8**, 138–149.

LARSÉN, O. (1945), Das Meron der Insekten, *Förh. K. fysiogr. Sällsk.*, **15**, 96–104.

—— (1950), Die Veränderungen im Bau der Heteropteren bei der Reduktion des Flugapparates, *Opusc. Ent.*, **15**, 17–51.

—— (1966), On the morphology and function of the locomotor organs of the Gyrinidae and other Coleoptera, *Opusc. Ent.*, *Suppl.*, **30**, 1–242.

LESTON, D. (1962), Tracheal capture in ontogenetic and phylogenetic phases of insect wing development, *Proc. R. ent. Soc. Lond.* (A), **37**, 135–144.

—— (1963), The origin of flight in insects, *Proc. R. ent. Soc. Lond.* (C), **28**, 23–32.

MANTON, S. M. (1972), The evolution of arthropodan locomotory mechanisms. Part 10: Locomotory habits, morphology and evolution of the hexapod classes, *Zool. J. Linn. Soc.*, **51**, 203–400.

MATSUDA, R. (1970), Morphology and evolution of the insect thorax, *Mem. ent. Soc. Canada*, **76**, 431 pp.

MITTELSTAEDT, H. (1950), Physiologie des Gleichgewichtssinnes bei fliegenden Libellen, *Z. vergl. Physiol.*, **2**, 422–463.

NACHTIGALL, W. (1962), Funktionelle Morphologie, Kinematik und Hydromechanik des Ruderapparates von *Gyrinus*, *Z. vergl. Physiol.*, **45**, 193–226.

—— (1965), Locomotion: swimming (hydrodynamics) of aquatic insects, In: Rockstein, M. (ed.), *The Physiology of Insecta*, **2**, 255–281.

—— (1966), Die Kinematik der Schlagflügelbewegungen von Dipteren. Methodische und analytische Grundlagen zur Biophysik des Insektenflugs, *Z. vergl. Physiol.*, **52**, 155–211.

—— (1974), *Insects in Flight*, Allen & Unwin, London, 153 pp.

NEEDHAM, J. G. (1903), A genealogic study of dragonfly wing venation, *Publ. U.S. nat. Mus.*, **26**, 703–764.

—— (1935), Some basic principles of insect wing venation, *Jl N.Y. ent. Soc.*, **43**, 113–129.

NEVILLE, A. C. (1960), Aspects of flight mechanics in anisopterous dragonflies, *J. exp. Biol.*, **37**, 631–656.

NORBERG, R. A. (1972), The pterostigma of insect wings as an inertial regulator of wing pitch, *J. Comp. Physiol.*, **83**, 9–22.

PHILPOTT, A. (1924), The wing-coupling apparatus in *Sabatinca* and other primitive genera of Lepidoptera, *Rep. Aust. Ass. Advmt Sci.*, **16**, 414–419.

—— (1925), On the wing-coupling apparatus of the Hepialidae, *Trans. ent. Soc. Lond.*, **1925**, 331–340.

PRINGLE, J. W. S. (1948), The gyroscopic mechanism of the halteres of Diptera, *Phil. Trans. R. Soc.*, *Ser. B.*, **233**, 347–384.

—— (1957), *Insect Flight*, Cambridge Univ. Press, Cambridge, 133 pp.

—— (1965), Locomotion: Flight, In: Rockstein, M. (ed.), *The Physiology of Insecta*, **2**, 283–329.

—— (1968), Comparative physiology of the flight motor, *Adv. Insect Physiol.*, **5**, 163–227.

RAINEY, R. C. (ed.) (1976), Insect Flight, *Symp. R. ent. Soc. Lond.*, **7**, 296 pp.

REID, J. A. (1941), The thorax of the wingless and short-winged Hymenoptera, *Trans. R. ent. Soc., Lond.*, **91**, 367–446.

SACKTOR, B. (1970), Regulation of intermediary metabolism, with special reference to the control mechanisms in insect flight muscle, *Adv. Insect Physiol.*, **7**, 267–347.

SANDER, K. (1956), Bau und Funktion des Sprungapparates von *Pyrilla perpusilla* Walker (Homoptera–Fulgoroidea), *Zool. Jb. (Anat.)*, **75**, 383–388.

SARKARIA, D. S. AND PATTON, R. L. (1949), Histological and morphological factors in the penetration of DDT through the pulvilli of several insect species, *Trans. Am. ent. Soc.*, **75**, 71–82.

SCHENKE, G. (1965), Schwimmhaarsystem und Rudern von *Notonecta glauca*, *Z. Morph. Ökol. Tiere*, **55**, 631–640.

SCHNEIDER, G. (1953), Die Halteren der Schmeissfliege (*Calliphora*) als Sinnesorgane und als mechanische Flugstabilisatoren, *Z. vergl. Physiol.*, **35**, 416–458.

SÉGUY, E. (1959), Introduction à l'étude morphologique de l'aile des insectes, *Mém. Mus. Hist. nat. Paris, N.S.*, **21A**, 1–248.

SHARPLIN, J. (1963, 1964), Wing base structure in Lepidoptera I–III, *Can. Ent.*, **95**, 1024–1050, 1121–1145; **96**, 943–949.

SMART, J. (1956), A note on insect wing veins and their tracheae, *Q. Jl microsc. Sci.*, **97**, 535–539.

SMITH, D. S. (1964), The structure and development of flightless Coleoptera: a light- and electron microscope study of the wings, thoracic exoskeleton and rudimentary flight muscles, *J. Morph.*, **114**, 107–184.

SNODGRASS, R. E. (1927), Morphology and mechanism of the insect thorax, *Smithson. misc. Collns*, **80**, 1–108.

—— (1929), The thoracic mechanism of a grasshopper and its antecedents, *Smithson. misc. Collns*, **82**, 1–111.

—— (1935), *Principles of Insect Morphology*, McGraw-Hill, London and New York, 667 pp.

STELLWAAG, F. (1916), Wie steuern die Insekten im Flug? *Naturwissenschaften*, **4**, 256–259, 270–272.

ŠULC, K. (1927), Das Tracheensystem von *Lepisma* (Thysanura) und Phylogenie der Pterygogenea, *Acta. Soc. sci. nat. Moravia*, **4**, 227–344.

TANNERT, W. (1958), Die Flügelgelenkung bei Odonaten, *Dtsch. ent. Z.*, (*N.F.*) **5**, 394–455.

TIEGS, O. W. (1955), The flight muscles of insects – their anatomy and histology, with some observations on the structure of striated muscles in general, *Phil. Trans. R. Soc., Ser. B*, **238**, 221–348.

TILLYARD, R. J. (1918a), The Panorpoid complex. i. The wing-coupling apparatus with special reference to the Lepidoptera, *Proc. Linn. Soc. N.S. Wales*, **43**, 286–319.

—— (1918b), The Panorpoid complex. ii. The wing trichiation and its relation to the general scheme of venation, *Proc. Linn. Soc. N.S. Wales*, **43**, 626–657.

VIGNON, P. (1929), Introduction à nouvelles recherches de morphologie comparée sur l'aile des insectes, *Arch. Mus. Hist. nat. Paris*, **4**, 89–125.

VOGEL, S. (1966–67), Flight in *Drosophila*. I–III, *J. exp. Biol.*, **44**, 567–578; **46**, 383–392; 431–443.

WADDINGTON, C. H. (1941), The genetic control of wing development in *Drosophila*, *J. Genet.*, **41**, 73–139.

WEBER, H. (1930), *Biologie der Hemipteren. Eine Naturgeschichte der Schnabelkerfe*, Springer, Berlin, 543 pp.

WEIS-FOGH, T. (1950), An aerodynamic sense organ in locusts, *Proc. 8th int. Congr. Ent., Stockholm, 1948*, 584–588.

—— (1956a), Biology and physics of locust flight. II. Flight performance of the desert locust (*Schistocerca gregaria*), *Phil. Trans. R. Soc., Ser. B*, **239**, 459–510.

—— (1956b), Biology and physics of locust flight. IV. Notes on sensory mechanisms in locust flight, *Phil. Trans. R. Soc., Ser. B*, **239**, 553–584.

—— (1972), Energetics of hovering flight in hummingbirds and in *Drosophila*, *J. exp. Biol.*, **56**, 79–104.

WEIS-FOGH, T. AND JENSEN, M. (1956), Biology and physics of locust flight. I. Basic principles in insect flight. A critical review, *Phil. Trans. R. Soc., Ser. B*, **239**, 415–458.

WHITTEN, J. M. (1962), Homology and development of insect wing tracheae, *Ann. ent. Soc. Am.*, **55**, 288–295.

WIGGLESWORTH, V. B. (1954), Growth and regeneration in the tracheal system of an insect, *Rhodnius prolixus* (Hemiptera), *Q. Jl microsc. Sci.*, **95**, 115–137.

—— (1963), The origin of flight in insects, *Proc. R. ent. Soc. Lond.* (C), **28**, 23–24.

—— (1973), Evolution of insect wings and flight, *Nature*, **246**, 127–129.

WILSON, D. M. (1966), Insect walking, *A. Rev. Ent.*, **11**, 103–122.

—— (1968), The nervous control of insect flight and related behaviour, *Adv. Insect Physiol.*, **5**, 289–338.

WOODRING, J. P. (1962), Oribatid (Acari) pteromorphs, Pterogasterine phylogeny, and evolution of wings, *Ann. ent. Soc. Am.*, **55**, 394–403.

YEAGER, J. F. AND HENDRICKSON, G. O. (1934), Circulation of the blood in wings and wing-pads of the cockroach, *Periplaneta americana* Linn, *Ann. ent. Soc. Am.*, **27**, 257–272.

Chapter 6

THE ABDOMEN

Segmentation of the Abdomen

The abdomen (Fig. 1) is composed of a series of segments which are more equally developed than in the other regions of the body (Matsuda, 1975). For the most part they retain their simple annular form, the terga and sterna are generally undivided shields, while the pleura are membranous and usually without differentiated sclerites. Each intersegmental sclerite is believed to have fused with the segmental plate behind it. Reduction or

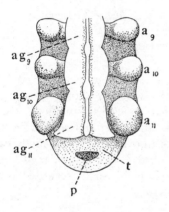

FIG. 41
Ventral view of last abdominal segments of young embryo of *Gryllotalpa*. From Heymons

a_9 to a_{11}, appendages of 9th to 11th segments and ag_9–ag_{11}, neuromeres of those segments; p, proctodaeum; t, telson.

special modification of certain segments is evident at the anterior and posterior ends of the abdomen, more especially in the latter region, and this specialization increases from the lower to the higher orders. The primitive number of abdominal segments as revealed by embryology is eleven, with a terminal non-segmental region or telson (Anderson, 1972). The telson is present in the embryos of certain insects (Fig. 41), but it rarely persists as a discrete region; it is evident, however, in the Protura, while traces are found in other insects. The 11th segment is present in the adults of the lower orders where its tergum is represented by the *epiproct* above the anus (often fused with the 10th tergum), while vestiges of its sternum are seen in the *paraprocts* which lie on either side of the anus (Fig. 42). The 10th segment is usually distinct and forms the terminal segment in the higher orders. The Protura differ from all other hemimetabolous insects in that the number of abdominal segments increases during post-embryonic development, the youngest instars having only eight segments and a telson. The Collembola are also exceptional in never having more than six abdominal segments, either in the embryo or the adult. In most insects the 1st abdominal segment, and more especially its sternum, is reduced or vestigial. Among Apocritan Hymenoptera this segment becomes fused with the metathorax during the change from the larva to the pupa and is known as the *propodeum, epinotum* or *median segment*.

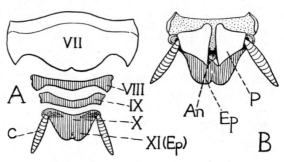

FIG. 42   Last abdominal segments of *Blatta*. A. Dorsal view. B. Ventral view

An, anus; *C*, cercus; *Ep*, epiproct; *P*, paraproct; VII–XI, terga. *After* Snodgrass.

In many Endopterygote insects, more especially those whose eggs are deposited within plant tissues or in other concealed situations, the distal abdominal segments become attenuated and often telescoped to form a retractile tube or *oviscapt* which is used as an ovipositor. This modification is particularly well exhibited in the Cerambycidae, Cecidomyidae, Trypetidae, Muscidae and other families. A true ovipositor is usually regarded as appendicular and is dealt with in the next section.

Appendages and Processes of the Abdomen

In the embryos of most insects evident rudiments of paired abdominal appendages appear at some stage during development. They are commonly present on each of the eleven segments in the less specialized Exopterygotes but do not occur on the telson (Fig. 41). A variable number of these appendages may become transformed into organs that are functional during post-embryonic life while the remainder disappear. The most conspicuous of the persistent appendages are the *cerci* of the 11th segment, which exhibit wide diversity of form and may even be transformed into forceps, as in the Japygidae and the earwigs. Among the Apterygota the retention of at least some abdominal appendages is a general feature. They are well exhibited in the Machilidae, where they are present in a reduced condition on the 2nd to 9th segments, and as cerci on the 11th segment. The reduced appendages each consist of a limb-base or *coxite* bearing a distal *stylus* which is sometimes regarded as the vestige of the shaft of a typical walking limb but which may conceivably be a 'coxal epipodite' homologous with the coxal styles of Symphylan or Machilid legs. In the Diplura the coxites fuse with the sterna and the styli arise directly from the composite plates so formed (Fig. 43). In most insects, however, the styli disappear so that the typical abdominal sternum of insects is probably a plate of compound origin – a 'coxosternum'. In adult Pterygota the cerci are retained in the Palaeopteran and Orthopteroid orders but pregenital abdominal appendages are absent and the only other abdominal appendages present are the *gonopods* which are

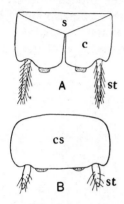

FIG. 43
Abdominal sterna and appendages of A, *Machilis* and B, *Campodea*

s, sternum; c, coxite; cs, coxosternum; st, stylus.

believed by many to enter into the formation of the *genitalia* or external reproductive organs. These are associated with the 8th and 9th segments in the female and with the 9th segment in the male. A completely developed gonopod consists of a coxite bearing a distal stylus, while on its medial border the coxite is produced into a tubular outgrowth or *gonapophysis* (Fig. 44). The term 'gonapophysis' is sometimes also used loosely to denote any genital appendage. The genital aperture is usually located in the membrane immediately behind the 8th or 9th segment.

Among immature Pterygote insects, the gills of mayfly nymphs, the abdominal prolegs of Lepidopteran and Symphytan larvae, the terminal appendages of Trichopteran larvae and the gills of Sialoid larvae have all been regarded as true abdominal appendages (Snodgrass, 1935), but this is not satisfactorily established (Pryor, 1951). On the one hand, the prolegs or gills develop from what seem to be the abdominal appendages of the polypod embryo (Friedman, 1934). On the other hand in the Lepidoptera and other Panorpoid orders the prolegs appear to have arisen independently in each group, they are absent or lack muscles in the more primitive Panorpoids, and only in the Lepidoptera do they become important locomotor structures (Hinton, 1955). Examples of what are certainly secondary abdominal processes include the gills of the Trichoptera and most Plecoptera, the pseudopods and creeping welts found in some Dipteran larvae (Hinton, 1955) and the copulatory organs on the 2nd abdominal segment of male Odonata.

External Genitalia

These include the characteristic ovipositor of many female insects and the diverse male copulatory organs. The morphological interpretation of these structures presents many problems, especially as there is a basic and long-standing controversy as to whether they are, at least in part, modified abdominal appendages serially homologous with the thoracic limbs, or whether they are simply secondary sternal processes. Many taxonomists have preferred to avoid a terminology based on uncertain homologies, so that a large number of special anatomical terms are now applied to the external genitalia of insects (Tuxen, 1970). In the above brief description of a gonopod we have, in effect, interpreted the genitalia as appendicular derivatives, with homologies existing between the various components of the male and

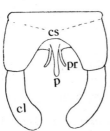
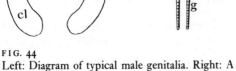
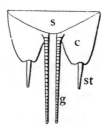
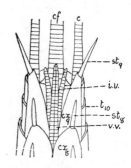

FIG. 44
Left: Diagram of typical male genitalia. Right: A pair of primitive gonopods

c, coxite; cl, clasper; cs, coxosternum; g, gonapophysis; p, penis; pr, paramere; s, sternum; st, stylus.

FIG. 45
Ventral view of the apex of the abdomen of a female *Machilis* showing genitalia

c, cercus; cf, median caudal filament; cx_8, cx_9, coxites of 8th and 9th sterna; st_8, st_9, styli; iv, vv, inner and ventral valves of ovipositor; t_{10}, 10th tergum. *After* Walker, *Ann. ent. Soc. Amer.* 15.

female organs. The matter is by no means settled, however, and further information and discussion can be found in recent summaries of the various theories by Michener (1944), Gustafson (1950), Nielsen (1957), Snodgrass (1957), Matsuda (1958), Gouin (1962), Scudder (1961, 1971) and Smith (1969). In general, the suggested homologies usually start from the structures found in the Thysanuran genitalia and depend also on evidence from postembryonic development, the structure of gynandromorphs, and the anatomical relations, musculature and innervation of the various parts. Scudder (1964), however, has emphasized the need to apply modern concepts of homology based on experimental studies of development; such an approach, combined perhaps with the use of genetic markers for homologous segmental regions may eventually help to resolve some of the persistent difficulties in this field (Sokoloff and Hoy, 1968; Fletcher, 1970).

The Female Genitalia consist typically of three pairs of valves which

collectively form the *ovipositor* or egg-laying organ. Their degree of development and co-adaptation varies according to the uses to which that organ is
subjected. In the Mallophaga and Siphunculata, for example, the ovipositor
is vestigial or absent; in the Dictyoptera its valves are small and free; in the
Tettigoniidae (Figs. 46 and 47) those of one side are held together by tongue
and groove joints and form, along with their counterparts of the opposite
side, an elongate and powerful egg-laying instrument; in most Hymenoptera
(Fig. 47) the ovipositor is greatly attenuated and modified for piercing or
stinging. A complete Pterygote ovipositor, according to the interpretation of
Scudder (1971), consists of (1) a pair of small coxites of the 8th abdominal
segment, each bearing (2) a gonapophysis, which forms an anterior (ventral

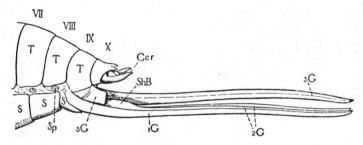

FIG. 46   Ovipositor of a long-horned grasshopper (*Conocephalus*)
VII–X, terga; s, sterna; *Cer*, cerci; *1G*, ventral valve; *2G*, inner valves; *ShB*, bulb-like
swelling formed by the union of the bases of *2G*; *3G*, dorsal valves, the left one is shown as if
cut off near its base. *After* Snodgrass, *U.S. Bur. Ent. Tech. Ser.* 18.

or first) ovipositor valve; (3) a pair of small coxites of the 9th segment, which
bear (4) the paired gonapophyses that form the inner (posterior or second)
ovipositor valves, as well as (5) a coxal extension or *gonoplac*, forming the
dorsal (lateral or third) ovipositor valve. In addition there is on each side a
characteristic sclerite (6) the *gonangulum* which is derived from the anterior
part of the coxite of the 9th abdominal segment and articulates primitively
with this coxite, the 9th abdominal tergum and the gonapophysis of the 8th
segment. Essentially the same structures are found among the Thysanura in
the Lepismatidae, though the coxites are unmodified there and bear styles.
In the Machilidae the gonangulum has not evolved, so that the appendages
of the genital segments are rather similar to those of the pregenital region of
the abdomen, with the gonapophyses corresponding in position to the eversible vesicles (see p. 435).

The Male Genitalia present a wide range of variation and are particularly valuable for separating the genera and species of many groups of
insects. When fully developed (Figs. 44 and 48) they consist essentially of a
pair of *claspers* which help to grip the female during copulation and between
which lies the *aedeagus*. The latter is composed of a *penis* which is usually
unpaired (but may be double, as in the Ephemeroptera and some

Dermaptera) and a pair of more lateral structures which are perhaps best referred to as *parameres*, though this term has been used in several different senses. The homologies of these parts have been the subject of much controversy. The claspers probably represent the coxites and styles of the 9th segment, though in some cases they seem to be composed only of the styles, the coxites having fused with the sternum. Snodgrass (1957) has argued that the claspers of the Hemiptera and the Endopterygote insects are non-

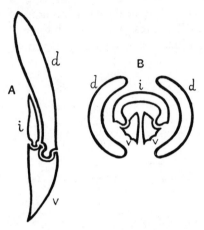

FIG. 47
Transverse sections of the ovipositor of: A, an Orthopteran (*Tettigonia*) *after* Dewitz; B, a Hymenopteran (*Sirex*) *after* Taschenberg. The method of interlocking of the valves is shown

d, dorsal valve; *i*, inner valve; *v*, ventral valve.

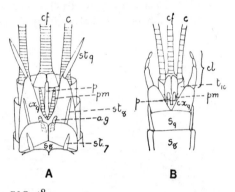

FIG. 48
Ventral view of the apex of the abdomen of A, *Machilis* and B, an Ephemerid showing male genitalia

ag, gonapophyses of 8th segment; *c*, cercus; *cf*, median caudal filament; *cl*, clasper; *cx₉*, coxite of 9th sternum; *p*, penis (paired in B); *pm*, paramere; *s₈ s₉*, 8th and 9th sterna; *st₇–st₉*, 7th to 9th styli; *t₁₀*, 10th tergum. *After* Walker, Ann. ent. soc. Am. 15.

appendicular parameres, formed from the paired phallic rudiments of the 10th abdominal segment. Part or all of the aedeagus is held by some authors to be of a secondary, non-appendicular nature but others consider it to be formed by the division of the gonapophyses of the 9th abdominal segment, the two median halves fusing during development to form the penis while the lateral halves constitute the parameres. Yet another view regards it as formed in a similar manner from the gonopods of the 10th abdominal segment.

Deviations from the typical structure of the genitalia of both sexes are described in the sections dealing with the anatomy of the various orders – they are sometimes supplemented by secondary formations or show various degrees of reduction. In addition to the general works cited above, recent studies of the morphology and development of the genitalia of particular groups of insects include the following: Qadri (1940), Snodgrass (1941), Neumann (1958), Dupuis (1955), Klier (1956), Scudder (1959), Davies (1961), Menees (1963), Salzer (1968), Smith, E. L. (1970) and Häfner (1971).

Literature on the Abdomen and External Genitalia

ANDERSON, D. T. (1972), The development of hemimetabolous insects, In: Counce, S. J. and Waddington, C. H. (eds), *Developmental Systems: Insects*, Academic Press, London and New York, I, 95–163.

DAVIES, R. G. (1961), The postembryonic development of the female reproductive system in *Limothrips cerealium* Haliday (Thysanoptera: Thripidae), *Proc. zool. Soc. Lond.*, 136, 411–437.

DUPUIS, C. (1955), Les genitalia des Hemiptères Hétéroptères (Genitalia externes des deux sexes; voies ectodermiques femelles). Revue de la morphologie. Lexique de la nomenclature. Index bibliographique analytique, *Mem. Mus. nat. Hist. nat.* (*A*), *Zool.*, 6, 183–278.

FLETCHER, L. W. (1970), Abdominal and genitalic homologies in the screw-worm *Cochliomyia hominovorax* (Dipt. Calliphoridae) established by a genetic marker, *Ann. ent. Soc. Am.*, 63, 490–495.

FRIEDMAN, N. (1934), Ein Beitrag zur Kenntnis der embryonalen Entwicklung der Abdominalfüsse bei den Schmetterlingsraupen, *Comment. biol.*, 4, 1–29.

GOUIN, F. J. (1962), Anatomie, Histologie und Entwicklungsgeschichte der Insekten und der Myriapoden. Das Abdomen der Insekten, *Fortschr. Zool.*, 15, 337–353.

GUSTAFSON, J. F. (1950), The origin and evolution of the genitalia of the Insecta, *Microentomology*, 15, 35–67.

HÄFNER, P. (1971), Muskeln und Nerven des Abdomens besonders des männlichen Geschlechtsapparates von *Haematopinus suis* (Anoplura), *Zool. Jb.* (*Anat.*), 88, 421–449.

HINTON, H. E. (1955), On the structure, function and distribution of the prolegs of the Panorpoidea, with a criticism of the Berlese–Imms theory, *Trans. R. ent. Soc. Lond.*, 106, 455–545.

KLIER, E. (1956), Zur Konstruktionsmorphologie des männlichen Geschlechtsapparates der Psocopteren, *Zool. Jb.* (*Anat.*), 75, 207–286.

MATSUDA, R. (1958), On the origin of the external genitalia of insects, *Ann. ent. Soc. Am.*, 51, 84–94.

—— (1975), *Morphology and Evolution of the Insect Abdomen*, New York, 568 pp.

MENEES, J. H. (1963), Embryonic and postembryonic homologies of insect genitalia as revealed in development of male and female reproductive systems of the European chafer, *Amphimallon majalis* Razoumowski (Coleoptera: Scarabaeidae), *Mem. Cornell Univ. agric. Exp. Sta.*, 381, 1–59.

MICHENER, C. D. (1944), A comparative study of the appendages of the eighth and ninth abdominal segments of insects, *Ann. ent. Soc. Am.*, 37, 336–351.

NEUMANN, H. (1958), Der Bau und die Funktion der männlichen Genitalapparate von *Trichocera annulata* Meig. und *Tipula paludosa* Meig., *Dtsch. ent. Zeitschr.*, (*N.F.*), 5, 235–298.

NIELSEN, A. (1957), On the evolution of the genitalia in male insects, *Ent. Medd.*, 28, 27–57.

PRYOR, M. G. M. (1951), On the abdominal appendages of larvae of Trichoptera, Neuroptera and Lepidoptera, and the origins of jointed limbs, *Q. Jl microsc. Sci.*, 92, 351–376.

QADRI, M. A. H. (1940), On the development of the genitalia and their ducts of Orthopteroid insects, *Trans. R. ent. Soc. Lond.*, 90, 121–175.

SALZER, R. (1968), Konstruktionsanatomische Untersuchungen des männlichen Postabdomens von *Calliphora erythrocephala* Meigen, *Z. Morph. Ökol. Tiere*, **63**, 155–238.

SCUDDER, G. G. E. (1959), The female genitalia of the Heteroptera: morphology and bearing on classification, *Trans. R. ent. Soc. Lond.*, **111**, 405–467.

—— (1961), The comparative morphology of the insect ovipositor, *Trans. R. ent. Soc. Lond.*, **113**, 25–40.

—— (1964), Further problems in the interpretation and homology of the insect ovipositor, *Can. Ent.*, **96**, 405–417.

—— (1971), Comparative morphology of insect genitalia, *A. Rev. Ent.*, **16**, 379–406.

SMITH, E. L. (1969), Evolutionary morphology of external insect genitalia. 1. Origin and relationships to other appendages, *Ann. ent. Soc. Am.*, **62**, 1051–1079.

—— (1970), Evolutionary morphology of the external insect genitalia. 2. Hymenoptera, *Ann. ent. Soc. Am.*, **63**, 1–27.

SNODGRASS, R. E. (1935), *Principles of Insect Morphology*, McGraw-Hill, New York, 667 pp.

—— (1941), The male genitalia of Hymenoptera, *Smithson. misc. Collns*, **95**, 96 pp.

—— (1957), A revised interpretation of the external reproductive organs of male insects, *Smithson. misc. Collns*, **135**, 60 pp.

SOKOLOFF, A. AND HOY, M. A. (1968), Mutations as possible aids in establishing genetic homologies in the sexes in *Tribolium castaneum*, *Ann. ent. Soc. Am.*, **61**, 550–553.

TUXEN, S. L. (ed.) (1970), *Taxonomist's Glossary of Genitalia in Insects*, 2nd edn, Munksgaard, Copenhagen, 359 pp.

Chapter 7

THE ENDOSKELETON

In certain regions of the body the integument becomes invaginated and greatly hardened, forming rigid processes which serve for the attachment of muscles and the support of certain other organs. This internal framework is termed the *endoskeleton* and its individual parts are known as *apodemes* or *phragmata*. They arise as invaginations of the body-wall between adjacent sclerites, or at the edge of a sclerite or segment. In some insects the invaginations remain open throughout life but, more usually, they become completely solid through the deposition of cuticular material.

The two most important parts of the endoskeleton are the tentorium and the endothorax, but endoskeletal structures may be developed in almost any part of the body where muscles are attached.

The Tentorium (Figs. 49, 50) (Hudson, 1945–51)

This name is given to the endoskeleton of the head and, in generalized Pterygote insects, it is composed of two or three pairs of apodemes which coalesce at their bases. The functions of the tentorium are (1) to afford a basis for the attachment of many of the cephalic muscles and, at the same time, to give rigidity to the head; (2) to lend support to the brain and fore intestine; (3) to strengthen the points of articulation of certain of the mouth-parts. The apodemes which enter into the formation of the cephalic endoskeleton are termed the *anterior, posterior* and *dorsal arms of the tentorium* according to their positions. The inner ends of these arms fuse with each other and the median skeletal part thus formed is termed the *body of the tentorium*. Despite earlier views to the contrary (e.g. Snodgrass, 1951), it has been claimed by Manton (1964) that an homologous system of tentorial apodemes occurs in the Myriapods, the apterygote insects and the Pterygota. Only in the Lepismatidae and the Pterygota, however, has the evolution of transverse biting movements of the mandibles led to the appearance of a strong rigid tentorium.

The Anterior Arms of the Tentorium – The invaginations which form these apodemes usually lie on either side of the frontoclypeal sulcus, when the latter is present, and just above the anterior articulations of the mandibles. In most insects they appear externally as mere slits but in many Diptera they form intracranial tunnels.

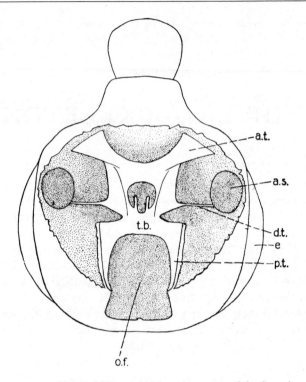

Head of *Blatta* with the greater part of the frontal
wall dissected away to show the tentorium

a.s, antennal socket; *a.t*, *d.t*, *p.t*, anterior, dorsal and
posterior arms of tentorium; *e*, compound eye; *o.f*, occipital foramen; *t.b*, body of tentorium.

The Posterior Arms of the Tentorium – These apodemes arise as
ingrowths from the ventral ends of the postoccipital sulci, generally in close
relation with the occipital foramen. In some prognathous insects they tend to
lie more forward on the ventral wall of the head (see p. 29).

The Dorsal Arms of the Tentorium – These arise not as invaginations
of the wall of the head but as outgrowths either from the body of the

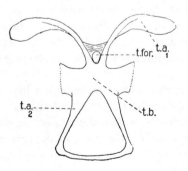

FIG. 50
Tentorium of a winged termite

t.a₁, *ta₂*, anterior and posterior arms; *t.b*,
body of tentorium; *t.for*, tentorial foramen.

tentorium or from the bases of the anterior arms. They pass upwards and outwards, often to become attached to the head wall near the antennae or eyes. They are generally present in Orthoptera, but in some cases (e.g. *Blatta*) they are tendon-like, while they are often undeveloped in other orders.

The Body of the Tentorium – This is a median plate which is often large and its shape varies to some extent in conformity with that of the head; thus, in the soldiers of many termites it is elongate, while in the workers it is a relatively narrow band.

When an insect moults, the dorsal arms of the tentorium are largely dissolved away by the moulting fluid, the tentorium splits medially, much of the central body is dissolved and the remainder is pulled out as four separate pieces, one from each tentorial pit (Sharplin, 1965).

The Endothorax (Figs. 51, 52)

This is composed of invaginations of the tergal, pleural and sternal regions of each thoracic segment. The several structures may be conveniently termed the *endotergites*, *endopleurites* and *endosternites* respectively.

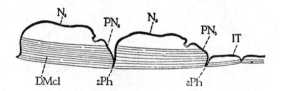

FIG. 51   Longitudinal section through the dorsal part of the meso- and metathorax and base of the abdomen of a stonefly (*Alloperla*)

DMcl, dorsal longitudinal muscles; *IT*, 1st abdominal tergum; N_2, mesonotum; N_3, metanotum; PN_2, PN_3, postnotum of meso- and metathorax; 2*Ph*, 3*Ph*, phragmata. *After* Snodgrass, *Proc. U.S. Nat. Mus.* 39.

The Endotergites or **Phragmata** (Snodgrass, 1935) arise as transverse infoldings of the intersegmental sclerites (see p. 43) and as their main function is to provide increased areas of attachment for the dorsal longitudinal muscles they are best developed in winged insects, especially those that fly actively. There are normally three phragmata associated with the intersegmentalia between the pro-, meso- and metathorax and the 1st abdominal segment, but owing to the varied transpositions of the intersegmental sclerites (p. 43) the relations of the phragmata to the definitive segmentation differ in different groups of insects. Thus, they are sometimes situated at the front of the segment, between the acrotergite and prescutum and sometimes at the back of a segment, behind the postnotum. Phragmata occupying these two positions have been referred to as pre- and postphragmata respectively and both may be carried by either the meso- or metathorax of some insects; no phragma is ever borne by the prothorax.

The Endopleurites or **Lateral Apodemes** are infoldings between the pleurites. In a typical wing-bearing segment of most insects there is a single apodeme on either side formed by infolding along the pleural sulcus and known as the *pleural ridge*. It terminates in the wing process above, the coxal process below, and often bears an inwardly projecting *pleural arm*. The endopleurites are well developed in the Odonata.

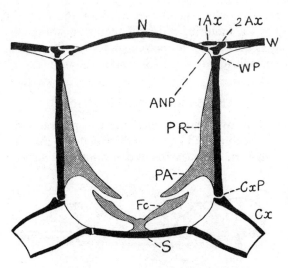

FIG. 52   Diagram of a section across a wing-bearing segment

ANP, anterior notal wing process; 1*Ax*, 2*Ax*, 1st and 2nd axillary sclerites; *Cx*, coxa; *CxP*, coxal process of pleuron; *Fc*, furca; *N*, notum; *PA*, pleural arm; *PR*, pleural ridge; *S*, sternum; *W*, wing; *WP*, pleural wing process. *After* Snodgrass, *Proc. U.S. Nat. Mus.* 36.

The Endosternites are typically formed of a pair of apophyses arising from pits on the eusterna between the basisternum and sternellum. Frequently, however, there is also an ingrowth of the sternum near their origin so that they are carried inwards and the whole forms a Y-shaped structure, the *furca*, whose internal arms may become fused with the pleural arms or are connected with them by short muscles. In many generalized insects there is also a median, unbranched apodeme or *spina* which arises from the intersegmental spinasternum. In the higher orders, the spina is lost or consolidated with the furca. In the Odonata the endosternites are paired, and are inclined so far inwards, towards the median line, that they almost meet over the nerve cord. In the honey bee those of the prothorax fuse to form a supraneural bridge, and the combined meso- and metathoracic endosternites together form a second bridge of a similar character.

The Abdominal Endoskeleton

In the abdomen apodemes are developed to provide firm origins for some of the more important muscles. Most of the terga usually present internal ridges or phragmata, as in the thorax, giving attachment to the chief longitudinal muscles. On the ventral aspect sternal apophyses are commonly present and in some cases highly developed. Specialized apodemes may be developed in connection with the ovipositor and the male copulatory organs.

Literature on the Endoskeleton

There is little special literature dealing with the endoskeleton. Reference may be made to the following and to the papers cited under sections dealing with the skeletomuscular systems in the main regions of the body.

HUDSON, G. B. (1945–51), Studies in the comparative anatomy and systematic importance of the Hexapod tentorium, I–IV, *J. ent. Soc. sth. Afr.*, 8, 71–90; 9, 99–110; 11, 37–49; 14, 3–23.

MANTON, S. M. (1964), Mandibular mechanisms and the evolution of Arthropoda, *Phil. Trans. R. Soc., Ser. B*, 247, 1–183.

SHARPLIN, J. (1965), Replacement of the tentorium of *Periplaneta americana* Linnaeus during ecdysis, *Can. Ent.*, 97, 947–951.

SNODGRASS, R. E. (1935), *Principles of Insect Morphology*, McGraw-Hill, New York, 667 pp.

—— (1951), *Comparative Studies on the Head of Mandibulate Arthropods*, Comstock Publ. Co., Ithaca, N.Y., 118 pp.

Chapter 8

THE MUSCULAR SYSTEM

Insect muscles are mostly translucent, colourless or grey, though the flight-muscles often show a yellowish or brown tinge. Both skeletal and visceral muscle fibres are cross-striated and some older reports of unstriated visceral muscles need reinvestigation. Insect muscles differ fundamentally from those of Annelids, both in histological structure and in not being incorporated into the body-wall to form a dermo-muscular tube. In most skeletal muscles, especially those of the appendages, one end of the muscle (its *origin*) is attached to a fixed skeletal region while the other end (its *insertion*) is attached to a movable part. Cuticular invaginations or apodemes, in the form of cords, bands or plate-like structures, may provide the true sites of attachment and therefore intervene between the muscle and the main structure on which it acts. For a general review of insect muscle see Usherwood (1975).

Histology of the Muscles (Figs. 53, 54)

The skeletal muscles of insects have a complex structure, in which one may distinguish (i) the fibrous contractile system, (ii) the mitochondria, (iii) the tracheal and nervous supply, and (iv) the membrane systems (Smith, 1961–66). Variations in the histology and ultrastructure of these components are associated with important functional differences between different groups of muscles. Essentially, however, a muscle is composed of a number of long *fibres*, each measuring from a few to a few hundred μm in diameter. Each of these is in turn subdivided into separate, smaller *fibrils* which are themselves composed of a highly organized array of *myofilaments* made up of the proteins actin and myosin (Fig. 53). The fibre is surrounded by an outer membrane, the *sarcolemma*, which encloses the nucleated sarcoplasm in which the fibrils are embedded. Each fibril is typically composed of alternating isotropic (I) and anisotropic (A) regions which, under appropriate microscopical illumination, appear as light and dark bands or disks. Because these appear at about the same level in adjacent fibrils they give to the whole fibre its characteristic cross-striated appearance. Crossing each isotropic region is a partition, the Z-disk, which thus divides the fibril into short units or *sarcomeres*, each comprising an anisotropic region (the A-band) and two half I-bands. The dense A-band is further traversed by a lighter H-band. These features, easily seen by light-microscopy, reflect the submicroscopic organization of the myofibrils.

An array of thin I-band actin filaments (each some 5 nm in diameter) extends from the Z-disk to the edge of the H-band in a relaxed fibre, while thicker myosin filaments (each about 15 nm in diameter) run throughout the A-band. Actin and myosin filaments are linked by temporary cross-bridges, each myosin filament usually being surrounded by 6 actin filaments (Garamvölgyi, 1965; Hagopian, 1966). According to the well-established theory of Huxley and Hanson, contraction of the fibril is due to the sliding of the actin and myosin filaments relative to each other (Hanson, 1956; Pringle, 1966; Osborne, 1967); the actin filaments move further into the A-disk while the myosin filaments thus approach the Z-disks.

The mitochondria of insect muscle vary greatly in size, shape and distribution, being most extensively developed in the flight-muscles, as one would expect from the extremely high metabolic rate of these actively contracting structures. They may be scattered randomly throughout the sarcoplasm or arranged between fibrils opposite the Z-disks; in the flight-muscles of the Odonata they form large slab-like structures and in the powerful fibrillar indirect flight-muscles of Hymenoptera and Diptera the giant mitochondria form conspicuous 'sarcosomes'. The tracheal supply of insect muscles also varies with their activity, visceral muscles being poorly supplied while flight-muscles are much more richly tracheated, with intracellular tracheoles penetrating the fibrils. Of considerable physiological importance are the membrane systems revealed by electron microscopy. The sarcolemma, about 7·5 nm thick, is a unit-membrane like that investing other cells. Transverse tubular invaginations from it form the so-called *T-system*, while close to these are the longitudinally arranged cisternae of a separate membrane system, the *sarcoplasmic reticulum*. The

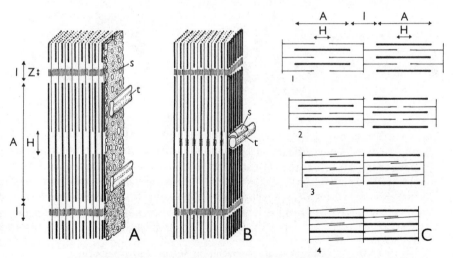

FIG. 53   Ultrastructure of insect muscle (*after* Smith and Huxley). A, synchronous muscle fibril showing extensive sarcoplasmic reticulum and wide I-band. B, asynchronous fibril showing reduced sarcoplasmic reticulum and narrow I-band. C, schematic representation of successive stages in the contraction of two sarcomeres, illustrating changes in the relative positions of the thick and thin filaments (respectively of myosin and actin)

s, sarcoplasmic reticulum; t, transverse membrane system (T-system). The relationships of the A, I, H and Z bands to the ultrastructural and molecular constitution of the fibril are indicated at the left and top right.

fibrillar asynchronous indirect flight-muscles of Diptera, Hymenoptera and Coleoptera have a reduced sarcoplasmic reticulum whereas the synchronous flight-muscles of other orders and the remaining skeletal muscles of all insects have both systems well developed. The T-system may provide a pathway along which the peripheral excitation of a fibre is conducted inwards, while the sarcoplasmic reticulum probably controls the contraction cycle through the activation of myosin ATP-ase by calcium ions. Further references to the ultrastructure of insect muscle are listed on p. 96 ff.

Visceral muscle fibres occur singly or in groups or larger lattice-like arrays around the gonads and their ducts and in the heart, diaphragms and gut wall. Their properties are in some ways intermediate between those of the skeletal muscles and the smooth muscles found in other phyla. Histologically they differ from skeletal muscle in having smaller fibres, linked by desmosomes and sometimes containing only one fibril. They also have a poor tracheolar supply, few mitochondria and a poorly developed T-system and sarcoplasmic reticulum. On the other hand, the fibres are divided into sarcomeres with A- and I-disks, though each myosin filament is usually surrounded by 12 orbital actin filaments (Davey, 1964; Smith, Gupta and Smith, 1966; Goldstein and Burdette, 1971).

Skeletal muscles are attached to the cuticle in a complicated way, only now being fully elucidated by electron microscopy (Fig. 54). At the junction

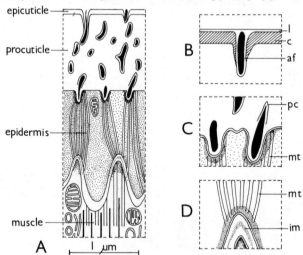

FIG. 54   Ultrastructure of attachment of muscle to integument (*after* Caveney, 1967). A, general diagram; B, relation of attachment fibre to epicuticle; C, junction of procuticle and epidermis, showing two hemidesmosomes; D, junction of epidermis and muscle, showing structure of a desmosome

af, muscle attachment fibre, running in pore canal; *c*, 'cuticulin' layer of epicuticle; *im*, intercellular matrix; *l*, lipid layer of epicuticle; *mt*, microtubules traversing epidermis from hemidesmosome to desmosome; *pc*, pore canal.

of muscle and epidermis the cells show regular interdigitation, the processes being lined with desmosomes. Within the epidermal cells microtubules connect the desmosomes of the interdigitating regions with cone-like depressions of the outer epidermal plasma membrane. Then, from each cone, an electron-dense muscle attachment fibre or *tonofibrilla* runs through the procuticle in a pore-canal and finally inserts on the epicuticle. The tonofibrillae are only slowly dissolved when the old cuticle is digested by moulting fluid. New tonofibrillae become attached to the epicuticle only when it is growing. For further details see Lai-Fook (1967) and Caveney (1969).

Arrangement of the Skeletal Muscles (Myology)

In general arrangement, the skeletal muscular system corresponds with the segmentation of the body and is shown in its least modified form in some apterygotes, the lower Pterygota and many larvae. There are usually very many muscles – up to 2000 or so in some Lepidopteran larvae – and the system is bilaterally symmetrical. Though detailed descriptive accounts of the muscular system of many insect species are available, the homologies of the muscles are sometimes difficult to decide and no uniform terminology exists. The older tendency to classify muscles according to their probable functions is giving way to a system of naming them from their positions, origins and insertions. Neither method is entirely satisfactory since apparently homologous muscles may change their sites of attachment during evolution and alter their functions. It seems rare, however, for muscles to transgress the limits of the segments to which they originally belonged. Despite the difficulties indicated above, general schemes are available in the works of Weber (1933), Snodgrass (1935), von Kéler (1963) and Matsuda (1965, 1970). Among the more recent myological studies are those by Alicata (1963), Chadwick (1959), Ewer (1967), Ewer and Nayler (1967), Manton (1964, 1972), Mickoleit (1962, 1968, 1969) and Parsons (1959, 1960, 1962, 1963).

A simplified list of the principal skeletal muscles is given below, based to some extent on conditions in the more generalized Orthopteroid insects.

A. Cephalic Muscles – The principal muscles of the head may be divided into (*a*) cervical muscles, (*b*) muscles of the mouthparts, and (*c*) muscles of the antennae.

(*a*) The *Cervical Muscles*. These control the movements of the head and are classified into levators, depressors, retractors and rotators according to their function. They take their origin from the prothorax and cervix and are inserted into the tentorium and epicranium.

(*b*) The *Muscles of the Mouthparts*. Associated with the labrum are (Cook, 1944):

1. The *labral compressors*, running between the dorsal and ventral surfaces of the labrum.

2. The *posterior labral* muscles which run from the tormal sclerites of the labrum to the wall of the head.

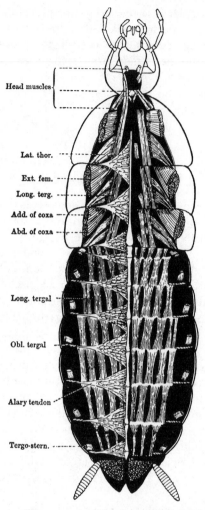

Head muscles

Lat. thor.

Ext. fem.
Long. terg.
Add. of coxa
Abd. of coxa

Long. tergal

Obl. tergal

Alary tendon

Tergo-stern.

FIG. 55
Muscles of the dorsal wall of a cockroach, with the heart and aliform muscles. *After* Miall and Denny

3. The *anterior labral* muscles (retractors) which run from the anterior margin of the labral base to the wall of the head.

The mandibular muscles (Snodgrass, 1950) include:

4. The *dorsal abductors* which originate on the upper lateral part of the epicranium and insert each on an apodeme connected with the outer, basal region of the mandible.

5. The *dorsal adductors* have an extensive origin on the posterodorsal part of the head and each inserts on an apodeme connected with the inner, basal region of the mandible.

6. The *ventral adductors* are present only in the Apterygotes and some lower Pterygotes.

The principal maxillary muscles are as follows:

7. *Dorsal basal* muscles arising on the dorsal part of the head and forming the *anterior* and *posterior rotators* of the cardo and the *cranial flexor of the lacinia*.

8. *Ventral basal muscles* which are inserted on the cardo and stipes and which

originate on the tentorium in most Pterygotes and on the tentorial apodemes in the Apterygotes.

9. *Stipital muscles.* These all originate on the stipes and include the *levator and depressor* of the palp, the *flexor* of the galea and the *stipital flexor of the lacinia.*

10. The *intrinsic palp muscles.*

The muscles of the labium are:

11. *Extrinsic labial muscles.* These arise on the tentorium or cranial wall and insert on the prementum. They correspond to the ventral basal muscles of the maxilla.

12. *Median labial muscles.* When present, these run from the back of the prementum to the postmentum and have no homologues in the maxilla.

13. *Labial salivary muscles.* There are usually two pairs, arising on the prementum and converging on the labial wall of the salivarium near the opening of the salivary duct.

14. *Muscles of the endites and palps.* From the prementum there run the *levator* and *depressor* muscles of the palps and a *flexor* of each glossa and paraglossa. These are homologues of the stipital muscles of the maxilla but it should be noted that the glossae possess no muscles corresponding to the cranial flexors of the laciniae.

15. *Intrinsic palp muscles.* Inserted on the suspensorium of the hypopharynx there are typically the *hypopharyngeal adductors of the mandible* and two pairs of *frontal* muscles (the so-called 'retractors of the mouth angles').

(c) *Muscles of the Antennae* (Imms, 1939). These are:

1. *Extrinsic antennal muscles.* A *levator* and, usually, two *depressors* are inserted on the base of the scape. They arise from the dorsal and anterior arms of the tentorium or from the dorsal arm alone or from the wall of the head.

2. *Intrinsic antennal muscles.* In all insects with normally developed antennae there is a pair of muscles arising in the scape and inserted on the base of the pedicel. In the Collembola and Diplura similar muscles also occur in all other segments except the last, but in the Thysanura and Pterygota the flagellar segments are devoid of muscles.

B. Thoracic Muscles – The muscles of a typical alate segment are set out below, following a simplified version of Snodgrass's (1935) scheme, but it should be noted that the prothorax differs in the absence of some flight-muscles.

(a) *Longitudinal.* As in the abdomen, these are divisible into *tergal* and *sternal* groups, the former being important indirect flight-muscles.

(b) *Dorsoventral.* Two main groups occur here, the *tergosternals*, which are the principal levators of the wing (acting antagonistically to the longitudinal tergals) and the *tergocoxal* muscles, which act as the tergal promotors and remotors of the leg.

(c) *Pleural.* Three sets of pleural muscles may be distinguished. The *tergopleural* muscles are very variable in development and include the axillary muscles; the *pleurosternals* are short fibres linking the pleural and sternal apophyses while the *pleurocoxals* act as abductors of the coxae.

(d) *Sternal.* These include two muscle groups. The *sternocoxals* are the sternal promotors and remotors of the leg while the *lateral intersegmental* muscle runs from the sternum to the pleuron or tergum of the succeeding segment and is best developed in larval forms.

(e) *Intrinsic Leg Muscles.* In addition to the various extrinsic leg-muscles mentioned above, whose function it is to move the whole limb, there are also muscles lying within the segments of the leg. They include the levator and depressor of the trochanter, tibia and tarsus and the levator of the pretarsus.

As well as these, there are the spiracular muscles and the epipleural muscles, the latter inserting on the subalare and basalare.

C.   Muscles of Flight – Although the flight of insects has been referred to on p. 64, the coordinated activities of the skeletomuscular system of flying insects need further brief discussion (Tiegs, 1955; Pringle, 1957, 1965, 1968). The flight movements are caused by three sets of muscles, the indirect, direct and accessory indirect flight-muscles (Figs. 56, 57). The *indirect muscles* are usually the largest in the body, and are attached to the thorax and not to the wing-bases. In most insects they consist of two groups of muscles: (1) a pair of dorsoventral muscles by whose contraction the tergal region of the thorax is depressed, with the result that the wings are forced upwards

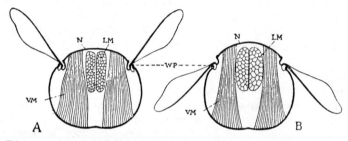

FIG. 56   Diagrammatic cross-section of thorax of a bee through pleural wing process *WP*

A, wings thrown upward by depression of tergum *N* caused by contraction of vertical muscles *VM*. B, wings thrown downward by elevation of tergum *N* caused by contraction of longitudinal muscles *LM*. *After* Snodgrass.

owing to the peculiar nature of their articulation with the thorax; (2) a pair of longitudinal muscles by whose contraction the tergal region becomes arched upwards which results in the wings being forced downwards. The rapid alternate contraction of these two groups of muscles consequently raises and lowers the wings by their action upon the dorsal wall of the thorax.

The *direct muscles* (Fig. 57) are typically the epipleural and axillary muscles and generally consist of four pairs: (*a*) the 1st anterior extensor arising usually from the sternal region and attached to the basalar sclerite (p. 54); (*b*) the 2nd anterior extensor arising from the rim of the coxa, just in front of its pleural articulation, and similarly inserted into the basalar sclerite; (*c*) the posterior extensor arising from the rim of the coxa, just behind its pleural articulation, and inserted into the subalar sclerite; (*d*) the flexor arising from the pleural ridge and inserted into the 3rd axillary sclerite. The accessory indirect flight-muscles are the pleurosternals, tergopleurals and lateral intersegmentals; they act by bracing the cuticular skeleton or by changing its elastic properties in various ways.

Fuller analysis of the skeletomuscular mechanisms that operate in flight shows that these are complicated and vary appreciably from one group of insects to another. *Schistocerca*, for example, has well-developed indirect flight-muscles with the usual functions, but their action in producing the downstroke of the wings is reinforced by the direct muscles attached to the basalar and subalar sclerites. These direct muscles also cause pronation and supination of the wings while the dorsoventral indirect flight-muscles also have an additional role as promotors and remotors of the legs. In the Odonata a different situation occurs: the direct muscles attached to the basalar and subalar sclerites are very large and mainly responsible for the downstroke, whereas the small dorsal longitudinal indirect flight-muscles are of little functional importance. In the bees and the higher Diptera only the indirect muscles of the mesothorax generate the power needed for flight; the metathoracic indirect

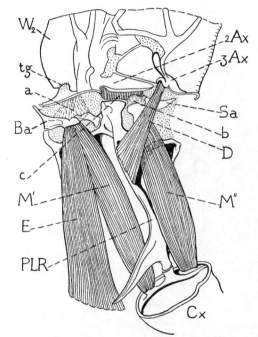

FIG. 57

Direct wing muscles of mesothorax of a grasshopper

$2Ax$, 2nd axillary sclerite and c, its ventral plate; $3Ax$, 3rd axillary sclerite; Ba, basalar sclerites; Cx, coxa; D, flexor muscle; E, 1st anterior extensor muscle; M', 2nd anterior extensor muscle; M'', posterior extensor muscle; PLR, pleural ridge; Sa, subalar sclerite; tg, tegula; W_2, fore wing, turned upward. Adapted from Snodgrass, 1930.

muscles are capable only of tonic contraction and act so as to control the amount of power transmitted from the mesothorax to the metathorax. The extent to which insect flight-muscles are developed is, as one would expect, correlated with the capacity for flight. Related species may differ appreciably in this respect in some groups, as in the Orthoptera discussed by Atzinger (1952). Alary polymorphism and associated flight-muscle polymorphism may also occur within a species, e.g. in some aquatic Heteroptera like the Corixidae (Young, 1965a, 1965b; Acton and Scudder, 1969) and others mentioned on p. 54.

D. The Abdominal Muscles – The principal muscles of a typical abdominal segment may be grouped into the following series, simplified and altered somewhat from Snodgrass (1935).

(a) *Longitudinal*. These may be divided into (a) *tergal* and (b) *sternal* longitudinal muscles. In each case they run between the intersegmental folds or antecostae of successive segments. Acting together, the two groups serve as retractors by telescoping the abdomen. Acting alone, the sternal muscles curve the abdomen downwards and the tergals straighten it or bend it upwards.

(*b*) *Lateral.* These typically run dorsoventrally and are both inter- and intrasegmental in position. They are usually *tergosternals*, but when distinct pleurites are present there may also be *tergopleural* and *sternopleural* muscles. By contraction they tend to compress the segment and are therefore important in respiratory movements.

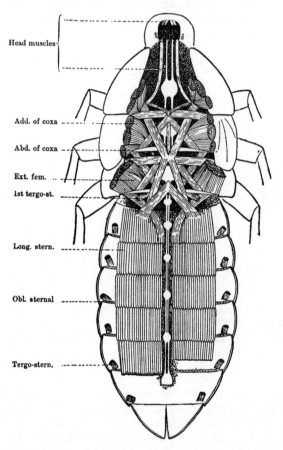

Head muscles
Add. of coxa
Abd. of coxa
Ext. fem.
1st tergo-st.
Long. stern.
Obl. sternal
Tergo-stern.

FIG. 58   Muscles of the ventral wall of a cockroach, with the nerve-cord. *After* Miall and Denny

(*c*) *Transverse.* These lie internal to the longitudinals on both ventral and dorsal sides and are better known as the muscles of the dorsal and ventral diaphragms (see p. 234).

In addition there are special muscles concerned with the movements of the genitalia, cerci and spiracles.

Physiology of Insect Muscles – Various aspects of the physiology of insect muscle, reviewed by Pringle (1966), Usherwood (1967, 1969, 1975), Aidley (1967) and others, can only be touched on here. In many respects the properties of insect skeletal muscles do not differ greatly from those of vertebrates, such features as absolute muscular power and the characteristics of the simple contraction (twitch) being similar in the two groups. Most insect muscles are phasically responsive (i.e. they contract and relax relatively quickly) but some skeletal and all visceral muscles undergo sustained tonic contractions. In some muscles both phasic and tonic fibres occur (Cochrane, Elder and Usherwood, 1972). Unlike vertebrates, insects have muscles that contain relatively few fibres and to achieve smooth contractions they are supplied by multiple nerve endings some 30–80 μm apart, each inducing local contraction. There is also a dual innervation, with a 'fast' axon supplying all or most of the fibres in a muscle while a 'slow' axon innervates a proportion of them. The motor end-plates have a complex structure (e.g. Osborne, 1970) and the neuromuscular transmitter substance of insects is probably L-glutamate. The muscles of the legs and abdomen, and in many insects all the flight muscles as well, respond synchronously to the nervous impulses that reach them. The duration of such responses would, however, be too great to allow the rapid succession of contractions needed to maintain the high frequency of the wing beat in many efficient flyers. The Diptera, Hymenoptera, Coleoptera and Hemiptera have therefore evolved the characteristic fibrillar type of asynchronous indirect flight-muscles (Cullen, 1974). Here the frequency of contraction is not determined by the central nervous system, but is directly controlled by the loading on the muscles. So long as they are stimulated by low-frequency nervous impulses, such muscles can respond directly to stretching. The resulting twitch is of a duration controlled by the conditions of loading and contraction causes direct stimulation of the antagonistic muscles. The oscillatory rhythm resulting from repetition of the cycle is therefore myogenic and its frequency depends on the mechanical properties of the cuticular thoracic wall. A similar mechanism operates in the rapid contractions of the tymbal muscle of cicadas.

Metabolism of Insect Muscle – Some of the metabolic events occurring in active muscular tissue, especially in flight-muscle, are now known in considerable biochemical detail (e.g. Sacktor, 1965, 1970). The oxygen consumption of an insect may rise a hundredfold when flight begins and if it is to continue for long periods a reserve of oxidizable respiratory material is needed. In Diptera and Hymenoptera, with asynchronous fibrillar flight-muscles, a respiratory quotient of unity during flight indicates that carbohydrates are the main substrate. In Lepidoptera, Homoptera and Orthoptera on the other hand, R.Q. values of about 0·7 occur and fat reserves are depleted. Some species, such as locusts and aphids, use glycogen and the disaccharide trehalose at first, but then consume fat during prolonged flight. Much more unusual is the occurrence of the amino-acid proline as an energy reserve for flight in *Glossina* (Bursell, 1966). Flight-muscle homogenates of *Hyalophora cecropia* and *Locusta migratoria* will actively oxidize fatty acids by mechanisms not unlike those occurring in vertebrate tissues (Gilbert,

1967) but much more is known of flight-muscle metabolism in the species that use carbohydrates. Here the processes of breakdown are qualitatively similar to those in vertebrate skeletal muscle. Enzymatically controlled equilibria involving various reserves such as glycogen in the fat-body, trehalose in the blood and both substrates in the muscle itself, lead to the appearance of glucose in the muscle. This undergoes anaerobic glycolysis, with concomitant formation of energy-rich adenosine triphosphate (ATP), through the Embden–Meyerhof pathway, followed by further degradation to carbon dioxide and water in the tricarboxylic acid (Krebs) cycle. Several of the dehydrogenases that participate in the tricarboxylic acid cycle have been identified in the mitochondria of insect flight-muscle and the cycle can also accomplish the terminal stages of lipid and amino-acid breakdown. There are important biochemical differences in the metabolism of different muscles; in *Locusta* flight-muscle, for example, lactic dehydrogenase is virtually absent and lactic acid is not an end-product of glycolysis, whereas in the leg-muscle lactic acid accumulates after exertion and is slowly removed by oxidation and conversion to glycogen.

Degeneration of Musculature – Degeneration of the flight-muscles occur after sexual maturity in many species, notably among queen ants and termites (pp. 617, 1237), where the process promotes reproduction by releasing amino acids that can be used in egg-formation. The alate forms of *Brevicoryne brassicae, Myzus persicae, Aphis fabae* and other aphids also undergo flight-muscle histolysis after settling on their host-plants. These changes and the accompanying hypertrophy of the fat-body and resumption of embryo development in the ovaries are caused by hormonal changes (Johnson, 1957, 1959). A comparable process is also found among Scolytid beetles in which, after moving to a new tree, the insects (of both sexes) do not fly and their flight-muscles are reduced to functionless ribbons. The fibrils become small, many mitochondria and most of the myofilaments disappear and little is eventually left beyond tracheoles, lysosomes and some granular sarcoplasm (e.g. Atkins and Farris, 1962; Bhakthan, Borden and Nair, 1970). Degeneration of the indirect flight-muscles also occurs in adult females of *Dysdercus*, where it coincides with oocyte growth and is under endocrine control (Edwards, 1969). Other interesting changes take place in the segmental muscles of the abdomen of *Rhodnius*, which undergo periodic regressive and regenerative changes associated with the moulting cycle (Wigglesworth, 1956; Toselli and Pepe, 1968).

Literature on the Muscular System

ACTON, A. B. AND SCUDDER, G. G. E. (1969), The ultrastructure of the flight muscle polymorphism in *Cenocorixa bifida* (Hung.) (Heteroptera, Corixidae), *Z. Morph. Tiere*, **65**, 327–335.

AIDLEY, D. J. (1967), The excitation of insect skeletal muscles, *Adv. Insect Physiol.*, **4**, 1–31.

ALICATA, P. (1963), Muscolatura e sistema nervoso del torace di *Eyprepocnemis plorans* (Charp.) e considerazioni sul sistema nervoso cervico-toracico degli Insetti, *Archo zool. ital.*, **47**, (1962), 263–337.

ASHURST, D. E. (1967), The fibrillar flight muscles of giant waterbugs: an electron-microscope study, *J. Cell Sci.*, **2**, 435–444.

ATKINS, M. D. AND FARRIS, S. H. (1962), A contribution to the knowledge of flight muscle changes in Scolytidae (Coleoptera), *Can. Ent.*, **94**, 25–32.

ATZINGER, L. (1952), Vergleichende Untersuchungen über die Beziehungen zwischen Ausbildung der Flügel, der Flugmuskulatur und des Flugvermögens bei Feldheuschrecken, *Zool. Jb.* (*Anat.*), **76**, 199–222.

BHAKTHAN, N. M. G., BORDEN, J. H. AND NAIR, K. K. (1970), Fine structure of degenerating and regenerating flight muscles in a bark beetle, *Ips confusus* (Col., Scolytidae), *J. Cell Sci.*, **6**, 807–819.

BIENZ-ISLER, G. (1968), Elektronenmikroskopische Untersuchungen über die imaginale Struktur der dorsolongitudinalen Flugmuskeln von *Antheraea pernyi* Guér. (Lepidoptera), *Acta anat.*, **70**, 416–433.

BURSELL, E. (1966), Aspects of the flight metabolism of tsetse flies (*Glossina*), *Comp. Biochem. Physiol.*, **19**, 809–818.

CAVENEY, S. (1969), Muscle attachment related to cuticle architecture in Apterygota, *J. Cell Sci.*, **4**, 541–559.

CHADWICK, L. E. (1959), Spinasternal musculature in certain insect orders, *Smithson. misc. Collns*, **137**, 117–156.

COCHRANE, D. G., ELDER, H. Y. AND USHERWOOD, P. N. R. (1972), Physiology and ultrastructure of phasic and tonic skeletal muscle-fibres in the locust *Schistocerca gregaria*, *J. Cell Sci.*, **10**, 419–441.

COOK, E. F. (1944), The morphology and musculature of the labrum and clypeus of insects, *Microentomology*, **9**, 1–35.

CULLEN, M. J. (1974), The distribution of asynchronous muscle in insects with particular reference to the Hemiptera: an electron microscope study. *J. Ent.*, (A), **49**, 17–41.

DAVEY, K. G. (1964), The control of visceral muscles in insects, *Adv. Insect Physiol.*, **2**, 219–245.

EDWARDS, F. J. (1969), Development and histolysis of the indirect flight muscles in *Dysdercus intermedius*, *J. Insect Physiol.*, **15**, 1591–1599.

EWER, D. W. (1967), The pterothoracic musculature of *Bullacris* Roberts and *Pneumora* Stål (Orthoptera, Pneumoridae), *J. ent. Soc. sth. Afr.*, **26**, 411–424.

EWER, D. W. AND NAYLER, L. S. (1967), The pterothoracic musculature of *Deropeltis erythrocephala*, a cockroach with a wingless female, and the origin of wing movements in insects. *J. ent. Soc. sth. Afr.*, **30**, 18–33.

GARAMVÖLGYI, N. (1965a), The arrangement of the myofilaments in the insect flight muscle. I, II, *J. Ultrastruc. Res.*, **13**, 409–434.

—— (1965b), Inter-Z bridges in the flight muscle of the bee, *J. Ultrastruc. Res.*, **13**, 435–443.

GILBERT, L. I. (1967), Lipid metabolism and function in insects, *Adv. Insect Physiol.*, **4**, 69–211.

GOLDSTEIN, M. A. AND BURDETTE, W. J. (1971), Striated visceral muscle of *Drosophila melanogaster*, *J. Morph.*, **134**, 315–334.

HAGOPIAN, M. (1966), The myofilament arrangement in the femoral muscle of the cockroach, *Leucophaea maderae* Fabricius, *J. Cell Biol.*, **28**, 545–562.

HAGOPIAN, M. AND SPIRO, D. (1967), The sarcoplasmic reticulum and its association with the T-system in an insect, *J. Cell Biol.*, **32**, 535–545.

—— (1968), The filament lattice of cockroach thoracic muscle, *J. Cell Biol.*, **36**, 433–442.

HANSON, J. (1956), Studies on the cross-striation of the indirect flight myofibrils of the blowfly *Calliphora*, *J. Biophys. Biochem. Cytol.*, **2**, 691–709.

IMMS, A. D. (1939), On the antennal musculature in insects and other Arthropods, *Q. Jl microsc. Sci.*, **81**, 273–320.

JOHNSON, B. (1957, 1959), Studies on the degeneration of the flight muscles of alate aphids. I. II, *J. Insect Physiol.*, **1**, 248–256; **3**, 367–377.

KÉLER, S. von (1963), *Entomologisches Wörterbuch mit besonderer Berücksichtigung der morphologischen Terminologie*, Akademie-Verlag, Berlin, 4th edn, 790 pp.

LAI-FOOK, J. (1967), The structure of developing muscle insertions in insects, *J. Morph.*, **123**, 503–508.

MANTON, S. M. (1964), Mandibular mechanisms and the evolution of Arthropods, *Phil. Trans. R. Soc., Ser. B*, **247**, 1–183.

—— (1972), The evolution of arthropodan locomotory mechanisms. Part 10: Locomotory habits, morphology and evolution of the hexapod classes, *Zool. J. Linn. Soc.*, **51**, 203–400.

MATSUDA, R. (1965), Morphology and evolution of the insect head, *Mem. Am. ent. Inst.*, **4**, 334 pp.

—— (1970), Morphology and evolution of the insect thorax, *Mem. ent. Soc. Canada*, **76**, 431 pp.

MICKOLEIT, G. (1962), Die Thoraxmuskulatur von *Tipula vernalis* Meigen. Ein Beitrag zur vergleichenden Anatomie des Dipterenthorax, *Zool. Jb. (Anat.)*, **80**, 213–244.

—— (1968), Zur Thoraxmuskulatur der Bittacidae, *Zool. Jb. (Anat.)*, **85**, 386–410.

—— (1969), Vergleichend-anatomische Untersuchungen an der pterothorakalen Pleurotergalmuskulatur der Neuropteria und Mecopteria (Insecta, Holometabola), *Z. Morph. Tiere*, **64**, 151–178.

MILL, P. J. AND LOWE, D. A. (1971), Ultrastructure of the respiratory and non-respiratory dorso-ventral muscles of the larva of a dragonfly, *J. Insect Physiol.*, **17**, 1947–1960.

OSBORNE, M. P. (1967), Supercontraction in the muscles of the blowfly larva: an ultrastructural study, *J. Insect Physiol.*, **13**, 1471–1482.

—— (1970), Structure and function of neuromuscular junctions and stretch receptors, In: Neville, A.C. (ed.), Insect Ultrastructure. *Symp. R. ent. Soc. Lond.*, **5**, 77–100.

PARSONS, M. C. (1959), Skeleton and musculature of the head of *Gelastocoris oculatus* (Fabricius) (Hemiptera–Heteroptera), *Bull. Mus. comp. Zool. Harv.*, **122**, 3–53.

—— (1960), Skeleton and musculature of the thorax of *Gelastocoris oculatus* (Fabricius) (Hemiptera–Heteroptera), *Bull. Mus. comp. Zool. Harv.*, **122**, 299–357.

—— (1962), Skeleton and musculature of the head of *Saldula pallipes* (F.) (Heteroptera: Saldidae), *Trans. R. ent. Soc. Lond.*, **114**, 97–130.

—— (1963), Thoracic skeleton and musculature of adult *Saldula pallipes* (F.) (Heteroptera: Saldidae), *Trans. R. ent. Soc. Lond.*, **115**, 1–37.

PASQUALI-RONCHETTI, I. (1969), The organization of the sarcoplasmic reticulum and T system in the femoral muscle of the housefly, *Musca domestica, J. Cell Biol.*, **40**, 269–273.

—— (1970), The ultrastructural organization of femoral muscles in *Musca domestica* (Dipt., Muscidae). *Tissue and Cell*, **2**, 339–354.

PRINGLE, J. W. S. (1957), *Insect Flight*, Cambridge Univ. Press, Cambridge, 133 pp.

—— (1965), Locomotion: Flight, In: Rockstein, M. (ed.), *The Physiology of Insecta*, **2**, 283–329.

—— (1966), The contractile mechanism of insect fibrillar muscle, *Progr. Biophys. Mol. Biol.*, **17**, 1–60.

—— (1968), Comparative physiology of the flight motor, *Adv. Insect Physiol.*, **5**, 163–227.

REGER, J. F. (1967), The organization of sarcoplasmic reticulum in direct flight muscle of the lepidopteran *Achalarus lyciades, J. Ultrastruct. Res.*, **18**, 595–599.

REGER, J. F. AND COOPER, D. P. (1967), A comparative study on the fine structure of the basalar muscle of the wing and the tibial extensor muscle of the leg of the lepidopteran *Achalarus lyciades, J. Cell Biol.*, **33**, 531–542.

SACKTOR, B. (1965), Energetics and respiratory metabolism of muscular contraction, In : Rockstein, M. (ed.), *The Physiology of Insecta*, **2**, 483–580.

—— (1970), Regulation of intermediary metabolism, with special reference to the control mechanisms in insect flight muscles, *Adv. Insect Physiol.*, **7**, 267–347.

SAITA, A. AND CAMATINI, M. (1967), Studio comparativo al microscopia elettronico di diversi tipi di fibre muscolari in *Cetonia aurata, R. Ist. lomb. Sci. Lett.*, (Ser. B), **101**, 521–541.

SHAFIQ, S. A. (1963), Electron microscopic studies on the indirect flight muscles of *Drosophila melanogaster*, I. Structure of the myofibrils, *J. Cell Biol.*, **17**, 351–362.

—— (1964), An electron microscopical study of the innervation and sarcoplasmic reticulum of the fibrillar flight muscle of *Drosophila melanogaster, Q. Jl microsc. Sci.*, **105**, 1–6.

SMITH, D. S. (1961a), The structure of insect fibrillar flight muscle, a study made with special reference to the membrane systems of the fibre, *J. biophys. biochem. Cytol.*, **10** Suppl., 123–158.

—— (1961b), The organization of the flight muscle in a dragonfly, *Aeshna* sp. (Odonata), *J. biophys. biochem. Cytol.*, **11**, 119–145.

—— (1963), The structure of flight muscle sarcosomes in the blowfly *Calliphora erythrocephala* (Diptera), *J. Cell Biol.*, **19**, 115–138.

—— (1965), The organization of flight muscle in an aphid, *Megoura viciae* (Homoptera). With a discussion on the structure of synchronous and asynchronous striated muscle fibres, *J. Cell Biol.*, **27**, 379–393.

—— (1966a), The organization of flight muscle fibres in the Odonata, *J. Cell Biol.*, **28**, 109–126.

—— (1966b), The structure of intersegmental muscle fibres in an insect, *Periplaneta americana* L., *J. Cell Biol.*, **29**, 449–459.

—— (1966c), The organization and function of the sarcoplasmic reticulum and T-system of muscle cells, *Prog. Biophys. Mol. Biol.*, **16**, 107–142.

SMITH, D. S. AND ALDRICH, H. C. (1971), Membrane systems of freeze-etched striated muscle, *Tissue and Cell*, **3**, 261–281.

SMITH, D. S., GUPTA, B. L. AND SMITH, U. (1966), The organization and myofilament array of insect visceral muscles, *J. Cell Sci.*, **1**, 49–57.

SMITH, D. S. AND SACKTOR, B. (1960), Disposition of membranes and the entry of haemolymph-borne ferritin in flight-muscle fibers of the fly *Phormia regina*, *Tissue and Cell*, **2**, 355–374.

SNODGRASS, R. E. (1935), *Principles of Insect Morphology*, McGraw-Hill, New York, 667 pp.

—— (1950), Comparative studies on the jaws of mandibulate arthropods, *Smithson. misc. Collns*, **116**, 85 pp.

TIEGS, O. W. (1955), The flight muscles of insects – their anatomy and histology, with some observations on the structure of striated muscles in general, *Phil. Trans. R. Soc., Ser. B*, **238**, 221–348.

TOSELLI, P. A. AND PEPE, F. A. (1968), The fine structure of the ventral intersegmental abdominal muscles of the insect *Rhodnius prolixus* during the molting cycle, I, II, *J. Cell Biol.*, **37**, 445–461; 462–481.

USHERWOOD, P. N. P. (1967), Insect neuromuscular mechanisms, *Am. Zool.*, **7**, 553–582.

—— (1969), Electrochemistry of insect muscle, *Adv. Insect Physiol.*, **6**, 205–278.

—— (ed.) (1975), *Insect Muscle*, Academic Press, London, 621 pp.

WALCOTT, B. AND BURROWS, M. (1969), The ultrastructure and physiology of the abdominal air-guide retractor muscles in the giant water-bug, *Lethocerus*, *J. Insect Physiol.*, **15**, 1855–1872.

WEBER, H. (1933), *Lehrbuch der Entomologie*, Jena, 726 pp.

WIGGLESWORTH, V. B. (1956), Formation and involution of striated muscle fibres during the growth and moulting cycles of *Rhodnius prolixus* (Hemiptera), *Q. Jl microsc. Sci.*, **97**, 465–480.

YOUNG, E. C. (1965a), The incidence of flight polymorphism in British Corixidae and description of the morphs, *J. Zool.*, **146**, 567–576.

—— (1965b), Flight muscle polymorphism in British Corixidae: Ecological observations, *J. Anim. Ecol.*, **34**, 335–390.

Chapter 9

THE NERVOUS SYSTEM

The General Nervous System

Introduction – The nervous system forms an elaborate connecting link between the *sense organs*, which respond to various external and internal stimuli, and *effector organs* such as muscles, glands and luminous structures, through which the insect reacts to the stimuli by coordinated behavioural changes (Schmitt, 1962; Roeder, 1963; Smith and Treherne, 1963; Gouin, 1965; Horridge, 1965; Treherne and Beament, 1965; Treherne, 1966, 1974; Hoyle, 1970). In addition, some nerve cells have a special secretory role (p. 273) and the integrity of the nervous system may be required for normal development and regeneration of other tissues (Finlayson, 1960; Nüesch, 1968). The nervous system is composed essentially of ramifying *neuroglial cells* with supporting and nutritive functions (Wigglesworth, 1960; Smith, 1967) and the more numerous *neurons*, which are highly specialized for the relatively rapid generation and conduction of electrochemical nervous impulses (Pipa, 1961; Landolt, 1965; Chiarodo, 1969; Sohal *et al.*, 1972). Processes of each neuron put it into contact by synaptic junctions with other nerve cells (Chiarodo *et al.*, 1970) or with sensory structures or effector organs. One or more of these processes forms a long conducting fibre or *axon* (Narahashi, 1963) ending in a group of fine, branching fibrils; the number of axons enables the neuron to be classified as uni-, bi- or multipolar. Functionally the nerve cells may be sensory (afferent) neurons, conveying impulses inwards from the sense organs (p. 123 ff.), motor or efferent neurons, conveying them outwards to the effector organs, and association neurons (interneurons) that link the other two types. Over most of the nervous system the neurons are grouped together into *ganglia* (p. 102), all with a more or less complex histological structure (Wigglesworth, 1959, and other references cited below). The axons of many neurons are organized into fibre tracts within the ganglion or run out, arranged together, as peripheral nerves (which also contain the incoming axons of afferent neurons). Neuroglial cells occur within and around the ganglia and nerves, forming an outer layer of cells, the perineurium, that secretes an external, more homogeneous, sheath-like neural lamella (Ashurst and Chapman, 1961; Madrell and Treherne, 1967; Lane, 1968, 1972). This contains neutral mucopolysaccharides and collagen fibres and is primarily a supporting structure, while the perineurium helps to form a physiological barrier between the ionic and other constituents of the blood and those of the nervous tissue (Treherne and Pichon, 1972). The axons are wrapped around by specialized neuroglial cells (Schwann cells) which form a thick myelin sheath in places. Individual axons vary greatly in diameter, the largest forming 'giant axons' adapted for very rapid nervous conduction over considerable distances (Seabrook, 1970b, 1971; Parnas and Dagan, 1971; Harris and Smyth, 1971).

The very large number of neurons (around a million in the brain of *Apis* according to Witthöft, 1967), and the multiplicity of possible connections among them means that the nervous pathways between and within the ganglia are extremely complicated. Some of the coarser ones have long been known from simple histological studies, but more recent mapping of the nerve tracts and individual axonal connections has required a variety of specialized techniques. These include modern silver impregnation methods, electron microscopy, tracing degenerating fibres after surgical transection of peripheral nerves, intraneuronal injection of dye-stuffs and the use of electrophysiological recording after experimental stimulation. The integration of such physiological and histological results is still actively in progress.

It is convenient to divide the insect nervous system into (*a*) the central nervous system, (*b*) the visceral nervous system, and (*c*) the peripheral nervous system, all three parts being connected with each other.

The Central Nervous System

This is the principal division of the nervous system and consists of a double series of ganglia joined by longitudinal and transverse strands of nerve fibres (Fig. 59). The longitudinal cords are termed *connectives* and join a pair of ganglia with those which precede and succeed it. The transverse fibres or *commissures* unite the two ganglia of a pair. Typically there is a pair of ganglia in each segment of the body, but the members of a pair are usually so closely united that they appear as a single ganglion, the commissure being no longer evident externally. The connectives may be separate and distinct throughout the body as in *Machilis* and *Corydalis*, or in the thorax only as in the Orthoptera, Coleoptera and many Lepidopteran larvae, but usually they are so closely approximated as to form a single longitudinal cord. In many cases the ganglia of adjacent segments coalesce to form *ganglionic centres*. Two of these are always present in the head, and varying degrees of coalescence of the thoracic and abdominal ganglia are revealed by a comparative study of the nervous system in different orders of insects.

Transverse sections through a typical ganglion show the outermost perineurium and neural lamella enclosing groups of neurons that form the ganglionic cortex and enclose a central neuropile. The neurons include the usually large, unipolar, motor neurons and the generally smaller interneurons, which may be massed locally into globuli. Some of the larger motor neurons are arranged in bilaterally symmetrical pairs and can be identified from one insect to another (Cohen and Jacklet, 1967). The neuropile is composed of axons, collaterals and dendrites, the latter often grouped locally into dense glomeruli.

The central nervous system is divisible into the brain or cerebral ganglion, the suboesophageal ganglion and the ventral nerve-cord.

1. **The Brain** (Figs. 60–63) lies just above the oesophagus between the supporting apodemes of the tentorium. It is the dorsal ganglionic centre of the head; its nerve-cells are almost entirely association neurons and it is formed by the coalescence of the first three neuromeres in the embryo. This threefold division is maintained in the completed organ which is divided into corresponding regions which are designated the *protocerebrum*, the *deutocerebrum* and the *tritocerebrum* respectively, though the division is not always apparent externally. Among general works on the structure of the brain are those by Jawłowski (1936), Hanström (1940), Ehnbohm (1948), Goossen (1949), Satija (1957, etc.), Guthrie (1961), Abraham (1967), Groth (1971) and Strausfeld (1976); various more specialized accounts of its component parts are listed below.

The Protocerebrum represents the fused ganglia of the acron and the preantennal segment. It forms the greater part of the brain and innervates the compound eyes and ocelli. The protocerebrum is divisible into (*a*) the protocerebral lobes and (*b*) the optic lobes.

(*a*) The *protocerebral lobes* are fused together along the mid line to form a bilobed ganglion. The two lobes are connected by a median commissural system termed the *central body*, towards which fibres converge from various parts of the brain. In addition to the central body there are two smaller commissures, the anterior and posterior dorsal. The *anterior dorsal commissure* passes in front of and above the central body. The *posterior dorsal commissure* is a⊤⊤-shaped fibre-tract lying behind the anterior commissure. The most conspicuous formations in the neuropile of the protocerebral lobes are the paired *mushroom bodies* or *corpora pedunculata* which are important association-centres (Landolt and Ris, 1966; Mancini and Frontali, 1967; Steiger, 1967; Frontali and Mancini, 1970; Schürmann, 1970, 1972; Pearson, 1971).

Each mushroom body consists essentially of one or two dorsal cup-like *calyces* of neuropile containing masses of neurons, the so-called calyx cells or *globuli cells*. The calyx (or joined calyces) then continue into a stalk or *pedunculus* of nerve fibres which ends deep in the brain by a complicated root system composed of a recurrent *alpha-lobe*, a *beta-lobe* and sometimes even a third *gamma-lobe*. In the Lepidoptera another separate group of globuli cells gives rise to a tract that leads to an additional tripartite *Y-lobe* (Pearson, *loc. cit.*). The axons of the calyx cells send branches to the stalk and the lobes of the root, where they form synapses with fibres from interneurons lying outside the corpora pedunculata. In the region of the brain between the mushroom bodies are four small *ocellar centres* from each of which an ocellar nerve takes its origin. The two outer nerves supply the paired dorsal ocelli, while the two inner nerves unite just outside the brain to form a single nerve supplying the median ocellus.

(*b*) The *optic lobes* (*optic ganglia* or *optic tract*) form a complex region of the brain and their degree of development is related to that of the compound eyes (e.g. Zuberi, 1963). Their immensely complicated histology was revealed in classical studies by Cajal and Sanchez (1915). This aspect, and the important role of the optic

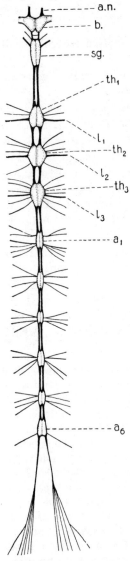

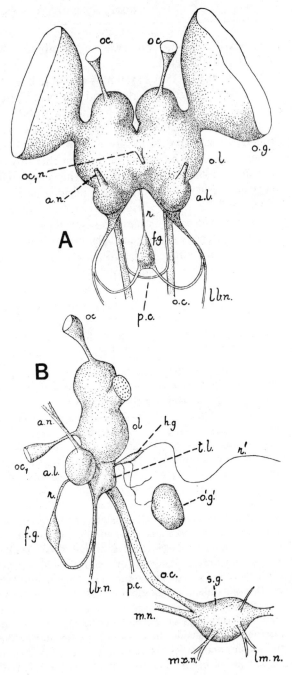

FIG. 59
Central nervous system of
Forficula

a.n, antennal nerve; *b*, brain;
sg, suboesophageal ganglion;
th₁–th₃, thoracic ganglia; *l₁–l₃*,
nerves to legs; *a₁–a₆*, 1st and
terminal abdominal ganglia.

FIG. 60
Brain and suboesophageal ganglion of a locust (*Melanoplus*)

A, frontal view; B, lateral view. *a.l*, antennary lobe; *a.n*, antennal
nerve; *f.g*, frontal ganglion; *h.g*, hypocerebral ganglion; *lb.n*, labral
nerve; *lm.n*, labial nerve; *m.n*, mandibular nerve; *mx.n*, maxillary
nerve; *oc*, lateral ocellus; *oc₁*, median ocellus; *oc₁n*, root of nerve to

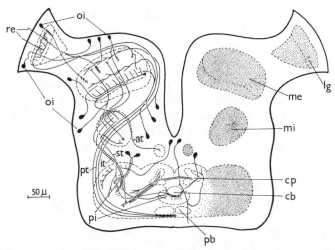

FIG. 61 Diagram to show neurons, nerve-tracts and centres of the protocerebrum and optic
lobes of *Gerris lacustris* (*after* Guthrie, 1962)

at, anterior optic tract; *cb*, central body; *cp*, corpora pedunculata; *it*, inferior median optic
tract; *lg*, lamina ganglionaris; *me*, medulla externa; *mi*, medulla interna; *oi*, optic interneurons;
pb, protocerebral bridge; *pi*, protocerebral interneurons; *pt*, posterior optic tract; *re*, retinal
elements; *st*, superior median optic tract.

lobe in processing the sensory input from the compound eye, still requires full
elucidation and has attracted many recent workers (e.g. Boschek, 1971; Campos-
Ortega and Strausfeld, 1972; Collett, 1970, 1971; Horridge and Meinertzhagen,
1970*a*, *b*; Strausfeld, 1970, 1971*a*, *b*; Strausfeld and Blest, 1970; Strausfeld and
Braitenberg, 1970, Trujillo-Cenoz, 1965; Varela and Porter, 1969; Varela, 1970;
Williams, 1975). Each lobe consists of three principal zones or tracts of nerve tissues
which are connected by a similar number of layers of nerve fibres (Figs. 62 and 86).
The *lamina ganglionaris* is the zone nearest the eye and is connected with the inner
ends of the ommatidia by the layer of *post-retinal fibres*. The middle zone is the
medulla externa and is connected with the lamina by the *external chiasma* which is
formed by the crossing of nerve fibres. The inner zone is the *medulla interna* or
lobula, united with the preceding zone by the *internal chiasma*. The fibres of the *optic
nerve* issue from the inner aspect of the *medulla interna* and divide into anterior and
posterior bundles, which pass to the centre of the protocerebrum. Studies of the
finer detail show that each major zone includes several types of neurons and a great
variety of nervous connections. Short axons from the receptor cells of each ommatidium join those from other ommatidia in the lamina to form a regular array of
optic 'cartridges'. The resulting projection of the retinal mosaic on to the lamina is
then further projected by other cells into an array of medullary cartridges, while
tangential nerve processes intersect the parallel, perpendicularly oriented neurons at
various characteristic levels. Links also occur between the two optic lobes and run
from the medial part of the protocerebrum to each medulla separately. Electrophysiological techniques have revealed units within the optic lobe mediating different

median ocellus; *o.c*, para-oesophageal connective; *o.g*, optic ganglion; *o'.g'*, oesophageal ganglion; *o.l*, optic
lobe; *p.c*, postoesophageal commissure; *r*, recurrent nerve (continued in B as the stomatogastric nerve *r'*);
s.g, suboesophageal ganglion; *t.l*, tritocerebral lobe. *After* Burgess, *2nd Rep. U.S. Ent. Comm.*

visual functions: some respond to movement of small parts of the visual field, others react to long, contrasting boundaries moving in a preferred direction; some subtend monocular visual fields, others cover the whole receptive area of both eyes, and so on.

The Deutocerebrum represents the fused ganglia of the antennal segment. It is chiefly composed of the paired *antennal* or *olfactory lobes* which are prominent swellings situated on the anteroventral aspect of the brain and innervate the antennae. For their detailed histology and ultrastructure see Schürmann and Wechsler (1969, 1970), Pareto (1972) and Masson (1972).

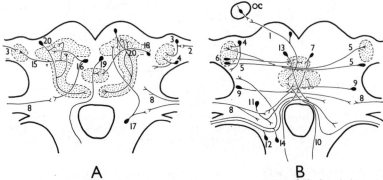

A   B

FIG. 62 Diagram of some major axon pathways in brain of *Periplaneta* showing links from corpora pedunculata (A) and central body (B) to main afferent association areas (*after* Bullock and Horridge, 1965, and Bretschneider, 1913)

oc, ocellus; *1*, second-order fibres from ocellar nerve; *2–4*, neurons of optic lobe; *5–6*, optic lobe to protocerebral bridge; *8*, antennal primary sensory fibres; *9, 11, 12*, deutocerebral commissure fibres; *14*, antennal motor fibres; *15–17*, fibres to calyx of corpora pedunculata; *18*, fibres between central body and calyx; *20*, globuli neurons of corpora pedunculata.

The so-called *dorsal lobe* is chiefly represented by a transverse fibrous tract situated above the antennal lobes and joining the latter together. Each half of the dorsal lobe is connected with the protocerebral lobe of the opposite side by a chiasma and the antennal lobe is connected with the mushroom body of its side and the central body by the *optico-olfactory chiasma*. Arising from the deutocerebrum are four pairs of nerves: the *antennal nerves* are the longest and most important, and are the sensory nerves of the antennae though the motor fibres mentioned next are also present in the same nerve; each has two roots, one of which is derived from the antennal lobe of its side and the other from the dorsal lobe. The *accessory antennal nerves* issue from the antennal lobes and are the motor nerves of the antennae. The *tegumentary nerves* are a pair of slender strands arising from the dorsal lobe and passing to the vertex.

The Tritocerebrum is formed by the ganglia of the third or intercalary segment of the head and is composed mainly of fibres and synapses between the brain, suboesophageal ganglia and frontal ganglion connectives (Willey, 1961). It is divided into two small widely separated lobes which are attached to the dorsal lobe of the deutocerebrum and receive nerve fibres from the latter. The tritocerebral lobes are joined together by means of the *postoesophageal commissure* which passes immediately behind the oesophagus. They also give origin to (1) the *para-oesophageal connectives* which unite the brain with the suboesophageal ganglion, and (2) the *labro-frontal nerves*. Each of the latter consists of two bundles of fibres, one of which passes to the labrum as the *labral nerve*, and the other forms the root of the frontal ganglion.

2. **The Suboesophageal Ganglion** is the ventral ganglionic centre of the head and is formed by the fusion of the ganglia of the mandibular, maxillary and labial segments. It gives off paired nerves supplying their respective appendages and may also connect with the corpora allata and contribute to the labral nerves (Willey, 1961).

3. **The Ventral Nerve-Cord** consists of a series of ganglia lying on the floor of the thorax and abdomen. They are united into a longitudinal chain by a pair of connectives which issue from the posterior border of the suboesophageal ganglion. The first three ganglia are situated one in each of the thoracic segments, and are known as the thoracic ganglia; the remainder lie in the abdomen and form the ganglia of that region.

The *thoracic ganglia* control the locomotor organs. Each ganglion gives off two pairs of principal nerves, one of which supplies the general musculature of the segment and the other innervates the muscles of the legs. In the meso- and metathorax an additional pair of nerves is present and controls the movements of the wings.

The *abdominal ganglia* are variable in number; in *Machilis* and in many larvae there are eight ganglia in the abdomen but as a rule there are fewer. The first abdominal ganglion frequently coalesces with that of the metathorax and the terminal ganglion is always composite, being formed by the fusion of at least three segmental nerve centres. Each abdominal ganglion gives off a pair of principal nerves to the muscles of its segment.

The histological structure of the ganglia of the ventral chain has been studied by Cohen and Jacklet (1967), Crossman *et al.* (1971), Huddart (1971a), Milburn and Bentley (1971), Pipa, Cook and Richards (1959), Seabrook (1968, 1970a), Young (1969) and others. Fielden (1963) has shown that the roots of the segmental nerves in *Anax imperator* arise from separate dorsal and ventral tracts in the neuropile; sensory activities predominate in the ventral tract and motor activity occurs almost entirely in the dorsal part. Further comparative studies are needed.

The Visceral Nervous System

The visceral or sympathetic nervous system is divided into (1) oesophageal sympathetic, (2) ventral sympathetic, and (3) caudal sympathetic systems.

1. **The Oesophageal Sympathetic (or Stomatogastric) Nervous System** is directly connected with the brain and innervates the fore and middle intestine, heart and certain other parts. It is dorsal in position, lying

above and at the side of the fore intestine and its structure is described in a wide variety of types by Cazal (1948), two common arrangements being shown in Fig. 63. Typically, a small triangular *frontal ganglion* lies above the oesophagus, a short distance in front of the brain. The frontal ganglion controls movements of the fore gut (Möhl, 1972). Its removal from immature stages reduces their growth and increases the osmolarity of the blood, apparently through hormonally controlled diuresis (Clark and Anstee, 1971; Penzlin, 1971). Anteriorly the frontal ganglion gives off a *frontal nerve* which passes to the clypeus, and a pair of lateral roots connect the frontal ganglion with the tritocerebrum. Posteriorly it emits a *recurrent nerve* which extends along the mid-dorsal line of the oesophagus and, passing just beneath the brain, expands a short distance behind the latter into a *hypocerebral ganglion*. The recurrent nerve leaves the hypocerebral ganglion as a median or paired lateral oesophageal nerve and passes back to the hinder region of the fore gut, where it (or each branch) terminates in a *ventricular* ganglion. This innervates the adjacent region of the fore and middle intestine. A pair of *corpora cardiaca* lies on the oesophagus just behind the brain, each joined with the hypocerebral ganglion. They are also connected with the protocere-

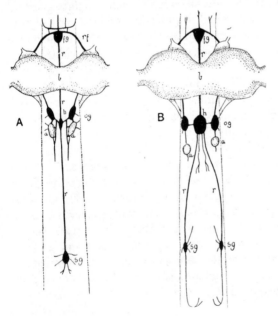

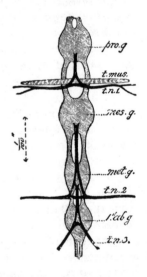

FIG. 63
Semi-diagrammatic figures of two prevalent types of sympathetic nervous system (in black)

A, with a single recurrent nerve and ventricular ganglion; B, with paired recurrent nerves and ganglia. The fore intestine is represented by the dotted lines. *a, a,* corpora allata; *b,* brain; *fg,* frontal ganglion; *h,* hypocerebral ganglion; *og,* corpus cardiacum (right); *r,* recurrent nerve; *rf,* root of frontal ganglion; *sg,* ventricular ganglion.

FIG. 64
Thoracic ganglia and portion of sympathetic nervous system of a *Chironomus* larva

pro.g, mes.g, met.g, thoracic ganglia; 1*ab.g,* 1st abdominal ganglion; *t.mus,* transverse muscle; $tn_1–tn_3$, sympathetic nerves. *After* Miall and Hammond.

brum, and include both nervous and endocrine structures (p. 274). Connected with them by nerves are the non-nervous *corpora allata*, with important endocrine functions (see p. 276). Among variations on this generalized Pterygote plan may be mentioned the connection of the frontal ganglion with the protocerebrum by a median *nervus cohnectivus*, the regression of the hypocerebral ganglion and varying degrees of fusion between the hypocerebral ganglion, the corpora cardiaca and the corpora allata, culminating in the condition found in the larvae of Cyclorrhaphan Diptera where all three fuse to form a composite structure known as Weismann's ring which encircles the aorta just behind the brain (p. 278).

2. **The Ventral Sympathetic Nervous System** (Fig. 64), when typically developed (Zawarsin, 1924), consists of a pair of transverse nerves associated with each ganglion of the ventral nerve-cord; each pair is connected with the ganglion preceding it by a median longitudinal nerve. The transverse nerves pass to the spiracles of their segment (Case, 1957; Miller, 1967). In many orders of insects, each transverse nerve is also connected in various ways with the so-called *perisympathetic system*, consisting of neurohaemal organs that release the products of neurosecretory cells in the ventral ganglia (Raabe, 1971).

3. **The so-called Caudal Sympathetic Nervous System** arises from the posterior compound ganglion of the ventral nerve-cord and supplies the reproductive system and posterior part of the gut; it is, however, difficult to draw a clear distinction between such a system and the visceral branches of the peripheral nervous system considered below.

The Peripheral Nervous System

Strictly speaking, this includes all the nerves radiating from the ganglia of the central and sympathetic nervous systems. The ultrastructure of the larger nerves resembles that of ganglia, with an outer lamella of collagen fibres in a mucopolysaccharide matrix secreted by perineurial cells, and with other glial cells lying most abundantly near the larger axons (Osborne, 1966; Huddart, 1971b). Micro-anatomical accounts of peripheral nerve distribution are generally fragmentary, dealing only with the innervation of particular organs but for more detailed descriptions of thoracic and abdominal innervation in selected species see Nüesch (1954, 1957), Wittig (1955), Pipa and Cook (1959), Schmitt (1962) and others. Of morphological interest is their evidence that the nervous system is not very much more conservative in its evolution than are the muscular and exoskeletal systems: muscles which seem to form an integral part of one segment may in part be innervated from the ganglia of other segments.

In some viscera, such as the salivary glands of *Periplaneta*, which are supplied from both the suboesophageal ganglion and the stomatogastric system, nerves reaching an acinus form a plexus on its surface and the axons then penetrate between the acinar cells to end in close contact with the gland-cell membranes by specially differentiated terminations (Whitehead, 1971). The junctions between peripheral nerve-endings and muscles are also specialized regions (Osborne, 1967), as are the stretch-receptors that form part of the proprioceptor sensory system (p. 130) and from which connections join the central nervous system. Various regions of the gut are elaborately innervated (e.g. Zawarsin, 1916; Orlov, 1924), with multipolar neurons in or on the intestinal wall. In the blowfly *Phormia* this system helps to regulate feeding: a group of 4–8 neurons are located in branches of the median abdominal nerve and respond to changes in the volume of the crop (Gelperin, 1971a, b). The reproductive system, too, has a complex innervation concerned in copulation and oviposition (Rehm, 1939; Degrugillier and Leopold, 1972; Grossman and Parnas, 1973). Lastly, the integument is supplied with a delicate plexus of nerve fibres and sensory neurons; these differentiate from epidermal cells and acquire connections distally with cuticular sensilla and proximally with the central nervous system (Zawarsin, 1912a, b; Rogosina, 1928; Wigglesworth, 1953; Finlayson, 1972, Hasenfuss, 1973).

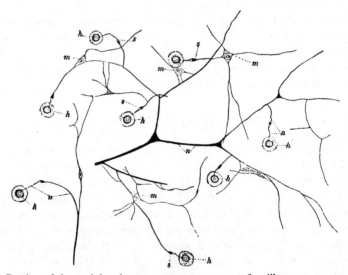

FIG. 65   Portion of the peripheral sensory nervous system of a silkworm

h, h, bases of sensory hairs; *s,* bipolar nerve cells; *m,* multipolar nerve cells; *n,* nerves. *After* Hilton, *Amer. Nat.* 36.

Modifications of the Nervous System

There are many degrees of cerebral development in insects, more or less clearly correlated with the complexity of the insect's sensory equipment or behaviour (Howse, 1975). Thus, the volume of the brain is 1/174th of the body volume in *Apis* and 1/280th in *Formica*, but only 1/3290th in

Melolontha and 1/4200th in *Dytiscus*. Again, the optic lobes are developed in proportion to the size of the eyes and the antennal lobes related to the development of the antennae and the senses mediated by them. Internally, the mushroom bodies attain great size and complexity in Hymenoptera with elaborate behaviour. Structural differences in the brains of drone, worker and queen bees appear to be correlated with the degree of development of the special instincts and activities of the three forms. Further, by comparing the size of the mushroom bodies to the brain as a whole it has been shown in the Hymenoptera that the sawflies come lowest in the scale and the social Hymenoptera highest, while the solitary bees occupy an intermediate position (Hanström, 1926; Pandazis, 1930; Gejvall, 1936; Goossen, 1949; and earlier workers; cf. also Zuberi, 1963, on termites).

In the ventral nerve-cord (Fig. 66), the most generalized condition occurs in the Thysanura, many larvae and some lower Pterygotes where the suboesophageal ganglion, three thoracic and eight abdominal ganglia are separately visible, the most posterior being a composite ganglion. Most Orthopteroid insects, Mecoptera, Trichoptera and Hymenoptera show only a little more concentration than this, but the metathoracic ganglion commonly fuses with

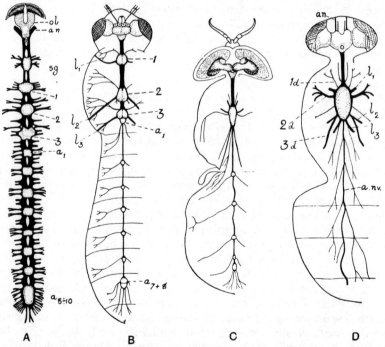

A   B   C   D

FIG. 66   Schematic figures of the central nervous system showing degrees of concentration, based upon various authors

A, *Machilis* (Oudemans); B, *Chironomus* (Brandt); C, *Stratiomys* (Kunckel d'Herculais); D, *Musca* (Hewitt). 1–3, thoracic ganglia; a_1, a_{7-8}, a_{8-10}, abdominal ganglia; *an*, antennal nerve; *ol*, optic lobe; *sg*, suboesophageal ganglion; l_{1-3}, nerves to legs; $1d-3d$, dorsal thoracic nerves; *a.nv*, abdominal nerves.

the first 1–3 abdominal ones and the 7th and subsequent abdominal ganglia form a compound centre. The other orders show various increasing degrees of fusion, often reaching extremes. Thus, in many Heteroptera, the suboesophageal and prothoracic ganglia are distinct but all the others have fused together. In many Sternorrhynchan Homoptera and higher Diptera, only the suboesophageal ganglion and a single compound thoracico-abdominal ganglion are to be seen while in the Coccoidea, Aphidoidea and some Coleopteran larvae all the ventral ganglia (including the suboesophageal) are united in a single centre. Data amplifying these generalizations will be found in the chapters dealing with individual insect orders.

Physiology of the Nervous System

A detailed treatment of this topic cannot be attempted here and reference should be made to the reviews and selected papers cited on p. 115 ff. The biophysical processes involved in the conduction of nervous impulses in insects seem to be essentially similar to those in other invertebrates (Narahashi, 1963). The resting potential of a giant axon of *Periplaneta* is about 80 mV; when the neuronal membrane is depolarized temporarily (mainly through an increase in its permeability to sodium ions) an action potential of 85–100 mV, lasting about 1 ms, is produced. The passage of such momentary changes of potential along an axon represents the propagation of the nervous impulse. The velocity of conduction varies somewhat; in the giant axons of *Periplaneta* it occurs at about 7 metres per second, while in the afferent nerves from the paraprocts of *Anax* nymphs it is about 2 m s^{-1} (see, for example, Chapman and Pankhurst, 1967).

Synaptic junctions between neurons lie in the ganglionic neuropile (especially the glomeruli), where associations occur between axon branches or dendrites. Interneuronal junctions between cell bodies (Landolt and Ris, 1966) or between axons and somata are rare or absent. The synaptic zones are without glial processes, the participating axon branches or dendrites are separated by a synaptic cleft about 20 nm wide, and the presynaptic knob contains mitochondria, glycogen granules and vesicles. The latter vary from about 30 to 70 nm in diameter with cores that may be electron-dense or electron-transparent (Boistel, 1968; Chiarodo, 1969; Chiarodo *et al.*, 1970; Schürmann, 1970, 1972; and others). In addition to the above chemical synapses, where the vesicles indicate the role of transmitter substances, other interneuronal contacts are formed by ephapses or tight junctions between the pre- and post-synaptic membranes; the ultrastructure of such contacts suggests that they are synapses across which electrical transmission occurs (e.g. Steiger, 1967; Huddart, 1971*b*). The identity of the chemical transmitters in the insect nervous system has been discussed extensively (see Pitman, 1971; Gerschenfeld, 1973; McDonald, 1975). Acetylcholine (ACh) and the enzymes that hydrolyse and synthesize it (acetylcholine esterase and choline acetylase) are present in relatively large quantities in insects (Smallman and Mansing, 1969) and it seems likely that acetylcholine is a transmitter at neuropilar axo-axonic synapses; γ-aminobutyric acid (GABA) may also play a part in synaptic transmission in the neuropile. Transmitter substances appear also to be released at neuromuscular junctions: there is evidence that glutamate is responsible for transmitting excitation from nerves to muscle (e.g. Beranek and Miller, 1969) and that GABA may act as an inhibitory transmitter at these sites.

In addition to the more or less complex pathways mapped histologically, several simpler pathways have now been analysed by electrophysiological recording techniques. For example, in *Calliphora* flight starts after a jump due to the contraction of muscles running from the tergum to the trochanter of the mid leg. Mulloney (1969) has shown that these muscles are activated by a motor neuron in the thorax which in turn is driven by two interneurons in the brain, from which descending paths synapse with the motor neurons in the thorax. A more intensively studied pathway is that leading from the cerci or paraprocts through the last abdominal ganglion up ascending giant axons in the ventral nerve cord. In the nymph of the dragonfly *Anax imperator* the giant fibres run the length of the cord (Fielden, 1960); they synapse with efferent fibres in the segmental nerves of the thoracic and abdominal ganglia and a rapid evasion response can be elicited by stimulating the afferent nerves from the paraprocts. In *Periplaneta* the cercal nerve tracts lead directly into the extensively ramifying dendritic processes of the interneurons from which giant axons run forwards through the thorax to activate the antennae and a general alarm system, though the accompanying leg movements are controlled by separate motor neurons with more slowly conducting pathways (Dagan and Parnas, 1970; Milburn and Bentley, 1971; Harris and Smyth, 1971. See also Seabrook, 1970*b*, 1971, for the giant fibre system of *Schistocerca*).

An important function of the nervous system is the central processing of information derived from the sense organs, especially from the compound eyes and auditory organs, which respond to stimuli with relatively elaborate spatial or temporal patterns. In the ventral nerve cord of *Locusta*, for example, electrophysiological recording has detected several types of symmetrically arranged 'acoustic interneurons' dealing with impulses from the auditory tympanal organs in the first abdominal segment (Kalmring, Rheinlaender and Rehbein, 1972; Kalmring, Rheinlaender and Römer, 1972). Three of these types depend on input from the organ of the same side, another responds to stimuli received by both sides, while a third type is unusual in simply measuring the repetition rate of the sound pattern. Comparable mechanisms have been found in several other insect species (e.g. Fraser Rowell and McKay, 1969; McKay, 1969; Roeder, 1969). The central processes involved in vision are likely to prove much more complicated (Horridge *et al.*, 1965) but it is already clear that electrophysiological units in the optic lobes of selected Orthoptera, Diptera and Lepidoptera respond differentially when a variety of visual stimuli is presented to the insect. Collett (1972), for example, found several directionally sensitive movement detectors in the brain of the hawkmoth *Sphinx ligustri*. These included (*a*) medial protocerebral neurons with large binocular fields, projecting to the optic lobe or to the ventral nerve cord; (*b*) movement detectors projecting from the medulla interna of one optic lobe to the opposite medial protocerebrum, and (*c*) optic lobe output cells which excite the binocular movement detectors centrifugal to the medulla. For other examples see Dingle and Fox (1966), Collett (1970), Mimura (1971) and Northrop and Guignon (1970).

The many interesting and sometimes elaborate forms of instinctive behaviour shown by insects have often been described and some have also been explained as simple taxes and kineses or interpreted in more complex ethological terms (Tinbergen, 1951; Baerends, 1959; Fraenkel and Gunn, 1961; Thorpe, 1963; Haskell, 1966; Evans, 1966; von Frisch, 1967; Sudd, 1967; Ewing and Manning, 1967; Manning, 1972). The neural mechanisms that underlie this behaviour, and especially those by which a simple act is incorporated into a larger biologically significant behaviour pattern, are now being studied increasingly, despite the many difficulties (Hoyle, 1970). Most attention has been paid to the nervous control of motor activity in respiration, walking, flight, stridulation and mating behaviour. In *Locusta*, for example, there is one neural pacemaker controlling spiracular opening and closing and another centre that controls ventilatory movements of the abdomen (Miller, 1966, 1967). The centres are synchronized by interneurons that run from the ventilation centre to other interneurons which are antecedent to the spiracular motor neurons in the ganglion of the same segment as the spiracle concerned.

Apart from simple evasive and defensive reactions, in which they play a dominant role, reflex arcs seem mainly to be involved in perfecting movements that are initiated by central mechanisms or in helping to maintain the nervous system in a state of adequate excitation. Studies on the nervous control mechanisms of insect flight (Wilson, 1968) suggest that a simple pattern of neural output to the flight-muscles, produced centrally, can support a form of flight, even in the absence of sensory input from wing receptors. Similarly, in a physiological preparation of *Periplaneta* in which all sensory input from the legs had been eliminated, centrally generated patterns of activity could be demonstrated in the motor neurons that control the coxal levator and depressor muscles (Pearson and Iles, 1970). Normal locomotion involves local reflexes, however, as is clearly shown in *Carausius* (Wendler, 1964). Here the proprioceptor hair-plates of the legs lead to a 'negative feedback' loop which moderates the exaggerated steps caused by a separate, centrally generated motor sequence.

The neural mechanisms involved in stridulation and epigamic behaviour of several species of Orthoptera have been intensively studied (Huber, 1965; Elsner and Huber, 1973). By ablation experiments and the local electrical stimulation of parts of the brain of *Gryllus campestris* it has been shown that sites near the mushroom bodies control the production of normal song-rhythms, while stimulation of the central body leads to abnormal songs through the changed temporal pattern of sound-pulses. In the grasshopper *Gomphocerippus rufus* elements of the male courtship pattern can be related to the activities of single motor units and sound production may be described quantitatively by reference to the efferent discharge to known groups of muscles (Elsner, 1968). It is interesting that in *Stenobothrus rubicundus* the same nervous mechanism can induce two different activities: thoracic motor neurons produce identical firing patterns in controlling leg and wing movements so that certain fundamental features of the wing-beat pattern recur in stridulatory movements of the leg (Elsner, 1974).

Some insects are known to show simple forms of learning and memory (Alloway, 1972). They can be trained to run simple mazes or to associate food with colours or other visual stimuli. Bees, ants and some other Aculeate Hymenoptera remember the visual landmarks they use in following routes to and from their nests. The neurophysiological basis of such reactions is unknown, but some analysis of very simple forms of learning has been made (Horridge, 1962; Hoyle, 1965). Headless *Periplaneta* and *Schistocerca* can be trained to flex their legs for long periods in order to avoid an electric shock. The response depends on continuous excitatory discharge to the coxal adductor muscles by only one or two neurons. If a spontaneous fall occurs in this background discharge, a further electric shock causes the frequency to rise again for several minutes or more; conversely, training to reduce the spontaneous discharge may also be accomplished. Simple learning processes of this kind can occur in preparations containing only a single isolated ganglion (Eisenstein, 1972).

Lastly one may note the role played by the nervous system in controlling circadian behavioural patterns. Circadian rhythms show persistent, stable periodicities of about 24 hours and may be observed in developmental, metabolic and behavioural events. Behavioural periodicity of this kind has been found in locomotor, feeding and reproductive activities and are all likely to be controlled through the central nervous system (Brady, 1974). The best studied example is the circadian rhythm of locomotor activity in *Periplaneta*, which is normally more active at night time than in daylight. In order to maintain this circadian rhythm it is only necessary that one protocerebral lobe and the ipsilateral medulla externa and lobula should be in nervous connection with the thorax. The simplest adequate interpretation of the evidence suggests that a driving circadian oscillator in the optic lobe is coupled electrically to the leg muscles via the protocerebrum, nerve cord connectives and thoracic ganglia, the whole system being modulated by sensory inputs which are integrated with the circadian oscillator output in the corpora pedunculata.

Literature on the Nervous System

ABRAHAM, A. (1967), Die Struktur der Gehirnzentren des Gelbrandkäfers (*Dytiscus marginalis* L.), *Z. mikrosk.-anat. Forsch.*, **76**, 435–465.

ALLOWAY, T. M. (1972), Learning and memory in insects, *A. Rev. Ent.*, **17**, 43–56.

ASHURST, D. E. AND CHAPMAN, J. A. (1961), The connective-tissue sheath of the nervous system of *Locusta migratoria*: an electron microscope study, *Q. Jl microsc. Sci.*, **102**, 463–467.

BAERENDS, G. P. (1959), Ethological studies of insect behaviour, *A. Rev. Ent.*, **4**, 207–234.

BERANEK, R. AND MILLER, P. L. (1969), The action of iontophoretically applied glutamate on insect muscle fibres, *J. exp. Biol.*, **49**, 83–93.

BOISTEL, J. (1968), The synaptic transmission and related phenomena in insects, *Adv. Insect Physiol.*, **5**, 1–64.

BOSCHEK, C. B. (1971), On the fine structure of the peripheral retina and lamina ganglionaris of the fly, *Musca domestica*, *Z. Zellforsch.*, **118**, 369–409.

BRADY, J. (1974), The physiology of insect circadian rhythms, *Adv. Insect Physiol.*, **10**, 1–115.

CAJAL, S. R. AND SANCHEZ, D. (1915), Contribución al conocimiento de los centros nervosos de los insectos, *Trab. Lab. Invest. biol. Univ. Madrid*, **13**, 1–164.

CAMPOS-ORTEGA, J. A. AND STRAUSFELD, N. J. (1972), The columnar organization of the second synaptic region of the visual system of *Musca domestica* L. I. Receptor terminals in the medulla, *Z. Zellforsch.*, **124**, 561–585.

CASE, J. F. (1957), The median nerves and cockroach spiracular function, *J. Insect Physiol.*, **1**, 85–94.

CAZAL, P. (1948), Les glandes endocrines rétrocérébrales des insectes (étude morphologique), *Bull. biol. Fr. Belg.*, *Suppl.*, **32**, 227 pp.; *Suppl.*, **33**, 9–18.

CHAPMAN, K. M. AND PANKHURST, J. H. (1967), Conduction velocities and their temperature coefficients in sensory nerve fibres of cockroach legs, *J. exp. Biol.*, **46**, 63–84.

CHIARODO, A. J. (1969), The fine structure of neurons and nerve fibres in the thoracic ganglion of the blowfly, *Sarcophaga bullata*, *J. Insect Physiol.*, **14**, 1169–1175.

CHIARODO, A. J., KISSEL, K. H. AND MACKEL, T. E. (1970), Structure of synaptic relations in the larval central nervous systems of the blowfly and the mealworm, *J. Insect Physiol.*, **16**, 361–371.

CLARKE, K. U. AND ANSTEE, J. H. (1971), Effects of the removal of the frontal ganglion on the mechanisms of energy-production in *Locusta*, *J. Insect Physiol.*, **17**, 717–732.

COHEN, M. J. AND JACKLET, J. W. (1967), The functional organization of motor neurons in an insect ganglion, *Phil. Trans. R. Soc.*, *Ser. B*, **252**, 561–572.

COLLETT, T. (1970), Centripetal and centrifugal visual cells in the medulla of the insect optic lobe, *J. Neurophysiol.*, **33**, 239–256.

—— (1971), Connections between wide-field monocular and binocular movement detectors in the brain of a hawk moth, *Z. vergl. Physiol.*, **75**, 1–31.

CROSSMAN, A. R., KERKUT, G. A., PITMAN, R. M. AND WALKER, R. J. (1971), Electrically excitable nerve cell-bodies in the central ganglia of two insect species, *Periplaneta americana*, and *Schistocerca gregaria*: investigations of cell geometry and morphology by intracellular dye-injection, *Comp. Biochem. Physiol.*, (*A*), **40**, 579–594.

DAGAN, D. AND PARNAS, I. (1970), Giant fibres and small-fibre pathways involved in the evasive response of the cockroach *Periplaneta americana* (Dict., Blattaria), *J. exp. Biol.*, **52**, 313–324.

DEGRUGILLIER, M. E. AND LEOPOLD, R. A. (1972), Abdominal peripheral nervous system of the adult female housefly and its role in mating behaviour and insemination, *Ann. ent. Soc. Am.*, **65**, 689–695.

DINGLE, H. AND FOX, S. S. (1966), Microelectrode analysis of light responses in the brain of the cricket (*Gryllus domesticus*), *J. cell. comp. Physiol.*, **68**, 45–59.

EHNBOHM, K. (1948), Studies on the central and sympathetic nervous system and some sense organs in the head of Neuropteroid insects, *Opusc. ent.*, *Suppl.*, **8**, 162 pp.

EISENSTEIN, E. M. (1972), Learning and memory in isolated insect ganglia, *Adv. Insect Physiol.*, **9**, 111–181.

ELSNER, N. (1968), Die neuromuskulären Grundlagen des Werbeverhaltens der Roten Keulenheuschrecke *Gomphocerippus rufus* (L.), *Z. vergl. Physiol.*, **60**, 308–350.

—— (1974), Neural economy: bifunctional muscles and common central pattern elements in leg and wing stridulation of the grasshopper *Stenobothrus rubicundus* Germ., (Orthoptera: Acrididae), *J. comp. Physiol.*, **89**, 227–236.

ELSNER, N. AND HUBER, F. (1973), Neurale Grundlagen artspezifischer Kommunikation bei Orthopteren, *Fortschr. Zool.*, **22**, 1–48.

EVANS, H. E. (1966), The behavior patterns of solitary wasps, *A. Rev. Ent.*, **11**, 123–154.

EWING, A. AND MANNING, A. (1967), The evolution and genetics of insect behaviour, *A. Rev. Ent.*, **12**, 471–494.

FIELDEN, A. (1960), Transmission through the last abdominal ganglion of the dragonfly nymph, *Anax imperator*, *J. exp. Biol.*, **37**, 832–844.

—— (1963), The localization of function in the root of an insect segmental nerve, *J. exp. Biol.*, **40**, 553–561.

FINLAYSON, L. H. (1960), A comparative study of the effects of denervation on the abdominal muscles of Saturniid moths during pupation, *J. Insect Physiol.*, **5**, 108–119.

—— (1972), Chemoreceptors, cuticular mechanoreceptors, and peripheral multiterminal neurones in the larva of the tsetse fly (*Glossina*), *J. Insect Physiol.*, **18**, 2265–2276.

FRAENKEL, G. S. AND GUNN, D. L. (1961), *The Orientation of Animals. Kineses, Taxes and Compass Reactions*, Dover Publ., New York, 2nd edn, 376 pp.

FRASER ROWELL, C. H. AND MCKAY, J. M. (1969), An acridid auditory interneurone. I. Functional connections and response to single sounds, *J. exp. Biol.*, **51**, 231–245.

FRISCH, K. von (1967), *The Dance Language and Orientation of Bees*, Harvard Univ. Press, Cambridge, Mass., 566 pp.

FRONTALI, N. AND MANCINI, G. (1970), Studies on the neuronal organization of cockroach corpora pedunculata. *J. Insect Physiol.*, **16**, 2293–2301.

GEJVALL, N. G. (1936), Untersuchungen über die relative und absolute Grösse der verschiedenen Gehirnzentren von *Apis mellifera*, *Förh. K. fysiogr. Sallsk. Lund*, **5**, 22–32.

GELPERIN, A. (1971*a*), Abdominal sensory neurons providing negative feedback to the feeding behaviour of the blowfly, *Z. vergl. Physiol.*, **72**, 17–31.

—— (1971*b*), Regulation of feeding, *A. Rev. Ent.*, **16**, 365–378.

GERSCHENFELD, H. M. (1973), Chemical transmission in invertebrate central nervous systems and neuro-muscular junctions, *Physiol. Rev.*, **53**, 1–119.

GOOSSEN, H. (1949), Untersuchungen an Gehirnen verschieden grosser, jeweils verwandter Coleopteren- und Hymenopteren-Arten, *Zool. Jb. (Allg. Zool.)*, **62**, 1–64.

GOUIN, F. J. (1965), Morphologie, Histologie und Entwicklungsgeschichte der Myriapoden und Insekten. III. Das Nervensystem und die neurocrinen Systeme, *Fortschr. Zool.*, **17**, 189–237.

GROSSMAN, Y. AND PARNAS, I. (1973), Control mechanisms involved in the regulation of the phallic neuromuscular system of the cockroach *Periplaneta americana*, *J. comp. Physiol.*, **82**, 1–21.

GROTH, U. (1971), Vergleichende Untersuchungen über die Topographie und Histologie des Gehirns der Dipteren, *Zool. Jb. (Anat.)*, **88**, 203–319.

GUTHRIE, D. M. (1961), The anatomy of the nervous system in the genus *Gerris* (Hemiptera–Heteroptera), *Phil. Trans. R. Soc., Ser. B*, **244**, 65–102.

HANSTRÖM, B. (1926), Untersuchungen über die relative Grösse der Gehirnzentren verschiedener Arthropoden unter Berücksichtigung der Lebensweise, *Z. mikr.-anat. Forsch.*, **7**: 135–190.

—— (1940), Inkretorische Organe, Sinnesorgane und Nervensystem des Kopfes einiger niederer Insektenordnungen, *K. svenska VetenskAkad. Handl.*, (3), **18**, 266 pp.

HARRIS, C. L. AND SMYTH, T. (1971), Structural details of cockroach giant axons revealed by injected dye, *Comp. Biochem. Physiol.*, (A), **40**, 295–303.

HASENFUSS, I. (1973), Vergleichend-morphologische Untersuchung der sensorischen Innervierung der Rumpfwand der Larven von *Rhyacophila nubila* Zett. (Trichoptera) und *Galleria mellonella* L. (Lepidoptera), *Zool. Jb. (Anat.)*, **90**, 1–54; 175–253.

HASKELL, P. T. (ed.) (1966), Insect Behaviour, *Symp. R. ent. Soc. Lond.*, **3**, 112 pp.

HORRIDGE, G. A. (1962), Learning leg position by the ventral nerve cord in headless insects, *Proc. R. Soc. (B)*, **157**, 33–52.

—— (1965), The Arthropoda, In: Bullock, T. H. and Horridge, G. A. (eds), *Structure and Function in the Nervous Systems of Invertebrates*, Freeman, San Francisco, **2**, 801–1270.

HORRIDGE, G. A. AND MEINERTZHAGEN, I. A. (1970*a*), The exact neural projection of the visual fields upon the first and second ganglia of the insect eye, *Z. vergl. Physiol.*, **66**, 369–378.

—— (1970*b*), The accuracy of the patterns of connexions of the first- and second-order neurons of the visual system of *Calliphora*, *Proc. R. Soc. (B)*, **175**, 69–82.

HORRIDGE, G. A., SCHOLES, J. H., SHAW, S. AND TUNSTALL, J. (1965), Extracellular recordings from single neurones in the optic lobe and brain of the locust, In: Treherne, J. E. and Beament, J. W. L. (eds), *The Physiology of the Insect Central Nervous System*, Academic Press, London, pp. 165–202.

HOWSE, P. E. (1975), Brain structure and behaviour in insects, *A. Rev. Ent.*, **20**, 359–379.

HOYLE, G. (1965), Neurophysiological studies on 'learning' in headless insects. In: Treherne, J. E. and Beament, J. W. L. (eds), *The Physiology of the Insect Central Nervous System*, London, pp. 203–232.

—— (1970), Cellular mechanisms underlying behaviour: neuroethology, *Adv. Insect Physiol.*, **7**, 349–444.

HUBER, F. (1965), Brain-controlled behaviour in Orthoptera, In: Treherne, J. E. and Beament, J. W. L. (eds), *The Physiology of the Insect Central Nervous System*, London, pp. 233–246.

HUDDART, H. (1971*a*), Ultrastructure of the prothoracic ganglion and connectives of the stick insect in relation to function. *J. Insect Physiol.*, **17**, 1451–1469.

—— (1971*b*), Ultrastructure of a peripheral nerve of *Locusta migratoria migratorioides* R. et F. (Orth., Acrididae), *Int. J. Insect Morphol. and Embryol.*, **1**, 29–41.

JAWŁOWSKI, H. (1936), Ueber den Gehirnbau der Käfer, *Z. Morph. Ökol. Tiere*, **32**, 67–91.

KALMRING, K., RHEINLAENDER, J. AND REHBEIN, H. (1972), Akustische Neuronen im Bauchmark der Wanderheuschrecke *Locusta migratoria*, *Z. vergl. Physiol.*, **76**, 314–332.

KALMRING, K., RHEINLAENDER, J. AND RÖMER, H. (1972), Akustische Neuronen im Bauchmark von *Locusta migratoria*. Der Einfluss der Schallrichtung auf die Antwortmuster, *J. comp. Physiol.*, **80**, 325–352.

LANDOLT, A. M. (1965), Elektronenmikroskopische Untersuchungen an der Perikaryenschicht der Corpora pedunculata der Waldameise (*Formica lugubris* Zett.) mit besonderer Berücksichtigung der Neuron-Glia-Beziehung, *Z. Zellforsch.*, **66**, 701–736.

LANDOLT, A. M. AND RIS, H. (1966), Electron microscope studies on soma-somatic interneuronal junctions in the corpus pedunculatum of the wood ant (*Formica lugubris* Zett.), *J. Cell Biol.*, **28**, 391–403.

LANE, N. J. (1968), The thoracic ganglia of the grasshopper, *Melanoplus differentialis*: fine structure of the perineurium and neuroglia with special reference to the intracellular distribution of phosphatases, *Z. Zellforsch.*, **86**, 293–312.

—— (1972), Fine structure of a Lepidopteran nervous system and its accessibility to peroxidase and lanthanum, *Z. Zellforsch.*, **131**, 205–222.

MADDRELL, S. H. P. AND TREHERNE, J. E. (1967), The ultrastructure of the perineurium in two insect species, *Carausius morosus* and *Periplaneta americana*, *J. Cell Sci.*, **2**, 119–128.

MANCINI, G. AND FRONTALI, N. (1967), Fine structure of the mushroom body neuropile of the brain of the roach, *Periplaneta americana*, *Z. Zellforsch.*, **83**, 334–343.

MANNING, A. W. G. (1972), *An Introduction to Animal Behaviour*, Arnold, London, 2nd edn.

MASSON, C. (1972), La système antennaire chez les fourmis. I. Histologie et ultrastructure du deutocérébron. Étude comparée chez *Camponotus vagus* (Formicinae) et *Mesoponera caffraria* (Ponerinae), *Z. Zellforsch.*, **134**, 31–64.

MCDONALD, T. J. (1975), Neuromuscular pharmacology of insects, *A. Rev. Ent.*, **20**, 151–166.

MCKAY, J. M. (1969), The auditory system of *Homorocoryphus* (Orth., Tettigoniidae), *J. exp. Biol.*, **51**, 787–802.

MILBURN, N. S. AND BENTLEY, D. R. (1971), On the dendritic topology and activation of cockroach giant interneurons, *J. Insect Physiol.*, **17**, 607–623.

MILLER, P. L. (1966), The regulation of breathing in insects. *Adv. Insect Physiol.*, **3**, 279–354.

—— (1967), The derivation of the motor command to the spiracles of the locust, *J. exp. Biol.*, **46**, 349–371.

MIMURA, K. (1971), Movement discrimination by the visual system of flies, *Z. vergl. Physiol.*, **73**, 105–138.

MÖHL, B. (1972), The control of foregut movements by the stomatogastric nervous system in the European house cricket *Acheta domesticus* L., *J. comp. Physiol.*, **80**, 1–28.

MULLONEY, B. (1969), Interneurons in the central nervous system of flies, and the start of flight, *Z. vergl. Physiol.*, **64**, 243–253.

NARAHASHI, T. (1963), The properties of insect axons, *Adv. Insect Physiol.*, **1**, 175–256.

NORTHROP, R. B. AND GUIGNON, E. F. (1970), Information processing in the optic lobes of the lubber grasshopper, *J. Insect Physiol.*, **16**, 691–713.

NÜESCH, H. (1954), Segmentierung und Muskelinnervation bei *Telea polyphemus*, *Revue suisse Zool.*, **61**, 420–428.

—— (1957), Die Morphologie des Thorax von *Telea polyphemus* Cr. (Lepid.). II. Nervensystem, *Zool. Jb. (Anat.)*, **75**, 615–642.

—— (1968), The role of the nervous system in insect morphogenesis and regeneration, *A. Rev. Ent.*, **13**, 27–44.

ORLOV, J. (1924), Die Innervation des Darmes der Insekten (Larven von Lamellicornien), *Z. wiss. Zool.*, **122**, 425–502.

OSBORNE, M. P. (1966), Ultrastructural observations on adult and larval nerves of the blowfly, *J. Insect Physiol.*, **12**, 501–507.

—— (1967), The fine structure of neuromuscular junctions in the segmental muscles of the blowfly larva, *J. Insect Physiol.*, **13**, 827–833.

PANDAZIS, G. (1930), Über die relative Ausbildung der Gehirnzentren bei biologisch verschiedenen Ameisenarten, *Z. Morph. Ökol. Tiere*, **18**, 114–169.

PARETO, A. (1972), Die zentrale Verteilung der Fühlerafferenz bei Arbeiterinnen der Honigbiene, *Apis mellifera* L., *Z. Zellforsch.*, **131**, 109–140.

PARNAS, I. AND DAGAN, D. (1971), Functional organizations of giant axons in the central nervous systems of insects: new aspects, *Adv. Insect Physiol.*, **8**, 95–144.

PEARSON, K. G. AND ILES, J. F. (1970), Discharge patterns of coxal levator and depressor motoneurons of the cockroach *Periplaneta americana* (Dict., Blattaria), *J. exp. Biol.*, **52**, 139–165.

PEARSON, L. (1971), The corpora pedunculata of *Sphinx ligustri* L. and other Lepidoptera: an anatomical study, *Phil. Trans. R. Soc., Ser. B*, **259**, 477–516.

PENZLIN, H. (1971), Zur Rolle des Frontalganglions bei Larven der Schabe *Periplaneta americana*, *J. Insect Physiol.*, **17**, 559–573.

PIPA, R. L. (1961), Studies on the hexapod nervous system. III. Histology and histochemistry of insect neuroglia, *J. comp. Neurol.*, **116**, 15–26.

PIPA, R. L. AND COOK, E. F. (1959), Studies on the hexapod nervous system. I. The peripheral distribution of the thoracic nerves of the adult cockroach, *Periplaneta americana*, *Ann. ent. Soc. Am.*, **52**, 695–710.

PIPA, R. L., COOK, E. F. AND RICHARDS, A. G. (1959), Studies on the hexapod nervous system. II. The histology of the thoracic ganglia of the adult cockroach, *Periplaneta americana* (L.), *J. comp. Neurol.*, **113**, 401–423.

PITMAN, R. M. (1971), Transmitter substances in insects: a review, *Comp. Gen. Pharmacol.*, **2**, 347–371.

RAABE, M. (1971), Neurosécrétion dans la chaine nerveuse ventrale des insectes et organes neurohémaux métamériques, *Arch. Zool. exp. gén.*, **112**, 679–694.

REHM, E. (1939), Die Innervation der inneren Organe von *Apis mellifica*, zugleich ein Beitrag zur Frage des sog. sympathetischen Nervensystems der Insekten, *Z. Morph. Ökol. Tiere*, **36**, 89–122.

ROEDER, K. D. (1963), *Nerve Cells and Insect Behaviour*, Harvard Univ. Press, Cambridge, Mass., 198 pp.

—— (1969), Acoustic interneurons in the brain of Noctuid moths, *J. Insect Physiol.*, **15**, 825–838.

ROGOSINA, M. (1928), Über das periphere Nervensystem der *Aeschna*-Larve, *Z. Zellforsch.*, **6**, 732–758.

SATIJA, R. C. (1957), A histological study of the brain, optic lobes and thoracic nerve cord of *Petrobius brevistylis* with special reference to the descending nervous pathways, *Res. Bull. Panjab Univ., Hoshiarpur (Zool.)*, **131**, 493–510.

SCHMITT, J. B. (1962), The comparative anatomy of the insect nervous system, *A. Rev. Ent.*, **7**, 137–156.

SCHÜRMANN, F. W. (1970), Über die Struktur der Pilzkörper des Insektenhirns. I. Synapsen im Pedunculus, *Z. Zellforsch.*, **103**, 365–381.

—— (1972), Über die Struktur der Pilzkörper des Insektenhirns. II. Synaptische Schaltungen im Alpha-Lobus des Heimchens *Acheta domesticus* L., *Z. Zellforsch.*, **127**, 240–257.

SCHÜRMANN, F. W. AND WECHSLER, W. (1969), Elektronenmikroskopische Untersuchung am Antennallobus des Deutocerebrum der Wanderheuschrecke *Locusta migratoria*, *Z. Zellforsch.*, **95**, 223–248.

—— (1970), Synapsen im Antennenhügel von *Locusta migratoria*, *Z. Zellforsch.*, **108**, 563–581.

SEABROOK, W. D. (1968), The structure of a pregenital abdominal ganglion of the desert locust *Schistocerca gregaria* (Forskål), *Can. J. Zool.*, **46**, 965–980.

—— (1970a), The structure of the terminal ganglionic mass of the locust, *Schistocerca gregaria* (Forskål) (Orth., Acrididae), *J. comp. Neurol.*, **138**, 63–85.

—— (1970b), The histology of the giant fibre system in the abdominal ventral nerve cord of the desert locust, *Can. Ent.*, **102**, 1163–1168.

—— (1971), An electrophysiological study of the giant fiber system of the locust *Schistocerca gregaria*, *Can. J. Zool.*, **49**, 555–560.

SMALLMAN, B. N. AND MANSING, A. (1969), The cholinergic system in insect development, *A. Rev. Ent.*, **14**, 387–408.

SMITH, D. S. (1967), The trophic role of glial cells in insect ganglia, In: Beament, J.

W. L. and Treherne, J. E. (eds) (1967), *Insects and Physiology*, Oliver & Boyd, London, pp. 189–198.

SMITH, D. S. AND TREHERNE, J. E. (1963), Functional aspects of the insect nervous system, *Adv. Insect Physiol.*, **1**, 401–484.

SOHAL, R. S., SHARMA, S. P. AND COUCH, E. F. (1972), Fine structure of the neural sheath, glia and neurons in the brain of the housefly, *Musca domestica, Z. Zellforsch.*, **135**, 449–459.

STEIGER, U. (1967), Über den Feinbau des Neuropils im Corpus pedunculatum der Waldameise. Elektronenmikroskopische Untersuchungen, *Z. Zellforsch.*, **81**, 511–536.

STRAUSFELD, N. J. (1970), Golgi studies on insects. Part II. The optic lobes of Diptera, *Phil. Trans. R. Soc., Ser. B.*, **258**, 135–223.

—— (1971*a*), The organization of the insect visual system (light microscopy). I. Projections and arrangements of neurons in the lamina ganglionaris of Diptera, *Z. Zellforsch.*, **121**, 377–441.

—— (1971*b*), The organization of the insect visual system (light microscopy). II. The projection of fibres across the first optic chiasma, *Z. Zellforsch.*, **121**, 442–454.

—— (1976), *Atlas of an Insect Brain*, Springer, Berlin, 230 pp.

STRAUSFELD, N. J. AND BLEST, A. D. (1970), Golgi studies on insects. Part I. The optic lobes of Lepidoptera, *Phil. Trans. R. Soc., Ser. B*, **258**, 81–134.

STRAUSFELD, N. J. AND BRAITENBERG, V. (1970), The compound eye of the fly (*Musca domestica*): connections between the cartridges of the lamina ganglionaris, *Z. vergl. Physiol.*, **70**, 95–104.

SUDD, J. M. (1967), *An Introduction to the Behaviour of Ants*, London, 208 pp.

THORPE, W. H. (1963), *Learning and Instinct in Animals*, London, 2nd edn.

TINBERGEN, N. (1951), *The Study of Instinct*, Clarendon Press, Oxford, 228 pp.

TREHERNE, J. E. (1966), *The Neurochemistry of Arthropods*, Cambridge Univ. Press, Cambridge, 156 pp.

—— (ed.) (1974), *Insect Neurobiology*, North-Holland, Amsterdam, 450 pp.

TREHERNE, J. E. AND BEAMENT, J. W. L. (eds) (1965), *The Physiology of the Insect Central Nervous System*, Academic Press, London, 277 pp.

TREHERNE, J. E. AND PICHON, Y. (1972), The insect blood-brain barrier, *Adv. Insect Physiol.*, **9**, 257–313.

TRUJILLO-CENÓZ, O. (1965), Some aspects of the structural organization of the intermediate retina of Dipterans, *J. Ultrastruct. Res.*, **13**, 1–33.

VARELA, F. G. (1970), Fine structure of the visual system of the honeybee (*Apis mellifera*). II. The lamina, *J. Ultrastruct. Res.*, **31**, 178–194.

VARELA, F. G. AND PORTER, K. R. (1969), Fine structure of the visual system of the honeybee (*Apis mellifera*). I. The retina, *J. Ultrastruct. Res.*, **29**, 236–259.

WENDLER, G. (1964), Laufen und Stehen der Stabheuschrecke *Carausius morosus*: Sinnesborstenfelder in den Beingelenken als Glieder von Regelkreisen, *Z. vergl. Physiol.*, **48**, 198–250.

WHITEHEAD, A. T. (1971), The innervation of the salivary gland in the American cockroach: light- and electron-microscopic observations, *J. Morph.*, **135**, 483–505.

WIGGLESWORTH, V. B. (1953), The origin of sensory neurones in an insect, *Rhodnius prolixus* (Hemiptera), *Q. Jl microsc. Sci.*, **94**, 93–112.

—— (1959), The histology of the nervous system of an insect, *Rhodnius prolixus*

(Hemiptera). I. The peripheral nervous system. II. The central ganglia, *Q. Jl microsc. Sci.*, **100**, 285–298; 299–313.

—— (1960), The nutrition of the central nervous system in the cockroach *Periplaneta americana* L. The role of perineurium and glial cells in the mobilization of reserves, *J. exp. Biol.*, **37**, 500–512.

WILLEY, R. B. (1961), The morphology of the stomodeal nervous system in *Periplaneta americana* (L.) and other Blattaria, *J. Morph.*, **108**, 219–247.

WILLIAMS, J. L. D. (1975), Anatomical studies of the insect central nervous system: a ground-plan of the midbrain and an introduction to the central complex in the locust, *Schistocerca gregaria* (Orthoptera), *J. Zool.*, **176**.

WILSON, D. M. (1968), The nervous control of insect flight and related behaviour, *Adv. Insect Physiol.*, **5**, 289–338.

WITTHÖFT, W. (1967), Absolute Anzahl und Verteilung der Zellen im Hirn der Honigbiene, *Z. Morph. Tiere*, **61**, 160–184.

WITTIG, G. (1955), Untersuchungen am Thorax von *Perla abdominalis* Burm. (Larve und Imago), *Zool. Jb. (Anat.)*, **74**, 491–570.

YOUNG, D. (1969), The motor neurons on the mesothoracic ganglion of *Periplaneta americana*, *J. Insect Physiol.*, **15**, 1175–1179.

ZAWARSIN, A. (1912a), Histologische Studien über Insekten. II. Das sensible Nervensystem der *Aeschna*-Larve, *Z. wiss. Zool.*, **100**, 245–286.

—— (1912b), Histologische Studien über Insekten. III. Über das sensible Nervensystem der Larven von *Melolontha vulgaris*, *Z. wiss. Zool.*, **100**, 447–458.

—— (1916), Quelques données sur la structure du système nerveux intestinal des insectes, *Rev. Zool. Russe.*, **1**, 176–180.

—— (1924), Histologische Studien über Insekten. V. Ueber die histologische Beschaffenheit des unpaaren ventralen Nervs der Insekten, *Z. wiss. Zool.*, **122**, 97–115.

ZUBERI, H. A. (1963), L'anatomie comparée du cerveau chez les Termites en rapport avec le polymorphisme, *Bull. biol. Fr. Belg.*, **97**, 147–207.

Chapter 10

THE SENSE ORGANS AND PERCEPTION

The sense organs or receptors (Carthy and Newell, 1968; Dethier, 1963; Horridge, 1965) are those structures whereby the energy of a stimulus arising outside or, less obviously, within the insect, is transformed into a nervous impulse which, after transmission to one of the central ganglia, usually results in a change in the behaviour of the insect or in the maintenance of some existing activity. Proof that a given organ is a receptor has, in many cases, been obtained by electrical recording of the nervous impulse which follows stimulation but some receptors have been studied only through observations on the insect's behaviour and occasionally a sensory function has been ascribed solely on the basis of structure and anatomical relations. The connection between a receptor and the effector organs which change or maintain activity is often represented as a relatively simple reflex arc (Fig. 67) though it is probable that even the simplest behavioural act involves the coordination of many variable conducting paths whose functional activity depends on the quality of the stimulus and the physiological state of the receptor and nervous system.

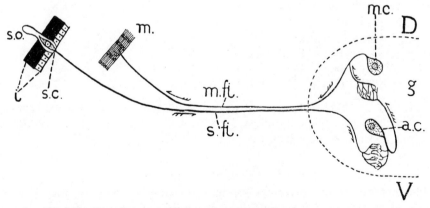

FIG. 67   Diagram of a reflex arc in the nervous system of an insect.
One half of a ganglion, *g*, of the ventral nerve-cord is represented in outline

D, dorsal aspect; *V*, ventral aspect. A motor (*m.fi*) and a sensory fibre (*s.fi*) of a lateral nerve are shown; *i*, integument; *s.o*, sense organ; *s.c*, sensory neuron; *m*, muscle; *a.c*, association neuron; *m.c*, motor neuron. (The course traversed by a stimulus, received by the sense organ, is represented by arrows.)

Sense organs are associated especially with the integument and each organ or *sensillum* (pl. *sensilla*) typically comprises: (*a*) a structure – cuticular and/or epidermal – through which the stimulus is amplified or directed or translated into other mechanical or chemical changes (not always adequately analysed as yet) and (*b*) one or more sensory neurons responding to the modified stimulus resulting from (*a*) by nervous activity. In the simplest cases, however, the neuron may end distally in fine ramifications among apparently unmodified epithelial cells or be produced into a simple process surmounted by cuticle which is hardly, if at all, specialized (Finlayson, 1972). The sensory neurons of insects are *primary sense cells* since they are produced centripetally into axons which enter a ganglion and end in contact with association neurons. Secondary sense cells, i.e. ectodermal structures innervated by deep-lying sensory neurons similar to those in the spinal ganglia of vertebrates, are apparently absent in the Insecta. So far as is known the nerve-impulses leaving the sensory neurons are qualitatively similar in all the different types of receptors and the specificity which these organs exhibit (i.e. the fact that some respond only to light, others to sound, etc.) depends mainly on the character of the transducing structures mentioned above. Essentially, therefore, the net result of stimulation is the production of a *generator potential* in the sensory neuron, leading to the development of *action potentials* that are conducted away along the afferent nerve fibre. *Tonic sensory neurons* produce action potentials throughout the duration of a stimulus, while *phasic receptors* fire at the beginning of a stimulus (*on-receptors*) or its end (*off-receptors*) or both (*on–off-receptors*).

It is convenient to consider insect sense organs under the following six main headings but a few unusual forms of sensory perception are discussed separately on p. 147.

    1. Mechanoreceptors.
    2. Auditory Organs.
    3. Chemoreceptors.
    4. Temperature Receptors.
    5. Humidity Receptors.
    6. Visual Organs.

It may be noted that auditory organs are actually only specialized types of mechanoreceptor while humidity perception may ultimately prove reducible to one of the other mechanisms.

1. Mechanoreceptors

The members of this diverse group have the common property that they are excited by processes involving the mechanical deformation of some part of the receptor (Dethier, 1963; McIver, 1975). As such they may mediate the sense of touch, including contact with solid objects or currents of air or water; they may also respond to mechanical stresses set up in the cuticle and so function as proprioceptors (Finlayson, 1968), including the specialized organs of equilibrium; and, as mentioned above, in certain cases they respond to changes in air pressure or the displacement of air-particles and so act as sound receptors. Four main structural types of mechanoreceptor may be distinguished:

(*a*) **Articulated Sensory Hairs** (Fig. 68). Each sensory hair or *trichoid sensillum* comprises the usual trichogen and tormogen cells which secrete respectively the hair and its socket (p. 15) together with a bipolar neuron which is produced distally into a specialized structure containing arrays of

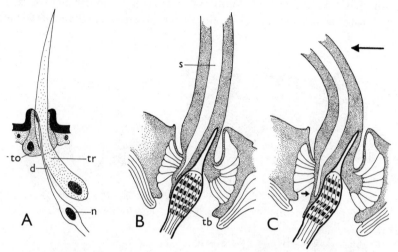

FIG. 68 Mechanoreceptor trichoid sensilla. A, histology of a mechanoreceptor hair from the prosternum of *Calliphora* (*after* Richter, 1964). B and C, ultrastructural diagrams of the socket-region, illustrating the mode of action of a mechanoreceptor hair (*after* Gaffal and Hansen, 1972); displacement of the hair shaft (indicated by the large arrow in C, above) results in a smaller displacement of the inner end of the hair, causing it to deform the tubular body in which the dendrite terminates, thus stimulating the neuron to discharge

d, dendrite; *n*, neuron; *s*, shaft of hair; *tb*, tubular body; *to*, tormogen cell; *tr*, trichogen cell.

microtubules and in close contact with the base of the hair (Richter, 1964; Guthrie, 1966; Gaffal and Hansen, 1972). Such hairs are widely distributed over the insect body, especially on the antennae, tarsi and cerci and are tactile organs, being stimulated to produce a nervous impulse on movement of the hair in its socket (Thurm, 1963, 1968).

Trichoid sensilla are often grouped together into 'hair plates' or 'bristle fields' where they act as proprioceptors and thus subserve many different biological functions. On the vertex of *Locusta* and *Schistocerca* they form a cephalic air-flow receptor that helps to regulate flight (Guthrie, 1966; Gewecke, 1972*b*), and in *Gryllus* males a pad of sensory hairs on each fore wing controls their stridulatory movements (Möss, 1971). In ants and other

Hymenoptera, bristle fields occur on the neck, antennae, petiole and legs, enabling the insects to orient themselves with respect to gravity (Markl, 1962, 1966); in *Apis* workers cervical hair plates and sensory hairs on the mandibles and antennae are needed for normal comb construction (Martin and Lindauer, 1966). Reflexes involving the trichoid sensilla of the tarsi and the coxal joints are important in the nervous control of posture and walking (Wendler, 1964; Runion and Usherwood, 1968).

(*b*) **Campaniform sensilla** – These are comparable to the innervated hairs in that two specialized epidermal cells and a bipolar neuron are associated with each sensillum (Fig. 69); structures intermediate between the

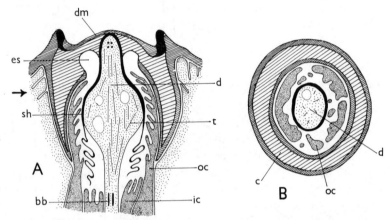

F I G. 69   Ultrastructure of a campaniform sensillum (*after* Chevalier, 1969). A, longitudinal section; B, transverse section at about level indicated by arrow at left-hand side

bb, basal body; *c*, cuticle; *d*, dendrite; *dm*, cuticular dome; *es*, extracellular space; *ic*, internal enveloping cell; *oc*, outer enveloping cell; *sh*, sheath; *t*, microtubules.

two forms occur in *Drosophila* mutants (Lees, 1942). A campaniform sensillum consists essentially of a dome-shaped region of relatively thin cuticle, often oval in surface view and set a little above or below the general level of the integument, with a denser rim of cuticle surrounding it. The distal process of the neuron is packed with microtubules, partly arranged into a cilium-like structure; its tip is covered with a dense secretion bound tightly to the cuticle of the dome (Moran *et al.*, 1971; Smith, 1969; Chevalier, 1969). Stresses in the cuticle are thought to be resolved into displacements of the dome which stimulate the distal process, though the detailed mechanism of transduction (as in other mechanoreceptors) is not fully explained. Campaniform sensilla occur on various parts of the body, including the cerci, wings and legs. In *Periplaneta* there is one at the base of each large tactile spine on the legs (Chapman, 1965) and in *Schistocerca* the campaniform sensilla near the base of the subcostal vein play a part in regulating the twisting of the wing during flight (Gettrup, 1966). Large numbers of campaniform sensilla occur arranged in parallel rows over the base of the haltere of Diptera (Chevalier, 1969; Smith, 1969) and are of major importance in flight control (see p. 65). In the ant *Atta cephalotes* three groups of campaniform sensilla at the junction of the trochanter and femur can perceive vibrations of the substrate occurring with frequencies between 0·05 and 4 kHz (Markl, 1970).

(c) **Chordotonal Organs (Scoloparia)** – These are usually compound structures composed of a number of specialized sensilla (*scolopophores* or *scolopidia*) which each contain a relatively conspicuous sensory body, the *scolops* or *scolopale* (Eggers, 1928; McFarlane, 1953; Howse, 1968). Each chordotonal organ consists typically of a spindle-shaped bundle of scolopophores attached at each end to the integument, but less frequently the distal attachment is absent and the organ ends freely in the body-cavity. The two

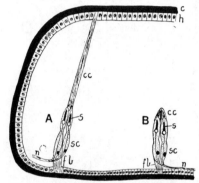

FIG. 70
Diagram of the two types of scolopophores

A, integumental; B, subintegumental. *c*, cuticle; *h*, epidermis; *cc*, cap cell; *s*, scolopale; *sc*, sensory cell; *fb*, connective tissue; *n*, nerve.

types may be distinguished respectively as *integumental* (*amphinematic*) and *subintegumental* (*mononematic*) chordotonal organs (Fig. 70) though the distinction is probably not very important.

The detailed structure of the scolopophore varies somewhat, even within a single chordotonal organ, but they show many common features (Corbière-Tichané, 1971; Young, 1970; Burns, 1974); we can take as an example those found in the subgenual organ of the fore leg of *Gryllus assimilis* (Friedman, 1972a). Each sensillum comprises three main cellular structures: a bipolar sensory neuron, a *scolopale cell* (or *envelope cell*) which surrounds the terminal dendrite of the neuron, and a more distally placed *interstitial cell* or *cap cell* (Fig. 71). The neuron, which has associated glial cells, contains granular endoplasmic reticulum, a Golgi apparatus and membrane-bounded vesicles varying from 55 to 105 nm in diameter. Its terminal dendrite is up to 50 μm long and about 2 μm in diameter, again containing granular endoplasmic reticulum with numerous mitochondria and a system of microtubules forming a ciliary structure. The scolopale cell encloses the dendrite from about the level of the ciliary root and its most characteristic inclusions are several masses of scolopale material containing microtubules about 20 nm in diameter; these masses fuse distally to form a continuous tube around the dendrite. At the apex of the sensillum the extracellular space between the scolopale cell and the dendrite is filled with a relatively homogeneous, electron-opaque substance – the cap material – containing some lacunae and surrounded by the interstitial cell. The sensilla are enclosed in connective tissue which, with the interstitial cells, provides attachment structures.

In addition to their occurrence in auditory organs (described below) groups of a few to very many chordotonal sensilla form organs with a wide distribution in the insect body. They are among the most important proprioceptors, responding to an increase in their longitudinal tension caused by displacement of the structures to which they are attached. In soft-bodied

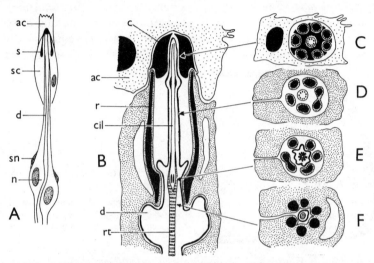

FIG. 71   A, Two scolopophores from the tibiotarsal chordotonal organ of
          Periplaneta (*after* Young, 1970). B, Ultrastructure (longitudinal
          section) of a scolopophore from the tympanal organ of *Locusta*
          (*after* Gray, 1960). C, D, E and F, Transverse sections at levels
          indicated by arrows

          ac, attachment cell; *c*, cap of scolopale; *cil*, cilium; *d*, dendrite; *n*, neuron; *r*,
          rod of scolopale; *rt*, root of cilium; *s*, scolopale; *sc*, scolopale cell; *sn*, nucleus
          of Schwann cell.

larvae, especially those of the Diptera, chordotonal organs are arranged segmentally and run in transverse, oblique and longitudinal directions, as described in *Drosophila* by Hertweck (1931). In adult insects they occur in the antennae and mouthparts and in the cervical region, where they record movements of the head. In the thorax of *Gryllus* pleurotergal scolopidia perceive vibrations of the substrate (Möss, 1971), while in other Orthoptera there are thoracic chordotonal organs that control the frequency with which the wings beat (Wilson and Gettrup, 1963). The aquatic Corixidae possess a thoracic structure (the organ of Hagemann) which apparently functions as a hydrostatic pressure receptor through chordotonal sensilla attached to a movable cuticular process that it contains (Popham, 1961). A large number of chordotonal organs have been described in the legs of many species

(Debaisieux, 1936, 1938), usually in association with the joints between segments. In *Periplaneta* downward and backward movements of the mid tarsus are perceived by a chordotonal organ made up of 26 bipolar neurons with 14 associated scolopales (Young, 1970), while in *Schistocerca, Locusta* and *Romalea* the femoral chordotonal organs of the hind leg provide information on the position, velocity and direction of movement of the tibia (Usherwood *et al.*, 1968). It is perhaps also through chordotonal organs in the fore and mid tarsi that *Notonecta* can localize its prey by responding to the vibrations that it sets up on the water surface (Markl and Wiese, 1969). Of widespread occurrence in the insect leg is the so-called *subgenual organ* which lies in the proximal part of the tibia and consists of an array of chordotonal sensilla suspended between the epidermis and a trachea. It responds to vibrations of the substrate on which the insect is standing and in *Periplaneta* it is sensitive to those with frequencies from 1 to 5 kHz and with threshold amplitudes as low as 10^{-7} to 10^{-10} cms (Autrum and Schneider, 1948; Schnorbus, 1971; Friedmann, 1972a). Some Hemiptera, Coleoptera and Diptera lack subgenual organs and are insensitive to such vibrations.

A highly specialized type of chordotonal organ in the antenna is known as *Johnston's organ*. It is located in the second antennal segment of most, if not all, Pterygota and of *Lepisma* (Eggers, 1928; Debauche, 1936) and consists of a variable number of radially arranged sensilla. These are attached at one end to the membrane between the 2nd and 3rd antennal segments and at the other to the wall of the 2nd segment, the axons from them running back and entering the antennal nerve (Fig. 72). The organ is developed to differing degrees in the different groups of insects and reaches its greatest complexity in male Chironomids and Culicids (Risler, 1955) where it has an auditory function (p. 137). In the hawkmoth *Manduca sexta* Johnston's organ consists of about 650 scolopidial units, each with a bipolar neuron whose dendrite contains numerous microtubules and is modified terminally into three ciliary-like structures (Van de Berg, 1971). In all cases the essential function of Johnston's organ is the perception of movement at the joint between the second antennal segment and the flagellum, as a consequence of which it can mediate responses to a variety of stimuli. It enables the antennae to act as gravity receptors (Bückmann, 1962) and in several species it has now been established that the antennae measure the rate at which air flows past the insect while it flies. In the Cyclorrhaphan Diptera the rate of air flow is assessed by the drag on the arista which in turn causes the third antennal segment to rotate around its axis of suspension, thus stimulating Johnston's organ (Burckhardt and Gewecke, 1965). A similar response occurs in *Apis* (Heran, 1959) and also in *Locusta*, where movement at the joint between the pedicel and flagellum is recorded by Johnston's organ as well as other chordotonal organs and campaniform sensilla in the antenna (Gewecke, 1972a). Different functions are performed by Johnston's organ in several aquatic insects. Thus, in *Dytiscus* and in *Aeshna* larvae it enables the antennae to be used as velocity receptors (Hughes, 1958) and in the whirligig beetle *Gyrinus* it perceives disturbances of the water surface on which the insect swims (de Wilde, 1941; Rudolf, 1967). It also forms part of an organ of equilibrium in *Notonecta, Ilyocoris, Plea* and *Corixa*. In all these Cryptocerate Heteroptera the small antenna is held against an air bubble at the back of the head; when the insect changes its position the bubble moves because of its buoyancy and displaces the antenna, so causing Johnston's organ to respond (Rabe, 1953).

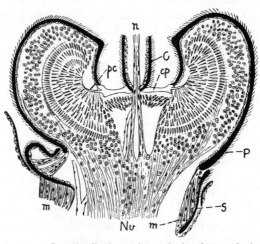

FIG. 72   Longitudinal section of the base of the antenna of a male Culicid (*Chaoborus*) showing Johnston's organ

s, scape; *p*, pedicel; *cp*, conjunctival plate and its process *pc*; *C*, base of clavola; *Nv*, antennal nerve; *n*, nerve to clavola; *m*, antennal muscles. *After* Child, 1894.

(*d*) **Multipolar Stretch Receptors** – Unlike the previously described mechanoreceptors, which contain bipolar (Type I) neurons and are associated with the cuticle, these consist of a multipolar (Type II) neuron associated with connective tissue or a muscle (Finlayson, 1968). Such receptors may form strands slung between intersegmental folds or between a fold and the dorsal body wall or between the latter and a nerve. Stretching the organ produces a linear increase in the frequency of the electrophysiological response (Finlayson and Loewenstein, 1958). They are of wide occurrence and may form segmentally arranged, serially homologous sets (Osborne and Finlayson, 1962, 1965; Whitten, 1963; Weevers, 1966).

Lastly we may mention a few examples of relatively complex or inadequately understood mechanoreceptor organs in some aquatic insects. Adults of *Nepa* have sense organs adjacent to three of the paired abdominal spiracles which apparently perceive relative pressure differences at the different spiracles, while in *Aphelocheirus* a pair of organs on the second abdominal sternum contains tactile hairs that respond to pressure changes at the air-water interface (Thorpe and Crisp, 1947). Statocyst-like organs occur in the terminal abdominal segments of some Limnobiine Tipulid larvae (von Studnitz, 1932) and Palmen's organ in the cephalic tracheal system of Ephemeropteran nymphs may perhaps also function as a statocyst (Wodsedalek, 1912).

2. Auditory Organs

Though noises may be transmitted through liquid and solid media, little is known of their reception by insects in such cases and it is convenient here to regard sound as consisting of air-disturbances of low intensity, irrespective of whether they fall within the range of human hearing (Pumphrey, 1940). Such disturbances produce both a local pressure increase and a displacement of air-particles away from the source of the sound, and it is probable that some insect auditory organs act as displacement receptors while others may be pressure receptors. They are, in fact, mechanoreceptors which differ from tactile sense organs only in that they respond to disturbances of much lower intensity and further study may well reveal receptors of an intermediate character. In spite of their underlying similarity, four types of sound receptors may be recognized (Haskell, 1961; Busnel, 1955; Autrum, 1963).

(a) **Tympanal Organs** – These are paired structures always composed of a thin cuticular membrane, the *tympanum*, associated with tracheal air-sacs and chordotonal sensilla. They occur in the Orthoptera (Acridoidea, Tettigonioidea and Grylloidea), the Cicadidae and some Lepidoptera (Noctuidae, Geometridae, Cymatophoridae, Uraniidae, Pyralididae, etc.).

In the Acridoidea there is a tympanal organ on each side of the first abdominal tergum, with an externally visible tympanum surrounded by a cuticular ring (Figs. 73, 74). Internally (Schwabe, 1906), a group of numerous scolopophores form a swelling known as Müller's organ, applied to the inner surface of each tympanum and connected by the auditory nerve to the metathoracic ganglion. Two sclerotized processes and a pyriform vesicle filled with a clear liquid are intimately associated with Müller's organ; they probably transmit the tympanal vibrations to the sensilla. The first abdominal spiracle, near the anterior margin of the tympanum, gives off an

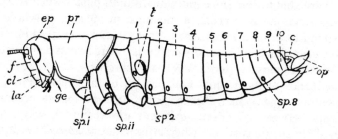

FIG. 73   Lateral view of a locust with wings and legs removed showing tympanum, *t*.
After Carpenter.

air-sac which is applied to the inner surface of that membrane. Two additional air-sacs arise from the ventral tracheal trunk on each side of the 2nd abdominal segment and lie internal to and in close contact with the other sac. The ultrastructure of the scolopophore is essentially similar to that of the mechanoreceptors discussed above (Gray, 1960). The cell body and axon of the bipolar neuron are enclosed in a Schwann cell and a fibrous sheath cell is wrapped around the basal part of the dendrite (Fig. 74); a total of 60–80 chordotonal units is arranged in four groups.

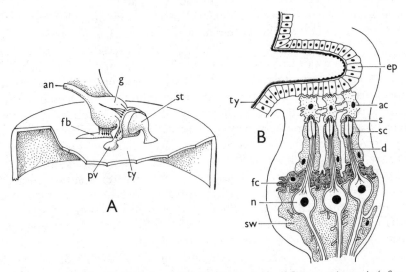

FIG. 74   Structure and histology of the auditory organ of *Locusta migratoria* (*after* Gray, 1960). A, Müller's organ on internal surface of tympanum; B, histology of ganglion, showing three scolopophores and associated structures

ac, attachment cell; *an*, auditory nerve; *d*, dendrite; *ep*, epidermis; *fb*, folded body; *fc*, fibrous sheath-cell; *g*, ganglion; *n*, neuron; *pv*, pyriform vesicle; *s*, scolopale; *sc*, scolopale cell; *st*, styliform body; *sw*, Schwann cell; *ty*, tympanic membrane.

The tympanal organs of the Acridoidea show directional sensitivity to sound and are able to recognize individual sound pulses when these occur at rates of up to 90–300 per second, depending on the species (Haskell, 1956). They respond over a wide frequency range, from below 1·5 kHz to over 80 kHz. The electrophysiological discharge is related to the intensity of the sound but the response to pure tones is asynchronous, i.e. it does not depend on the frequency of the sound. This has led to the theory (Pumphrey and Rawdon-Smith, 1939) that insects with such tympanal organs do not discriminate between sounds of different pitch, though when the sound varies regularly in intensity (i.e. when it undergoes *amplitude modulation*) the responses synchronize with the modulation frequency; the frequency of the 'carrier wave' is immaterial provided it falls within the auditory range of the species. There is, however, more recent evidence that the Acridoid tympanum can, to some extent at any rate, distinguish different frequencies (Horridge, 1961). This seems to be partly due to differences in the sensitivity of the four groups of sensilla and partly to the existence in the tympanum of spatially distinct centres of resonance (Michelsen, 1971). Further, in *Chorthippus* Howse *et al.* (1971) have shown that the normal response in the auditory nerve does not follow the amplitude modulation envelope but rather depends on the existence in the stridulatory song of ultrasonic *transients* (i.e. very rapid changes in sound intensity).

In the Tettigonioidea and Grylloidea there is often a pair of tympanal organs at the base of each fore tibia (Figs. 75–77). In many genera they are easily seen, but in others each organ is concealed by a cuticular fold and comes to lie in a cavity which communicates with the exterior by a slit-like opening (Fig. 77).

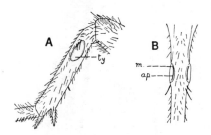

FIG. 75
Left fore tibia of *Acheta domesticus* seen from the outside showing tympanum, *ty* B. Portion of fore tibia of *Tettigonia viridissima*, frontal view

m, membrane covering tympanum; *ap*, aperture into tympanal chamber.

These organs attain great complexity of structure and most of what is known concerning them is due to the researches of Schwabe (1906) and others. In *Decticus verrucivorus* the tympanal organs are of the concealed type (Figs. 76 and 77). The trachea supplying the leg is greatly modified and on entering the tibia it becomes inflated and divides into an anterior and a posterior branch, which reunite below the auditory organ. Each trachea is closely applied to the tympanum of its side, which thus has air on both its aspects: the open air on the outer surface, and the air of the trachea on its inner surface. It is noteworthy that these tracheae communicate with the exterior by a special orifice on either side, in close proximity to the prothoracic spiracle, and these orifices are only present in species with tympanal organs. In a transverse section of the tibia (Fig. 77) it will be observed that the two tracheae occupy the area between the tympana. There is an extensive outer chamber in the leg (above the tracheae, as seen in the figure) and a corresponding inner chamber below. The outer chamber contains the supratympanal organ together with leucocytes and adipose cells. The *supratympanal organ* is placed a short distance above the tympana, and is composed of a number of scolopophores whose cap cells are attached to the integument of the leg. Immediately below this organ, on its outer side, there is a smaller sensory structure which is termed the *intermediate organ*; it is composed of scolopophores of the subintegumental type. On the outer face

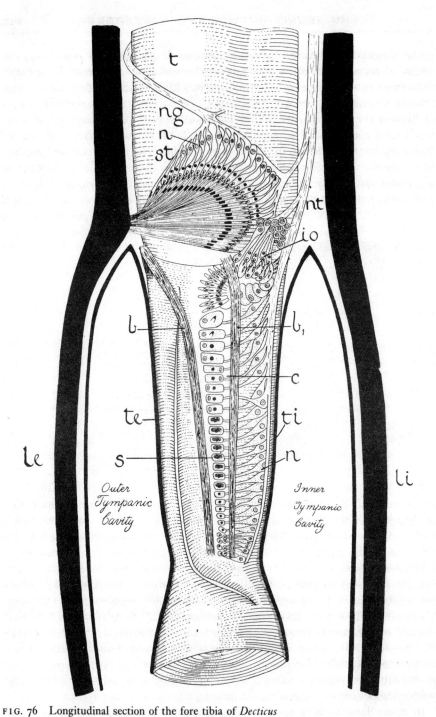

FIG. 76   Longitudinal section of the fore tibia of *Decticus*

c, crista acustica with its supporting bands *b* and *b*$_1$; *io*, intermediate organ; *le*, *li*, outer and inner aspects of tibia; *n*, nerve cells; *ng*, subgenual branch of crural nerve; *nt*, tympanal nerve; *s*, scolopalia; *st*, supratympanal organ; *t*, main trachea; *te*, *ti*, outer and inner tympana. Redrawn from Schwabe, *Zoologica*, 1906.

of the anterior trachea is a third chordotonal organ – the *crista acustica* (*organ of Siebold*). It is an elongated ridge or crest composed of a large number of subintegumental scolopophores which gradually decrease in size towards the distal extremity of the tibia. There are two principal nerves in the tibia – the tibial nerve and the tympanal nerve – both arising from the prothoracic ganglion. The supratympanal organ is supplied by a branch from each of those nerves, while the two remaining organs are innervated by the tympanal nerve. Rather surprisingly, Friedmann (1972*b*) has found that in *Gryllus assimilis* the ultrastructure of the cells in the tympanal organ is quite unlike those of the subgenual and intermediate organs. The tympanal

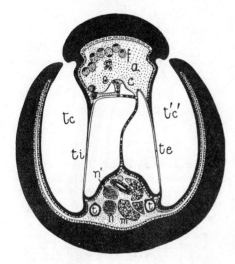

FIG. 77
Transverse section of the fore tibia of *Decticus* passing through the crista acustica (*c*)

a, anterior blood space; *f*, fat-body; *m*, muscles; *n*, tarsal nerve; *n'*, tibial nerve; *t*, tracheae; *tc*, *t'c'*, inner and outer tympanic cavities; *ti*, *te*, inner and outer tympana. Redrawn from Schwabe.

cells here are apparently specialized epidermal cells with arborescent processes and they lack all association with scolopophores. They do, however, contain scolopale-like structures and longitudinally arranged microtubules but their innervation is not understood. Further comparisons of Tettigonioid and Grylloid tympanal organs are needed since Schwabe's account has been supported by Michel (1974). In general the known physiology of the Tettigonioid tympanal organs is similar to that of the Acridoidea. The organs show directional sensitivity, their range of response extends to ultrasonic frequencies, they show asynchronous discharge and they respond to transients in the stimulus. In *Homorocoryphus* a trachea from the prothoracic spiracle conducts sound to the rear of the tympanum and the insects are more sensitive to sound directed at the open spiracles than at the tympanic openings on the tibia (Lewis, 1973).

In many families of Lepidoptera conspicuous tympanal organs occur at each side of the metathorax or at the base of the abdomen, their structure differing in the different families that possess them (Eggers, 1919; von Kennel and Eggers, 1933; Gohrbrandt, 1937; Fig. 78). Each organ is lodged in a cavity formed by invagination of the cuticle and its external opening may be guarded by an integumentary fold. A tympanum, tracheal air-sacs and two chordotonal sensilla are present. The latter have the typical ultra-structure (Ghiradella, 1971) and are connected by the tympanal nerve to the metathoracic ganglion. The physiology of these auditory organs is best known in some Noctuidae, where they respond to frequencies from 3 to 240 kHz, the two sensilla differing in sensitivity (Roeder and Treat, 1957; Adams, 1971). Their function is to detect the cries of predatory bats; sonic

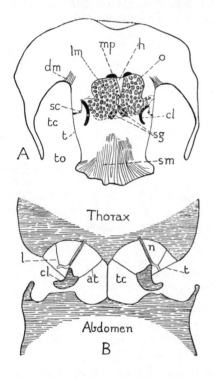

FIG. 78
Tympanal organs of Lepidoptera. A. Vertical section across base of abdomen of a Geometrid moth. B. A diagrammatic horizontal section across base of thorax and abdomen of a Noctuid moth

at, accessory tympanum; *cl*, cuticular lamella; *dm*, dorsal muscle of tympanum; *h*, heart; *l*, ligament supporting chordotonal organ; *lm*, longitudinal muscles; *mp*, mesophragma; *n*, chordotonal nerve; *o*, oesophagus; *sc*, scolopophore; *sg*, salivary gland; *sm*, sternal muscles; *t*, tympanum; *tc*, tympanic chamber; *to*, external opening of tympanic chamber. Adapted from Eggers and v. Kennel.

and ultrasonic bat cries cause electrical activity in the auditory nerve and the organs also enable the moths to detect the direction of a sound in three-dimensional space (Payne *et al.*, 1966). Interneurons in the brain are concerned with central processing of the auditory information (Roeder, 1969).

Tympanal organs are also found in other insects but have been less extensively studied. They occur in both sexes of the Cicadidae at the base of the abdomen, where a group of about 1500 chordotonal sensilla is stretched across an auditory cavity bounded by a transparent tympanum, the 'mirror' (Vogel, 1923; Pringle, 1954). In *Chrysopa* small swellings near the base of the radial vein of each fore wing are bounded ventrally by thin cuticle and contain chordotonal organs. Although the cavity of the organ is filled with blood rather than by a tracheal air-sac, there is experimental evidence that they perceive ultrasonic vibrations between 13 and 120 kHz and can detect sound pulses delivered at up to 150 per second (Miller, 1970, 1971). Chordotonal organs in the mesothorax of *Notonecta*, *Nepa* and *Corixa* also react to air-borne sound in the range from about 100 to 10000 Hz (Larsén, 1957; Arntz, 1972; Prager, 1973).

(b) **Auditory Hairs** – Minnich (1925, 1936) has shown in many Lepidopteran larvae that responses to sound are mediated by hair sensilla which react to low frequencies. Pumphrey and Rawdon-Smith (1936) found that hairs on the cerci of some Orthoptera act as displacement receptors of sound vibrations having a frequency of less than about 3000 cycles per second and that, at least in the lower parts of the range, the impulses in the cercal nerves synchronize with the frequency of the stimulus. See also Haskell (1956).

(c) **Johnston's Organ** – Roth (1948) has shown that in males of *Aedes aegypti*, sound waves can set in motion the hairs on the antennal flagellum and this in turn causes movements of the whole flagellum which stimulate the sensilla of Johnston's organ. In this and other species the males can hear in the frequency range 150 to 550 Hz with the maximum sensitivity around 380 Hz, a frequency that corresponds to the flight tone of the females. Males do not perceive their own more highly pitched flight sounds nor do females respond to sound in the above range, their less plumose antennae being mainly chemoreceptive (Tischner, 1953; Tischner and Schief, 1955; see also Risler, 1953, for histological details).

(d) **Pilifer of Choerocampine Hawkmoths** – A unique auditory organ, sensitive to ultrasonic frequencies, has been found in the head of several species of Sphingidae belonging to the subfamily Choerocampinae. In these insects sound perception depends on contact between the median wall of the second segment of the labial palp and the distal lobe of the pilifer, which contains the sensory transducer. When stimulated experimentally at the optimum frequency of 30 to 70 kHz the pilifer lobe responds to displacements as minute as 0·02–0·1 nm (Roeder, 1972).

3. Chemoreceptors

The physiology of the chemical senses and the receptors involved have been reviewed by Hodgson (1958, 1965), Dethier (1963), Schneider (1969, 1971), Schneider and Steinbrecht (1968), Lewis (1970) and others. Many different structural types of receptor are encountered (Slifer, 1970) and they are often clustered together so that extirpation experiments cannot easily reveal the chemical stimuli to which they respond. Electrophysiological recordings from individual sensilla (or even individual sense-cells) are now being made increasingly often and provide the best evidence of their function, but inferences from the ultrastructure of the sensilla and from the results of behavioural experiments can also now be drawn with increasing confidence.

Despite considerable variation in detail, the chemoreceptor sensilla of insects show many common structural features (Fig. 79). They are apparently derived from unspecialized clothing hairs and their cuticular parts are secreted by equivalents of the trichogen and tormogen cells, though other epidermal cells may also contribute (Ernst, 1972a). Forming part of each sensillum is a group of bipolar neurons (occasionally a single neuron, sometimes up to 60 or more). The perikaryon of the neuron is normal, with Golgi bodies, endoplasmic reticulum and numerous mitochondria. Where the dendritic process of the neuron leaves the cell body it has a ciliary ultrastructure and is often surrounded by a 'cuticular' sheath of uncertain composition. Distal to the ciliary region the dendrite contains only

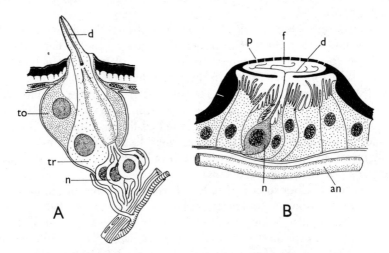

FIG. 79   A, Section through a basiconic sensillum from the hypopharynx of *Apis mellifera* (*after* Galic, 1971). B, Section through a plate-organ (probably olfactory) from the antenna of the aphid *Megoura viciae* (*after* Slifer, Sekhon and Lees, 1964)

an, antennal nerve; *c*, cuticle; *d*, dendrite; *f*, fluid-filled chamber; *n*, sensory neuron; *p*, cuticular pore; *to*, tormogen cell; *tr*, trichogen cell.

microtubules and small vesicles and it may undergo extensive branching in the fluid-filled vacuole left after retraction of the trichogen cell. The cuticle surmounting olfactory sensilla is pierced by many complex pores through which the chemical stimulus reaches the nerve endings; contact chemoreceptors, perceiving taste stimuli, have one or two larger pores near their tip. The exact nature of the pores, the tubules by which they extend inwards, and their relations with the dendritic branches are still under active investigation (Richter, 1962; Slifer and Sekhon, 1964a; Steinbrecht, 1973). There is evidence that in some sensilla the individual sensory neurons respond to different chemical stimuli and in several cases a single sensory hair is known to act both as a mechanoreceptor and a chemoreceptor. For detailed accounts of sensillar ultrastructure see Adams *et al.* (1965), Ernst (1969,

1972*a*), Hansen and Heumann (1971), Hawke and Farley (1971), I-Wu and Axtell (1971), Lewis (1970), Zacharuk and Blue (1971) and many papers by Slifer and by Slifer and Sekhon, as well as others listed below.

According to the gross structure of the cuticular parts it is possible to distinguish several main types of chemoreceptor sensilla, as well as a few less common forms. The recent classification of Slifer (1970) lays primary emphasis on the thickness of the sensillar cuticle, but the more familiar scheme based on the shape of the sensillum is retained here for the major divisions (Fig. 79).

(*a*) **Sensilla trichoidea** – These are hair-like structures which may be thin-walled or thick-walled (when they are sometimes known as *sensilla chaetica* and may be difficult to distinguish superficially from mechanoreceptors or non-innervated clothing hairs). They are the most abundant and widespread type of chemoreceptor, occurring on various parts of the body. Many examples are found on the labella and tarsi of Diptera, where they mediate taste (Adams and Forgash, 1966; Dethier and Hanson, 1968; Gothilf *et al.*, 1971; Owen, 1963; and Wilczek, 1967). Others occur on the maxillary palps of Acridids (Blaney, Chapman and Cook, 1971) where they play a part in food-plant selection, and on the antennae of many species, where they have an olfactory function (e.g. Lacher, 1967; Myers and Brower, 1969; Borden and Wood, 1966; Jefferson *et al.*, 1970). They are the most numerous olfactory sensilla on the antennae of *Bombyx mori* (Schneider and Kaissling, 1957) and are also found on the ovipositor of *Musca*, where they are presumably gustatory (Hooper *et al.*, 1972).

(*b*) **Sensilla basiconica** and **styloconica** – These are peg-like or cone-like organs, distinguished from the trichoid sensilla by the more thick-set appearance of the projecting portion. Like the trichoid sensilla they can have a thin or a thick cuticular wall (which may be grooved or variously sculptured) and they respond to various chemical stimuli. Examples are known from the maxillary palps of Lepidopteran larvae, where they respond to solutions of sugars, alcohols, amino acids and other plant constituents (Schoonhoven, 1969; Dethier and Kuch, 1971). Others, on the ovipositor of *Phormia*, react to inorganic ions (Wallis, 1962). On the maxillary palps of female Culicidae they are sensitive to carbon dioxide (McIver 1972) and they are widely distributed as antennal olfactory receptors (e.g. Slifer and Sekhon, 1964*b*; Morita and Yamashita, 1961; Steinbrecht, 1970; Zacharuk, Yin and Blue, 1971).

(*c*) **Sensilla coeloconica** – These are less common and differ from thin-walled basiconic sensilla in that the cuticular peg is sunk below the general surface of the cuticle, whence their alternative name of pit-peg organs. In some the pit is very deeply sunk and connected with the surface by a more or less elongate tube – the so-called *sensilla ampullacea*. Coeloconic sensilla were described from the antennae of Coleoptera, Lepidoptera and Diptera by older authorities. More recently they have been studied on the antennae of *Apis mellifera*, where some of them are carbon dioxide receptors (Lacher, 1964) and on the antennae of *Hippelates* (Dubose and Axtell, 1968), *Anopheles* (Steward and Atwood, 1963), *Bombyx* (Schneider and Kaissling, 1957), *Leucophaea* (Schafer, 1971) and *Lasius* (Dumpert, 1972).

(*d*) **Sensilla placodea** – This rather distinctive form of sensillum appears as a thin, elongate, oval or circular plate of cuticle which is flat or raised into a slight dome. Pores are present over the surface of the plate or around its periphery. Sensilla placodea are best known from the antennae of *Apis* (Slifer and Sekhon, 1961) where they are very numerous and act as olfactory receptors (Lacher and Schneider, 1963). On the antennae of Aphidoidea there are a small number of specialized sensilla placodea in which an outer, perforated cuticular plate is separated from an inner plate by a space containing many fine dendritic branches (Slifer *et al.*, 1964).

(*e*) **Atypical sensilla** – There is undoubtedly far greater diversity of structure among chemoreceptor sensilla than is indicated above. Among others which have attracted detailed attention may be mentioned the following: (i) the *sensilla coelosphaerica* found on the last antennal segment of *Nicrophorus*; they are olfactory receptors in which the usual pore tubules are replaced by a complicated network of filaments (Ernst, 1972*b*); (ii) the elaborately looped *circumfila* found on Cecidomyid antennae, which are essentially thin-walled chemoreceptors with the pores lying among a labyrinth of fine surface ridges (Slifer and Sekhon, 1972); (iii) the much-folded sensory plaque organs of the Fulgorid *Pyrops candelaria*, which have numerous bipolar neurons arranged in groups and may have evolved from a cluster of basiconic sensilla (Lewis and Marshall, 1970); (iv) the three pairs of tufted sensilla found on the thorax of *Calliphora* larvae and which are probably chemoreceptors, as are also the papilla-like sensilla occurring among the pseudotracheae on the labellar lobes of the adult (Peters, 1961, 1965).

Because of the high specificity of chemoreceptor sensilla, and even of their constituent neurons, it might be most logical to regard them as mediating a great variety of individual senses. It is, however, convenient to distinguish two main groups, the olfactory senses (smell) and the gustatory senses (taste). The differences between these seem relatively clear in terrestrial insects but need further elucidation in aquatic species and those inhabiting such moist environments as the soil or plant and animal tissues.

Olfactory Senses (Smell) – The sense of smell in insects is stimulated by low concentrations of the vapour phase of a great variety of substances which are relatively volatile at ordinary temperatures. The olfactory receptors are trichoid, basiconic, coeloconic and placoid sensilla, usually if not always provided with many cuticular pores. Olfactory stimuli have been intensively studied through the search for attractant and repellent substances in economic entomology (Hocking, 1960; Jacobson, 1966; Beroza, 1970), but while chemical and behavioural aspects of the problem have received great attention much less is known of the detailed mechanism of perception and very little of the processes of transduction. Measurements of insect olfactory acuity vary considerably in accuracy, but the olfactory thresholds of many simple chemical substances are of the same order for the honeybee as for man (Dethier, 1963). A few examples of threshold concentrations, measured in molecules per cc of air for various insect species and substances are: *Nicrophorus*, skatol, $4 \cdot 17 \times 10^{14}$; *Pieris rapae*, benzaldehyde, $3 \cdot 28 \times 10^{14}$; *Apis mellifera*, nerol, $3 \cdot 2 \times 10^9$. Much greater sensitivity is shown towards certain highly specific biological attractants, such as bombykol, the female sex-attractant of *Bombyx mori*, a single molecule of which is enough to elicit a nervous response in an antennal sense-cell (Steinbrecht, 1970). Many factors such as age, sex, nutritional status and previous conditioning, as well as temperature, humidity and the rate of air-flow, are known to affect olfactory thresholds and in some cases a reversal of response may occur naturally during the life of the insect or can be induced experimentally. For example, the newly-emerged females of *Pimpla ruficollis* – a parasite of the Pine-shoot Moth, *Evetria buoliana* – are repelled by the essential oils of pine, but when

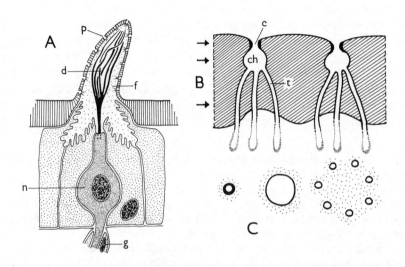

FIG. 80 Fine structure of olfactory basiconic sensillum from antenna of *Nicrophorus* (*after* Ernst, 1969). A, Longitudinal section; B, Section of cuticle to show details of two pores; C, Pore structure in tangential sections at levels indicated by arrows to left of B

c, pore channel; *ch*, pore chamber; *d*, dendrite; *f*, fluid-filled cavity of sensillum; *g*, glia cell; *p*, pore; *t*, pore-tubule.

sexually mature they become attracted to them and return to the pines where their hosts are available for oviposition (Thorpe and Caudle, 1938). Again, Thorpe (1939), by larval conditioning, has induced *Drosophila* adults to react positively to the normally repellent odour of peppermint oils.

There is electrophysiological evidence that excitatory odours induce dendritic polarization of the sense-cell (the receptor potential) followed by nervous impulses in the axon. Inhibitory odours hyperpolarize the cell and depress the impulses. Among the sense-cells of olfactory receptors there seem to be two major kinds (Boeckh *et al.*, 1965). *Odour specialist cells* respond identically to highly specific, biologically important odours like the pheromones discussed below. *Odour generalist cells*, on the other hand, differ in the spectra of their responses to a variety of odours and perhaps provide a peripheral basis for central nervous discrimination of odours. Attempts to relate the olfactory properties of substances to their physico-chemical characteristics have not yet been very successful, though there are several interesting theories, such as that of Wright (1966) who postulates a limited number of 'primary odours' whose attractiveness to insects can be inferred from the infra-red spectra of the various compounds tested.

One of the most remarkable features of olfactory responses is their considerable specificity and the role played by specific attractant or communicating stimuli in the life of the insect. For many such substances the term 'pheromone' was introduced by Karlson and Butenandt (1959) and they have become the subject of active study, reviewed by Butler (1967), Beroza (1970), Wood et al. (1970), Jacobson (1972), Shorey (1973) and Birch (1974). A pheromone may be defined as a substance secreted externally by an individual and received by other conspecific individuals, in which it elicits some specific behaviour pattern or developmental process. Such substances are generally olfactory stimulants and may be classified by their biological functions into the following groups:

(a) *Trail-marking pheromones*. Odour trails have been described in a few species of termites and in many ants. They are laid on the ground by successful foragers returning to the nest and consist usually of short-lived, undirected streaks of material secreted from the gut or various epidermal glands. Comparable pheromones are secreted by the mandibular glands of Meliponinae workers and male *Bombus* spp. to mark stationary objects on their flight-paths.

(b) *Sex attractants*. These are now among the most fully investigated pheromones and though best known from the substances released by the abdominal glands of some female moths to attract males that lie down-wind, female sex attractants have also been found in at least nine other orders of insects (Jacobson, 1972). Many have been identified chemically, some can now be synthesized, and those from economically important insects have been used in their control. Examples of female sex attractants include bombykol (*trans*-10,*cis*-12-hexadecadien-1-ol) from *Bombyx mori*, disparlure (*cis*-7,8-epoxy-2-methyloctadecane) from *Porthetria dispar*, propylure (10-propyl-*trans*-5,9-tridecadien-1-ol) from *Pectinophora gossypiella*, and muscalure (tricos-9-ene) from *Musca domestica* (Gribble et al., 1973). Male insects may also produce sex pheromones that attract females and induce them to mate. These include the substances produced by the tergal glands of some male Blattaria (Roth, 1969), those from the hair-pencils, coremata and androconia of various Lepidoptera, and from comparable structures in various species of Hemiptera, Coleoptera, Hymenoptera, Mecoptera and Diptera. Female cotton boll weevils, *Anthonomus grandis*, are attracted by a mixture of terpenoids secreted by the males.

(c) *Olfactory markers and surface pheromones*. These attract other individuals to the secreting insect or to sites which they frequent, but they are not epigamic in function. They include the caste-recognition scents and colony odours of social Hymenoptera and the secretions of the Nassanoff glands in the abdomen of *Apis* spp. The latter include geraniol, nerolic and geranic acid and attract other workers to newly-located foraging areas.

(d) *Assembly and aggregation pheromones*. These induce the formation of large, temporary or persistent aggregations of a species and include the pheromone that causes hibernation aggregations of the Coccinellid beetle *Semadalia undecimnotata* (Hodek, 1960). The cohesion of a swarm of honeybees is due to the production of 9-hydroxydec-*trans*-2-enoic acid from the mandibular glands of the queen (Butler *et al.*, 1964). In several species of Scolytidae attractant pheromones are secreted by the male or female according to the species and attract both sexes, though sometimes unequally. The situation is complicated as the species are attracted to a tree initially by olfactory stimuli, several pheromones are involved, and it is debatable whether they should be regarded as sex attractants or aggregation pheromones (see reviews cited above).

(*e*) *Alarm pheromones* (Maschwitz, 1964; Blum, 1969). These include a variety of 20 or more relatively simple compounds, of which 4-methyl-3-heptanone and 6-methyl-5-hepten-2-one are the most widespread, though citral, limonene and formic acid also function in this way. They have been found in many Formicidae but also in *Apis*, *Trigona* and a few Isoptera, and are produced by the mandibular glands and various abdominal glands according to the species concerned. Alarm pheromones cause the insects perceiving them to run or fly more actively and to show aggressive behaviour, but their effects vary considerably and may depend on the concentration of the pheromone.

(*f*) *Morphogenetic pheromones.* Used in a broad sense, this term denotes pheromones which affect development, including the postembryonic and postmetamorphic growth of the reproductive system needed to attain sexual maturity. Mature males of *Schistocerca gregaria*, for example, secrete material that promotes the sexual maturation of other males (Norris, 1954; Loher, 1960). In *Apis mellifera* 9-oxodec-*trans*-2-enoic acid, together with an unidentified scent, make up the 'queen substance' that is secreted by the mandibular glands of the queen bee and inhibits the development of the ovary rudiments of the workers. It also has behavioural effects on the workers, preventing them from constructing queen cells and rearing new queens. In the Isoptera, the extensive morphogenetic changes that make up caste-differentiation appear to be controlled by several pheromones of unknown composition (Lüscher, 1961; see also p. 278).

Allied to the pheromones are other biologically important odours produced by one species, but of benefit mainly to the other species that perceives them. These are the 'kairomones' of Brown *et al.* (1970) and they are included in the oviposition, food and host odours which attract insects to the plants, animals and other sites on which they feed or lay their eggs.

(*a*) *Oviposition attractants.* The deposition of eggs in places suitable for their further development is often the result of the female insect being attracted by scents from the oviposition site. Many parasitic insects, for example, are attracted to their hosts in this way (Thorpe and Jones, 1937; Vinson, 1976). Insects with saprophagous larvae are attracted by odours such as ammonia and skatol arising from the bacterial decomposition of organic materials. Among phytophagous species, females of *Hylemyia antiqua* are attracted to and oviposit in sand treated with *n*-propyl mercaptan, an odorous constituent of onion plants on which oviposition normally occurs and on which the larvae develop (Matsumoto and Thorsteinson, 1968). The attraction of adult females of the moth *Chilo suppressalis* to rice plants or plant extracts and their oviposition on the plant is related to the concentration of *p*-methyl acetophenone present, though other attractants may also be involved (Saito and

Munakata in Wood *et al.*, 1970). In neither of the last two examples do the adults feed on the plants concerned.

(*b*) *Food-plant attractants.* Gustatory responses play a major role in food-plant selection, as indicated below, but olfactory attraction also occurs. Phytophagous species are attracted more or less specifically by the smell of essential oils and other non-nutritive substances, locusts are attracted by the smell of grass, and the larvae of *Hylemyia* and *Chilo* in the experiments cited above were attracted by the same plant odours as were the adult insects. The aggregation of *Dendroctonus ponderosae* and other Scolytid beetles depends not only on the attractant pheromones mentioned above but also on the release of volatile substances in the resinous exudate from attacked pine trees (Pitman and Vité, 1969). Experiments on the caterpillars of *Manduca sexta* have shown that each receptor cell of the antennal basiconic sensilla has only a limited specificity and that the specificity patterns of different cells overlap so that the larva can recognize a greater variety of volatile plant constituents than if each receptor were sensitive to only one odour (Schoonhoven and Dethier, 1966; see also Schoonhoven, 1968, 1973). In many insects it must be remembered that olfactory responses to plant stimuli operate only over distances of a few centimetres in still air; they may then be more important in discrimination than in attraction from a distance, and in some insects olfactory stimuli may play little part in host-plant selection. In the aphids, for example, plants seem to be discovered by random dispersal, aided by the effects of physical features on air-movement. After the insect has alighted it discriminates between hosts and unsuitable plants through responses that are more likely to be visual or gustatory than olfactory (Kennedy and Stroyan, 1959). For further information on the whole subject of olfactory responses in host-plant selection, see reviews by Thorsteinson (1960) and various contributions to De Wilde and Schoonhoven (1969) and van Emden (1973).

(*c*) *Attraction to animal hosts.* There is much behavioural evidence that olfactory stimuli play some part in enabling blood-sucking insects such as Culicidae and *Glossina* to locate their hosts from a distance, but there is relatively little precise information on the chemical and physiological basis of the responses (Dethier, 1957; Hocking, 1971). Some species of mosquitoes can respond to the presence of calves or an equivalent source of carbon dioxide at distances of 15 metres or more, though other species do not respond at half this distance (Gillies and Wilkes, 1972). A combination of L-lactic acid and carbon dioxide attracts female *Aedes aegypti* (Acree *et al.*, 1968) and a rather miscellaneous variety of substances are attractive to other mosquitoes, though none equals the odours from the host animal itself. In *Glossina morsitans* carbon dioxide and host odours are attractive under laboratory conditions (Turner, 1971) and in the forest-inhabiting species *G. medicorum* an olfactory response involving up-wind orientation is probably an important factor in host-location (Chapman, 1961).

Gustatory Senses (Taste) – These are defined by responses to relatively high concentrations of non-volatile stimulatory substances which usually come into contact with the receptors in aqueous solution. For this reason the gustatory sensilla are sometimes known as contact chemoreceptors and they mediate important responses in feeding, the rejection of unpalatable substances and in oviposition by adults with endoparasitic larvae. The receptors include trichoid, basiconic and styloconic sensilla as well as special forms like the interpseudotracheal papilla of Diptera; a list is given by Hansen and

Heumann (1971). Contact chemoreceptors are commonly distributed on the tarsi, the maxillary and labial palps (including the labella of Diptera) and less often on the antennae. Concentrations of such sensilla are found on the mouthparts and the walls of the pre-oral food cavity, e.g. Thomas (1966), Galić (1971), and Moulins (1971). In blood-sucking Diptera they form functionally important groups in the wall of the cibarium and elsewhere (e.g. Rice et al., 1973) and in Hemiptera they are localized in a special gustatory organ (Wensler and Filshie, 1969). There are wide differences in insect taste-thresholds according to the substance, species and receptor investigated (Dethier, 1963). In general sugars are acceptable; some – such as fructose, glucose, sucrose and maltose – are perceived at low concentrations while others like galactose, mannose and arabinose stimulate only at high concentrations. A wide variety of substances, including acids, salts, alcohols, esters, amino acids and oils, are all rejected if the concentration is sufficiently high, but dilute solutions of acids and salts are sometimes preferred to water or dilute sugar solutions.

In the well-studied gustatory hairs on the labella of *Phormia* there are five sense-cells in each sensillum. One of these is a mechanoreceptor, but the other four comprise two 'salt cells', sensitive respectively to chlorides and the sodium salts of fatty acids (Dethier and Hanson, 1968), a 'water cell' and a 'sugar cell'. Evidence is now accumulating for a similar specialization among the individual sense-cells of other gustatory receptors and the biophysical basis of transduction is under active study (e.g. Rees, 1968, 1970). The complexity of gustatory perception is illustrated by studies on the responses of the styoloconic sensilla on the maxilla of Lepidopteran larvae (e.g. Dethier and Kuch, 1971; Schoonhoven, 1973). Each of these two sensilla has 4 chemoreceptor cells and in most species some of the cells are individually sensitive to salts, amino acids, inositol and sucrose. Others are more specifically sensitive to substances characteristic of particular food-plants, such as mustard-oil glucosides, populin or anthocyanins, while still further cells respond to various unrelated deterrent substances. There are also additional sources of diversity between species and receptor cells that need to be considered. Those cells that are generally sensitive to a certain substance may differ quantitatively in their thresholds and sensitivity spectra; inhibitory and synergistic effects on the receptor occur through interactions between the stimuli; and no doubt these various peripheral processes are further modified to determine feeding behaviour through the central integrative action of the nervous system. Food-plant selection and discrimination between different foods seem to be complex processes in which varying roles are played by nutritionally important constituents of the plant and by non-nutritional 'secondary plant substances' with specific attractant or deterrent properties (Fraenkel, 1969). In a simplified experimental analysis of feeding in *Schistocerca*, Haskell and Mordue (1969) found that contact receptors on the palps detected sucrose, which stimulates feeding, while a deterrent (azadirachtin) was perceived by receptors near the apex of the labrum. Continued feeding on acceptable material depends on the stimulation of receptors on the labrum, clypeus and hypopharynx. Among the specific gustatory attractants and feeding stimulants of various insects are the mustard-oil glucosides of cabbage (Nayer and Thorsteinson, 1963), hypericin from *Hypericum hirsutum* (Rees, 1969), and various essential oils, alkaloids, catalposides and other compounds. For fuller accounts of the role of gustatory physiology in food-plant selection and feeding see de Wilde and Schoonhoven (1969) and Schoonhoven (1968), and for the special case of plant-sucking Hemiptera see Miles (1958) and Mittler (1967). Less is known of the part played by taste receptors in controlling the feeding behaviour of blood-sucking Diptera, though there is little doubt of their importance (Dethier, 1957; Hocking, 1971; Rice *et al.*, 1973). In mosquitoes, for example, labellar receptors are sensitive to sugars, water and deterrent substances while labral receptors respond to blood. The ingestion of liquids is, however, controlled by cibarial receptors which are stimulated by blood, sugar and deterrents (Salama, 1966). Gustatory senses are also involved in the oviposition behaviour of parasitic insects and in the discrimination they show in finding hosts. The Braconid *Orgilus lepidus*, for example, can distinguish between healthy and previously parasitized hosts through contact chemoreceptors on the medial and lateral stylets of its ovipositor (Hawke *et al.*, 1973). Even more remarkable is the Cynipid hyperparasite of aphids, *Charips victrix*, which apparently uses sensilla at the tip of its ovipositor to discriminate among the parasitic larvae lying in the aphid haemocoele (Gutierrez, 1970).

4. Temperature Receptors (Herter, 1953)

Heat may be transferred to or from an insect by radiation, convection or conduction and also by evaporation and condensation. Only the first three methods are important as stimuli to behavioural changes and experimental assessments of their relative importance have been made in some cases (Fraenkel and Gunn, 1940). Responses to radiant heat alone have been shown in few insects. *Schistocerca gregaria* displays a postural response to the sun's rays which, though partly a visual response, seems principally to depend on radiant energy (Fraenkel, 1930). Specialized areas on the body of Acrididae have been identified as thermoreceptors (Slifer, 1953), but this now seems unlikely (Makings, 1964). Blood-sucking insects are often attracted by the heat of their mammalian hosts or by artificial bodies of similar temperature, probably responding through antennal receptors to convective transfer of heat in the air (Wigglesworth and Gillett, 1934; Wigglesworth, 1941). Temperature receptors on the antennae, maxillary palps and tarsi of several insects are discussed by Gebhart (1953) and there is now electrophysiological evidence of cold receptors in a few species. In *Periplaneta americana* they occur on the arolia and tarsal pulvilli (Kerkut and Taylor, 1957), and also on the antennae, where a few small trichoid sensilla on the ventral side of the flagellar segments respond both to the temperature at a given instant and to the rate of temperature change (Loftus, 1968, 1969). The third antennal segment of Lepidopteran larvae also contain receptors, perhaps styloconic sensilla, that are very sensitive to falls in temperature (Schoonhoven, 1967). Many experimenters have exposed insects in an apparatus providing a gradient of temperature and found that they tend to congregate in a preferred zone (e.g. mainly from 24–32° C. for *Stomoxys calcitrans*; Nieschulz, 1934). Krumbiegel (1932) found that different races of

Carabus nemoralis have preferenda correlated with the temperatures characteristic of the districts which they inhabit and in some species the zone depends on the previous history of the specimens. Not all experiments with a temperature gradient are satisfactory since the method fails to distinguish between convective and conductive transfer and humidity variations within the apparatus are not always eliminated.

5. Humidity Receptors

Many insects react to differences in humidity and detailed behavioural studies have been made for some species, e.g. Gunn and Pielou, 1940; Lees, 1943; Willis and Roth, 1950; Syrjamaki, 1962; Madge, 1965; Youdeowei, 1967. Some insects orientate by the vapour from a distant source of water while others, by avoiding high or low humidities, tend to congregate in a preferred zone. This may be similar to the humidity of their natural habitat but behaviour is also affected by the insects' water-balance (Roth and Willis, 1951*a*). In *Stomoxys calcitrans* the probing responses induced by a rapid increase in relative humidity may be more important in normal feeding behaviour than any olfactory response to host odours (Gatehouse, 1970). Inferences from the behaviour of experimentally treated insects have led to humidity receptors being identified on the antennae (e.g. Andersen and Ball, 1959), sometimes as basiconic, trichoid or placoid sensilla (Roth and Willis, 1951*b*) or, in *Pediculus*, as peculiar tuft-like sensilla (Wigglesworth, 1941). Electrophysiological recording has shown that basiconic sensilla on the antennae and maxillary palps of *Aedes aegypti* respond to water vapour (Kellogg, 1970), as do three receptor cells near the median basiconic sensillum on the third antennal segment of *Manduca sexta* caterpillars (Dethier and Schoonhoven, 1968). The mode of action of humidity receptors is not clear.

6. Perception of Miscellaneous Stimuli

Responses to a few unusual stimuli have been reported occasionally. Species of the Buprestid beetle *Melanophila* are attracted to forest fires, apparently through the action of infra-red receptors consisting of some 70 sensilla in a pit near the mid-coxal cavities. They are most sensitive to radiation with wavelengths from 2·4 to 4 μm and are perhaps modified temperature receptors (Evans, 1966). A possible role in the detection of infra-red radiation has also been ascribed on structural grounds to unusual placoid sensilla on the antenna of the Braconid *Coeloides brunneri* (Richerson *et al.*, 1972). In *Camponotus* (Formicidae), Martinsen and Kimeldorf (1972) have found behavioural responses to X-rays and in *Calliphora* and *Musca* the direction of landing after flight is apparently related to the axis of the horizontal component of the earth's magnetic field (Becker and Speck, 1964). The earth's magnetic field has also been found to affect the direction of the 'waggle dance' performed by worker honeybees on the vertical comb (Lindauer and Martin, 1968).

7. Vision and Visual Organs

Responses to light (though not necessarily to the same wavelengths as are perceived by the human eye) are mediated by (*a*) *Dermal receptors*, (*b*) *Dorsal ocelli*, (*c*) *Lateral ocelli* or *stemmata* and (*d*) *Compound eyes*. Typically, the imago possesses compound eyes and dorsal ocelli though the latter may be absent; lateral ocelli occur only in Endopterygote larvae (but see p. 157). Reduction or loss of photoreceptors is common among species which live habitually in dark situations (e.g. endoparasitic and cavernicolous insects and those inhabiting the nests of termites or ants or burrowing in soil or plant tissues). It is also characteristic of many ectoparasites (Mallophaga, Siphunculata, Siphonaptera, most Pupiparan Diptera).

(*a*) **Dermal Light Sense** – Several insects (Lepidopteran larvae, *Periplaneta, Tenebrio* larvae) react to light even after the eyes and ocelli have been removed or covered with opaque material. The general body surface appears to be sensitive to light and localized receptors have not been identified. In *Schistocerca* and *Locusta*, rhythmic deposition of cuticle can be uncoupled from the 'circadian clock' that controls it through a direct, long-term effect of light – especially the wavelengths between 435 and 520 nm – on the epidermal cells (Neville, 1967).

(*b*) **Dorsal Ocelli** – The structure and functions of the dorsal ocelli have been reviewed by Goodman (1970). When typically developed there are three dorsal ocelli on the frons and vertex of the head, though the median one may be missing. In the Blattaria they are usually absent or represented by a pair of reduced, though light-sensitive, fenestrae. In the Siphonaptera Wachmann (1972*a*) has shown that the only visual organs have a structure quite unlike the ommatidia of a compound eye; they may perhaps be the paired dorsal ocelli, as Hanström (1927) first suggested. As a general rule the ocelli are absent in apterous insects and present in winged forms (Götze, 1927; Kalmus, 1945) though a few strong fliers like the Sphingidae and the Tabanidae also lack dorsal ocelli. They vary considerably in their detailed structure in different groups of insects (Ruck and Edwards, 1964; Munchberg, 1966) but the following major features can be recognized (Figs. 81, 82): (i) *The cornea*. This is usually a thickened, more transparent region of cuticle that surmounts the ocellus externally to form a lens. In the Ephemeroptera (Fig. 81) the cornea is convex but not thickened and the lens is formed from a mass of polyhedral cells lying beneath the corneagen layer. (ii) The *corneagen layer* consists of modified epidermal cells; they are colourless and transparent and secrete the lens, sometimes forming a distinctive layer of vitreous cells, apparently dioptric in function. (iii) *The retina*. This commonly consists of some 500–1000 primary sense-cells which form a

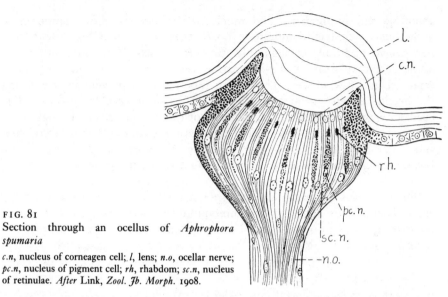

FIG. 81
Section through an ocellus of *Aphrophora spumaria*

c.n, nucleus of corneagen cell; *l*, lens; *n.o*, ocellar nerve; *pc.n*, nucleus of pigment cell; *rh*, rhabdom; *sc.n*, nucleus of retinulae. *After Link, Zool. Jb. Morph.* 1908.

shallow cup and are sometimes arranged into groups of 2–5 cells, the retinulae. Proximally the retinal cells form short axons synapsing with second-order neurons whose cell-bodies lie in the pars intercerebralis of the brain. Some part of the surface of each retinal cell is specialized as a light-sensitive rhabdomere, composed of closely-packed microvilli which measure, in *Schistocerca*, about 0·5 μm in length. Within the retinal cells there may be conspicuous rough endoplasmic reticulum as well as multivesicular bodies

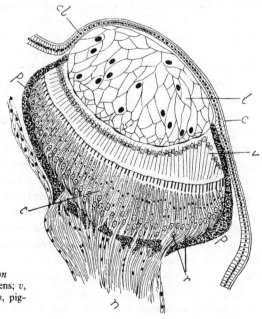

FIG. 82
Section of the median ocellus of *Cloeon*
c, cuticle; *cl*, corneagen layer; *l*, cellular lens; *v*, vitreous layer; *r*, retinulae; *t*, tapetum; *p*, pigment; *n*, ocellar nerve. *After Hesse, 1901.*

about 50 nm in diameter and much larger 'onion bodies' composed of numerous concentric membranes. (iv) *Pigment cells.* These are absent from *Periplaneta* and vary considerably in other species. They may invest the whole ocellus or form an iris-like ring of cells; pigment migration has been reported in a few species. Their main function is to prevent light entering the ocellus except through the lens. (v) *Central nervous connections.* The ocellar nerves consist of the short axons of the retinal cells and longer ones from protocerebral neurons. Partial decussation occurs in the protocerebrum and tracts appear to run to the calyces of the corpora pedunculata with further descending connections to the thoracic ganglia in the ventral nerve-cord.

The functions of the dorsal ocelli are not entirely clear. Their visual fields are overlapped by those of the compound eyes and they are unlikely to mediate the perception of form. This is because the principal focal plane of their lens falls below the level of the retinal layer and because of the convergence of the very many retinal cells on to a much smaller number of ocellar nerve fibres. Despite claims to the contrary, the dorsal ocelli are probably unable to recognize polarized light and they do not seem to be implicated in colour vision or in the entrainment of circadian rhythms. On the other hand, there is electrophysiological evidence that they signal the level of light intensity and changes in that level; the electrical response of a stimulated ocellus is a complicated process resulting from the presence of up to four different components (Ruck, 1961). Occluding the ocelli usually leads the insect to move less rapidly, especially in bright light. This has favoured the view – not now generally accepted – that the dorsal ocelli may be 'stimulatory organs' serving to raise the excitatory level of the insect with respect to other visual stimuli perceived through the compound eyes. Another view (Jander and Barry, 1968) is that the ocelli and compound eyes interact in mediating the phototactic behaviour of insects. In *Locusta* and *Gryllus*, for example, they act together in promoting more accurate directional orientation: in dim light the dorsal ocelli and eyes interact synergistically while in brighter light they behave antagonistically.

(c) **Lateral Ocelli (Stemmata)** – The lateral ocelli are, with very few exceptions, the only eyes present in insect larvae. As their name implies, they are located on the sides of the head where they occupy positions corresponding with those of the compound eyes of the imagines. The number of lateral ocelli is variable and not always constant in the same species; in some groups there is a single ocellus present on either side, while in others there may be six, seven or more ocelli. They differ from the dorsal ocelli in being innervated from the optic lobes of the brain, and also in that a crystalline refractive body may sometimes be developed beneath the corneal lens and that pigment granules are sometimes absent. The histological structure is very varied, but four main types may be considered (Fig. 83):

(i) In the Tenthredinidae (Corneli, 1924) the single lateral ocellus is not unlike the typical dorsal ocellus in structure (see also Meyer-Rochow, 1974). The cuticle

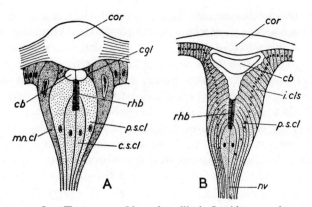

FIG. 83   Two types of lateral ocelli. A. Lepidopteran larva.
B. *Dytiscus* larva (*after* Snodgrass, 1935)

cb, crystalline body; cgl, corneagen cell; cor, corneal lens;
c.s.cl, central retinal cells; i.cls, pigmented iris cells; mn.cl,
mantle cell; nv., ocellar nerve; p.s.cl., peripheral retinal
cells; rhb, rhabdom.

forms a lens-like cornea secreted by a thick underlying layer of corneagen cells
(modified epidermis). Beneath the latter is a retina composed of many retinulae,
each made up of four cells whose apposed rhabdomeres form a rhabdom. A broadly
similar type of structure is found in *Tipula* (Constantineanu, 1930), *Cicindela*
(Friedrichs, 1931) and *Acilius*.

(ii) In *Dytiscus* (Günther, 1912), *Euroleon* (Jokusch, 1967) and *Sialis*, a lens-like
crystalline body is secreted beneath the cornea but the structure otherwise resembles
the first type.

(iii) In the Lepidoptera and Trichoptera, each larval ocellus has a cornea and
crystalline body but only seven retinal cells are present, forming a single retinula
with their apposed rhabdomeres constituting a single axial rhabdom. The resulting
organ is strikingly similar to each ommatidium of a compound eye (see below).

(iv) In the larvae of several Nematoceran Diptera the lateral ocelli may be simplified (Constantineanu, 1930). There are few retinulae and a corneal lens is reduced
or absent but the corneagen cells of some species have a vitreous appearance recall-
ing that of a more or less degenerate crystalline body. In the unpigmented ocellus of
Chironomus even these vitreous cells are absent and a single retinula lies directly
beneath unmodified cuticle. In *Aedes aegypti* larvae the retinular cells have the kind
of ultrastructure expected in insect eyes, with rough endoplasmic reticulum, rhabdomeric microvilli and a zonula adhaerens junction between contiguous cells
(White, 1967, 1968). Pinocytosis occurs at the base of the microvilli and protein
taken up in this way is sequestered in multivesicular bodies, themselves later transformed into lamellar structures. The larvae of Cyclorrhaphan Diptera have a pair of
very simple photoreceptors which are probably degenerate lateral ocelli. In *Musca*
there is a small group of light-sensitive cells on each side of the pharyngeal sclerites
and invisible externally (Bolwig, 1946). They are most sensitive to green light and
apparently unable to perceive red. See also Strange (1961).

The physiology of the lateral ocelli of Lepidoptera has been studied by
Dethier (1942, 1943) who showed that both cornea and crystalline body can

form a more or less distinct, inverted image which, irrespective of the distance of the object, falls somewhere on the rhabdom. Acting together, the six pairs of ocelli form twelve visual fields with little or no overlap and so provide a very coarse mosaic of intensities. By moving the head from side to side as it advances, the larva can examine a larger field and is at least capable of orientating itself towards the boundary between light and dark parts of its environment (Hundertmark, 1936). The behaviour of caterpillars seeking food and pupation sites suggests that they have some colour vision (Götz, 1936) and in *Bombyx mori* larvae the spectral sensitivity curves of the lateral ocelli, obtained by electrophysiological recording, show peaks in the near ultra-violet and green regions that suggest a two-receptor system of colour vision (Ichikawa, 1969).

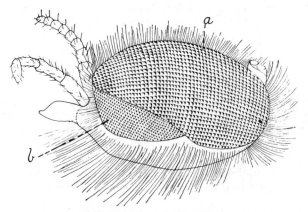

FIG. 84   Head of *Bibio marci* (male), showing divided eye
a, upper division of eye; b, lower division.

(*d*) **Compound Eyes** – The principal feature distinguishing compound eyes from ocelli is that in the former the cornea is divided into a number of separate facets, whereas there is only a single facet to each ocellus. Compound eyes are aggregations of separate visual elements known as ommatidia, each corresponding with a single facet of the cornea. Like lateral ocelli, they are innervated from the optic lobes of the brain (Fig. 86).

The number and size of the facets of the compound eye vary greatly. In workers of the ant *Ponera punctatissima*, each eye is a single facet; there are 6–9 facets in the same caste of *Solenopsis fugax*, while among other ants the number varies between about 100 and 600 in the workers, 200 and 830 in the females, and 400 and 1200 in the males. In *Musca* the eye consists of about 4000 facets, in some Lepidoptera from 12 000 to 17 000 and in Odonata between 10 000 and 28 000 or more. In most insects the facets are very closely packed together and hexagonal but where they are fewer and less closely compacted, they are circular. The facets are not always of equal dimensions over the whole area of the eye. Thus, in males of *Tabanus* they are often larger over the anterior and upper parts of the eye, but the two fields are not sharply demarcated. In males of other Diptera such as *Bibio* and *Simulium*, the two areas of different-sized facets are very distinctly separated, each eye appearing double (Fig. 84). An extreme condition is attained among certain Coleoptera (*Gyrinus*, several Cerambycidae, Fig. 85) and Ephemeroptera (*Cloeon*), where the two parts of the eye are separate so that the insect appears to possess two pairs of compound eyes. In *Cloeon* the anterior division of each eye is elevated on a pillar-like outgrowth of the head, while the posterior division is normal. Modern general accounts of the compound eyes and visual processes of insects will be found in works by Bernhard (1966), Horridge (1968, 1974), Mazokhin-Porschnyakov (1969) and Wehner (1972).

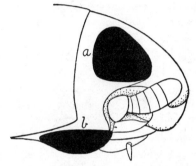

FIG. 85
Head of *Gyrinus natator*, showing divided eye (right)

a, upper division of eye; *b*, lower division.

The structure of an ommatidium (Figs. 86, 87). The main structural components found in almost all ommatidia are as follows:

(i) The cornea. This is the outermost transparent layer of cuticle forming the external facet and acting as a lens. It is usually more or less biconvex and may be uniformly refractive or composed of paraboloidal laminae whose refractive index decreases from the centre outwards (e.g. Meyer-Rochow, 1972–73). In some insects hairs arise between the facets and if these are experimentally removed from the eye of *Apis*, the insect is unable to measure accurately the angle between the directions of the sun and a feeding site (Neese, 1965). Externally the cornea is quite smooth in many insects (e.g. Coleoptera, Orthoptera, Hymenoptera), while in others such as the Diptera and Lepidoptera its outer surface is provided with many ultramicroscopic cone-like projections (the corneal nipple array). These lead to a less abrupt discontinuity in refractive index between the air and the main bulk of the cornea, thus reducing reflection at the surface and increasing the transmission of light through the cornea (Bernhard *et al.*, 1965, 1970; Gemne, 1971).

(ii) The corneagen layer. The part of the epidermis that extends beneath the cornea is known as the corneagen layer. It normally consists of two cells in each ommatidium, but in some adults these can only be seen with difficulty. In other cases the corneagen cells are said to be absent and the cornea is secreted by the crystalline cone cells.

(iii) The crystalline cone cells. Beneath the corneagen layer or cornea lie four cells which, in eucone eyes, secrete the crystalline cone. The cells may contain glycogen inclusions and microtubules are also present. In some insects the crystalline cone cells are produced proximally into six processes which end deep in the ommatidium and may perhaps provide support or facilitate the transport of metabolites (Brammer, 1970; Burton and Stockhammer, 1969; Horridge and Giddings, 1971).

(iv) The primary pigment cells (primary iris cells). These are densely pigmented, commonly two in number, disposed in a circlet around the crystalline cone and corneagen cells.

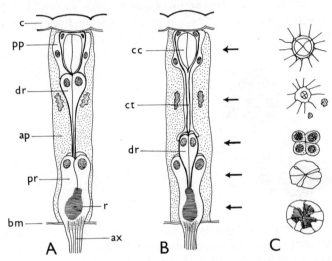

FIG. 86 Histology of an ommatidium from compound eye of *Archichauliodes guttiferus* (Neuroptera) (*after* Walcott and Horridge, 1971). A, Longitudinal section of ommatidium from dark-adapted eye. B, The same, from light-adapted eye. C, Transverse sections through B at levels indicated by arrows

ap, accessory pigment cell; *ax*, eight axons leaving each ommatidium; *bm*, basement membrane; *c*, cornea; *cc*, crystalline cone; *ct*, crystalline tract, which varies in length according to state of light/dark adaptation; *dr*, four distal retinula cells; *pp*, primary pigment cells; *pr*, proximal retinula cells; *r*, rhabdom formed by contributions from the eight retinular cells.

(v) The retinula forms the basal portion of the ommatidium, consisting of a group of more or less elongate visual cells, each continued into an axon, the post-retinal fibre, that communicates with the central nervous system. The retinular cells show considerable variation in number, size, arrangement and ultrastructural details, as shown below and in the next section. There are commonly eight cells but *Ephestia* has 9–12 (Fischer and Horstmann, 1971) and *Erebus* has 7 (Fernández-Moran, 1958). In some cases they are similar and radially arranged, as in *Mastotermes* (Horridge and Giddings, 1971); in others they may be more or less stratified, as in the Odonata (Ninomiya *et al.*, 1969), or they consist of central and peripheral groups as in *Aedes*

(Brammer, 1970). The cells commonly contain pigment granules or lipid droplets as well as multivesicular bodies and most of the usual organelles (rough and smooth endoplasmic reticulum, Golgi elements, mitochondria, free ribosomes and longitudinally arranged microtubules). Their most striking feature is the way in which a more or less extensive portion of the cell surface is produced into large numbers of microvilli which collectively form a rhabdomere. Some cells may bear an isolated rhabdomere but usually the rhabdomeres occupy contiguous regions of adjoining cells and so fuse into a central rod-like rhabdom. The rhabdomeric microvilli are highly orientated, they are thought to contain the visual pigment rhodopsin, on which the primary photoreceptor process depends, and they may show signs of pinocytosis at their bases. The region of retinular cytoplasm adjacent to the rhabdomere may show some specialization; in *Locusta* the rhabdom of the dark-adapted eye is surrounded by endoplasmic reticular cisternae which are replaced in the light-adapted condition by closely crowded mitochondria (Horridge and Barnard, 1965).

(vi) The secondary pigment cells (secondary iris cells). These are usually rather numerous, often elongate, pigment-containing cells that surround the retinula and primary pigment cells like a sleeve, isolating each ommatidium optically from its neighbours. Such screening pigment may, however, be absent around part or all of the ommatidium or may migrate according to the incident light in some species (see below).

The proximal extremities of the ommatidia rest on a fenestrated basement membrane through whose perforations pass the post-retinal fibres and often fine tracheae. The latter, as they enter further into the eye, become arranged parallel to the long axes of the ommatidia and in some nocturnal insects they form a reflecting layer, the tapetum, causing the eyes to glow when illuminated in the dark.

The types of compound eye. The older histologists Grenacher and Kirchhoffer distinguished four main types of compound eye according to the presence and nature of the crystalline cone: (i) In eucone eyes each ommatidium contains a hard, refractive, conical body secreted within the crystalline cone cells and forming part of the dioptric apparatus. Such eyes occur in the Thysanura, Ephemeroptera, Odonata, Orthoptera, some Hemiptera, Chrysopidae, Trichoptera, Lepidoptera, Hymenoptera and some Coleoptera (Carabidae, Cicindelidae, Dytiscidae and Scarabaeidae). (ii) In the pseudocone eyes of Brachyceran and Cyclorrhaphan Diptera the cone cells are filled with a transparent viscous liquid. (iii) In acone eyes the long, transparent 'crystalline cone cells' do not secrete any refractive material. They occur in the Dermaptera, some Hemiptera, some Nematoceran Diptera and some Coleoptera (Staphylinidae, Histeridae, Silphidae, Coccinellidae and Curculionidae). (iv) Exocone eyes are those in which the crystalline body is replaced by a cone-shaped extension of the inner surface of the cornea, lying distal to the unmodified 'crystalline cone cells', as in the Dermestidae, Elateridae, Byrrhidae and Cantharidae.

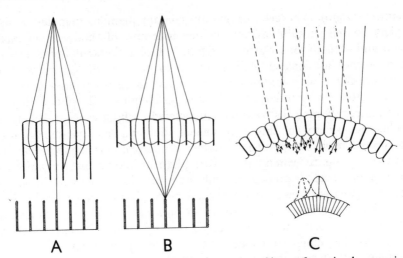

FIG. 87   A and B show the classical interpretation of image-formation by apposition (A) and superposition (B). In A the only light rays to stimulate a
given rhabdom are those passing through the dioptric unit directly above
it; lateral rays are absorbed by pigment before reaching the retina. In B
the pigment is retracted and light rays passing through lateral ommatidia
reach the rhabdom immediately opposite the light source. C shows an
interpretation of image-formation by superposition in the eye of
Hesperiid butterflies (*after* Horridge, Giddings and Stange, 1972). A
parallel beam of light (full lines) is focused as a spot on the receptor layer
with a normal distribution of light intensity. Movement of the stimulus
gives a new beam (broken lines) with corresponding displacement of the
light distribution on the receptor layer

A second major classification of compound eyes is the functional division
made by Exner (1891) into apposition and superposition eyes, depending on
the way in which an image is formed (see below). Structurally this difference
is based on the extent of the screening pigment between adjacent ommatidia,
but it now seems that true, functional superposition eyes occur only in the
atypical Hesperiidae (Horridge *et al.*, 1972) so that Exner's distinction has
lost much of its force (Horridge, 1971, 1972). Instead, several types of eye
may be recognized on structural grounds, though the functional significance
of many of these variations is not yet fully clear (Horridge, 1969) and in
some cases the dark- and light-adapted eyes differ histologically as well as
optically. In typical apposition eyes the rhabdom runs from the apex of the
cone to the basement membrane, as in *Apis* and many other insects (Perrelet,
1970; Skrzipek and Skrzipek, 1971; Meyer-Rochow, 1972; Horridge and
Giddings, 1971). As a rule the rhabdomeres of adjacent retinular cells are in
contact throughout their length to form a closed rhabdom, but in the
Diptera, with an open-rhabdom type of eye, there is a central axial space
between the component rhabdomeres (e.g. Boschek, 1971). The rhabdomeres may also occur at two or three different levels, as in *Photuris*
(Horridge, 1969) and in *Dytiscus*, where the retina consequently has a tiered

structure; the proximal retina of *Dytiscus* seems to be concerned with gathering light and the distal retina with form-perception (Horridge *et al.*, 1970). The relations of the cone with the retinula cells and the rhabdom are also of great functional importance. In the dark-adapted eye of *Archichauliodes* four of the retinular cells reach the tip of the cone (though the rhabdomeres are more proximal in position). In light-adapted eyes, however, a crystalline thread, 100 μm long runs from the cone to the retinula. Similar crystalline-thread eyes occur in the Neuroptera, lower Coleoptera and some Lepidoptera (Walcott and Horridge, 1971), while in many other cases a permanent crystalline thread is present (Horridge, 1969). In all such cases the crystalline thread guides light from the cone to the deeper receptor cells. Where threads are absent and an optically homogeneous region lies between the cones and the retinulae one has the typical clear-zone eye, in which the light reaching a single ommatidium will have been refracted through several facets (Horridge, 1971, 1972).

Atypical eyes. Certain insects possess lateral 'eyes' which have a distinctly ocelliform structure and whose phylogenetic relationship to compound eyes or to ocelli is in doubt. Examples are found in the Mallophaga and Siphunculata (Wundrig, 1936; Wigglesworth, 1941) where there is a single facet on each side and in male Strepsiptera (Wachmann, 1972b) where about 50 facets are grouped together. In both cases, each corneal facet surmounts a large number of retinulae. On the other hand, in the Collembola and *Lepisma*, where each facet is associated with a single retinula, the structure is very reminiscent of the ocelli of Lepidopterous larvae. The lateral 'ocelli' of some Aphidoidea and male Coccoidea may also be mentioned here.

Visual pigments and light perception. Light falling on the rhabdoms induces nervous activity in the retinular cells through photochemical reactions that are associated with systems of visual pigments, the rhodopsins and their breakdown products the metarhodopsins (Langer, 1967; Gogala *et al.*, 1970, Schwemer *et al.*, 1971). These pigments, which are similar to those found in the eyes of vertebrates and cephalopod molluscs, are located in the microvilli of the rhabdomeres. Rhodopsin is a conjugated protein, of which the chromophore is an aldehyde of Vitamin A. In *Ascalaphus* ultraviolet light causes the conversion of rhodopsin to metarhodopsin with the initiation of a generator potential in the retinular cell. The rhodopsin is then rapidly regenerated under the influence of light of longer wavelengths with a further increase in the generator potential. In the eyes of the moth *Deilephila* two similar pigment systems are found, respectively sensitive to ultraviolet and to green light.

The optical mechanism of the compound eye. The classical view of the way in which the insect compound eye works is the mosaic theory of vision proposed originally by Johannes Müller in 1829. According to this interpretation each ommatidium acts independently, subtending a very small part of the total visual field which it perceives as a minute spot of more or less bright light. The juxtaposition of large numbers of these elementary units then produces a larger erect image of the whole field, rather like a newspaper photograph consists of very many minute dots which differ individually only in their light or dark colour. Subsequent investigations have shown many reasons why the mosaic theory cannot be retained in its simple form. Each ommatidium commonly receives light from a wider angle than it subtends geometrically and the field of vision of adjoining ommatidia may overlap considerably. Some insects have eyes with a resolving power appreciably greater than would be expected from their inter-ommatidial angle. The projection of the ommatidial cells on to the lamina ganglionaris and deeper levels of the optic lobe is an elaborate procedure which may involve the coupling of ommatidia or the separation of the inputs of the individual retinular cells of each ommatidium, the latter process apparently facilitated by the separation of rhabdomeres in the open-rhabdom type of eye. Lastly, the migration of pigment or the development of crystalline threads in the light-adapted eye permits various optical mechanisms in different light conditions, while a tiered retina allows the eye to respond in more than one way, according to various qualities in the light stimulus.

The first to extend the mosaic theory of insect vision was Exner (1891) who distinguished between apposition images, formed in the eyes of diurnal insects rather as the mosaic theory requires, and superposition images formed in eyes where the distribution of pigment is restricted so that the ommatidia are no longer isolated optically from each other. Light entering several different ommatidia may then be focused through their cones on to the rhabdom of a single ommatidium directly opposite the source; when many receptors are stimulated in this way, a well-defined, erect image results. Such a mechanism has been demonstrated in the eyes of Hesperiid butterflies (Horridge, Giddings and Stange, 1972) where a parallel beam of light falling on the eye reaches a single receptor via a circular area of eye-surface that subtends an angle of about 30° at the centre of the eye. One would expect this to be an adaptation to vision in poor light, but in fact the Hesperiidae are diurnal fliers, showing excellent daytime vision; their well-focused clear-zone eye lacks light guides and its pigment is in the position normally found in dark-adapted eyes. In nocturnal insects, light is apparently admitted to the dark-adapted eye through each facet with a Gaussian distribution of intensity and then spreads out from the cone as another Gaussian distribution, unrelated to the first except in total energy. The direction of arrival of a ray on the receptor is therefore only minimally related to its direction of origin, the ray-paths do not agree with Exner's theory of a superposition eye, there is no sharply focused image, and the eye has gained sensitivity at the expense of acuity. This type of unfocused clear-zone eye, the focused eye of the Hesperiidae and intermediate forms have been brought into a single mathematical theory (Diesendorf and Horridge, 1973), but arguments in favour of Exner's superposition image are still current (Kunze, 1970). The complexity of image perception is further indicated by another special case, the open-rhabdom eye of Cyclorrhaphan Diptera (Kirschfeld, 1971, 1972) where a process of 'neural superposition' takes place. Here the optical axes of the seven separate rhabdoms of an ommatidium diverge, but six receptor cells from six different ommatidia converge on to a single optical cartridge in the lamina ganglionaris of the optic lobe. Moreover the rhabdomeres of these six receptors are stimulated by the same point in the visual field so that it is the cartridge rather than any single ommatidium which may be said to 'look' at the point concerned!

An alternative theory of insect vision emphasizes the role of diffraction images in the eye (Burtt and Catton, 1966). The visual acuity of *Locusta* and *Calliphora* is 3–4 times that expected from a single ommatidium on the simple mosaic theory (cf. Burtt and Catton, 1969 and earlier works). This, it is argued, can best be explained by the existence of a series of diffraction images (of which Exner's superposition image is only one) that are formed at successively deeper levels in the eye. One or more of these are thought to be used by the insect, at least to detect small movements or resolve fine patterns. The theory of diffraction images in the insect eye has also been expressed in a mathematical model (Rogers, 1962) though the interpretation has not won general support among insect physiologists (e.g. Horridge, 1971).

Form perception and pattern recognition. Whatever the means by which images are formed in the compound eye, there is little doubt of the biological importance of form perception and pattern recognition in many insects. Most of the research into this subject is based on inferences from the behaviour of the insect, much of it on the honeybee, *Apis mellifera* (but cf. Tsuneki, 1961). Early experimental work on bees showed that their ability to distinguish between different geometrical figures depends largely on 'figural intensity', that is, the amount of contour per unit area of figure (Hertz, 1934, 1937; Wolf and Zerrahn-Wolf, 1937). Thus, simple triangles, squares or circles can be distinguished from patterns of stripes, dots or radially arranged lines and bees visit spontaneously the complex figures more often than the simple ones; there is no discrimination between patterns of approximately the same complexity. Other features of a pattern, such as the angle of inclination of stripes, may be perceived without reference to figural intensity (Wehner, 1971), but seem to be less important. Despite claims to the contrary, bees are unable to recognize generalized shapes as such, e.g. as triangles irrespective of their size, orientation and details of outline (Anderson, in Wehner, 1972). An essential feature of form discrimination by figural intensity is the more varied stimulation of ommatidia that occurs when a complex image moves across the eye. Behavioural reactions to moving patterns (optomotor responses) have therefore been used to develop theories of form perception (e.g. McCann and Maginitie, 1965; Reichardt, 1973). Under natural conditions many insects respond in relatively complex ways to specific forms of special biological importance, such as the food-plant response of *Carausius* (Jander and Heinrichs, 1970). Simple forms of binocular vision enable predacious insects such as the larva of *Aeshna* to locate their prey (Baldus, 1926) and in the mantid *Stagmatoptera biocellata* a specialized region of the eye, the so-called fovea, estimates the striking distance (Maldona and Barrós-Pita, 1970). The water-cricket *Velia caprai* reacts to visual stimuli from its prey when these are accompanied by vibrations of the water surface; it responds best to complex, changing stimuli and depends particularly on discrimination by two horizontal rows of ommatidia in the ventro-median part of the eye (Meyer, 1971). Little is known of the central mechanisms involved in perceiving form and movement, but recordings from the optic lobe show several kinds of electrophysiological units responding selectively to displacement of stimuli in the visual field (e.g. Kaiser and Bishop, 1970; Mimura, 1971). There is evidence that the effective field of vision of neurons in the optic lobe of *Calliphora* is much narrower than that of receptor cells, suggesting that lateral inhibition occurs, with consequent sharpening of the visual image (Zettler and Järvilehto, 1972).

Colour vision. The ability of insects to distinguish between different colours was first demonstrated by von Frisch (1913), who trained honeybees to associate sugar-solutions with differently coloured backgrounds having the same intensity of reflected light. Several other experimental methods have also been employed to establish and investigate colour vision (Burkhardt, 1964). They include; (a) the use of optomotor responses to a moving background of alternately coloured stripes of uniform intensity (Moller-Racke, 1952); (b) the use of colour-adaptation techniques in which insects adapted to light of one wavelength will respond to equally bright light of another colour; and (c) the study of electrophysiological responses by the eye to lights of different wavelengths but of equal energy-content, now extended to recordings from single retinular cells (e.g. Burkhardt, 1962; Autrum and Kolb, 1968; Bennett et al., 1967; Bruckmoser, 1968). Though the results are not always easy to interpret, wavelength discrimination of some kind has now been demonstrated in many insects from the orders Orthoptera, Hemiptera, Diptera, Lepidoptera, Coleoptera and Hymenoptera. As a rule most insects respond to ultraviolet radiation at wavelengths around 350 nm and also to light in the range 430–500 nm, but few if any seem sensitive to red. An elaborate form of colour vision occurs in the honeybee *Apis mellifera* (Daumer, 1956), which can distinguish at least six major categories of colour – yellow, blue-green, blue, violet, ultraviolet and 'bee-purple' (a mixture of yellow and ultraviolet). The colour vision of *Apis* may be represented by a 'colour circle' of six sectors, like that of man but displaced some 100 nm towards the ultraviolet end of the spectrum (Fig. 88). As in man, opposite shades of the bee's colour circle are complementary and blend to give the appearance of white. Discrimination is poor in the orange-green region but good for blue-green, violet and bee-purple. Since ultraviolet light is reflected differentially from natural objects they may reveal to the insect's eye 'latent patterns' which the human eye cannot see. Thus, flowers which seem uniformly coloured to us may have pollen-guides or nectar-guides clearly defined by ultraviolet reflection. These and the similar patterns which will show up on the wings of insects or the bodies of other animals may be of considerable behavioural importance in feeding, courtship and prey-recognition or in the use of warning or protective coloration (Daumer, 1958).

Many measurements have now been made of the sensitivity of single retinular cells to a range of monochromatic lights of equal energy-content. The resulting spectral sensitivity curves have established the existence of a small number of different receptor cell types, each with its own peak of sensitivity at a particular wavelength. In *Calliphora*, for example, the seven retinular cells of each ommatidium fall into three classes (Burkhardt, 1962). All are highly sensitive to ultraviolet but five of the seven cells also have a peak at about 490 nm (the 'green cells'), while the two others have their respective peaks at about 470 nm ('blue cells') and at 521 nm ('yellow-green cells'). In *Apis* the drones have three cell types with maxima at 340, 450 and 530 nm while workers have cells with maxima at 340, 430, 460 and 530 nm (Autrum and Zwehl, 1964). Other comparable examples are known from *Heliconius* (Struwe, 1972a, 1972b), though *Periplaneta* has only two types of receptors with maxima at 365 and 407 nm (Mote and Goldsmith, 1970). In some insects different regions of the eye show different responses. In the Gyrinid beetle *Dineutes* the ventral part of the eye is more sensitive to ultraviolet (Bennett, 1967) and in *Ascalaphus* the electroretinogram of the lateral region has a subsidiary peak in the green (Gogala, 1967). The retinular pigments responsible for colour vision, including ultraviolet sensitivity, are thought to be the rhodopsins; the different spectral sensitivities are probably associated with the coupling of the chromophore to different proteins. The ultraviolet-fluorescent substances often found in insect eyes are not connected with colour vision (Kay, 1969).

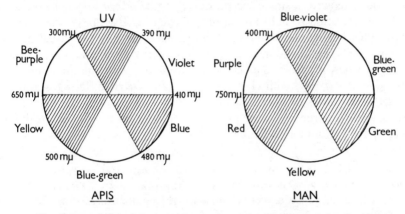

FIG. 88   Colour circle of man and honey bee (*Apis mellifera*) (*after* Burkhardt, 1964, and Daumer, 1956). Cross-hatched areas denote primary colours, white areas the secondary colours. Pairs of complementary colours lie at the opposite ends of diameters of the circle. A mixture of appropriate quantities of two complementary colours will appear indistinguishable from white light

While the existence of two or more retinular cell types with different spectral sensitivities is sufficient to explain colour vision, little is known of the central mechanisms accompanying discrimination. Bishop (1969) was unable to detect responses to different wavelengths by interneurons in the lobula (medulla interna) of the optic lobe of *Calliphora* and *Drosophila*. On the other hand, single-unit recordings from high-order visual neurons in the protocerebrum of butterflies have shown various patterns of electrical activity associated with different wavelengths (Swihart, 1970, 1972), suggesting that the nervous processing of colour information occurs in the brain proper rather than in the optic lobes.

Perception of polarized light. Light reaching insects from the sky is plane-polarized. The ability of *Apis mellifera* to recognize the plane of polarization was first established by von Frisch in 1948 and has since been very fully investigated in this species (von Frisch, 1968; von Frisch and Lindauer, 1956; Lindauer, 1967). The same faculty has now been found in many Arthropods, including such insects as *Trigona*, *Andrena*, *Vespa* and several species of ants, as well as some Coleoptera, Diptera and Heteroptera

(Stockhammer, 1959). In most cases the moving insects are able to orientate at a constant angle to the plane of polarization and are thus able to supplement their normal capacities for navigating by landmarks, scent trails or the direction of the sun. The reactions of honeybees are based on a direct perception of the plane of polarization and do not depend on the bee recognizing a pattern of brightness produced in its surroundings by the incident light or its reflection (von Frisch, Lindauer and Daumer, 1960). The mechanism by which the plane of polarization is detected lies in the individual retinular cells; the dioptric apparatus is optically isotropic but the action potentials from single receptor cells of *Calliphora* vary with the direction of polarization of light falling on the eye (e.g. Burkhardt and Wendler, 1960). This is apparently due to dichroic absorption by the visual pigment rhodopsin in the rhabdomeric microvilli, the light being most strongly absorbed when its plane of polarization coincides with the direction of the long axes of the microvilli (Langer, 1965). In *Gerris*, different electrophysiological responses to polarized light in different parts of the eye are correlated with differences in the orientation of the microvilli (Schneider and Langer, 1969; Bohm and Täuber, 1971).

Literature on the Sense Organs and Perception

ACREE, F., TURNER, R. B., GOUCK, H. K., BEROZA, M. AND SMITH, N. (1968), *l*-Lactic acid: a mosquito attractant isolated from humans, *Science*, **161**, 1346–1347.

ADAMS, J. R. AND FORGASH, A. J. (1966), The location of the contact chemoreceptors of the stable fly, *Stomoxys calcitrans* (Diptera: Muscidae), *Ann. ent. Soc. Am.*, **59**, 133–141.

ADAMS, J. R., HOLBERT, P. E. AND FORGASH, A. J. (1965), Electron microscopy of the contact chemoreceptors of the stable fly, *Stomoxys calcitrans* (Diptera: Muscidae), *Ann. ent. Soc. Am.*, **58**, 909–917.

ADAMS, W. B. (1971), Intensity characteristics of the noctuid acoustic receptor, *J. gen. Physiol.*, **58**, 562–579.

ANDERSEN, L. W. AND BALL, H. J. (1959), Antennal hygroreceptors of the milkweed bug, *Oncopeltus fasciatus* (Dallas) (Hemiptera, Lygaeidae), *Ann. ent. Soc. Am.*, **52**, 279–284.

ARNTZ, B. (1972), The hearing capacity of water bugs, *J. comp. Physiol.*, **80**, 309–311.

AUTRUM, H. (1963), Anatomy and physiology of sound receptors in invertebrates, In: Busnel, R. G. (ed.) *Acoustic Behaviour of Animals*, Elsevier, Amsterdam, pp. 412–433.

AUTRUM, H. AND KOLB, G. (1968), Spektrale Empfindlichkeit einzelner Sehzellen der Aeschniden, *Z. vergl. Physiol.*, **60**, 450–477.

AUTRUM, H. AND SCHNEIDER, W. (1948), Vergleichende Untersuchungen über den Erschütterungssinn der Insekten, *Z. vergl. Physiol.*, **31**, 77–78.

AUTRUM, H. AND ZWEHL, V. VON (1964), Die spektrale Empfindlichkeit einzelner Sehzellen des Bienenauges, *Z. vergl. Physiol.*, **48**, 357–384.

BALDUS, K. (1926), Experimentelle Untersuchungen über die Entfernungslokalisation der Libellen (*Aeschna cyanea*), *Z. vergl. Physiol.*, **3**, 475–505.

BECKER, G. AND SPECK, U. (1964), Untersuchungen über die Magnetfeld-Orientierung von Dipteren, Z. vergl. Physiol., 49, 301–340.

BENNETT, R. R. (1967), Spectral sensitivity studies on the whirligig beetle, Dineutes ciliatus, J. Insect Physiol., 13, 621–633.

BENNETT, R. R., TUNSTALL, J. AND HORRIDGE, G. A. (1967), Spectral sensitivity of single retinula cells of the locust, Z. vergl. Physiol., 55, 195–206.

BERNHARD, C. G. (ed.) (1966), The Functional Organization of the Compound Eye, Pergamon Press, Oxford, 591 pp.

BERNHARD, C. G., GEMNE, G. AND SÄLLSTRÖM, Z. (1970), Comparative ultra-structure of corneal surface topography in insects with considerations on phylogeny and function, Z. vergl. Physiol., 67, 1–25.

BERNHARD, C. G., MILLER, W. H. AND MØLLER, A. R. (1956), The insect corneal nipple array. A biological, broad-band impedance transformer that acts as an antireflection coating, Acta physiol. scand., 63 (Suppl. 43), 1–79.

BEROZA, M. (ed.) (1970), Chemicals Controlling Insect Behaviour, Academic Press, New York, 170 pp.

BIRCH, M. C. (ed.) (1974), Pheromones, North-Holland, Amsterdam, 495 pp.

BISHOP, L. G. (1969), A search for color-encoding in the responses of a class of fly interneurons, Z. vergl. Physiol., 64, 355–371.

BLANEY, W. M., CHAPMAN, R. F. AND COOK, A. G. (1971), The structure of the terminal sensilla on the maxillary palps of Locusta migratoria (L.), and changes associated with moulting (Orth., Acrididae), Z. Zellforsch., 121, 48–68.

BLUM, M. S. (1969), Alarm pheromones, A. Rev. Ent., 14, 57–80.

BOECKH, J., KAISSLING, K.-E. AND SCHNEIDER, D. (1965), Insect olfactory receptors, Cold Spring Harb. Symp. quant. Biol., 30, 263–280.

BOHM, H. AND TÄUBER, U. (1971), Beziehung zwischen der Wirkung polarisierten Lichtes auf das Elektroretinogramm und der Ultrastruktur des Auges von Gerris lacustris L., Z. vergl. Physiol., 72, 32–53.

BORDEN, J. H. AND WOOD, D. L. (1966), The antennal receptors and olfactory response of Ips confusus (Coleoptera: Scolytidae) to male sex attractant in the laboratory, Ann. ent. Soc. Am., 59, 253–261.

BOSCHEK, C. B. (1971), On the fine structure of the peripheral retina and lamina ganglionaris of the fly Musca domestica, Z. Zellforsch., 118, 369–409.

BRAMMER, J. D. (1970), The ultrastructure of the compound eye of a mosquito, Aedes aegypti L., J. exp. Zool., 175, 181–196.

BROWN, W. L., JR., EISNER, T. AND WHITTAKER, R. H. (1970), Allomones and kairomones: transspecific chemical messengers, Bioscience, 20, 21.

BRUCKMOSER, P. (1968), Die spektrale Empfindlichkeit einzelner Sehzellen des Rückenschwimmers Notonecta glauca L. (Heteroptera), Z. vergl. Physiol., 59, 187–204.

BÜCKMANN, D. (1962), Das Problem des Schweresinnes bei den Insekten, Naturwissenschaften, 49, 28–33.

BURKHARDT, D. (1962), Spectral sensitivity and other response characteristics of single visual cells in the arthropod eye, Symp. Soc. exp. Biol., 16, 86–109.

—— (1964), Colour discrimination in insects, Adv. Insect Physiol., 2, 131–173.

BURKHARDT, D. AND GEWECKE, M. (1965), Mechanoreception in Arthropoda: the chain from stimulus to behavioral pattern, Cold Spring Harb. Symp. quant. Biol., 30, 601–614.

BURKHARDT, D. AND WENDLER, L. (1960), Ein direkter Beweis für die Fähigkeit einzelner Sehzellen des Insektenauges, die Schwingungsrichtung polarisierten Lichtes zu analysieren, *Z. vergl. Physiol.*, **43**, 687–692.

BURNS, M. D. (1974), Structure and physiology of the locust femoral chordotonal organ, *J. Insect Physiol.*, **20**, 1319–1339.

BURTON, P. R. AND STOCKHAMMER, K. A. (1969), Electron microscopic studies of the compound eye of the toadbug, *Gelastocoris oculatus*, *J. Morph.*, **127**, 233–258.

BURTT, E. T. AND CATTON, W. T. (1966), Image formation and sensory transmission in the compound eye, *Adv. Insect Physiol.*, **3**, 1–52.

—— (1969), Resolution of the locust eye measured by rotation of radial striped patterns, *Proc. R. Soc.*, **173** (B), 513–529.

BUSNEL, R. G. (ed.) (1955), *Colloque sur l'Acoustique des Orthoptères*, Paris, Inst. Nat. Rech. Agr., 448 pp.

BUTLER, C. G. (1967), Insect pheromones, *Biol. Rev.*, **42**, 42–87.

BUTLER, C. G., CALLOW, R. K. AND CHAPMAN, J. R. (1964), 9-Hydroxydec-*trans*-2-enoic acid, a pheromone stabilizing honeybee swarms, *Nature*, **201**, 733.

CARTHY, J. D. AND NEWELL, G. E. (eds) (1968), Invertebrate receptors, *Symp. zool. Soc. Lond.*, **23**, 341 pp.

CHAPMAN, K. M. (1965), Campaniform sensilla on the tactile spines of the legs of the cockroach, *J. exp. Biol.*, **42**, 191–203.

CHAPMAN, R. F. (1961), Some experiments to determine the methods used in host finding by tsetse fly G. *medicorum*, *Bull. ent. Res.*, **52**, 83–97.

CHEVALIER, R. L. (1969), The fine structure of campaniform sensilla on the halteres of *Drosophila melanogaster*, *J. Morph.*, **128**, 443–464.

CONSTANTINEANU, M. J. (1930), Der Aufbau der Sehorgane bei den in Süsswasser lebenden Dipterenlarven und bei Puppen und Imagines von *Culex*, *Zool. Jb. (Anat.)*, **52**, 253–346.

CORBIÈRE-TICHANÉ, G. (1971) Ultrastructure des organes chordotonaux des pièces céphaliques chez la larve du *Speophyes lucidulus* Delar. (Coléoptère cavernicole de la sous-famille des Bathysciinae), *Z. Zellforsch.*, **117**, 275–302.

CORNELI, W. (1924), Von dem Aufbau des Sehorgans der Blattwespenlarven und der Entwicklung des Netzauges, *Zool Jb. (Anat.)*, **46**, 473–608.

DAUMER, K. (1956), Reizmetrische Untersuchungen des Farbensehens der Bienen, *Z. vergl. Physiol.*, **38**, 413–478.

—— (1958), Blumenfarben, wie sie die Bienen sehen, *Z. vergl. Physiol.*, **41**, 49–110.

DEBAISIEUX, P. (1936, 1938), Organes scolopidiaux des pattes d'insectes. I, II, *Cellule*, **44**, 271–314; **47**, 77–202.

DEBAUCHE, H. (1936), Étude cytologique et comparée de l'organe de Johnston des insectes, *Cellule*, **45**, 77–148.

DETHIER, V. G. (1942–43), The dioptric apparatus of lateral ocelli. I, II, *J. cell. comp. Physiol.*, **19**, 301–303; **22**, 115–126.

—— (1957), The sensory physiology of blood-sucking arthropods, *Exp. Parasitol.*, **6**, 68–122.

—— (1963), *The Physiology of Insect Senses*, Methuen, London, 266 pp.

DETHIER, V. G. AND HANSON, F. E. (1968), Electrophysiological responses of the chemoreceptors of the blowfly to sodium salts of fatty acids, *Proc. natn. Acad. Sci., U.S.A.*, **60**, 1296–1303.

DETHIER, V. G. AND KUCH, J. H. (1971), Electrophysiological studies of gustation in lepidopterous larvae. I. Comparative sensitivity to sugars, amino acids and glycosides, *Z. vergl. Physiol.*, **72**, 343–363.

DETHIER, V. G. AND SCHOONHOVEN, L. M. (1968), Evaluation of evaporation by cold and humidity receptors in caterpillars, *J. Insect Physiol.*, **14**, 1049–1054.

DIESENDORF, M. O. AND HORRIDGE, G. A. (1973), Two models of the partially focussed clear zone compound eye, *Proc. R. Soc. (B)*, **183**, 141–158.

DUBOSE, W. P. AND AXTELL, R. C. (1968), Sensilla on the antennal flagella of *Hippelates* eye gnats, *Ann. ent. Soc. Am.*, **61**, 1547–1561.

DUMPERT, K. (1972), Bau und Verteilung der Sensillen auf der Antennengeissel von *Lasius fuliginosus* (Latr.) (Hymenoptera, Formicidae), *Z. Morph. Tiere*, **73**, 95–116.

EGGERS, F. (1919), Das thoracale bitympanale Organ einer Gruppe der Lepidoptera Heterocera, *Zool. Jb. (Anat.)*, **41**, 273–376.

—— (1928), Die stiftführenden Sinnesorgane. Morphologie und Physiologie der chordotonalen und der tympanalen Sinnesapparate der Insekten, *Zool. Baust.*, **2**(1), 354 pp.

EMDEN, H. F. VAN (ed.) (1973), Insect/Plant Relationships, *Symp. R. ent. Soc. Lond.*, **6**, 215 pp.

ERNST, K. D. (1969), Die Feinstruktur von Riechsensillen auf der Antenne des Aaskäfers *Necrophorus* (Coleoptera), *Z. Zellforsch.*, **94**, 72–102.

—— (1972a), Die Ontogenie der basiconischen Riechsensillen auf der Antenne von *Necrophorus* (Coleoptera), *Z. Zellforsch.*, **129**, 217–236.

—— (1972b), Sensillum coelosphaericum, die Feinstruktur eines neuen olfaktorischen Sensillentyp, *Z. Zellforsch.*, **132**, 95–106.

EVANS, W. G. (1966), Morphology of the infra-red sense organs of *Melanophila acuminata* (Buprestidae: Coleoptera), *Ann. ent. Soc. Am.*, **59**, 873–877.

EXNER, S. (1891), *Die Physiologie der facettierten Augen von Krebsen und Insekten*, Deuticke, Leipzig and Wien, 206 pp.

FERNÁNDEZ-MORÁN, H. (1958), Fine structure of the light receptors in the compound eyes of insects, *Exp. Cell. Res.*, **1958**, Suppl. **5**, 586–644.

FINLAYSON, L. H. (1968), Proprioceptors in the invertebrates, *Symp. zool. Soc., Lond.*, **23**, 217–249.

—— (1972), Chemoreceptors, cuticular mechanoreceptors, and peripheral multiterminal neurones in the larva of the tsetse fly (*Glossina*), *J. Insect Physiol.*, **18**, 2265–2275.

FINLAYSON, L. H. AND LOWENSTEIN, O. (1958), The structure and function of abdominal stretch receptors in insects, *Proc. R. Soc. (B)*, **148**, 433–449.

FISCHER, A. AND HORSTMANN, G. (1971), Der Feinbau des Auges der Mehlmotte, *Ephestia kuehniella* Zeller (Lepidoptera, Pyralididae), *Z. Zellforsch.*, **116**, 275–304.

FRAENKEL, G. S. (1930), Die Orientierung von *Schistocerca gregaria* zu strahlender Wärme, *Z. vergl. Physiol.*, **13**, 300–313.

—— (1969), Evaluation of our thoughts on secondary plant substances, *Entomologia exp. appl.*, **12**, 473–486.

FRAENKEL, G. S. AND GUNN, D. L. (1940), *The Orientation of Animals. Kineses, Taxes and Compass Reactions*, Clarendon Press, Oxford, 352 pp.

FRIEDMAN, M. H. (1972a), A light and electron microscopic study of sensory organs and associated structures in the foreleg of the cricket, *Gryllus assimilis*, *J. Morph.*, **138**, 263–328.

—— (1972b), An electron microscopic study of the tympanal organ and associated structures in the foreleg tibia of the cricket *Gryllus assimilis*, *J. Morph.*, **138**, 329–348.

FRIEDRICHS, H. F. (1931), Beiträge zur Morphologie und Physiologie der Sehorgane der Cicindelinen (Col.), *Z. Morph. Ökol. Tiere*, **21**, 1–172.

FRISCH, K. VON (1913), Ueber den Farbensinn der Bienen und die Blumenfarben, *Sber. Ges. Morph., München*, **28** (1912), 50–59.

—— (1967), *The Dance Language and Orientation of Bees*, Harvard Univ. Press, Cambridge, Mass. 580 pp.

FRISCH, K. VON AND LINDAUER, M. (1956), The 'language' and orientation of the honey bee, *A. Rev. Ent.*, **1**, 45–58.

FRISCH, K. VON, LINDAUER, M. AND DAUMER, K. (1960), Über die Wahrnehmung polarisierten Lichtes durch das Bienenauge, *Experientia*, **16**, 289–301.

GAFFAL, K. P. AND HANSEN, K. (1972), Mechanorezeptive Strukturen der antennalen Haarsensillen der Baumwollwanze *Dysdercus intermedius* Dist, *Z. Zellforsch.*, **132**, 79–94.

GALIĆ, M. (1971), Die Sinnesorgane an der Glossa, dem Epipharynx und dem Hypopharynx der Arbeiterin von *Apis mellifica* L. (Insecta, Hymenoptera), *Z. Morph. Tiere*, **70**, 201–228.

GATEHOUSE, A. G. (1970), The probing response of *Stomoxys calcitrans* to certain physical and olfactory stimuli, *J. Insect Physiol.*, **16**, 61–74.

GEBHARDT, H. (1953), Die Lage der wichtigsten Thermorezeptoren bei einigen Insekten, *Zool. Jb. (Allg. Zool.)*, **63**, 558–592.

GETTRUP, E. (1966), Sensory regulation of wing twisting in locusts, *J. exp. Biol.*, **44**, 1–16.

GEMNE, G. (1971), Ontogenesis of corneal surface ultrastructure in nocturnal Lepidoptera, *Phil. Trans. R. Soc. Ser. B*, **262**, 343–363.

GEWECKE, M. (1972a), Bewegungsmechanismus und Gelenkrezeptoren der Antennen von *Locusta migratoria* L. (Insecta, Orthoptera), *Z. Morph. Tiere*, **71**, 128–149.

—— (1972b), Antennen und Stirnscheitelhaare von *Locusta migratoria* L. als Luftströmungs-Sinnesorgane bei der Flugsteuerung, *J. comp. Physiol.*, **80**, 57–94.

GHIRADELLA, H. (1971), Fine structure of the Noctuid moth ear. I. The transducer area and connections to the tympanic membrane in *Feltia subgothica* Haworth, *J. Morph.*, **134**, 21–46.

GILLIES, M. T. AND WILKES, T. J. (1972), The range of attraction of animal baits and carbon dioxide for mosquitoes. Studies in a freshwater area of West Africa, *Bull. ent. Res.*, **61**, 389–404.

GOGALA, M. (1967), Die spektrale Empfindlichkeit der Doppelaugen von *Ascalaphus macaronius* Scop. (Neuroptera, Ascalaphidae), *Z. vergl. Physiol.*, **57**, 232–243.

GOGALA, M., HAMDORF, K., AND SCHWEMER, J. (1970), UV-Sehfarbstoff bei Insekten, *Z. vergl. Physiol.*, **70**, 410–413.

GOHRBRANDT, I. (1937), Das Tympanalorgan der Drepaniden und der Cymatophoriden, *Z. wiss. Zool.*, **149**, 537–600.

GOODMAN, L. J. (1970), The structure and function of the insect dorsal ocellus, *Adv. Insect Physiol.*, **7**, 97–195.

GOTHILF, S., GALUN, R. AND BAR-ZEEV, M. (1971), Taste reception in the Mediterranean fruit fly: electrophysiological and behavioural studies, *J. Insect Physiol.*, **17**, 1371–1384.

GÖTZ, B. (1936), Beiträge zur Analyse des Verhaltens von Schmetterlingsraupen beim Aufsuchen des Futters und des Verpuppungsplatzes, *Z. vergl. Physiol.*, **23**, 429–503.

GÖTZE, G. (1927), Untersuchungen an Hymenopteren über das Vorkommen und die Bedeutung der Stirnaugen, *Zool. Jb.* (*Allg. Zool.*), **44**, 211–268.

GRAY, E. G. (1960), The fine structure of the insect ear, *Phil. Trans. R. Soc., Ser. B*, **243**, 75–94.

GRIBBLE, G., SANSTEAD, J. AND SULLIVAN, J. (1973), One-step synthesis of the housefly sex attractant (Z)-tricos-9-ene (Muscalure), *J. chem. Soc. Chemical Communications*, **1973**, 735–736.

GUNN, D. L. AND PIELOU, D. P. (1940), The humidity behaviour of the mealworm beetle, *Tenebrio molitor* L. III. The mechanism of the reaction, *J. exp. Biol.*, **17**, 307–316.

GÜNTHER, H. (1912), Die Sehorgane der Larve und Imago von *Dytiscus marginalis*, *Z. wiss. Zool.*, **100**, 60–115.

GUTHRIE, D. M. (1966), The function and fine structure of the cephalic airflow receptor in *Schistocerca gregaria*, *J. Cell Sci.*, **1**, 463–470.

GUTIERREZ, A. P. (1970), Studies on host selection and host specificity of the aphid hyperparasite *Charips victrix* (Hym., Cynipidae). 6. Description of sensory structures, and a synopsis of host selection and host specificity, *Ann. ent. Soc. Am.*, **63**, 1705–1709.

HANSEN, K. AND HEUMANN, H.-G. (1972), Die Feinstruktur der tarsalen Schmeckhaare der Fliege *Phormia terranovae* Rob.-Desv., *Z. Zellforsch.*, **117**, 419–442.

HANSTRÖM, B. (1927), Das Gehirn und die Sinnesorgane der Aphanipteren, *Ent. Tidskr.*, **48**, 154–160.

HASKELL, P. T. (1956), Hearing in certain Orthoptera. I, II, *J. exp. Biol.*, **33**, 756–766; 767–776.

—— (1961), *Insect Sounds*, Witherby, London, 189 pp.

HASKELL, P. T. AND MORDUE, A. J. (1969), The role of mouthpart receptors in the feeding behaviour of *Schistocerca gregaria*, *Entomologia exp. appl.*, **12**, 591–610.

HAWKE, S. D. AND FARLEY, R. D. (1971), Antennal chemoreceptors of the desert burrowing cockroach, *Arenivaga* sp, *Tissue and Cell*, **3**, 649–664.

HAWKE, S. D., FARLEY, R. D. AND GREANY, P. D. (1973), The fine structure of sense organs in the ovipositor of the parasitic wasp, *Orgilus lepidus* Muesebeck, *Tissue and Cell*, **5**, 171–184.

HERAN, H. (1959), Wahrnehmung und Regelung der Fluggeschwindigkeit bei *Apis mellifica* L., *Z. vergl. Physiol.*, **42**, 103–163.

HERTER, K. (1953), *Der Temperatursinn der Insekten*, Duncker & Humbolt, Berlin, 378 pp.

HERTWECK, H. (1931), Anatomie und Variabilität des Nervensystems und der Sinnesorgane von *Drosophila melanogaster* (Meigen), *Z. wiss. Zool.*, **139**, 559–663.

HERTZ, M. (1934), Zur Physiologie des Formen- und Bewegungssehens. III, *Z. vergl. Physiol.*, **21**, 604–615.

—— (1937), Beitrag zum Farbensinn und Formensinn der Bienen, *Z. vergl. Physiol.*, **24**, 413–421.

HOCKING, B. (ed.) (1960), Smell in insects: a bibliography with abstracts, *Canad. Defence Res. Board, Tech. Rept.*, **8**, 266 pp.

—— (1971), Blood-sucking behavior of terrestrial arthropods, *A. Rev. Ent.*, **16**, 1–26.

HODEK, I. (1960), Hibernation-bionomics in Coccinellidae (6th contribution to the ecology of Coccinellidae), *Acta Soc. ent. Csl.*, **57**, 1–20.

HODGSON, E. S. (1958), Chemoreception in Arthropods, *A. Rev. Ent.*, **3**, 19–36.

—— (1965), The chemical senses and changing viewpoints in sensory physiology, *Viewpoints in Biol.*, **4**, 83–124.

HOOPER, R. L., PITTS, C. W. AND WESTFALL, J. A. (1972), Sense organs on the ovipositor of the face fly, *Musca autumnalis, Ann. ent. Soc. Am.*, **65**, 577–586.

HORRIDGE, G. A. (1961), Pitch discrimination in locusts, *Proc. R. Soc.* (B), **155**, 218–231.

—— (1965), In: Bullock, T. H. and Horridge, G. A. (eds), *Structure and Function in the Nervous Systems of Invertebrates*, **2**, pp. 1005–1113.

—— (1968), *Interneurons. Their origin, action, specificity, growth and plasticity*, Freeman, London and San Francisco, 436 pp.

—— (1969), The eye of the firefly *Photuris*, *Proc. R. Soc.* (B), **171**, 445–463.

—— (1971), Alternatives to superposition images in clear-zone compound eyes, *Proc. R. Soc.* (B), **179**, 97–124.

—— (1972), Further observations on the clear-zone eye of *Ephestia*, *Proc. R. Soc.* (B), **181**, 157–173.

—— (ed.) (1974), *The Compound Eye and Vision of Insects*, Clarendon Press, Oxford, 500 pp.

HORRIDGE, G. A. AND BARNARD, P. B. T. (1965), Movement of palisade in locust retinula cells when illuminated, *Q. Jl Microsc. Sci.*, **106**, 131–135.

HORRIDGE, G. A. AND GIDDINGS, C. (1971), The ommatidium of the termite *Mastotermes darwiniensis, Tissue and Cell*, **3**, 463–476.

HORRIDGE, G. A., GIDDINGS, C. AND STANGE, G. (1972), The superposition eye of skipper butterflies, *Proc. R. Soc.* (B), **182**, 457–495.

HORRIDGE, G. A., WALCOTT, B. AND IOANNIDES, A. C. (1970), The tiered retina of *Dytiscus*: a new type of compound eye, *Proc. R. Soc.* (B), **175**, 83–94.

HOWSE, P. E. (1968), The fine structure and functional organization of chordotonal organs, *Symp. zool. Soc. Lond.*, **23**, 167–198.

HOWSE, P. E., LEWIS, D. B. AND PYE, J. D. (1971), Adequate stimulus of the insect tympanic organ, *Experientia*, **27**, 598–600.

HUNDERTMARK, A. (1936), Helligkeits- und Farbenunterscheidungsvermögen der Eiraupen der Nonne (*Lymantria monacha* L.), *Z. vergl. Physiol.*, **24**, 42–57.

HUGHES, G. M. (1958), The coordination of insect movements. III. Swimming in *Dytiscus, Hydrophilus* and a dragonfly nymph, *J. exp. Biol.*, **35**, 567–583.

ISHIKAWA, S. (1969), The spectral sensitivity and the components of the visual system in the stemmata of silkworm larvae, *Bombyx mori* L. (Lep., Bombycidae), *Appl. Ent. Zool.*, **4**, 87–99.

I-WU CHU AND AXTELL, R. C. (1971), Fine structure of the dorsal organ of the housefly larva, *Musca domestica* L., *Z. Zellforsch.*, **117**, 17–34.

JACOBSON, M. (1966), Chemical insect attractants and repellents, *A. Rev. Ent.*, **11**, 403–422.

—— (1972), *Insect Sex Pheromones*, Academic Press, New York, 382 pp.

JANDER, R. AND BARRY, C. K. (1968), Die phototaktische Gegenkopplung von Stirnocellen und Facettenaugen in der Phototropotaxis der Heuschrecken und Grillen (Saltatoria: *Locusta migratoria* und *Gryllus bimaculatus*), *Z. vergl. Physiol.*, **57**, 432–458.

JANDER, R. AND HEINRICHS, I. (1970), Das strauchspezifische visuelle Perceptor-

System der Stabheuschrecke (*Carausius morosus*), *Z. vergl. Physiol.*, **70**, 425–447.

JEFFERSON, R. N., RUBIN, R. E., MCFARLAND, S. U. AND SHOREY, H. H. (1970), Sex pheromones of noctuid moths. XXII. The external morphology of the antennae of *Trichoplusia ni*, *Heliothis zea*, *Prodenia ornithogalli* and *Spodoptera exigua*, *Ann. ent. Soc. Am.*, **63**, 1227–1238.

JOKUSCH, B. (1967), Bau und Funktion eines larvalen Insektenauges: Untersuchungen am Ameisenlöwen (*Euroleon nostras* Fourcroy, Plannip., Myrmel.), *Z. vergl. Physiol.*, **56**, 171–198.

KAISER, W. AND BISHOP, L. G. (1970), Directionally selective motion-detecting units in the optic lobe of the honey bee, *Z. vergl. Physiol.*, **67**, 403–413.

KALMUS, H. (1945), Correlations between flight and vision, and particularly between wings and ocelli, in insects, *Proc. R. ent. Soc. Lond.* (A), **20**, 84–96.

KARLSON, P. AND BUTENANDT, A. (1959), Pheromones (ectohormones) in insects, *A. Rev. Ent.*, **4**, 39–58.

KAY, R. E. (1969), Fluorescent materials in insect eyes and their possible relationship to ultra-violet sensitivity, *J. Insect Physiol.*, **15**, 2021–2038.

KELLOG, F. E. (1970), Water vapour and carbon dioxide receptors in *Aedes aegypti*, *J. Insect Physiol.*, **16**, 99–108.

KENNEDY, J. S. AND STROYAN, H. L. G. (1959), Biology of aphids, *A. Rev. Ent.*, **4**, 139–160.

KENNEL, J. VON AND EGGERS, F. (1933), Die abdominalen Tympanalorgane der Lepidopteren, *Zool. Jb. (Anat.)*, **57**, 1–104.

KERKUT, G. A. AND TAYLOR, B. J. R. (1957), A temperature receptor in the tarsus of the cockroach, *Periplaneta americana*, *J. exp. Biol.*, **34**, 486–493.

KIRSCHFELD, K. (1971), Aufnahme und Verarbeitung optischer Daten im Komplexauge von Insekten, *Naturwissenschaften*, **58**, 201–209.

KRUMBIEGEL, J. (1932), Untersuchungen über physiologische Rassenbildung. Ein Beitrag zum Problem der Artbildung und der geographischen Variationen, *Zool. Jb. (Syst.)*, **63**, 183–280.

KUNZE, P. (1970), Verhaltensphysiologische und optische Experimente zur Superpositionstheorie der Bildentstehung in Komplexaugen, *Verh. dt. zool. Ges.*, **1970**, 234–238.

LACHER, V. (1964), Elektrophysiologische Untersuchungen an einzelnen Rezeptoren für Geruch, Kohlendioxyd, Luftfeuchtigkeit und Temperatur auf den Antennen der Arbeitsbiene und der Drohne (*Apis mellifica* L.), *Z. vergl. Physiol.*, **48**, 587–623.

—— (1967), Elektrophysiologische Untersuchungen an einzelnen Geruchsrezeptoren auf den Antennen weiblicher Moskitos (*Aedes aegypti* L.), *J. Insect Physiol.*, **13**, 1461–1470.

LACHER, V. AND SCHNEIDER, D. (1963), Electrophysiologischer Nachweis der Riechfunktion von Porenplatten (Sensilla placodea) auf den Antennen der Drohne und der Arbeitsbiene (*Apis mellifica* L.), *Z. vergl. Physiol.*, **47**, 274–278.

LANGER, H. (1965), Nachweis dichroitischer Absorption des Sehfarbstoffes in den Rhabdomeren des Insektenauges, *Z. vergl. Physiol.*, **51**, 258–263.

—— (1967), Grundlage der Wahrnehmung von Wellenlänge und Schwingungsebene des Lichtes, *Zool. Anz.*, *Suppl.*, **30**, 195–233.

LARSÉN, O. (1957), Truncale Scolopalorgane in den pterothorakalen und den beiden ersten abdominalen Segmenten der aquatilen Heteropteren, *Acta Univ. lund.* (N.F.) **53** (1), 1–68.

LEES, A. D. (1942), Homology of the campaniform organs on the wing of *Drosophila melanogaster*, *Nature*, **150**, 375.

—— (1943), On the behaviour of wireworms of the genus *Agriotes* Esch. (Coleoptera, Elateridae), *J. exp. Biol.*, **20**, 43–60.

LEWIS, C. T. (1970), Structure and function in some external receptors, *Symp. R. ent. Soc. Lond.*, **5**, 59–76.

LEWIS, C. T. AND MARSHALL, A. T. (1970), The ultrastructure of the sensory plaque organs of the antennae of the Chinese lantern fly, *Pyrops candelaria* (Hem., Hom., Fulgoridae), *Tissue and Cell*, **2**, 375–385.

LINDAUER, M. (1967), Recent advances in bee communication and orientation, *A. Rev. Ent.*, **12**, 439–470.

LINDAUER, M. AND MARTIN, H. (1968), Die Schwereorientierung der Bienen unter dem Einfluss des Erdmagnetfeldes, *Z. vergl. Physiol.*, **60**, 219–243.

LOFTUS, R. (1968), The response of the antennal cold receptor of *Periplaneta americana* to rapid temperature changes and to steady temperature, *Z. vergl. Physiol.*, **59**, 413–455.

—— (1969), Differential thermal components in the response of the antennal cold receptor of *Periplaneta americana* to slowly changing temperature, *Z. vergl. Physiol.*, **63**, 415–433.

LOHER, W. (1960), The chemical acceleration of the maturation process and its hormonal control in the male of the desert locust, *Proc. R. Soc.* (B), **153**, 380–397.

LÜSCHER, M. (1961), Social control of polymorphism in termites, In: Kennedy, J. S. (ed.), Insect polymorphism. *Symp. R. ent. Soc. Lond.*, **1**, 57–67.

MADGE, D. S. (1965), The responses of cotton stainers (*Dysdercus fasciatus* Sign.) to relative humidity and temperature, and the location of the hygroreceptors, *Entomologia exp. appl.*, **8**, 135–152.

MAKINGS, P. (1964), 'Slifer's patches' and the thermal sense in Acrididae (Orthoptera), *J. exp. Biol.*, **41**, 473–497.

MALDONA, H. AND BARROS-PITA, J. C. (1970), A fovea in the praying mantis eye. I. Estimation of catching distance, *Z. vergl. Physiol.*, **67**, 58–78.

MARKL, H. (1962), Borstenfelder an den Gelenken als Schweresinnesorgane bei Ameisen und anderen Hymenopteren, *Z. vergl. Physiol.*, **45**, 475–569.

—— (1966), Schwerkraftdressuren an Honigbienen. 2. Die Rolle der schwererezeptorischen Borstenfelder verschiedener Gelenke für die Schwerekompassorientierung, *Z. vergl. Physiol.*, **53**, 353–371.

—— (1970), Die Verständigung durch Stridulationssignale bei Blattschneiderameisen. III. Die Empfindlichkeit für Substratvibrationen, *Z. vergl. Physiol.*, **69**, 6–37.

MARKL, H. AND WIESE, K. (1969), Die Empfindlichkeit des Rückenschwimmers *Notonecta glauca* L. für Oberflächenwellen des Wassers, *Z. vergl. Physiol.*, **62**, 413–420.

MARTIN, H. AND LINDAUER, M. (1966), Sinnesphysiologische Leistungen beim Wabenbau der Honigbiene, *Z. vergl. Physiol.*, **53**, 372–404.

MARTINSEN, D. L. AND KIMELDORF, D. J. (1972), The prompt detection of ionizing radiations by carpenter ants, *Biol. Bull.*, **143**, 403–419.

MASCHWITZ, U. (1964), Gefahrenalarmstoffe und Gefahrenalarmierung bei sozialen Hymenoptera, *Z. vergl. Physiol.*, **47**, 596–655.

MATSUMOTO, Y. AND THORSTEINSON, A. J. (1968), Olfactory response of larvae of the onion maggot, *Hylemyia antiqua* Meigen (Diptera: Anthomyiidae) to organic sulphur compounds, *Appl. Ent. Zool.*, **3**, 107–111.

MAZOKHIN-PORSCHNYAKOV, G. A. (1969), *Insect Vision*, Plenum Press, New York, 306 pp.

MCCANN, G. D. AND MAGINITIE, G. F. (1965), Optomotor response studies of insect vision, *Proc. R. Soc.* (B), **163**, 369–401.

MCFARLANE, J. E. (1953), The morphology of the chordotonal organs of the antenna, mouthparts and legs of the lesser migratory grasshopper, *Melanoplus mexicanus* (Saussure), *Can. Ent.*, **85**, 81–103.

MCIVER, S. B. (1972), Fine structure of pegs on the palps of female Culicine mosquitoes, *Can. J. Zool.*, **50**, 571–576.

—— (1975), Structure of cuticular mechanoreceptors of arthropods, *A. Rev. Ent.*, **20**, 381–397.

MEYER, H. W. (1971), Visuelle Schlüsselreize für die Auslösung der Beutefanghandlung beim Bachwasserläufer *Velia caprai* (Hemiptera, Heteroptera). I. Untersuchung der räumlichen und zeitlichen Reizparameter mit formverschiedenen Attrappen, *Z. vergl. Physiol.*, **72**, 260–297.

MEYER-ROCHOW, V. B. (1972), The eyes of *Creophilus erythrocephalus* F. and *Sartallus signatus* Sharp (Staphylinidae: Coleoptera): light-, interference-, scanning electron- and transmission electron microscope examinations, *Z. Zellforsch.*, **133**, 59–86.

—— (1973), The dioptric system of the eye of *Cybister* (Dytiscidae: Coleoptera), *Proc. R. Soc.* (B), **183**, 159–178.

—— (1974), Structure and function of the larval eye of the sawfly, *Perga*, *J. Insect Physiol.*, **20**, 1565–1591.

MICHEL, K. (1974), Das Tympanalorgan von *Gryllus bimaculatus* Degeer (Saltatoria, Gryllidae), *Z. Morph. Tiere*, **77**, 285–315.

MICHELSEN, A. (1971), The physiology of the locust ear. I–III, *Z. vergl. Physiol.*, **71**, 49–62; 63–101; 102–128.

MILES, P. W. (1958), Contact chemoreception in some Heteroptera, including chemoreception internal to the stylet food canal, *J. Insect Physiol.*, **2**, 338–347.

MILLER, L. A. (1970), Structure of the green lacewing tympanal organ (*Chrysopa carnea*, Neuroptera), *J. Morph.*, **131**, 359–382.

—— (1971), Physiological responses of green lacewings (*Chrysopa*: Neur., Chrysopidae) to ultrasound, *J. Insect Physiol.*, **17**, 491–506.

MIMURA, K. (1971), Movement discrimination by the visual system of flies, *Z. vergl. Physiol.*, **73**, 105–138.

MINNICH, D. E. (1925), The reactions of larvae of *Vanessa antiopa* Linn. to sounds, *J. exp. Zool.*, **42**, 443–469.

—— (1936), The responses of caterpillars to sounds, *J. exp. Zool.*, **72**, 439–453.

MITTLER, T. E. (1967), Gustation of dietary amino-acids by the aphid *Myzus persicae*, *Entomologia exp. appl.*, **10**, 87–96.

MOLLER-RACKE, I. (1952), Farbensinn und Farbenblindheit bei Insekten, *Zool. Jb.* (*Physiol.*), **63**, 237–274.

MORAN, D. T., CHAPMAN, K. M. AND ELLIS, R. A. (1971), The fine structure of cockroach campaniform sensilla, *J. Cell Biol.*, **48**, 155–173.

MORITA, H. AND YAMASHITA, S. (1961), Receptor potentials recorded from sensilla basiconica on the antenna of the silkworm larvae, *Bombyx mori*, *J. exp. Biol.*, **38**, 851–861.

MÖSS, D. (1971), Sinnesorgane im Bereich des Flügels der Feldgrille (*Gryllus campestris* L.) und ihre Bedeutung für die Kontrolle der Singbewegung und die Einstellung der Flügellage, *Z. vergl. Physiol.*, **73**, 53–83.

MOTE, M. I. AND GOLDSMITH, T. H. (1970), Spectral sensitivities of color receptors in the compound eye of the cockroach *Periplaneta, J. exp. Zool.*, **173**, 137–146.

MOULINS, M. (1971), Ultrastructure et physiologie des organes épipharyngiens et hypopharyngiens (chemiorécepteurs cibariaux) de *Blabera craniifer* Burm. (Insecte, Dictyoptère), *Z. vergl. Physiol.*, **73**, 139–166.

MUNCHBERG, P. (1966), Zum morphologischen Bau und zur funktionellen Bedeutung der Ocellen der Libellen, *Beitr. Ent.*, **16**, 221–249.

MYERS, J. AND BROWER, L. P. (1969), A behavioural analysis of the courtship pheromone receptors of the queen butterfly, *Danaus gilippus berenice, J. Insect Physiol.*, **15**, 2117–2130.

NAYAR, J. K. AND THORSTEINSON, A. J. (1963), Further investigations into the chemical basis of insect–host plant relationships in an oligophagous insect, *Plutella maculipennis* (Curtis) (Lepidoptera: Plutellidae), *Can. J. Zool.*, **41**, 923–929.

NEESE, V. (1965), Zur Funktion der Augenborsten bei der Honigbiene, *Z. vergl. Physiol.*, **49**, 543–585.

NEVILLE, A. C. (1967), A dermal light sense influencing skeletal structure in locusts, *J. Insect Physiol.*, **13**, 933–939.

NIESCHULZ, O. (1934), Ueber die Vorzugtemperatur von *Stomoxys calcitrans, Z. angew. Ent.*, **21**, 224–238.

NINOMIYA, N., TOMINAGA, Y. AND KUWABARA, M. (1969), The fine structure of the compound eye of a damsel-fly, *Z. Zellforsch.*, **98**, 17–32.

NORRIS, M. J. (1954), Sexual maturation in the desert locust (*Schistocerca gregaria* Forskål) with special reference to the effects of grouping, *Anti-Locust Bull.*, **18**, 1–44.

OSBORNE, M. P. AND FINLAYSON, L. H. (1962), The structure and topography of stretch receptors in representatives of seven orders of insects, *Q. Jl Microsc. Sci.*, **103**, 227–242.

—— (1965), An electron microscope study of the stretch receptor of *Antheraea pernyi* (Lepidoptera, Saturniidae), *J. Insect Physiol.*, **11**, 703–710.

OWEN, W. B. (1963), The contact chemoreceptor organs of the mosquito and their function in feeding behaviour, *J. Insect Physiol.*, **9**, 73–87.

PAYNE, R. S., ROEDER, K. D. AND WALLMAN, J. (1966), Directional sensitivity of the ears of Noctuid moths, *J. exp. Biol.*, **44**, 17–31.

PERRELET, A. (1970), The fine structure of the retina of the honeybee drone, *Z. Zellforsch.*, **108**, 530–562.

PETERS, W. (1961), Die sogenannten Fuss-stummelsinnesorgane der Larven von *Calliphora erythrocephala* Mg. (Diptera), *Zool. Jb. (Anat.)*, **79**, 339–346.

—— (1965), Die Sinnesorgane an den Labellen von *Calliphora erythrocephala* Mg. (Diptera), *Z. Morph. Ökol. Tiere*, **55**, 259–320.

PITMAN, G. B. AND VITÉ, J. P. (1969), Aggregation behaviour of *Dendroctonus ponderosae* (Coleoptera: Scolytidae) in response to chemical messengers, *Can. Ent.*, **101**, 143–149.

POPHAM, E. J. (1961), The function of Hagemann's organ in *Corixa punctata* (Illig.) (Hemiptera: Corixidae), *Proc. R. ent. Soc. Lond. (A)*, **36**, 119–125.

PRAGER, J. (1973), Die Hörschwelle des mesothorakalen Tympanalorgans von *Corixa punctata* Ill. (Heteroptera, Corixidae), *J. Comp. Physiol.*, **86**, 55–58.

PRINGLE, J. W. S. (1954), A physiological analysis of cicada song, *J. exp. Biol.*, **31**, 525–560.

PUMPHREY, R. J. (1940), Hearing in insects, *Biol. Rev.*, **15**, 107–132.

PUMPHREY, R. J. AND RAWDON-SMITH, A. F. (1936), Hearing in insects: the nature of the response of certain receptors to auditory stimuli, *Proc. R. Soc.* (B), **121**, 18–27.

—— (1939), 'Frequency discrimination' in insects: a new theory, *Nature*, **143**, 806–807.

RABE, W. (1953), Beiträge zum Orientierungsproblem der Wasserwanzen, *Z. vergl. Physiol.*, **35**, 300–325.

REES, C. J. C. (1968), The effect of aqueous solutions of some 1:1 electrolytes on the electrical response of the type 1 ('salt') chemoreceptor cell in the labella of *Phormia*, *J. Insect Physiol.*, **14**, 1331–1364.

—— (1969), Chemoreceptor specificity associated with choice of feeding site by the beetle, *Chrysolina brunsvicensis* on its foodplant, *Hypericum hirsutum*. *Entomologia exp. appl.*, **12**, 565–583.

—— (1970), The primary processes of reception in the type 3 ('water') receptor cell of the fly, *Phormia terranovae*, *Proc. R. Soc.* (B), **174**, 469–490.

REICHARDT, W. (1973), Musterinduzierte Flugorientierung. Verhaltensversuche an der Fliege *Musca domestica*, *Naturwissenschaften*, **60**, 122–138.

RICE, M. J., GALUN, R. AND MARGALIT, J. (1973), Mouthpart sensilla of the tsetse fly and their function. III. Labrocibarial sensilla, *Ann. trop. Med. Parasit.*, **67**, 109–116.

RICHERSON, J. V., BORDEN, J. H. AND HOLLINGDALE, J. (1972), Morphology of a unique sensillum placodeum on the antenna of *Coeloides brunneri* (Hym., Braconidae), *Can. J. Zool.*, **50**, 909–913.

RICHTER, S. (1962), Unmittelbarer Kontakt der Sinneszellen cuticularer Sinnesorgane mit der Aussenwelt. Eine licht- und elektronenmikroskopische Untersuchung der chemorezeptorischen Antennensinnesorgane der *Calliphora*-Larven. *Z. Morph. Ökol. Tiere*, **52**, 171–196.

—— (1964), Die Feinstruktur des für die Mechanorezeption wichtigen Bereichs der Stellungshaare auf dem Prosternum von *Calliphora erythrocephala* Mg. (Diptera), *Z. Morph. Ökol. Tiere*, **54**, 202–218.

RISLER, H. (1953), Das Gehörorgan der Männchen von *Anopheles stephensi* Liston (Culicidae), *Zool. Jb.* (*Anat.*), **73**, 165–186.

—— (1955), Das Gehörorgan der Männchen von *Culex pipiens* L., *Aedes aegypti* L. und *Anopheles stephensi* Liston (Culicidae), eine vergleichend morphologische Untersuchung, *Zool. Jb.* (*Anat.*), **74**, 478–490.

ROEDER, K. D. (1969), Acoustic interneurons in the brain of noctuid moths, *J. Insect Physiol.*, **15**, 825–838.

—— (1972), Acoustic and mechanical sensitivity of the distal lobe of the pilifer in Choerocampine hawkmoths, *J. Insect Physiol.*, **18**, 1249–1264.

ROEDER, K. D. AND TREAT, A. E. (1957), Ultrasonic reception by the tympanic organ of Noctuid moths, *J. exp. Zool.*, **134**, 127–157.

ROGERS, G. L. (1962), A diffraction theory of insect vision. II. Theory and experiments with a simple model eye, *Proc. R. Soc.* (B), **157**, 83–98.

ROTH, L. M. (1948), A study of mosquito behaviour. An experimental laboratory study of the sexual behaviour of *Aedes aegypti* (Linnaeus), *Am. Midl. Nat.*, **40**, 265–352.

—— (1969), The evolution of male tergal glands in the Blattaria, *Ann. ent. Soc. Am.*, **62**, 176–208.

ROTH, L. M. AND WILLIS, E. R. (1951*a*), Hygroreceptors in Coleoptera, *J. exp. Zool.*, **117**, 451–487.

—— (1951*b*), The effects of desiccation and starvation on the humidity behaviour and water balance of *Tribolium confusum* and *Tribolium castaneum*, *J. exp. Zool.*, **118**, 337–361.

RUCK, P. (1961), Electrophysiology of the insect dorsal ocellus. I–III, *J. gen. Physiol.*, **44**, 605–627; 629–639; 641–657.

RUCK, P. AND EDWARDS, G. A. (1964), The structure of the insect dorsal ocellus. I. General organization of the ocellus in dragonflies, *J. Morph.*, **115**, 1–5.

RUDOLF, P. (1967), Zum Ortungsverfahren von *Gyrinus substriatus* Steph. (Taumelkäfer), *Z. vergl. Physiol.*, **56**, 341–375.

RUNION, H. I. AND USHERWOOD, P. N. R. (1968), Tarsal receptors and leg reflexes in the locust and grasshopper, *J. exp. Biol.*, **49**, 421–436.

SALAMA, H. S. (1966), The function of mosquito taste-receptors, *J. Insect Physiol.*, **12**, 1051–1060.

SCHAFER, R. (1971), Antennal sense-organs of the cockroach *Leucophaea maderae*, *J. Morph.*, **134**, 91–104.

SCHNEIDER, D. (1969), Insect olfaction: deciphering system for chemical messages, *Science, N.Y.*, **163**, 1031–1037.

—— (1971), Molekulare Grundlagen der chemischen Sinne bei Insekten, *Naturwissenschaften*, **58**, 194–200.

SCHNEIDER, D. AND KAISSLING, K.-E. (1957), Der Bau der Antenne des Seidenspinners *Bombyx mori* L. II. Sensillen, cuticulare Bildungen und innerer Bau, *Zool. Jb. (Anat.)*, **76**, 223–250.

SCHNEIDER, D. AND STEINBRECHT, R. A. (1968), Checklist of insect olfactory sensilla, *Symp. zool. Soc. Lond.*, **23**, 279–297.

SCHNEIDER, L. AND LANGER, H. (1969), Die Struktur des Rhabdoms im 'Doppelauge' des Wasserläufers *Gerris lacustris*, *Z. Zellforsch.* **99**, 538–559.

SCHNORBUS, H. (1971), Die subgenualen Sinnesorgane von *Periplaneta americana*: Histologie und Vibrationsschwellen, *Z. vergl. Physiol.*, **71**, 14–48.

SCHOONHOVEN, L. M. (1967), Some cold receptors in larvae of three Lepidoptera species, *J. Insect Physiol.*, **13**, 821–826.

—— (1968), Chemosensory bases of host plant selection, *A. Rev. Ent.*, **13**, 115–135.

—— (1969), Gustation and foodplant-selection in some Lepidopterous larvae, *Entomologia exp. appl.*, **12**, 555–564.

—— (1973), Plant recognition by Lepidopterous larvae, In: van Emden, H. (1973) (ed.), Insect/Plant Relationships. *Symp. R. ent. Soc. Lond.*, **6**, 87–99.

SCHOONHOVEN, L. M. AND DETHIER, V. G. (1966), Sensory aspects of host-plant discrimination by Lepidopterous larvae, *Arch. néerl. Zool.*, **16**, 497–530.

SCHWABE, J. (1906), Beiträge zur Morphologie und Histologie der tympanalen Sinnesapparate der Orthopteren, *Zoologica, Stuttgart*, Hft. **50**, 154 pp.

SCHWEMER, J., HAMDORF, K. AND GOGALA, M. (1971), Der UV-Sehfarbstoff der Insekten: Photochemie *in vitro* und *in vivo*, *Z. vergl. Physiol.*, **75**, 174–188.

SHOREY, H. H. (1973), Behavioral responses to insect pheromones, *A. Rev. Ent.*, **18**, 349–380.

SKRZIPEK, K.-H. AND SKRZIPEK, H. (1971), Die Morphologie der Bienenretina (*Apis mellifica* L.) in elektronenmikroskopischer und lichtmikroskopischer Sicht, Z. Zellforsch., 119, 552–576.

SLIFER, E. H. (1953), The pattern of specialized heat-sensitive areas on the body of Acrididae (Orthoptera). Parts I, II, Trans. Am. ent. Soc., 79, 37–68; 69–97.

—— (1970), The structure of Arthropod chemoreceptors, A. Rev. Ent., 15, 121–142.

SLIFER, E. H. AND SEKHON, S. S. (1961), Fine structure of the sense organs on the antennal flagellum of the honey bee, *Apis mellifera* Linnaeus, J. Morph., 109, 351–381.

—— (1964a), Fine structure of the sense organs on the antennal flagellum of a flesh fly, *Sarcophaga argyrostoma* R.-D. (Diptera: Sarcophagidae), J. Morph., 114, 185–208.

—— (1964b), The dendrites of thin-walled olfactory pegs of the grasshopper (Orthoptera, Acrididae), J. Morph., 114, 393–410.

—— (1972), Circumfila and other sense organs on the antenna of the sorghum midge (Diptera, Cecidomyiidae), J. Morph., 133, 281–302.

SLIFER, E. H., SEKHON, S. S. AND LEES, A. D. (1964), The sense organs on the antennal flagellum of aphids (Homoptera), with special reference to the plate organs, Q. Jl Microsc. Sci., 105, 21–29.

SMITH, D. S. (1969), The fine structure of haltere sensilla in the blowfly, *Calliphora erythrocephala* (Meig.), with scanning electron-microscopic observations on the haltere surface, Tissue and Cell, 1, 443–484.

STEINBRECHT, R. A. (1970), Zur Morphometrie der Antenne des Seidenspinners, *Bombyx mori* L.: Zahl und Verbreitung der Riechensensillen (Insecta, Lepidoptera), Z. Morph. Ökol. Tiere, 68, 93–126.

—— (1973), Der Feinbau olfaktorischer Sensillen des Seidenspinners (Insecta, Lepidoptera). Rezeptorfortsätze und reizleitender Apparat, Z. Zellforsch., 139, 533–565.

STEWARD, C. C. AND ATWOOD, C. E. (1963), The sensory organs of the mosquito antenna, Can. J. Zool., 41, 577–594.

STOCKHAMMER, K. (1959), Die Orientierung nach der Schwingungsrichtung linear polarisierten Lichtes und ihre sinnesphysiologischen Grundlagen, Ergebn. Biol., 21, 23–56.

STRANGE, P. H. (1961), The spectral sensitivity of *Calliphora* maggots, J. exp. Biol., 38, 237–248.

STRUWE, G. (1972a) Spectral sensitivity of single photoreceptors in the compound eye of a tropical butterfly (*Heliconius numata*), J. comp. Physiol., 79, 197–201.

—— (1972b), Spectral sensitivity of the compound eye in butterflies (*Heliconius*), J. comp. Physiol., 79, 191–196.

STUDNITZ, G. VON (1932), Die statische Funktion der sog. 'pelotaktischen' Organe ('Schlammsinnesorgane') der Limnobiidenlarven, Zool. Jb. (Allg. Zool.), 50, 419–446.

SWIHART, S. L. (1970), The neural basis of colour vision in the butterfly *Papilio troilus*, J. Insect Physiol., 16, 1623–1636.

—— (1972), The neural basis of colour vision in the butterfly *Heliconius arato*, J. Insect Physiol., 18, 1015–1025.

SYRJAMAKI, J. (1962), Humidity perception in *Drosophila melanogaster*, Ann. Soc. zool. bot. fenn. Vanamo, Helsinki, 23(3), 1–74.

THOMAS, J. G. (1966), The sense organs of the mouthparts of the desert locust, *J. Zool., Lond.*, **148**, 420–448.

THORPE, W. H. (1939), Further studies on pre-imaginal olfactory conditioning in insects, *Proc. R. Soc.* (B), **27**, 424–433.

THORPE, W. H. AND CAUDLE, H. B. (1938), A study of the olfactory responses of insect parasites to the food plant of their host, *Parasitology*, **30**, 523–528.

THORPE, W. H. AND CRISP, D. J. (1947), Studies on plastron respiration. III, *J. exp. Biol.*, **24**, 310–328.

THORPE, W. H. AND JONES, F. G. W. (1937), Olfactory conditioning in a parasitic insect and its relation to the problem of host selection, *Proc. R. Soc.* (B), **124**, 56–81.

THORSTEINSON, A. J. (1960), Host selection in phytophagous insects, *A. Rev. Ent.*, **5**, 193–218.

THURM, U. (1963), Die Beziehungen zwischen mechanischen Reizgrössen und stationären Erregungszuständen bei Borstenfeldsensillen von Bienen, *Z. vergl. Physiol.*, **46**, 351–382.

—— (1968), Steps in the transducer process of mechanoreceptors, *Symp. zool. Soc. Lond.*, **23**, 199–216.

TISCHNER, H. (1953), Über den Gehörsinn von Stechmücken, *Acustica*, **3**, 335–343.

TISCHNER, H. AND SCHIEF, A. (1955), Fluggeräusch und Schallwahrnehmung bei *Aedes aegypti* L. (Culicidae), *Verh. dtsch. zool. Ges.*, **18**, 453–460.

TSUNEKI, K. (1953), On colour vision in two species of ants, with special emphasis on their relative sensitivity to various monochromatic lights, *Jap. J. Zool.*, **11**, 187–221.

—— (1961), Colour vision and figure discriminating capacity of the solitary diplopterous wasp, *Odynerus frauenfeldi*, Saussure, *Mem. Fac. lib. Arts, Fukui Univ.* (2), *Nat. Sci. No* 11, 103–160.

TURNER, D. A. (1971), Olfactory perception of live hosts and carbon dioxide by the tsetse fly *Glossina morsitans orientalis* Vanderplank, *Bull. ent. Res.*, **61**, 75–96.

USHERWOOD, P. N. R., RUNION, H. I. AND CAMPBELL, J. I. (1968), Structure and physiology of a chordotonal organ in the locust leg, *J. exp. Biol.*, **48**, 305–323.

VAN DE BERG, J. S. (1971), Fine-structural studies of Johnston's organ in the tobacco hornworm moth, *Manduca sexta* (Johannson), *J. Morph.*, **133**, 439–456.

VINSON, S. B. (1976), Host selection by insect parasitoids, *A. Rev. Ent.*, **21**, 109–133.

VOGEL, R. (1923), Über ein tympanales Sinnesorgan, das mutmassliche Hörorgan der Singzikaden, *Z. ges. Anat., München*, **67**, 190–231.

WACHMANN, E. (1972a), Das Auge des Hühnerflohes *Ceratophyllus gallinae* (Schrank) (Insecta, Siphonaptera), *Z. Morph. Tiere*, **73**, 315–324.

WACHMANN, E. (1972b), Zum Feinbau des Komplexauges von *Stylops* spec. (Insecta, Strepsiptera), *Z. Zellforsch.*, **123**, 411–424.

WALCOTT, B. AND HORRIDGE, G. A. (1971), The compound eye of *Archichauliodes* (Megaloptera), *Proc. R. Soc.* (B), **179**, 65–72.

WALLIS, D. I. (1962), The sense organs on the ovipositor of the blowfly, *Phormia regina* Meigen, *J. Insect Physiol.*, **8**, 453–467.

WEEVERS, R. DE G. (1966), The physiology of a Lepidopteran muscle receptor. I–III, *J. exp. Biol.*, **44**, 177–194; **44**, 195–208; **45**, 229–249.

WEHNER, R. (1971), The generalization of directional visual stimuli in the honey bee, *Apis mellifera*, *J. Insect Physiol.*, **17**, 1579–1591.

—— (ed.) (1972), *Information processing in the visual systems of Arthropods*, Springer, Berlin, 334 pp.

WENDLER, G. (1964), Laufen und Stehen der Stabheuschrecke *Carausius morosus*: Sinnesborstenfelder in den Beingelenken als Glieder von Regelkreisen, *Z. vergl. Physiol.*, **48**, 198–250.

WENSLER, R. J. AND FILSHIE, J. B. (1969), Gustatory sense organs in the food canal of aphids, *J. Morph.*, **129**, 473–492.

WHITE, R. H. (1967), The effect of light and light deprivation upon the ultrastructure of the larval mosquito eye. II. The rhabdom, *J. exp. Zool.*, **166**, 405–425.

—— (1968), The effect of light and light deprivation upon the ultrastructure of the larval mosquito eye. III. Multivesicular bodies and protein uptake, *J. exp. Zool.*, **169**, 261–278.

WHITTEN, J. (1963), Observations on the cyclorrhaphan peripheral nervous system: muscle and tracheal receptor organs and independent peripheral type II neurons associated with lateral segmental nerves, *Ann. ent. Soc. Am.*, **56**, 755–763.

WIGGLESWORTH, V. B. (1941), The sensory physiology of the human louse *Pediculus humanus corporis* de Geer (Anoplura), *Parasitology*, **33**, 67–109.

WIGGLESWORTH, V. B. AND GILLETT, J. D. (1934), The function of the antennae in *Rhodnius prolixus* (Hemiptera) and the mechanism of orientation to the host, *J. exp. Biol.*, **11**, 120–139.

WILCZEK, M. (1967), The distribution and neuroanatomy of the labellar sense organs of the blowfly, *Phormia regina* Meigen, *J. Morph.*, **122**, 175–201.

WILDE, J. DE (1941), Contribution to the physiology of the Johnston organ and its part in the behaviour of *Gyrinus*, *Arch. néerl. Physiol.*, **28**, 530–542.

WILDE, J. DE AND SCHOONHOVEN, L. M. (eds) (1969), Insect and host plant, (Proceedings of the 2nd International Symposium 'Insect and Host Plant' Wageningen, the Netherlands). *Entomologia exp. appl.*, **12**, 471–810.

WILLIS, E. R. AND ROTH, L. M. (1950), Humidity reactions of *Tribolium castaneum* (Herbst), *J. exp. Zool.*, **115**, 561–587.

WILSON, D. M. AND GETTRUP, E. (1963), A stretch reflex controlling wingbeat frequency in grasshoppers, *J. exp. Biol.*, **40**, 171–185.

WODSEDALEK, J. E. (1912), Palmen's organ and its function in nymphs of the Ephemeridae. *Heptagenia interpunctata* (Say) and *Ecdyonurus maculipennis* (Walsh), *Biol. Bull.*, **22**, 253–272.

WOLF, E. AND ZERRAHN-WOLF, G. (1937), Flicker and the reactions of bees to flowers, *J. gen. Physiol.*, **20**, 511–518.

WOOD, D. L., SILVERSTEIN, R. M. AND NAKAJIMA, M. (eds) (1970), *Control of Insect Behaviour by Natural Products*, Academic Press, New York, 346 pp.

WRIGHT, R. H. (1966), Primary odours and insect attraction, *Can. Ent.*, **98**, 1083–1093.

WUNDRIG, G. (1936), Die Sehorgane der Mallophagen, nebst vergleichenden Untersuchungen an Liposceliden und Anopluren, *Zool. Jb. (Anat.)*, **62**, 45–110.

YOUDEOWEI, A. (1967), The reactions of *Dysdercus intermedius* (Heteroptera: Pyrrhocoridae) to moisture, with special reference to aggregation, *Entomologia exp. appl.*, **10**, 194–210.

YOUNG, D. (1970), The structure and function of a connective chordotonal organ in the cockroach leg, *Phil. Trans. R. Soc., Ser. B*, **256**, 401–428.

ZACHARUK, R. Y. AND BLUE, S. G. (1971), Ultrastructure of the peg and hair sensilla on the antenna of larval *Aedes aegypti* (L.), *J. Morph.*, **135**, 433–445.
ZACHARUK, R. Y., YIN, L. R. AND BLUE, S. G. (1971), Fine structure of the antenna and its sensory cone in larvae of *Aedes aegypti* (L.), *J. Morph.*, **135**, 273–298.
ZETTLER, F. AND JÄRVILEHTO, M. (1972), Lateral inhibition in an insect eye, *Z vergl. Physiol.*, **76**, 233–244.

Chapter 11

THE SOUND- AND LIGHT- PRODUCING ORGANS

Sound-producing Organs

Sounds of different kinds and intensities are produced by a number of species in all the main orders of insects; for general reviews see Alexander (1967), Dumortier (1963*a*, *b* and *c*), Frings and Frings (1958, 1960), Haskell (1961), Michelsen and Nocke (1974) and Bennet-Clark (1975). In some cases, the sound-producing organs are similar in the two sexes (as in many Coleoptera) but often they are confined to, or more strongly developed in, the male (e.g. most Orthoptera and Cicadidae). The biological significance of the sounds which are produced is not always clear but in some species they facilitate mating by attracting the sexes or stimulating the female and in other instances they may express sexual rivalry between males, subserve species-recognition (thus helping to keep members of the same species together), communicate warnings of danger or have a defensive function. A single species may make several sounds, each with its own function.

The methods by which sounds are produced may be classified under the following main headings.

(*a*) By tapping part of the body against an external object.
(*b*) By friction of one part of the body against another.
(*c*) By vibration of the wings or thoracic wall.
(*d*) By vibration of a special membrane through muscular action.
(*e*) By emission of air.

A. Sounds Produced by Tapping Part of the Body against an External Object – The best-known example of this type of sound-production occurs in the Death-watch Beetle, *Xestobium rufovillosum*, and allied Anobiidae where what is thought to be a sexual call is made by tapping the head against the walls of the burrow (Darwin, 1909). Some male Plecoptera rap the end of the abdomen against the substrate (Rupprecht, 1968) and the Psocopteran *Trogium pulsatorium* (the 'Lesser Death-watch') also makes a sound in this way (Pearman, 1928). Again, the pupae of a few

Hesperiidae and Lycaenidae produce what is probably a defensive sound by knocking the body against the walls of their cells or the substrate (Hinton, 1948). Finally, the soldiers of many species of termites (e.g. *Bellicositermes*) respond to slight mechanical disturbances of their nest by hammering their heads against the nest-structure, sometimes in rhythmic unison (Emerson, 1930). The resulting vibrations are transmitted through the substrate and probably act as a warning signal (Howse, 1964).

B. Sounds Produced by Friction of one Part of the Body against Another – Most of the sounds emitted by insects are produced by this method, the parts concerned being known as stridulating organs. Practically every external part of the body which is subjected to friction on an adjoining part has given rise to a stridulating organ in one or other insect.

Stridulating organs are possessed by representatives of several orders of insects, particularly the Orthoptera, Coleoptera and Hemiptera, but it is in the first mentioned order that they are best known (e.g. Faber, 1953; Haskell, 1957–60; Jacobs, 1953, and others cited below). In many of the Acridoidea, Tettigonioidea and Grylloidea the males are capable of vigorous stridulation; outside these three groups fewer other Orthoptera stridulate and the faculty is less often present in the females (Ragge, 1955; Spooner, 1968). Among the Acrididae the sounds are often produced by a row of pegs on the inner side of each hind femur being worked against the outer surface of each tegmen (Fig. 89 and see Evans, 1952; Loher, 1957; Weih, 1951; Otte, 1970).

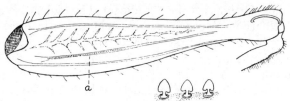

FIG. 89   Hind femur of an Acridid
a, row of pegs, three of which are shown greatly enlarged.

In some others (Oedipodinae) a row of prominences on a secondary tegminal vein is scraped by a ridge on the inner side of each hind femur. In the Tettigoniidae and Gryllidae the sound is produced when a row of teeth on the tegminal vein Cu_2 (the file) is scraped by a sclerotized part of the margin of the opposite tegmen, thus throwing into vibration certain areas of the tegmen (Stärk, 1958; Bailey, 1970; Bailey and Broughton, 1970; Nielsen and Dreisig, 1970; Bennet-Clark, 1970, and Nocke, 1971). Yet another type of stridulatory apparatus occurs in the Gryllacrididae where ridges on the side of the 2nd or 3rd abdominal segments are rubbed by the posterior femora and several other mechanisms peculiar to individual species of Orthoptera have been described (e.g. by Kevan, 1955). The sounds produced are often very characteristic of the species and each form may produce several types of sound including a normal song and a number of epigamic songs (Faber,

1929–36; Jacobs, 1950). Analysis of Orthopteran stridulation by electrical methods (Pierce, 1948 and much recent work) reveals the complex character of the sound, e.g. in *Gryllus assimilis* the normal song consists of a series of chirps at intervals of about $\frac{1}{3}$ of a second, each chirp comprising 2–6 very short pulses during which sound-waves with a frequency of about 5 kHz undergo amplitude modulation. In *Locusta migratoria* the greatest intensity of sound is emitted at about 12·5 kHz (Evans, *l.c.*) while artificial stimulation of the tegmen of *Homorocoryphus* produces an almost pure tone of 15 kHz (Bailey, *l.c.*). The songs of Orthoptera are generally characteristic of the species and sometimes allow morphologically very similar sibling species to be distinguished (e.g. Alexander and Thomas, 1959; Madsen *et al.*, 1970; Walker, 1964, 1969*a*, *b*). In a few well-studied cases the epigamic songs have been shown to ensure conspecific matings and thus maintain reproductive isolation in sympatric species (e.g. Perdeck, 1957).

Among Coleoptera, stridulatory organs have been reported from widely different parts of the body in a variety of species, both from larvae (e.g. *Geotrupes*) and adults (Gahan, 1900; Dudich, 1920–21). One part of the exoskeleton forms a file-like area (*pars stridulans*) which is rubbed by an adjacent region (*plectrum*) when the two undergo relative movement. The organs are generally inconspicuous and equally developed in both sexes; in a few species they are confined to one sex, the female in *Phanopate* but more usually the male (e.g. *Xenoderus, Cryptorrhynchus lapathi*). Stridulation sometimes has a sexual significance in beetles. The weevil *Rhynchaenus fagi* stridulates during courtship (Claridge, 1968) and in several species of the Scolytid *Ips* the females stridulate when entering the entrance tunnels made by the males; if they are stopped experimentally from stridulating the male prevents them from entering (Barr, 1969).

Among Hemiptera there is similarly a great diversity of stridulatory organs of this type, especially among Heteroptera (Leston, 1956), where both sexes or the male alone can produce sounds. The peculiar strigil of male Corixidae is apparently used in some unknown way to produce a sound in *Micronecta*, but in those other male Corixids which stridulate the noise is made by the passage of an area of spines on the femur over the angular side-margins of the face (von Mitis, 1935; Finke, 1968; Jansson, 1973).

Several Lepidoptera are known to be capable of stridulation (Hannemann, 1956). They usually produce a hissing or rustling sound and four mechanisms may be noted: (*a*) friction between the fore and hind wings, as in *Cidaria dotata* males (Hampson, 1894); (*b*) friction between legs and wings, as in some male Agaristinae (Hampson, 1892) and fertilized females of *Parnassius mnemosyne* (Jobling, 1936); (*c*) friction between adjacent parts of the thoracic wall (e.g. both sexes of the Arctiid *Rhodogastria*, where the sound is accompanied by the emission of froth through reflex bleeding and is probably defensive (Carpenter, 1938)); (*d*) friction between part of the 3rd abdominal sternum and an infolding of the adjacent sternopleural region, as in some male Lymantriids.

In the Hymenoptera stridulating organs are common among certain ants and vary in structure in different species and in the castes of the same species (Fig. 90). The organ commonly consists of a file on the first gastral tergum which is scraped by the posterior edge of the preceding tergum. Workers of *Atta cephalotes* produce sounds which are perceived by other workers only when conducted through the soil (with maximum intensity in the frequency range 1–3 kHz). In this way workers that have been accidentally buried can attract others that dig them out (Markl, 1967, 1968).

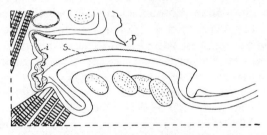

FIG. 90   Stridulating organ of *Myrmica rubra* (= *laevinodis*) in median section
p, edge of postpetiole forming a 'scraper'; *s*, stridulatory surface on first gastral segment; *i*, intersegmental membrane. *After* Janet.

C. Sounds Produced by Vibration of the Wings or Thoracic Wall – Many insects make a humming or buzzing sound during flight, the frequency of the emitted note appearing to some observers to be the same as that of the wing-beat. Sotavalta (1947) gives numerous examples, but in a more detailed physical analysis of the flight-tone of *Drosophila funebris* Williams and Galambos (1950) found that though each cycle of wing-movement corresponds to one of sound, there are, in addition to the fundamental frequency, prominent harmonics which result in wide departures from a simple sinusoidal type of vibration (see also Sotavalta, 1963). In some Diptera and Hymenoptera there is, in addition to the main wing-beat tone and its harmonics, an independent sound of higher frequency occurring on each half-cycle of the wing-beat and perhaps due to vibration of the thoracic exoskeleton (Esch and Wilson, 1967). This sound can also be produced when the wings are folded and in *Apis* a colony about to swarm will do so only after its members have received such a sound-signal from certain bees (Esch, 1967*b*). *Melipona* workers returning after a successful foraging flight can convey information on the site of a suitable food source by producing sounds, again with closed wings, at frequencies of 300–800 Hz, as compared with the normal flight tone of 200 Hz (Esch, 1967*a*). The males of various species of *Drosophila* emit characteristic courtship songs by special wing vibrations. These consist of regularly repeated pulses of sound and allow the females to recognize conspecific males (Bennet-Clark and Ewing, 1968; Ewing and Bennet-Clark, 1968).

D. Sounds Produced by the Vibration of a Special Membrane Exerted by Muscular Action – Sound-producing organs of this type are characteristic of the Auchenorrhynchan Homoptera and among the Cicadidae there is found one of the most complex kinds of sound-producing organs known. These structures are met with in the males, the females being either silent or only possessing rudiments of the apparatus. The great volume of sound emitted by the cicadas marks them out as being the noisiest representatives of the Insecta.

The apparatus comprises a pair of shell-like drums or tymbals situated at the base of the abdomen and vibrating by the action of powerful muscles

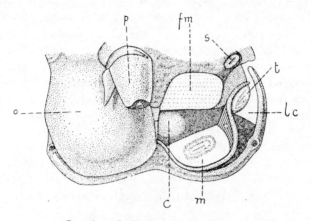

FIG. 91   Sound-producing apparatus of a cicada with the operculum of one side removed

c, ventral cavity; *fm*, folded membrane; *lc*, lateral cavity; *m*, tympanal organ; *o*, operculum, that of the other side removed; *p*, base of leg; *s*, spiracle; *t*, tymbal. *After* Carlet, *Ann. Sci. nat.*, Paris, 1887.

(Pringle, 1954, 1957). In *Magicicada septendecim* the true sound organs are freely exposed, but in many other cicadas the drums are covered by overlapping *opercula*, a pair of large plates which are backward extension of the metathoracic epimera and situated on the ventral side of the body, where they overlap the base of the abdomen (Fig. 91). On removing an operculum a pair of cavities containing the external parts of the sound-producing apparatus is disclosed. The larger of these cavities is ventral, and the smaller is lateral in position. Their walls contain three specialized areas of membrane which are known respectively as the *tymbal*, the *folded membrane* and the *mirror*. The *tymbal* is a crisp, plaited membrane surrounded by a sclerotized ring; it forms part of the inner wall of the lateral cavity, and is somewhat shell-like in appearance with its convex surface bulging outwards. The *folded membrane* is in the anterior wall of the ventral cavity, while the *mirror* is a tense, transparent membrane in the posterior wall of that cavity. In close association with the whole apparatus there is an extensive air-chamber which opens to the exterior by the third pair of spiracles. The sound is produced by the rapid in and out movement of the tymbal, which is brought about by a powerful muscle arising from the 2nd abdominal sternum and attached to the inner face of the tymbal. When the muscle contracts the tymbal is pulled inwards; on relaxation of the muscle the tymbal regains its former position by virtue of the elasticity of its cuticular ring. This method of sound production has been compared to the pushing in and out of the bottom of a tin vessel, which makes a cracking sound.

Electrophysiological analysis shows the patterns of sound produced by cicadas to be relatively complex (Pringle, *op. cit.*, Aidley, 1969, Young, 1972). The tymbal muscles may contract simultaneously or alternately and they may be synchronous or asynchronous with respect to the nerve-impulses reaching them. The sound waves, at frequencies from 850 to several thousand Hz, are amplitude-modulated to produce pulses repeated at the rate of 40 to over 1000 per second. In *Abricta* and *Magicicada* the strongly-developed ribs of the tymbal buckle separately during the 'in' movement, so that each muscle contraction produces 7–9 sound pulses. Among Australian cicadas the enlarged abdomen of *Cystosoma saundersi* acts as a resonating chamber, while in *Arunta perulata* the enlarged opercula impart strong directional properties to the sound. Cicada songs differ appreciably from one species to another (see also Alexander and Moore, 1958) and their function is apparently to bring the individuals together and to attract females to males (Myers, 1929), though distress songs have also been described.

Ossiannilsson (1949) has found that males in many other Auchenorrhynchan families (Delphacidae, Cixiidae, Cercopidae, Membracidae, Cicadellidae) have small stridulatory organs essentially similar to those of cicadas and more weakly developed organs even occur in some female Cercopids and Cicadellids. The sounds produced by these insects are always of low intensity and not always audible to the unaided ear, though several types of song (courtship, rivalry, distress, etc.) are said to be distinguishable. See also Strübing (1962, 1965) and Claridge and Howse (1968).

A pair of sound-producing tymbals also occurs in the Arctiid moth *Melese laodamia*, where each organ is a swollen, air-filled structure formed externally from the metathoracic katepisternum, with 15–20 striae on its outer face. It can be compressed by the basalar muscles, when the striae act as an array of 'microtymbals', each producing a brief pulse of sound. The overall result is a pattern of ultrasonic sounds generated between 30 and 90 kHz with a pulse repetition frequency of 1200 per second (Blest, Collett and Pye, 1963). Similar organs occur in some other Arctiids which are also known to be distasteful and to exhibit a protective display when stimulated (Blest, 1964).

E. Sounds Associated with the Emission of Air – Older authors considered that the humming of bees and flies was due to the vibration of lamellae just inside the thoracic spiracles but this idea has now been abandoned. More recently, the 'piping' of newly-emerged queens of *Apis mellifera* has been ascribed to the emission of air from the spiracles (Woods, 1956) though it is perhaps more likely to be due to thoracic vibration of the kind described above in section C (Simpson, 1964). The cry of the Death's Head Hawkmoth, *Acherontia atropos*, has puzzled many investigators but is probably due, as Prell (1920) first showed, to the emission of air from the pharynx near the base of the proboscis. Two sounds are emitted: a low note accompanied by vibration of the epipharynx and a high-pitched whistle that occurs when air passes out freely (Busnel and Dumortier, 1959).

The Light-Producing or Photogenic Organs

Certain insects are self-luminous owing to the possession of special photogenic organs; others owe their luminous properties to the presence of light-producing bacteria, or through having ingested luminous food (McElroy, 1964). True luminous insects are almost confined to the order Coleoptera and more particularly to various genera of Lampyridae and Cantharidae, notably *Lampyris, Luciola, Phosphaenus, Photuris, Photinus, Phengodes* as well as the Elaterid genera *Pyrophorus* (the 'cucujos') and *Photophorus*. Outside the Coleoptera, the Mycetophilids *Arachnocampa luminosa* and *Ceroplatus testaceus* have photogenic powers (Stammer, 1932; Gatenby, 1960). Their luminous larvae spin webs in which they catch prey, presumably attracted by the light. In *Arachnocampa* the light of older female pupae and adult females attracts the males (which are also luminous, though their light does not seem to be used epigamically); see Richards, A. M. (1960). Other examples of luminescence from the Collembola, Fulgoridae and a few other groups are not sufficiently understood (Harvey, 1957).

In the Lampyridae the luminosity is known in some species to extend to all the development stages, and is a character of their cytoplasm. In the egg the luminous substance is diffused, but in the postembryonic phases it is localized in the photogenic organs. The latter structures vary greatly in size, shape and position in different species and according to the sex and developmental stage. In *Phengodes* and *Phrixothrix*, for example, there are 11 or 12 segmentally arranged pairs of organs while in many species (e.g. *Lampyris noctiluca* females) they are found principally on the 6th and 7th abdominal sterna and in *Pyrophorus* they occur on the pronotum and anterior abdominal sterna. They may be equally developed in both sexes, as in *Luciola* and the Elateridae, or the females, which are sometimes apterous and larviform, may be the sole or principal luminous sex.

According to the species concerned, the light may be emitted as a continuous glow, an intermittent or pulsating glow of variable intensity or periodic short flashes whose characteristics are influenced by environmental factors e.g. Seliger *et al.* (1964). The spectacular synchronous flashing of Lampyrids (Buck and Buck, 1968; Baldaccini *et al.*, 1969) is not fully explained. The light occupies wavelengths over the range 5000 to 6500 Å and is usually yellowish-green though in *Pyrophorus* the abdominal organs emit a reddish light in flight and the thoracic ones a green light when the insect is not flying. The light is usually thought to assist in the attraction of the sexes for mating (e.g. Papi, 1969) and two main systems of communication have been distinguished (Lloyd, 1966, 1971). In some species one sex (usually the female) is sedentary and emits a species-specific light-signal that attracts the opposite sex (e.g. *Lampyris noctiluca*, *Phengodes laticollis*). In others the

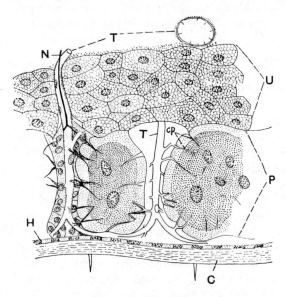

FIG. 92   Luminous organ of *Photinus*

C, cuticle; *cp*, tracheoles; H, epidermis; N, nerve; P,
photocyte layer; T, tracheae; U, reflector layer. *After*
Williams, 1916.

flying male emits a specific signal, the female responds by another signal and the male then flies towards her, as in *Photuris* and *Photinus*. In the synchronously flashing fireflies of S.E. Asia (*Pteroptyx*; Ballantyne and McLean, 1970) aggregations of synchronizing males attract both males and females, after which the females may respond and attract the males.

The organs are located beneath areas of transparent cuticle and their histological structure shows various degrees of complexity from those of *Phengodes*, where the photogenic structures (photocytes) are large independent cells similar to oenocytes, to those where a compact organ (Fig. 92) is composed of masses of photocytes arranged in rosette-like fashion around tracheae from which short trunks run into tracheal end-cells. From the latter arise numerous tracheoles which penetrate between the photogenic cells while the inner surface of the organ is covered by a layer of cells containing inclusions thought to be urates. These form the so-called reflecting layer, the function of which is uncertain.

The photocytes are packed with granules about 2·5 μm in diameter and the organ contains specialized nerve terminals. The axoplasm of these possess granules which are structurally identical with synaptic vesicles and neurosecretory material and the nerves end in association with the membranes of the photocytes and the tracheolar and tracheal end-cells. For details see Buck (1948), Smith (1963), Peterson and Buck (1968) and Peterson (1970).

As in other luminous animals and plants, the reactions leading to luminescence begin when an enzyme, *luciferase*, in the form of a complex with adenosine triphosphate, combines with reduced *luciferin* to form a further complex which is itself oxidized by molecular oxygen with the emission of light. The oxidized product is very stable but luciferase can be freed from it by pyrophosphate, thus providing material to start the cycle again (McElroy and Seliger, 1963; McElroy, Seliger and White, 1970). The flashing that occurs in luminous organs of complex histological structure is subject to nervous control, perhaps depending ultimately on the release of pyrophosphate in the photocytes (Carlson, 1969).

Literature on Sound- and Light-Production

On Sound-Production:

AIDLEY, D. J. (1969), Sound production in a Brazilian cicada, *J. exp. Biol.*, **51**, 325–337.

ALEXANDER, R. D. (1967), Acoustical communication in arthropods, *A. Rev. Ent.*, **12**, 495–526.

ALEXANDER, R. D. AND MOORE, T. E. (1958), Studies on the acoustical behaviour of the 17-year cicada, *Ohio J. Sci.*, **58**, 107–127.

ALEXANDER, R. D. AND THOMAS, E. S. (1959), Systematic and behavioral studies on the crickets of the *Nemobius fasciatus* group (Orthoptera, Nemobiinae), *Ann. ent. Soc. Am.*, **52**, 591–605.

BAILEY, W. J. (1970), The mechanics of stridulation in bush crickets (Tettigoniidae, Orthoptera). 1. The tegminal generator, *J. exp. Biol.*, **52**, 495–505.

BAILEY, W. J. AND BROUGHTON, W. B. (1970), The mechanics of stridulation in bush crickets (Tettigoniidae, Orthoptera). 2. Conditions for resonance in the tegminal generator, *J. exp. Biol.*, **52**, 507–517.

BARR, B. A. (1969), Sound production in Scolytidae (Coleoptera) with emphasis on the genus *Ips*, *Can. Ent.*, **101**, 636–672.

BENNET-CLARK, H. C. (1970), The mechanism and efficiency of sound production in mole crickets, *J. exp. Biol.*, **52**, 619–652.

—— (1975), Sound production in insects, *Sci. Progr.*, **62**, 263–283.

BENNET-CLARK, H. C. AND EWING, A. W. (1968), The wing mechanism involved in the courtship of *Drosophila*, *J. exp. Biol.*, **49**, 117–128.

BLEST, A. D. (1964), Protective display and sound production in some New World Arctiid and Ctenuchid moths, *Zoologica, N.Y.*, **49**, 161–181.

BLEST, A. D., COLLETT, T. S. AND PYE, J. D. (1963), The generation of ultrasonic signals by a New World Arctiid moth, *Proc. Roy. Soc.* (B), **158**, 196–207.

BUSNEL, R. G. (1963), *Acoustic Behaviour of Animals*, Elsevier, London and New York, 933 pp.

BUSNEL, R. G. AND DUMORTIER, B. (1959), Vérification par les méthodes d'analyse acoustique des hypothèses sur l'origine du cri du Sphinx *Acherontia atropos* (Linné), *Bull. Soc. ent. Fr.*, **64**, 44–58.

CARPENTER, G. D. H. (1938), Audible emission of defensive froth by insects, *Proc. zool. Soc. Lond.* (A), **108**, 243–252.

CLARIDGE, L. C. (1968), Sound production in species of *Rhynchaenus* (=*Orchestes*) (Coleoptera: Curculionidae), *Trans. R. ent. Soc. Lond.*, **120**, 287–296.

CLARIDGE, M. F. AND HOWSE, P. E. (1968), Songs of some British *Oncopsis* species (Hemiptera: Cicadellidae), *Proc. R. ent. Soc. Lond.* (A), **43**, 57–61.

DARWIN, C. (1909), *The Descent of Man and Selection in relation to Sex*, Murray, London, 301 pp.

DUDICH, E. (1920–22), Über den Stridulationsapparat einiger Käfer, *Ent. Blätter*, **16**, 146–161; **17**, 136–140; 145–155.

DUMORTIER, B. (1963*a*), Morphology of sound emission apparatus in Arthropoda, In: Busnel, R. G. (ed.), *Acoustic Behaviour of Animals*, pp. 277–345.

—— (1963*b*), The physical characteristics of sound-emissions in Arthropoda, In: Busnel, R. G. (ed.), *Acoustic Behaviour of Animals*, pp. 346–373.

—— (1963*c*), Ethological and physiological study of sound emissions in Arthropoda, In: Busnel, R. G. (ed.), *Acoustic Behaviour of Animals*, pp. 583–654.

EMERSON, A. E. (1930), Communication among termites, *Trans. 4th int. Congr. Ent.*, **2**, 722–727.

ESCH, H. (1967*a*), Die Bedeutung der Lauterzeugung für die Verständigung der stachellosen Bienen, *Z. vergl. Physiol.*, **56**, 199–220.

—— (1967*b*), The sounds produced by swarming honey bees, *Z. vergl. Physiol.*, **56**, 408–411.

ESCH, H. AND WILSON, D. (1967), The sounds produced by flies and bees, *Z. vergl. Physiol.*, **54**, 256–267.

EVANS, E. J. (1952), The stridulation noise of locusts, *Proc. R. ent. Soc. Lond.* (A), **27**, 39–42.

EWING, A. W. AND BENNET-CLARK, H. C. (1968), The courtship songs of *Drosophila*, *Behaviour*, **31**, 288–301.

FABER, A. (1929–32), Die Lautäusserungen der Orthopteren. I, II, *Z. Morph. Ökol. Tiere*, **13**, 745–803; **26**, 16–93.

—— (1936), Die Lautäusserungen und Bewegungsäusserungen der Oedipodinae, *Z. wiss. Zool.*, **149**, 1–85.

—— (1953), Laut- und Gebärdensprache bei Insekten: Orthoptera. Teil 1, Vergleichende Darstellung von Ausdrucksformen als Zeitgestalten und ihre Funktionen, *Ges. Fr. Mus. Naturk.*, **1953**, 198 pp.

FINKE, C. (1968), Lautäusserungen und Verhalten von *Sigara striata* und *Callicorixa praeusta* (Corixidae Leach, Hydrocorisae Latr.), *Z. vergl. Physiol.*, **58**, 398–422.

FRINGS, H. AND FRINGS, M. (1958), Uses of sounds by insects, *A. Rev. Ent.*, **3**, 87–106.

FRINGS, M. AND FRINGS, H. (1960), *Sound production and sound perception by Insects. A Bibliography*, Pennsylvania State Univ. Press, Philadelphia, 108 pp.

GAHAN, C. J. (1900), Stridulating organs in Coleoptera, *Trans. ent. Soc. Lond.*, **1900**, 433–452.

HAMPSON, G. F. (1892), On stridulation in certain Lepidoptera, and on the distortion of the hindwings in the males of certain Ommatophorinae, *Proc. zool. Soc. Lond.*, **1892**, 188–189.

—— (1894), [No title], *Proc. ent. Soc. Lond.*, **1894**, xiii–xiv.

HANNEMANN, H. I. (1956), Über ptero-tarsale Stridulation und einige andere Arten der Lauterzeugung bei Lepidopteren, *Dt. ent. Z.*, **3**, 14–27.

HASKELL, P. T. (1957), Stridulation and associated behaviour in certain Orthoptera. I. Analysis of the stridulation of, and behaviour between, males, *Br. J. Anim. Behav.*, **5**, 139–148.

—— (1958), Stridulation and associated behaviour in certain Orthoptera. 2. Stridulation of females and their behaviour with males, *Anim. Behav.*, **6**, 27–42.

—— (1960), Stridulation and associated behaviour in certain Orthoptera. 3. The influence of gonads, *Anim. Behav.*, **8**, 76–81.

—— (1961), *Insect Sounds*, Witherby, London, 189 pp.

HINTON, H. E. (1948), Sound production in Lepidopterous pupae, *Entomologist*, **81**, 254–269.

HOWSE, P. E. (1964), The significance of the sound produced by the termite *Zootermopsis angusticollis* (Hagen), *Anim. Behav.*, **12**, 284–300.

JACOBS, W. (1950), Vergleichende Verhaltensstudien an Feldheuschrecken, *Z. Tierpsychol.*, **7**, 169–216.

—— (1953), Verhaltensbiologische Studien an Feldheuschrecken, *Z. Tierpsychol.*, Beiheft 1, 228 pp.

JANSSON, A. (1973), Stridulation and its significance in the genus *Cenocorixa* (Hemiptera, Corixidae), *Behaviour*, **46**, 1–36.

JOBLING, B. (1936), On the stridulation of the females of *Parnassius mnemosyne* L., *Proc. R. ent. Soc. Lond.* (A), **11**, 66–68.

KEVAN, D. K. MCE. (1955), Méthodes inhabituelles de production de son chez les Orthoptères, In: Busnel, R.-G. (ed.), Colloque sur l'Acoustique des Orthoptères. Fasc. hors série des *Ann. Epiphyt.*, 448 pp.

LESTON, D. (1956), Stridulatory mechanisms in terrestrial species of Hemiptera Heteroptera, *Proc. zool. Soc. Lond.*, **128**, 369–386.

LOHER, W. (1957), Untersuchungen über den Aufbau und die Entstehung der Gesänge einiger Feldheuschreckenarten und der Einfluss von Lautzeichen auf das akustische Verhalten, *Z. vergl. Physiol.*, **39**, 313–356.

MADSEN, C., VICKERY, V. R. AND NOWOSIELSKI, J. (1970), Variation in song patterns of Antipodean *Teleogryllus* species (Orthoptera: Gryllidae) and a proposed phenetic classification, *Can. J. Zool.*, **48**, 797–801.

MARKL, H. (1967), Die Verständigung durch Stridulationssignale bei Blattschneiderameisen. I. Die biologische Bedeutung der Stridulation, *Z. vergl. Physiol.*, **57**, 299–330.

—— (1968), Die Verständigung durch Stridulationssignale bei Blattschneiderameisen. II. Erzeugung und Eigenschaften der Signale, *Z. vergl. Physiol.*, **60**, 103–150.

MICHELSEN, A. AND NOCKE, H. (1974), Biophysical aspects of sound communication in insects, *Adv. Insect Physiol.*, **10**, 247–296.

MITIS, H. von (1935), Zur Biologie der Corixiden. Stridulation, *Z. Morph. Ökol. Tiere*, **30**, 479–495.

MYERS, J. G. (1929), *Insect Singers: A Natural History of Cicadas*, Routledge, London, 304 pp.

NIELSEN, E. T. AND DREISIG, H. (1970), The behavior of stridulation in Orthoptera Ensifera, *Behaviour*, 37, 205–252.

NOCKE, H. (1971), Biophysik der Schallerzeugung durch die Vorderflügel der Grillen, *Z. vergl. Physiol.*, 74, 272–314.

OSSIANNILSSON, F. (1949), Insect drummers, *Opusc. ent.*, Suppl., 10, 145 pp.

OTTE, D. (1970), A comparative study of communicative behavior in grasshoppers, *Misc. Publs Mus. Zool. Univ. Mich.*, 141, 168 pp.

PEARMAN, J. V. (1928), On sound production in the Psocoptera and on a presumed stridulatory organ, *Entomologist's mon. Mag.*, 64, 179–186.

PERDECK, A. C. (1957), The isolating value of specific song patterns in two sibling species of grasshoppers (*Chorthippus brunneus* Thunb. and *C. biguttulus* L.), *Behaviour*, 12, 1–75.

PIERCE, G. W. (1948), *The Songs of Insects*, Cambridge, Mass., 329 pp.

PRELL, H. (1920), Die Stimme des Totenkopfes (*Acherontia atropos* L.), *Zool. Jb.* (*Syst.*), 42, 235–272.

PRINGLE, J. W. S. (1954), A physiological analysis of cicada song, *J. exp. Biol.*, 31, 525–560.

—— (1957), The structure and evolution of the organs of sound-production in cicadas, *Proc. Linn. Soc. Lond.*, 167, 144–159.

RAGGE, D. R. (1955), Le problème de la stridulation des femelles Acridinae (Orthoptera, Acrididae), *Colloque sur l'Acoustique des Orthoptères*. Fascicule hors de série des *Ann. Epiphyt.*, 448 pp.

RUPPRECHT, R. (1968), Das Trommeln der Plecopteren, *Z. vergl. Physiol.*, 59, 38–71.

SIMPSON, J. (1964), The mechanism of honey-bee queen piping, *Z. vergl. Physiol.*, 48, 277–282.

SOTAVALTA, P. (1947), The flight-tone (wing-stroke frequency) of insects, *Acta ent. fenn.*, 4, 1–117.

—— (1963), The flight-sounds of insects, In: Busnel, R.-G. (1960) (ed.), *Acoustic Behaviour of Animals*, Elsevier, pp. 374–390.

SPOONER, J. D. (1968), Pair-forming acoustic systems of phaneropterine katydids (Orthoptera, Tettigoniidae), *Anim. Behav.*, 16, 197–212.

STÄRK, A. A. (1958), Untersuchungen am Lautorgan einiger Grillen- und Laubheuschrecken-Arten, zugleich ein Beitrag zum Rechts-Links-Problem, *Zool. Jb.* (*Anat.*), 77, 9–50.

STRÜBING, H. (1962), Paarungsverhalten und Lautäusserung von Kleinzikaden, demonstriert an Beispielen aus der Familie der Delphacidae, *Verh. XI int. Kongr. Ent.*, 3, 12–14.

—— (1965), Das Lautverhalten von *Euscelis plebejus* Fall. und *Euscelis ohausi* Wagn. (Homoptera–Cicadina), *Zool. Beitr.* (N.F.), 11, 289–341.

WALKER, T. J. (1964), Cryptic species among sound-producing ensiferan Orthoptera (Gryllidae and Tettigoniidae), *Q. Rev. Biol.*, 39, 345–355.

—— (1969a), Systematics and acoustic behaviour of United States crickets of the genus *Orocharis* (Orthoptera, Gryllidae), *Ann. ent. Soc. Am.*, 62, 752–762.

—— (1969b), Systematics and acoustic behavior of United States crickets of the genus *Cyrtoxipha* (Orthoptera: Gryllidae), *Ann. ent. Soc. Am.*, 62, 945–952.

WEIH, A. S. (1951), Untersuchungen über das Wechselsingen (Anaphonie) und über das angeborene Lautschema einiger Feldheuschrecken. Z. Tierpsychol., 8, 1–41.
WILLIAMS, C. M. AND GALAMBOS, R. (1950), Oscilloscopic and stroboscopic analysis of the flight sounds of Drosophila, Biol. Bull., 99, 300–307.
WOODS, E. F. (1956), Queen piping, Bee World, 37, 185.
YOUNG, D. (1972), Neuromuscular mechanism of sound production in Australian cicadas, J. comp. Physiol., 79, 343–362.

On Light-Production:

BALDACCINI, N. E., FIASCHI, V. AND PAPI, F. (1969), Rhythmic synchronous flashing in a Bosnian firefly, Monitore zool. ital., 3, 239–245.
BALLANTYNE, L. A. AND MCLEAN, M. R. (1970), Revisional studies on the firefly genus Pteroptyx Olivier (Coleoptera: Lampyridae: Luciolinae: Luciolini), Trans. Am. ent. Soc., 96, 223–305.
BUCK, J. B. (1948), The anatomy and physiology of the light organ in fireflies, Ann. N.Y. Acad. Sci., 49, 397–482.
BUCK, J. AND BUCK, E. (1968), Mechanism of rhythmic synchronous flashing of fireflies, Science, N.Y., 159, 1319–1327.
CARLSON, A. D. (1969), Neural control of firefly luminescence, Adv. Insect Physiol., 6, 51–96.
GATENBY, J. B. (1960), The Australasian mycetophilid glow-worms, Trans. R. Soc. N.Z., 88, 577–593.
HARVEY, E. N. (1957), A History of Luminescence, Am. Phil. Soc., Philadelphia, 692 pp.
LLOYD, J. E. (1966), Studies on the flash communication system in Photinus fireflies, Misc. Publ. Mus. Zool. Univ. Mich., 130, 1–95.
—— (1971), Bioluminescent communication in insects, A. Rev. Ent., 16, 97–122.
MCELROY, W. D. (1964), Insect bioluminescence, In: Rockstein, M. (ed.), The Physiology of Insecta, 1, 463–508.
MCELROY, W. D. AND SELIGER, H. H. (1963), The chemistry of light emission, Adv. Enzymol., 25, 119–166.
MCELROY, W. D., SELIGER, H. H. AND WHITE, E. H. (1970), Mechanism of bioluminescence, chemiluminescence and enzyme function in the oxidation of firefly luciferin, Photochem. Photobiol., 10, 153–170.
PAPI, F. (1969), Light emission, sex attraction and male flash dialogues in a firefly, Luciola lusitanica (Charp.), Monitore zool. ital., 3, 135–184.
PETERSON, M. K. (1970), The fine structure of the larval firefly light organ, J. Morph., 131, 103–115.
PETERSON, M. K. AND BUCK, J. (1968), Light organ fine structure in certain Asiatic fireflies, Biol. Bull., 135, 335–348.
RICHARDS, A. M. (1960), Observations on the New Zealand glow-worm Arachnocampa luminosa (Skuse 1890), Trans. R. Soc. N.Z., 88, 559–574.
SELIGER, H. H., BUCK, J. B., FASTIE, W. G. AND MCELROY, W. D. (1964), Flash patterns in Jamaican fireflies, Biol. Bull., 127, 159–172.
SMITH, D. S. (1963), The organization and innervation of the luminescent organ in a firefly, Photuris pennsylvanica (Coleoptera), J. Cell Biol., 16, 323–359.
STAMMER, H. J. (1932), Zur Biologie und Anatomie der leuchtenden Pilzmückenlarve von Ceroplatus testaceus Dalm. (Diptera, Fungivoridae), Z. Morph. Ökol. Tiere, 26, 135–146.

Chapter 12

THE ALIMENTARY CANAL, NUTRITION AND DIGESTION

The Alimentary Canal

The alimentary canal is a tube of very variable length; in some cases it is about equal to that of the body, while in others it is much longer and convoluted. The shortest and simplest type is found in many larvae, notably those of the Lepidoptera, Hymenoptera and Diptera-Nematocera; in the Apterygotes, Dermaptera and certain Orthoptera this condition is maintained throughout life. In Homoptera, and larval Diptera Cyclorrhapha, it attains its greatest length and number of convolutions and is often several times longer than the whole insect. As a rule, the great length of digestive canal occurs in insects which feed on juices, rather than on the more solid tissues of animals and plants. Exceptions, however, are found in the larval Hymenoptera, whose food is entirely liquid, and whose digestive canal is a straight, simple tube. Morphologically the alimentary canal is divisible into three primary regions according to their method of embryonic origin (Fig. 93). The *fore gut* arises as an anterior ectodermal invagination (stomodaeum); the hind gut as a similar posterior invagination (proctodaeum); and the *mid gut*, which ultimately connects the two, develops as what is probably an endodermal sac (mesenteron) (p. 339). These differences in embryonic origin result in marked histological differences in the structure of the mid gut, as compared with the other regions. Both the fore and hind gut, being invaginations of the body-wall, resemble the latter in their essential histology and are lined with cuticle.

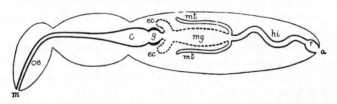

F I G. 93   Diagram of the digestive system of an insect

    The ectodermal parts are represented by heavy lines and the endodermal parts by broken lines.

    m, mouth; *oe*, oesophagus; *c*, crop; *g*, gizzard; *ec*, enteric caeca; *mg*, mid gut; *mt*, Malpighian tubules (probably endodermal); *hi*, hind gut; *r*, rectum; *a*, anus.

(*a*) **The Fore Gut** – The following layers, passing from within outwards, are generally recognizable in the walls of the fore gut (Fig. 94). 1. The *intima* or innermost lining, which is a cuticular layer directly continuous with the cuticle of the body-wall. 2. The *epithelial layer*, continuous with the epidermis and secreting the intima; it is often extremely thin. 3. The *basement membrane*, bounding the outer surface of the epithelium. 4. The longitudinal muscles. 5. The *circular muscles*. 6. The *peritoneal membrane* which consists of apparently structureless connective tissue and is often difficult to detect. Small amounts of connective tissue may also lie among the muscles. The fore gut is divisible into the following regions:

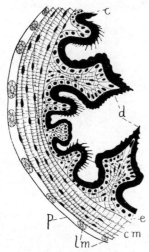

FIG. 94
Transverse section of the wall of the gizzard of a Tettigoniid (*Decticus albifrons*)

c, cuticular lining; *d*, teeth; *e*, epithelium; *cm*, circular muscles; *lm*, longitudinal muscles; *p*, peritoneal membrane. *After* Bordas.

The *preoral food cavity* is the space lying between the mouthparts and the labrum and is not, strictly speaking, a part of the gut. In insects with mandibulate mouthparts this space is divided by the hypopharynx into an anterior or dorsal *cibarium* and a posterior or ventral *salivarium*. The cibarium, whose walls are connected to the post-clypeus by the cibarial dilator muscles, may form only a small pouch for the temporary storage of food or is modified into a sucking-pump as in the Thysanoptera, Hemiptera and others. The salivarium may also undergo modification to form the salivary syringe of the Hemiptera and the silk-regulator of Lepidopterous larvae.

The *pharynx* is the region between the mouth and the oesophagus. It is normally provided with dilator muscles which run from its dorsal surface to the frontal region of the head-capsule and are separated from the cibarial dilator muscles by the frontal ganglion of the stomatogastric nervous system. These muscles are best developed where the pharynx participates in the formation of a well-developed sucking-pump (Lepidoptera, Hymenoptera, Neuroptera, Dytiscidae).

The *oesophagus* is a simple, straight tube passing from the hinder region of the head into the fore part of the thorax. It is very variable in length and the inner walls are longitudinally folded.

The *crop* is present in many insects and is usually a dilation of the hinder portion of the oesophagus. It is extremely variable in form, and functions mainly as a food reservoir, though digestion occurs when its contents are

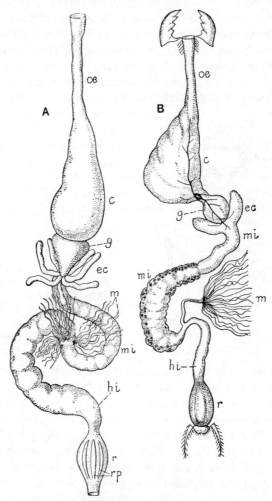

FIG. 95   A. Alimentary canal of *Periplaneta americana*.
B. Alimentary canal of *Nemobius sylvestris* (Gryllidae)

oe, oesophagus; *c*, crop; *g*, gizzard; *ec*, enteric caeca; *m*, Malpighian tubules; *mi*, mid gut; *hi*, hind gut; *r*, rectum; *rp*, rectal papillae. *After* Bordas, *Ann. Sci. nat., 8th ser.,* 5.

mixed with salivary enzymes and some lipids may be absorbed there (Eisner, 1955). In *Periplaneta* its movements are under nervous control and emptying depends on the osmotic pressure of its contents (Davey and Treherne, 1963). The walls of the crop are thin and its muscular coat weakly developed. In

most Orthoptera and Dictyoptera (Fig. 95), it is very capacious and constitutes the major portion of the fore gut. In a few insects it is developed as a
lateral dilation of one side of the oesophagus as in *Gryllotalpa*, certain
Isoptera and the larvae of *Myrmeleon* and the Curculionidae. Among various
sucking insects this dilation becomes greatly pronounced and connected with
the oesophagus by a slender tube. Such a food-reservoir is present in most
Diptera (Fig. 98) and also in the larvae of some of the Cyclorrhapha and in
the higher Lepidoptera (Fig. 99).

The *gizzard* or *proventriculus* is situated behind the crop and is best
developed in the Orthopteroid orders and the Coleoptera (Judd, 1948; Thiel,
1936; Balfour-Browne, 1944) (Fig. 95). It is also found in the Mecoptera,
Odonata, Isoptera and various Hymenoptera but is reduced to a valve in the
honey bee and most Diptera. The dominant features in its structure are the
great development of the cuticular lining into prominent denticles, and
increased thickness of its muscles. At the junction of the fore and mid gut,
there is often a *cardiac* or *oesophageal valve* formed by the wall of the fore gut
being prolonged into the cavity of the stomach and then reflected upon itself
and passing forwards to unite with the stomach-wall (Fig. 97). It shows
varying degrees of complexity and probably prevents or reduces regurgita-
tion of food from the mid gut.

(*b*) **The Mid Gut** – This region is also termed the *stomach* or *ventriculus*
and its shape and capacity vary greatly. It may be sac-like, coiled and tubular,
or divided into two or more well-defined regions as in the Heteroptera and
many Cyclorrhapha (Fig. 98). Histologically the wall of the stomach exhibits
the following structure (Fig. 96). Internally it is lined by the *mid gut epithelium*, the outer ends of whose cells rest upon a *basement membrane*, the latter
is followed by an inner layer of *circular muscles* and an outer layer of *longitudinal muscles*. The outermost coat of the mid gut is a thin *peritoneal
membrane*. Both muscle layers are composed of striated fibres and their
positions are the reverse of those in the fore gut. The structure of the mid
gut epithelium requires more detailed mention. Three main types of cells
may be distinguished: (*a*) *Columnar (cylinder) cells*, (*b*) *Regenerative cells* and
(*c*) *Calyciform (goblet) cells*. The first-mentioned are those which are at times
actively involved in the secretion of enzymes and the absorption of the
products of digestion. In well-studied examples such as *Calliphora* (De
Priester, 1971), and the larvae of *Hyalophora* (Anderson and Harvey, 1966)
and *Ephestia* (Smith *et al.*, 1969) they have a complex ultrastructure. The
cytoplasm is divided basally into compartments by deep infoldings of the
plasma membrane and produced towards the lumen in an array of microvilli
(Fig. 96). These increase the surface area greatly and are covered by indistinct material that forms the *glycocalyx*; they make up the 'brush border' of
the older microscopists. Large amounts of rough endoplasmic reticulum are
present in the columnar cells, with microtubules and mitochondria, the latter
arranged longitudinally in the basal compartments but randomly scattered
elsewhere. Golgi structures are widely distributed but pinocytosis does not

seem to be important. Laterally the plasma membranes of adjacent cells may be linked by simple or septate desmosomes or are separated locally to form intercellular spaces. In *Lucilia* larvae the columnar cells comprise ultrastructurally distinct lipophilic and cuprophilic cells, the former with large inclusions of storage fat and a lamellate border (Waterhouse and Wright, 1960), while in *Periplaneta* there are histological differences between the so-called secretory and absorptive cells (Threadgold and Gresson, 1962). In larval Lepidoptera and Trichoptera and in some other insects the mid gut

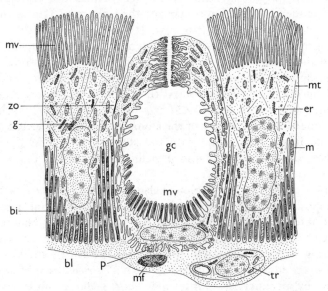

FIG. 96   Ultrastructure of mid gut epithelium of *Hyalophora cecropia* larva (*after* Anderson and Harvey, 1966). The figure shows a goblet cell, with a columnar cell on each side of it

bi, basal infoldings of columnar cell, associated with parallel arrays of mitochondria; *bl*, basal lamina; *er*, rough endoplasmic reticulum; *g*, Golgi body of columnar cell; *gc*, goblet cell cavity, opening into lumen by canal between apical cytoplasmic processes; *m*, mitochondrion; *mf*, muscle fibre; *mt*, microtubule; *mv*, microvilli; *p*, podocyte-like processes at base of goblet cell; *tr*, tracheolar cell; *zo*, zona occludens (junction between columnar and goblet cells).

epithelium contains *goblet cells* with less rough endoplasmic reticulum, less regular microvilli and a highly folded plasma membrane that encloses the invaginated or vacuole-like goblet chamber. They may be involved in active transport of potassium ions across the gut wall (Anderson and Harvey, *op. cit.*). The replacement or regenerative cells, which are absent in a few groups, may be scattered singly or in pairs beneath the columnar cells, or grouped into clusters (nidi) there, or variously arranged in crypt-like outpocketings of the mid gut. Their function is to renew the other epithelial cells when these are destroyed through secretion or in the large-scale processes of degeneration which accompany moulting or pupation in certain species.

In many insects the surface area of the mid gut is increased by sac-like diverticula – the *enteric* or *gastric caeca* (Fig. 95). These are usually situated near the oesophagus and vary in number. In certain Dipterous larvae and in the Gryllidae and Tettigoniidae two large caeca are present; in Dictyoptera and larval Culicidae there are eight, while in the larvae of Scarabaeidae they are more numerous and are disposed in an anterior, a middle, and a posterior annular series. Among various predacious Coleoptera they are represented

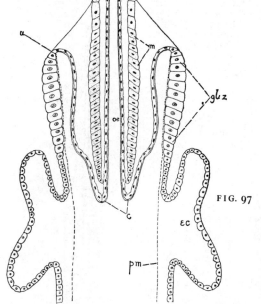

FIG. 97 Cardiac valve of a Dipteran larva (Nematocera) with the adjacent region of the mid gut, seen in longitudinal section

oe, oesophagus; *c*, cuticular intima; *m*, muscles; *u*, point of union of fore and mid gut; *glz*, zone of columnar gland cells which secrete the peritrophic membrane *pm*; *ec*, enteric caecum.

by numerous villiform processes, and in some orders (e.g. Collembola, Lepidoptera) caeca are generally wanting.

In the larvae of certain groups of insects the mid gut is closed posteriorly. In these instances the food is liquid and there is little solid residuum. This condition is prevalent in most larvae of the Hymenoptera Apocrita, and in those of the Neuroptera Plannipennia, *Glossina*, and other viviparous Diptera. The elaborate modifications of the mid gut which occur in the filter-chamber of many Homoptera are discussed on p. 693.

For further information on the histology and ultrastructure of the mid gut, see Baccetti (1960), Forbes (1964), Hecker *et al.* (1971*a*, *b*), and others cited below in connection with digestion and absorption.

Food particles in the mid gut are generally separated from the epithelium by being enclosed in a tubular sheath, the *peritrophic membrane*. This consists of a mucoprotein in which chitin fibrils are arranged irregularly or in hexagonal or orthogonal arrays (Peters, 1968, 1969; Platzer-Schultz and Welsch, 1969). Peritrophic membranes protect the mid gut from abrasion by food particles and are said to be absent from most insects feeding on a liquid diet, e.g. the Hemiptera except Corixidae (Sutton, 1951), and many adult Lepidoptera and blood-sucking insects. However, they occur in *Cicadella* (Gouraton and Maillet, 1965), mosquitoes and *Glossina* (Moloo *et al.*, 1970; Freeman, 1973) and they may have been overlooked elsewhere (Waterhouse, 1953*b*). The peritrophic membrane has been considered to arise in two ways (Wigglesworth, 1930; Waterhouse, 1953*a*): (*a*) In some Lepidoptera, Diptera

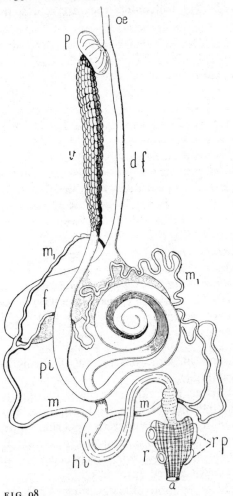

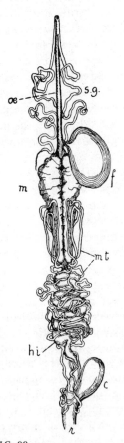

FIG. 98
Alimentary canal of *Calliphora*

oe, oesophagus; *p*, proventriculus; *v*, mid gut; *df*, duct of food-reservoir *f*; *pi*, proximal intestine; *m₁*, Malpighian tubules which unite to form a common duct (*m*) on either side; *hi*, hind gut; *r*, rectum; *rp*, rectal papillae; *a*, anus. Adapted from Lowne.

FIG. 99
Digestive system of *Sphinx ligustri* (imago)

oe, oesophagus; *sg*, salivary gland; *f*, crop; *m*, mid gut; *mt*, Malpighian tubules; *hi*, hind gut; *c*, caecum; *r*, rectum. *After* Newport.

(Fig. 97) and Dermaptera it is secreted by cells near the junction of fore and mid gut and extruded in tubular form by a muscular 'press' in this region; (*b*) in other insects it arises by delamination from part or all of the general surface of the mid gut and a series of concentric membranes, formed by successive delaminations, is often present. In *Apis* (Kusmenko, 1940) and *Periplaneta* (Lee, 1968) a membrane formed predominantly in the anterior part of the mid gut is supplemented by delamination further back.

(*c*) **The Hind Gut** – This consists of the same layers as the fore gut except that its circular muscles are developed to a varying degree both inside and outside the layer of longitudinal muscles. The commencement of the hind gut is normally marked by a pyloric valve and the insertion of the Malpighian tubules (see p. 248). In most insects three regions are recognizable: the *ileum, colon* and *rectum*. The cuticular lining of the ileum and colon is often thrown into folds and provided with hairlike or spiny projections; among certain Scarabaeid larvae the latter are highly developed and assume an arborescent form. The ileum may be very long as in *Dytiscus* and *Nicrophorus*, short as in many other insects, or it may be undifferentiated from the colon, as in many Orthoptera and Hemiptera. Among Lepidoptera, certain Coleoptera and others, a caecum arises from the colon; it is sac-like in *Sphinx ligustri* (Fig. 99) and many other Lepidoptera, while in *Dytiscus* it forms a tube nearly equal to the abdomen in length. The *rectum* is a more or less globular or pyriform chamber, generally provided with a variable number of inwardly projecting papillae. These (often called 'rectal glands') are composed either of a single layer of tall epithelial cells (Thysanura, Odonata, Orthoptera, Phasmida, etc.) or of two cell-layers, with or without a lumen (Neuroptera, Hymenoptera, Lepidoptera, Diptera). They are generally six in number, occasionally four (most Diptera, Thysanoptera) or very many (some Trichoptera, Lepidoptera). They do not occur in Hemiptera and the larvae of Endopterygotes and are found, among Coleoptera, only in the Silphidae and some Adephagan families (Palm, 1949; Reichenbach-Klinke, 1952). They are often richly tracheated though it seems unlikely that they have a respiratory function; they are, in fact, closely concerned with important processes of salt and water absorption that occur in the hind gut and their ultrastructure reflects these functions (Baccetti, 1962; Berridge, 1970; Berridge and Gupta, 1967; Gupta and Berridge, 1966; Noirot and Noirot-Timothée, 1971; Wall and Oschman, 1973). Two main systems are involved in the processes of hind gut absorption: the rectal papillae already mentioned and a structurally more elaborate cryptonephric system such as that of some beetles, in which the ends of the Malpighian tubules are closely applied to the rectum and enclosed with it in a perinephric space (Saini, 1964; Ramsay, 1964, 1971; Grimstone *et al.*, 1968; see also p. 248). In both cases the rectum can conserve water by absorbing it actively from the faeces into the haemolymph or the lumen of the Malpighian tubules (Phillips, 1964; Stobbart, 1968). The process is accompanied by the absorption of inorganic · ions (potassium, sodium and chloride) and the biophysical mechanisms are under active study. In *Schistocerca* neurosecretory substances control excretion partly by restricting the absorption of water through the rectal wall (Mordue, 1969).

Other important functions may also be exercised by the hind gut of some groups of insects. The rectal pouch of many Isoptera contains symbiotic Protozoa (p. 628) while in Anisopteran dragonfly nymphs there is a complex system of rectal gills (p. 508), and in Dytiscid larvae the capacious hind gut takes in water to split the old cuticle at ecdysis. Females of the Scolytid beetle *Trypodendron lineatum* produce an attractant pheromone from secretory cells near the junction of ileum and rectum (Schneider and Rudinsky, 1969) and in larvae of the Pyraustid moth *Ostrinia nubilalis* the cells of the ileum undergo cyclical changes perhaps indicating the secretion of a diapause-controlling hormone, *proctodone* (Hassemer and Beck, 1969; Alexander and Fahrenbach, 1969).

Nutrition

This large subject has been reviewed many times and only an outline of our present knowledge is given here. For full discussions of the various aspects see especially Lipke and Fraenkel, 1956; Legay, 1958; Friend, 1958; House, 1958, 1961; Dadd, 1963, 1970*a*, 1973; Gordon, 1968; Auclair, 1969; Haydak, 1970; and Rodriguez, 1973. Insects eat a great variety of foods (Brues, 1946) though the normal diets of many species are strictly limited to one or a few materials. The nutritional requirements for growth, development and reproduction are now known in biochemical terms for many representative species, largely through the use of synthetic diets (Altman and Dittmer, 1968; Vanderzant, 1974). The substances thus shown to be necessary are as follows:

(i) *Water and mineral salts:* Though normally available in the food, water may also be obtained by an insect in dry surroundings through the oxidation of respiratory substrates (Fraenkel and Blewett, 1944). Substantial amounts of potassium, magnesium and phosphates are needed by all insects but much less sodium, calcium and chloride; traces of zinc, iron, manganese and copper are sometimes also required.

(ii) *Sources of energy:* These are normally taken as carbohydrates, the different ones differing appreciably in the extent to which they can be utilized, but fats and amino acids can also provide energy on oxidation and form the major source in Dipteran larvae.

(iii) *Proteins or amino acids:* The amino acids needed for tissue replacement and growth are normally obtained by digestion of the dietary proteins. An external supply of the following is essential for most insects: arginine, histidine, *iso*-leucine, leucine, lysine, methionine, phenylalanine, threonine, tryptophan and valine. These are very much the same as the amino acids essential for vertebrates and it is generally only the L-amino acids that can be utilized.

(iv) *Water-soluble growth factors:* Seven B-complex vitamins are usually needed for normal insect growth: thiamin (B_1), riboflavin (B_2), nicotinamide, pyridoxine (B_6), pantothenate, folic acid and biotin. Some other more specific requirements have also been established: Tenebrionid beetles need carnitine (B_T) while ribonucleic acids and ascorbic acid (vitamin C) are needed by some other species; it is uncertain whether cobalamin (B_{12}) is required by insects.

(v) *Fat-soluble vitamins:* In general the fat-soluble vitamins of vertebrates are not required by insects though a few species need α-tocopherol (vitamin E) and carotenoids related to vitamin A. There is no evidence that calciferol or related compounds are required by insects.

(vi) *Lipogenic factors:* Dietary inositol and choline (water-soluble growth factors originally included in the B-complex but with special roles in lipid metabolism) are needed by many insects in larger quantities than the true vitamins.

(vii) *Fatty acids:* Apart from their role as energy sources, the poly-unsaturated (di- and trienoic C_{18}) fatty acids are essential for the development of many Lepidoptera and Orthoptera, more particularly in connection with wing-expansion and successful emergence from the pupal cuticle.

(viii) *Sterols:* Insects need an external source of sterols (Clayton, 1964; Robbins *et al.*, 1971) and this is probably the greatest nutritional distinction between them and the vertebrates. Cholesterol as such seems essential for some carrion-feeding Dermestidae but many other insects can apparently transform metabolically the phytosterols that they ingest to produce cholesterol; in only a few cases are there specific requirements for other sterols.

It should be noted that many insects depend on micro-organisms for their supply of essential nutrients. In some cases the bacteria, fungi, etc., are found in the material on which the insect feeds; in others a more intimate association occurs, the micro-organisms living in the gut or in special organs (mycetomes) of the insect and sometimes undergoing transmission from parent to progeny (Buchner, 1965; Brooks, 1963; Koch, 1967). As examples of such symbionts may be cited the yeast-like *Actinomyces rhodnii* which occurs in the gut of *Rhodnius prolixus* (Brecher and Wigglesworth, 1944) and the yeasts of *Stegobium paniceum* and *Lasioderma serricorne* which synthesize B-vitamins and a sterol (Pant and Fraenkel, 1950). See also under Digestion, below.

Digestion and Absorption

Digestion includes the processes whereby the food materials are broken down into smaller molecular forms such as monosaccharide sugars and amino acids which are then absorbed through the wall of the gut (Waterhouse, 1957; Dadd, 1970b). Such changes are catalysed by the digestive enzymes which, apart from those produced by the salivary glands (p. 266), are secreted by the columnar cells of the mid gut. Two types of secretory activity have been distinguished by the older histologists: *merocrine secretion*, in which globules of material are extruded from the luminal surface of the columnar cells, and *holocrine secretion*, in which the cell breaks down completely to release its contents. There is, however, considerable uncertainty as to the correct interpretation of the cytological observations. On the one hand, older descriptions of secretory changes in the mid gut cells of insects (e.g. by Duspiva, 1939, and Schönfeld, 1958) have been extended and, at least in a broad sense, confirmed by some ultrastructural findings (e.g. Bertram and Bird, 1961; Threadgold and Gresson, 1962; Stäubli *et al.*, 1966; Gander, 1968). On the other hand it has been claimed that there are often no visible signs of secretory activity, or that the changes are degenerative, or that they represent absorptive processes or perhaps even the synthesis of proteins used in vitellogenesis (Day and Powning, 1949; Khan and Ford, 1962; Smith *et al.*, 1969; de Priester, 1971; and see p. 297).

The enzymes produced are, broadly speaking, adapted to the diet of the species, the most abundant enzyme catalysing the breakdown of the predominant dietary constituent. There is also some experimental evidence that the secretion of digestive enzymes is controlled through neuroendocrine factors from the brain and that foods of different composition, fed to a given species, tend to evoke preferentially the secretion of enzymes able to digest them most effectively. Three main types of digestive enzymes have been recorded in insects:

(a) *Carbohydrases* – These catalyse the hydrolysis of the more complex carbohydrates and include the polysaccharases – of which the amylases, acting upon starch, are widely distributed in insects – and the glycosidases. The latter control the breakdown of, among other things, the disaccharide and trisaccharide sugars and include α-glucosidases which are responsible for the digestion of maltose, sucrose, melezitose, etc., α-galactosidases, dealing with raffinose and melibiose and a β-galactosidase with lactose as its substrate.

(b) Lipases – These catalyse the hydrolysis of dietary fats and are of low specificity, though different esters are attacked at different rates (Gilbert, 1967).

(c) *Proteases* are responsible for the degradation of proteins. The first type are the endopeptidases which catalyse the breakdown of proteins or peptones to polypeptides and which, in insects, are almost invariably of the tryptic type, acting best in alkaline media. The exopeptidases complete protein digestion by facilitating the hydrolysis of peptides to amino acids. All three types of exopeptidases have been reported from insects – the carboxypeptidases and aminopeptidases catalyse attack on the peptide chains at different points while the dipeptidases are responsible for the breakdown of dipeptides. Endopeptidases appear to occur mainly in the lumen of the gut and exopeptidases in the epithelium, suggesting that absorption may begin before hydrolysis of the protein is complete. There is also evidence that some proteases comprise a mixture of several electrophoretically distinct enzymes all acting on the same substrate.

The digestion of a few foods presents some unusual features indicated below:

(a) *Keratin* – Many insects live on wool or fur, but the processes of digestion are best known for the larvae of *Tineola*. Keratin is unusual in that the disulphide linkages which it contains are resistant to attack and even withstand the *in vitro* action of a protease from the gut of *Tineola* larvae. *In vivo*, however, the extremely low redox potential of the gut contents causes a reduction of the disulphide linkages to −SH (sulphydryl) groups and the protease – which differs from trypsin in not being inhibited by −SH groups – catalyses the breakdown (Day, 1951; Powning, Day and Irzykiewicz, 1951).

(b) *Wood, etc.* – The woody tissues of plants include the complex polysaccharides lignin, cellulose, the hemicelluloses and starch as well as the simpler carbohydrates, proteins, etc., of the cell-contents. Lignin is apparently never digested and most insects – even phytophagous ones – cannot digest cellulose though they may be able to utilize some of the contents of unbroken cells. A smaller number of species is able to break down cellulose either by secreting a cellulase (e.g. most Cerambycid larvae, *Xestobium rufovillosum* and *Ctenolepisma*) or through the presence in the gut of symbiotic celluloclastic bacteria (Scarabaeid larvae; Wiedemann, 1930) or

Protozoa (*Cryptocercus*, Isoptera – see p. 628). Some other wood-borers, such as *Phymatodes* larvae, cannot deal with cellulose but possess hemi-cellulases while yet further xylophagous species utilize only the cell-contents and polysaccharides below the hemi-celluloses (Parkin, 1940).

(*c*) *Wax* – A considerable proportion of the wax ingested by *Galleria* larvae is utilized and though the mechanism of digestion is uncertain it is likely that bacteria play a role (Florkin *et al.*, 1949; Niemierko, 1959).

(*d*) *Collagen* – This is normally unaffected by endopeptidases of the tryptic type but *Lucilia* larvae secrete a collagenase (Hobson, 1931).

Apart from a little uptake of partially hydrolysed lipids from the crop of *Periplaneta* (Eisner, 1955), insects seem to absorb the products of digestion entirely through the cells of the mid gut (Treherne, 1962, 1967; Berridge, 1970). Water, inorganic ions and dissolved nutrients pass in mainly through the walls of its anterior region, including the gastric caeca. The uptake of water concentrates the amino acids in the lumen and thus creates more pronounced gradients for their diffusion. Similar gradients for monosaccharides arise through their conversion to trehalose immediately after entering the haemocoele. In both these processes an important part is probably played by the well-developed extracellular space that is formed through extensive basal infolding of the mid gut cells. Also of interest is the fact that in the posterior region of the mid gut fluid is secreted into the lumen from the haemolymph, so that a forwardly directed cycle of fluid movement occurs. At the same time food particles are moving posteriorly within the peritrophic

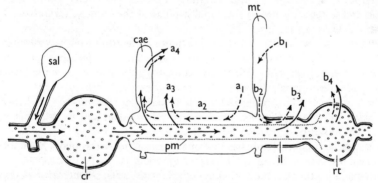

FIG. 100   Diagram illustrating fluid movement within the alimentary canal (*after* Berridge, 1970). Solid arrows denote movement of exogenous fluids, broken arrows that of endogenous fluid. Endogenous fluid movement comprises two cycles. In the absorptive cycle, fluid is transported into the posterior region of the mid gut (a_1) and moves forward (a_2) to be absorbed by the anterior mid gut (a_3) and gastric caeca (a_4). In the excretory cycle, fluid is secreted by the Malpighian tubules (b_1) and moves into the hind gut (b_2), where it is absorbed by the ileum or colon (b_3) or by the rectum (b_4)

cae, gastric caecum; *cr*, crop; *il*, ileum; *mt*, Malpighian tubule; *pm*, peritrophic membrane; *rt*, rectum; *sal*, salivary gland. Food particles are denoted by small open circles.

membrane, which means that digestion can proceed along the whole mid gut with the soluble digestive products being swept forwards to the absorptive sites while the residues accumulate posteriorly (Fig. 100).

Literature on the Alimentary Canal, Nutrition and Absorption

ALEXANDER, N. J. AND FAHRENBACH, W. H. (1969), Fine structure of endocrine hindgut cells of a lepidopteran, *Ostrinia nubilalis* (Hübn.), *Z. Zellforsch*, **94**, 337–345.

ALTMANN, P. L. AND DITTMER, D. S. (eds.) (1968), Nutrient requirements and utilization: insects, In: *Metabolism*, Fedn Am. Soc. exp. Biol, Bethseda, pp. 148–167.

ANDERSON, E. AND HARVEY, W. R. (1966), Active transport by the *Cecropia* midgut. 2. Fine structure of the midgut epithelium, *J. Cell Biol.*, **31**, 107–134.

AUCLAIR, J. L. (1969), Nutrition of plant-sucking insects on chemically defined diets, *Entomologia exp. appl.*, **12**, 623–641.

BACCETTI, B. (1960), Ricerche sull'ultrastruttura dell'intestino degli insetti. I. L'ileo di un Ortottero adulto. II. La cellula epiteliale del mesentero in un Ortottero, un Coleottero e un Dittero adulti, *Redia*, **45**, 263–278; **46**, 158–165.

—— (1962), Ricerche sull'ultrastruttura dell'intestino degli insetti. IV. Le papille rettali in un Ortottero adulto, *Redia*, **47**, 105–118.

BALFOUR-BROWNE, F. (1944), The proventriculus of the Coleoptera (Adephaga) and other insects – a study in evolution, *Jl R. microsc. Soc.*, **64**, 68–117.

BERRIDGE, M. (1970), A structural analysis of intestinal absorption, In: Neville, A. C. (ed.), Insect Ultrastructure, *Symp. R. ent. Soc. Lond.*, **5**, 135–151.

BERRIDGE, M. L. AND GUPTA, B. L. (1967), Fine-structural changes in relation to ion and water transport in the rectal papillae of the blowfly, *Calliphora, J. Cell Sci.*, **2**, 89–112.

BERTRAM, D. S. AND BIRD, R. G. (1961), Studies on mosquito-borne viruses in their vectors. I. The normal fine structure of the mid gut epithelium of the adult female *Aedes aegypti* L. and the functional significance of its modification following a blood meal, *Trans. R. Soc. trop. Med. Hyg.*, **55**, 404–423.

BRECHER, G. AND WIGGLESWORTH, V. B. (1944), The transmission of *Actinomyces rhodnii* Erikson in *Rhodnius prolixus* Stål (Hemiptera) and its influence on the growth of the host, *Parasitology*, **35**, 220–224.

BROOKS, M. A. (1963), Symbiosis and aposymbiosis in Arthropods. Symbiotic associations, *Symp. Soc. gen. Microbiol.*, **13**, 200–231.

BUCHNER, P. (1965), *Endosymbiosis of Animals with Plant Micro-organisms*, Interscience, New York, 909 pp.

CLAYTON, R. B. (1964), The utilization of sterols by insects, *J. Lipid Res.*, **5**, 3–19.

DADD, R. H. (1963), Feeding behaviour and nutrition in grasshoppers and locusts, *Adv. Insect Physiol.*, **1**, 47–109.

—— (1970a), Arthropod nutrition, In: Florkin, M. and Scheer, B. T. (eds.), *Chemical Zoology*, V (A), 35–95.

—— (1970b), Digestion in insects, In: Florkin, M. and Scheer, B. T. (eds.), *Chemical Zoology*, V (A), 117–145.

—— (1973), Insect nutrition: current developments and metabolic implications, *A. Rev. Ent.*, **18**, 381–420.

DAVEY, K. G. AND TREHERNE, J. E. (1963), Studies on crop function in the cockroach (*Periplaneta americana* L.). I–III, *J. exp. Biol.*, **40**, 763–773; 775–780; **41**, 513–524.

DAY, M. F. (1951), Studies on the digestion of wool by insects. I, III, *Aust. J. sci. Res. Ser. B*, **4**, 42–48; 64–72.

DAY, M. F. AND POWNING, R. F. (1949), A study of the process of digestion in certain insects, *Aust. J. sci. Res. Ser. B*, **2**, 175–215.

DUSPIVA, F. (1939), Untersuchungen über die Verteilung der proteolytischen Enzyme sowie der Sekret- und Resorptionszellen im Darm von *Dytiscus marginalis*, *Protoplasma*, **32**, 211–250.

EISNER, T. (1955), The digestion and absorption of fats in the foregut of the cockroach *Periplaneta americana* (L.), *J. exp. Zool.*, **130**, 159–182.

FLORKIN, M., LOZET, F. AND SARLET, H. (1949), Sur la digestion de la cire d'abeille par la larve de *Galleria mellonella* Linn., et sur l'utilisation de la cire par une bactérie isolée à partir du contenu intestinal de cette larve, *Arch. int. Physiol.*, **57**, 71–88.

FORBES, A. R. (1964), The morphology, histology and fine structure of the gut of the Green Peach Aphid, *Myzus persicae* (Sulzer) (Homoptera: Aphididae), *Mem. ent. Soc. Canada*, **36**, 1–74.

FRAENKEL, G. AND BLEWETT, M. (1944), The utilisation of metabolic water in insects, *Bull. ent. Res.*, **35**, 127–139.

FREEMAN, J. C. (1973), The penetration of the peritrophic membrane of the tsetse flies by trypanosomes, *Acta tropica*, **30**, 347–355.

FRIEND, W. G. (1958), Nutritional requirements of phytophagous insects, *A. Rev. Ent.*, **3**, 57–74.

GANDER, E. (1968), Zur Histochemie und Histologie des Mitteldarmes von *Aedes aegypti* und *Anopheles stephensi* im Zusammenhang mit der Blutverdauung, *Acta tropica*, **25**, 133–175.

GILBERT, L. I. (1967), Lipid metabolism and function in insects, *Adv. Insect Physiol.*, **4**, 69–211.

GORDON, H. T. (1968), Quantitative aspects of insect nutrition, *Am. Zool.*, **8**, 131–138.

GOURATON, J. AND MAILLET, P. - L. (1965), Sur l'existence d'une membrane péritrophique chez un insecte suceur de sève, *Cicadella viridis* L. (Homoptera, Jassidae), *C.r. Acad. Sci., Paris*, **261**, 1102–1105.

GRIMSTONE, A. V., MULLINGER, A. M. AND RAMSAY, J. A. (1968), Further studies on the rectal complex of the mealworm *Tenebrio molitor* L. (Coleoptera, Tenebrionidae), *Phil. Trans. R. Soc. (B)*, **253**, 343–382.

GUPTA, B. L. AND BERRIDGE, M. J. (1966), Fine-structural organisation of the rectum in the blowfly, *Calliphora erythrocephala* (Meig.) with special reference to connective tissue, tracheae and neurosecretory innervation in the rectal papillae, *J. Morph.*, **120**, 23–82.

HASSEMER, M. AND BECK, S. D. (1969), Ultrastructure of the ileum of the European corn borer, *Ostrinia nubilalis*, *J. Insect Physiol.*, **15**, 1791–1802.

HAYDAK, M. H. (1970), Honey bee nutrition, *A. Rev. Ent.*, **15**, 143–156.

HECKER, H., FREYVOGEL, T. A., BRIEGEL, H. AND STEIGER, R. (1971a), The ultrastructure of midgut epithelium in *Aedes aegypti* (L.) (Insecta, Diptera) males, *Acta tropica*, **28**, 275–290.

HECKER, H., FREYVOGEL, T. A., BRIEGEL, H. AND STEIGER, R. (1971*b*), Ultrastructural differentiation of the midgut epithelium in female *Aedes aegypti* (L.) (Insecta, Diptera) imagines, *Acta tropica*, **28**, 80–104.

HOBSON, R. P. (1931), On an enzyme from blow-fly larvae (*Lucilia sericata*) which digests collagen in alkaline solution, *Biochem. J.*, **25**, 1458–1463.

HOUSE, H. L. (1958), Nutritional requirements of insects associated with animal parasitism, *Expl Parasit.*, **7**, 555–609.

—— (1961), Insect nutrition, *A. Rev. Ent.*, **6**, 13–26.

JUDD, W. W. (1948), A comparative study of the proventriculus of Orthopteroid insects with reference to its use in taxonomy, *Can. J. Res. (Ser. D)*, **26**, 93–161.

KHAN, M. R. AND FORD, J. B. (1962), Studies on digestive enzyme production and its relationship to the cytology of the midgut epithelium in *Dysdercus fasciatus* Sign. (Hemiptera, Pyrrhocoridae), *J. Insect Physiol.*, **8**, 597–608.

KOCH, A. (1967), Insects and their endosymbionts, In: Henry, S. M. (ed.), *Symbiosis*, Academic Press, New York, **2**, 1–106.

KUSMENKO, S. (1940), Ueber die postembryonale Entwicklung des Darmes der Honigbiene und die Herkunft der larvalen peritrophischen Hüllen, *Zool. Jb.* (*Anat.*), **66**, 463–530.

LEE, R. F. (1968), The histology and histochemistry of the anterior mid-gut of *Periplaneta americana* L. (Dictyoptera: Blattidae) with reference to the formation of the peritrophic membrane, *Proc. R. ent. Soc. Lond.*, A, **43**, 122–134.

LEGAY, J. M. (1958), Recent advances in silkworm nutrition, *A. Rev. Ent.*, **3**, 75–86.

LIPKE, H. AND FRAENKEL, G. (1956), Insect nutrition, *A. Rev. Ent.*, **1**, 17–44.

MOLOO, S. K., STEIGER, R. AND HECKER, H. (1970), Ultrastructure of the peritrophic membrane formation in *Glossina* Wiedemann, *Acta tropica*, **27**, 378–383.

MORDUE, W. (1969), Hormonal control of Malpighian tube and rectal function in the desert locust, *Schistocerca gregaria*, *J. Insect Physiol.*, **15**, 273–285.

NOIROT, C. AND NOIROT-TIMOTHÉE, C. (1971), Ultrastructure du proctodeum chez le Thysanoure *Lepismodes inquilinus* Newman (= *Thermobia domestica* Packard). I. La region anterieure (iléon et rectum). II. Le sac anal, *J. Ultrastruct. Res.*, **37**, 119–137, 335–350.

PALM, N. B. (1949), The rectal papillae in insects, *Acta Univ. lund.*, **45** (8), 29 pp.

PANT, N. C. AND FRAENKEL, G. (1950), The function of the symbiotic yeasts of two insect species, *Lasioderma serricorne* F., and *Stegobium (Sitodrepa) paniceum* L., *Science, N.Y.*, **112**, 498–500.

PARKIN, E. A. (1940), The digestive enzymes of some wood-boring beetle larvae, *J. exp. Biol.*, **17**, 364–377.

PETERS, W. (1968), Vorkommen, Zusammensetzung und Feinstruktur peritrophischer Membranen im Tierreich, *Z. Morph. Tiere*, **62**, 9–57.

—— (1969), Vergleichende Untersuchungen der Feinstruktur peritrophischer Membranen von Insekten, *Z. Morph. Tiere*, **64**, 21–58.

PHILLIPS, J. E. (1964), Rectal absorption in the desert locust, *Schistocerca gregaria* Forskål. I–III, *J. exp. Biol.*, **41**, 15–38; 39–67; 69–80.

PLATZER-SCHULTZ, I. AND WELSCH, U. (1969), Zur Entstehung und Feinstruktur der peritrophischen Membran der Larven von *Chironomus strenzkei* Fittkau (Diptera), *Z. Zellforsch.*, **100**, 594–605.

POWNING, R. F., DAY, M. F. AND IRZYKIEWICZ, H. (1951), Studies on the digestion of wool by insects, II. *Aust. J. sci. Res. Ser. B*, **4**, 49–63.

PRIESTER, W. DE (1971), Ultrastructure of the midgut epithelial cells in the fly *Calliphora erythrocephala*, *J. Ultrastruct. Res.*, **36**, 783–805.

RAMSAY, J. A. (1964), The rectal complex of the mealworm *Tenebrio molitor* L. (Coleoptera: Tenebrionidae), *Phil. Trans. R. Soc. Ser. B*, **248**, 279–314.

—— (1971), Insect rectum, *Phil. Trans. R. Soc. Ser. B*, **262**, 251–260.

REICHENBACH-KLINKE, H. H. (1952), Die Rektalpapillen der Insekten insbesondere der Käfer, und ihre Beziehung zur Stammesgeschichte, *Zool. Jb. (Anat.)*, **72**, 230–250.

ROBBINS, W. E., KAPLANIS, J. A., SVOBODA, J. A. AND THOMPSON, M. J. (1971), Steroid metabolism in insects, *A. Rev. Ent.*, **16**, 53–72.

RODRIGUEZ, J. G. (1973) (ed.), *Insect and mite nutrition*, North-Holland, Amsterdam, 717 pp.

SAINI, R. S. (1964), Histology and physiology of the cryptonephridial systems of insects, *Trans. R. ent. Soc. Lond.*, **116**, 347–392.

SCHNEIDER, I. AND RUDINSKY, J. A. (1969), The site of pheromone production in *Trypodendron lineatum* (Coleoptera: Scolytidae): bioassay and histological studies of the hind gut, *Can. Ent.*, **101**, 1181–1186.

SCHÖNFELD, C. (1958), Histophysiologische Untersuchungen zur Verdauungstätigkeit der Mückenlarve *Chaoborus (Corethra)*, *Zool. Jb. (Allg. Zool.)*, **67**, 337–364.

SMITH, D. S., COMPHER, K., JANNERS, M., LIPTON, C. AND WHITTLE, L. W. (1969), Cellular organization and ferritin uptake in the midgut epithelium of a moth, *Ephestia kühniella*, *J. Morph.*, **127**, 41–71.

TÄUBLI, W., FREYVOGEL, T. A. AND SUTER, J. (1966), Structural modification of the endoplasmatic reticulum of midgut epithelial cells of mosquitoes in relation to blood intake, *J. Microscopie*, **5**, 189–204.

STOBBART, R. H. (1968), Ion movements and water transport in the rectum of the locust, *Schistocerca gregaria*, *J. Insect Physiol.*, **14**, 269–275.

SUTTON, M. F. (1951), On the food, feeding mechanism and alimentary canal of Corixidae (Hemiptera Heteroptera), *Proc. zool. Soc. Lond.*, **121**, 465–499.

THIEL, H. (1936), Vergleichende Untersuchungen an den Vormagen von Käfern, *Z. wiss. Zool.*, **147**, 395–432.

THREADGOLD, L. T. AND GRESSON, R. A. R. (1962), Electron microscopy of the epithelial cells of the mid-gut and hepatic caeca of *Blatta orientalis*, *Proc. R. Soc. Edinb.*, B, **68**, 162–170.

TREHERNE, J. E. (1962), The physiology of absorption from the alimentary canal in insects, *Biol. Viewpts*, **1**, 201–241.

—— (1967), Gut absorption, *A. Rev. Ent.*, **12**, 43–58.

VANDERZANT, E. S. (1974), Development, significance, and application of artificial diets for insects, *A. Rev. Ent.*, **19**, 139–160.

WALL, B. J. AND OSCHMAN, J. L. (1973), Structure and function of rectal pads in *Blattella* and *Blaberus* with respect to the mechanism of water uptake, *J. Morph.*, **140**, 105–118.

WATERHOUSE, D. F. (1953a), Occurrence and endodermal origin of the peritrophic membrane in some insects, *Nature*, **172**, 676–677.

—— (1953b), The occurrence and significance of the peritrophic membrane, with special reference to adult Lepidoptera and Diptera, *Aust. J. Zool.*, **1**, 299–318.

—— (1957), Digestion in insects, *A. Rev. Ent.*, **2**, 1–18.

WATERHOUSE, D. F. AND WRIGHT, M. (1960), The fine structure of the mosaic mid gut epithelium of blowfly larvae, *J. Insect Physiol.*, **5**, 230–239.

WIEDEMANN, J. F. (1930), Die Zelluloseverdauung bei Lamellicornierlarven, *Z Morph. Tiere*, **19**, 228–258.

WIGGLESWORTH, V. B. (1930), The formation of the peritrophic membrane in insects, with special reference to the larvae of mosquitoes, *Q. Jl. Microsc. Sci.*, **73**, 593–616.

Chapter 13

THE RESPIRATORY SYSTEM

In the vast majority of insects respiration takes place by means of internal air-tubes known as *tracheae*. These ramify through the organs of the body and its appendages, the finest branches being termed *tracheoles*. The air generally enters the tracheae through paired, usually lateral, openings termed *spiracles*, which are segmentally arranged along the thorax and abdomen. More rarely the spiracles are closed or wanting, respiration in such cases being cutaneous. In the immature stages of many aquatic insects special respiratory organs known as *gills* or *branchiae* are present, and these may or may not co-exist with open spiracles. The respiratory organs of insects are always derived from ectoderm: the tracheae are developed from solid ingrowths or tubular invaginations of that layer and the gills arise as hollow outgrowths. Histologically, both are composed of a layer of cuticle, the epidermis and usually a basement membrane, all directly continuous with similar layers forming the general body-wall. All or most of the cuticular lining of the tracheo-spiracular system is usually shed at ecdysis. A tracheal system is absent in most Collembola, some Protura and some endoparasitic Hymenopteran and Dipteran larvae.

The Spiracles

Number and Position of the Spiracles – The spiracles are derived from the mouths of the ectodermal invaginations which give rise to the tracheal system. They are normally placed on the pleura of the thoracic and abdominal segments, but their exact position is very variable. In the abdomen of most insects they lie in the soft membrane between the terga and sterna, sometimes towards the front or back of their segments. In many insects, particularly on the thorax, the spiracles occupy an intersegmental position, being situated just in front of each of the segments to which they are generally referred; or they may have moved from the pleura to the side margins of the terga, as in the abdominal spiracles of *Apis* and *Musca*.

In the developing embryo the spiracles appear as a series of ingrowths lying to the outer side of the rudiments of the appendages. Twelve pairs are said to be present in the embryo of *Leptinotarsa*, situated on each of the thoracic and the first nine abdominal segments. In the embryos of most insects, however, the prothoracic pair is wanting and the pair on the 9th abdominal segment is likewise absent. The resulting number – two thoracic and eight abdominal pairs – is the maximum found in the postembryonic stages of any insect apart from some Diplura and it is probable that their primitive position was intersegmental, the most anterior pair lying between the pro- and mesothorax and the last pair between the 7th and 8th abdominal segments (Keilin, 1944). Deviations from the number and arrangement of spiracles in such a primitive system have evolved through the migration of the spiracle to an adjacent segment (usually the posterior one) and the reduction of some or all of the spiracles. These either become closed or remain visible as small scars or are lost completely (see non-functional spiracles, below).

According to the number and arrangement of functional spiracles it is possible to classify respiratory systems as follows (De Gryse, 1926; Keilin, 1944; Hinton, 1947):

1. *The Holopneustic Respiratory System.* This is the most primitive arrangement found in living insects, 10 pairs of functional spiracles being present, on the first 8 abdominal segments, the metathorax and either the prothorax or the mesothorax. It is characteristic of the nymphs and imagines of many orders and of the larvae of the Bibionidae (Diptera) and some Hymenoptera.

2. *Hemipneustic Respiratory Systems.* These forms of respiratory system are prevalent among insect larvae and are derived from the holopneustic type through one or more pairs of spiracles becoming non-functional. The following terms indicate the different distributions of spiracles:

Peripneustic. Spiracles in a row along each side of the body. In typical examples the prothoracic and abdominal spiracles are open, that of the metathorax being closed. This condition is found in the terrestrial larvae of the orders Neuroptera, Mecoptera, Lepidoptera, of many Hymenoptera Symphyta, and of many Coleoptera; among Diptera it is prevalent in larvae of some Bibionidae and most Mycetophilidae and Cecidomyiidae.

Amphipneustic. Only the prothoracic and the posterior abdominal spiracles are open. This type is common among larval Diptera.

Propneustic. Only the prothoracic spiracles are open. A comparatively rare condition exhibited for example in the pupae of some Dipteran families.

Metapneustic. Only the last pair of abdominal spiracles are open. The prevalent type in larval Culicidae and Tipulidae and in *Hypoderma* among the Oestridae; also found in the first larval instar of most Cyclorrhapha and in the aquatic larvae of certain Coleoptera (Dytiscidae, Helodidae, etc.).

The last three types may together be denoted as oligopneustic systems and represent adaptive modifications to life in a liquid or semi-liquid medium.

3. *The Apneustic Respiratory System.* Here none of the spiracles are functional, air entering the closed tracheal system by diffusion through the general body surface or the gills (q.v.). Like the oligopneustic types, the apneustic system is an adaptation to life submerged in fluids and is therefore characteristic of aquatic and endoparasitic forms. It occurs, for example, in the nymphs of the Ephemeroptera and Odonata, the larvae of the

Trichoptera, of such Dipteran families as the Blephariceridae, Simuliidae, Chironomidae and Ceratopogonidae and of some Coleoptera (e.g. Elmidae, Haliplidae, Hygrobiidae). It is also found in some larval instars of endo-parasitic Hymenoptera and Tachinidae.

In all the above types of respiratory system, the total of functional and non-functional spiracles is equal to 10 pairs. In contrast to them, the term *hypopneustic* is used to denote systems in which one or more pairs of spiracles have disappeared completely. For example, the Mallophaga and Siphunculata have 1 thoracic and 6 abdominal pairs; the Thysanoptera have 2 pairs of thoracic and 2 pairs of abdominal spiracles; in the Hemiptera Sternorrhyncha their number is very variable and is reduced to 2 pairs in

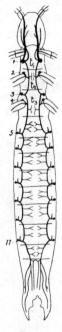

FIG. 101
Tracheal system of
Japyx

t_1, t_2, t_3, thoracic segments; 1–4, thoracic spiracles; 5, 11, abdominal spiracles. *After* Grassi.

many Coccoidea. Among Coleoptera, the Scarabaeoidea and Curculionoidea have from 1 to 3 of the hindmost abdominal spiracles wanting (Ritcher, 1969, 1969*a*). The Diptera usually exhibit a reduction in the number of abdominal spiracles and, among the Cyclorrhapha, a sexual difference is evident in this respect, the females often having 5 pairs and the males 6 or 7 pairs. Among parasitic Hymenoptera reduction is frequently evident and in the Chalcidoidea there are commonly only 3 pairs which are situated on the thorax, propodeum and 8th abdominal segment, respectively.

The Diplura and those Collembola with a tracheal system exhibit atypical arrangements of spiracles (see pp. 447, 467, Fig. 101).

Structure of the Spiracles – In general, the spiracles not only permit gaseous respiratory exchange but are also a major site of water loss and, at ecdysis, are the apertures through which the old tracheal lining is pulled out.

They show many adaptations to these diverse functions but a typical functional spiracle includes not only the external opening, and the annular sclerite or *peritreme* which surrounds it, but also the *atrium* or vestibule into which the opening leads, together with the *closing apparatus*. The latter consists of one or more muscles with associated cuticular parts and, by closing the spiracular aperture, prevents excessive loss of water-vapour. The atrium is a specialized region leading from the spiracular opening; it lacks taenidia and its walls are variously sculptured or provided with hairs, trabeculae and similar cuticular outgrowths. These help to reduce water-loss and prevent the entry of dust. Closely connected with the spiracles are frequently *peristigmatic glands* which secrete a hydrophobe material preventing the wetting of those organs. The structure of the spiracles presents an enormous range of variety among different groups of insects: it is also usually different in the thoracic and abdominal spiracles of the same insect and may be greatly

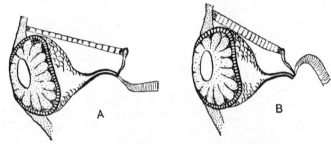

FIG. 102 Spiracle and occluding apparatus of *Trichodectes*, semi-diagrammatic
A, open; B, closed. *After* Harrison, *Parasitology*, 1915.

modified in different instars. It will, therefore, be readily appreciated that their classification is a matter of much difficulty (Bergold, 1935; Hassan, 1944; Keilin, 1944; Lotz, 1961; Hinton, 1967a, 1967b; Ritcher, 1969a, 1969b; Tonapi, 1957; Bhatnagar, 1972).

The most generalized type of spiracle is devoid of lips and closing apparatus and is little more than a simple crypt as in *Sminthurus*. No special chamber or atrium is developed and the spiracle opens directly into the tracheae.

In most Hemiptera, more especially in the abdomen, the spiracles are simple apertures surrounded by a peritreme. A well developed atrium is present and between the latter and the trachea is the closing apparatus (absent from the Hydrocorisae). This type of spiracle is also found in the Mallophaga and Siphunculata (Webb, 1946), Siphonaptera and in other insects (Fig. 102).

In the Acrididae the thoracic spiracles each have a slit-like opening guarded by two external valves or lips (Figs. 103 and 104, A). The metathoracic spiracles have movable lips (*a, p*) united by a ventral lobe (*n*): they open by their own elasticity but are closed by an occlusor muscle (*sm*) arising from a process (*o*) on the margin of the mesocoxal cavity. The abdominal spiracle

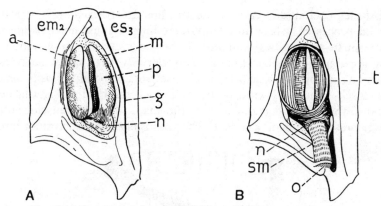

FIG. 103   Metathoracic spiracle of an Acridid (*Dissosteira*)

A, outer view. B, inner view. *em₂*, mesepimeron; *es₃*, metepisternum; *g*, intersegmental fold; *m*, membrane; *t*, trachea. Further explanation in the text. Adapted from Snodgrass, 1929.

(Fig. 104, B, C, D) have no projecting lips, the integument being inflected to form two hardened walls of the atrium – one wall (*v*) being movable and the other (*d*) fixed. The movable wall is prolonged into a process or manubrium (*q*) to which the occlusor (*sm1*) and opening muscles (*sm2*) are attached. In

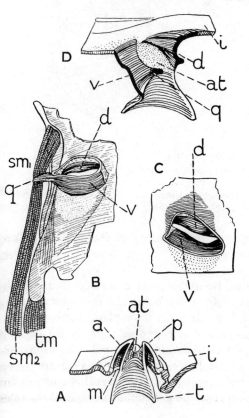

FIG. 104

Spiracles of an Acridid (*Dissosteira*)

A, D, sections through metathoracic and 1st abdominal spiracles respectively. B, inner view, and C, outer view of 1st abdominal spiracle. *at*, atrium; *i*, integument; *tm*, tympanal muscles. Further explanation in the text. Adapted from Snodgrass, 1929.

the spiracles of Lepidopteran larvae the lips are fringed with repeatedly branched processes, whose finest divisions form a most efficient guarding mechanism to the tracheal system. At the inner end of the atrium is the closing apparatus. This consists of a cuticular bow that partly encircles the trachea, on the opposite side of which is a sclerotized band; a closing lever or rod is closely connected with the band. The occlusor muscle is attached at one end to the bow and at the other to the lever; when the muscle contracts the lever presses the band against the bow, thus closing the entrance into the

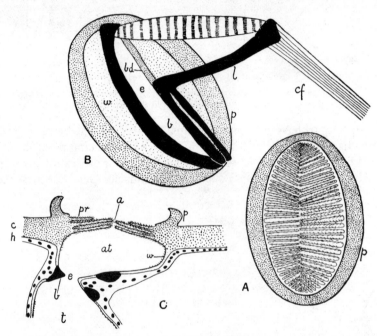

FIG. 105   Spiracle of a Lepidopterous larva (*Sphingidae*)

A, seen from the outside showing fringed processes of the lips; B, seen from the inside, lips omitted; C, sectional view. *a*, spiracular aperture; *at*, atrium; *b*, bow; *bd*, band; *c*, cuticle; *cf*, elastic fibre which opens spiracle; *e*, entrance into trachea; *h*, epidermis; *l*, lever; *p*, peritreme; *pr*, fringed processes of lips; *w*, wall of atrium; *t*, trachea.

trachea. The latter is opened partly by the elasticity of the cuticular parts which regain their former position, and partly by an antagonistic muscle or elastic fibre (Fig. 105).

In the larvae of the Scarabaeoidea the spiracles are described as *cribriform* or *multiforous* (Lotz, 1962; Hinton, 1967a). Each spiracle is almost circular and comprises a crescentic spiracular plate or sieve plate, supported by internal trabeculae and almost completely surrounding a projecting central bulla, around which, in turn, runs a curved 'slit' (Fig. 106). Gas exchange occurs only through minute circular, oval or elongate apertures in the sieve plate. The curved slit around the bulla represents the external opening of an ecdysial tube (p. 217); this collapses after the cuticle has been shed, so that its original orifice gives rise to a closed scar. An internal closing apparatus may also be present, as in the Lucanidae.

Larvae of the Elateridae, Cleridae, Nitidulidae and other Coleoptera have *biforous spiracles*, each with two contiguous openings which are more or less slit-like and separated by a partition wall. Each opening communicates either by a tubular passage with a common atrium or opens directly into the trachea (Fig. 106).

In Dipteran larvae the spiracles are without a closing apparatus (Keilin, 1944). In the third stage larvae of the higher Cyclorrhapha the posterior spiracles consist of a pair of cuticular plates. Each plate is surrounded by a

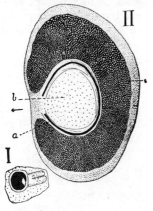

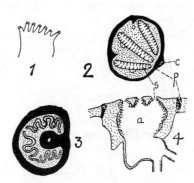

FIG. 106
I. Biforous spiracle of a Clerid larva. *After* Böving and Champlain. II. Abdominal spiracle of the larva of *Melolontha melolontha*

a, spiracular opening; *b*, bulla; *s*, sieve-plate. The arrow is directed anteriorly.

FIG. 107
Spiracles of larval Diptera

1, anterior spiracle of *Musca domestica*; 2, posterior spiracle of *Calliphora erythrocephala*; 3, posterior spiracle of *Musca domestica*; 4, vertical section through spiracle of *Calliphora*; sclerotized parts only shown; *a*, atrium; *c*, stigmatic scar; *p*, peritreme; *s*, spiracular slit.

peritreme and bears as a rule three openings which may be pyriform (*Muscina*) or in the form of straight slits (*Calliphora*) or sinuous slits (*Musca*). Each opening is traversed by a number of fine cuticular rods resembling a grating, and all three openings communicate with a common atrium. Just internal to the openings there is a system of branched cuticular trabeculae which, with the grating, form an efficient barrier to the entrance of foreign particles. The walls of the atrium are also lined with fibrous processes and form the so-called *felt chamber* which probably reduces water-loss in the absence of a closing mechanism. The anterior spiracles each consist of a variable number of digitate processes whose apices are perforated by openings. Each opening communicates with a small atrium and the atria of each spiracle all join with the main tracheal trunk of their side (Fig. 107). In the larvae of *Oestrus*, *Hypoderma* and other Oestridae instead of three openings to each spiracle there are multiple pores. In *Glossina* there are about 500 of these pores to a side; they form the sculpturing on the lobe-like posterior abdominal spiracles (Bursell, 1955). The pores are connected by tubular continuations with a tripartite felt chamber. A similar arrangement is found in the larva of *Hippobosca* except that the pores are much less numerous, while in *Melophagus* there are only three to each lobe.

The shedding of the old spiracles at ecdysis and the formation of new ones takes place in three different ways (Keilin, 1944; Hinton, 1947). In the most primitive type, where the new spiracular aperture is sufficiently wide to allow withdrawal of the old tracheae and atrial apparatus, the new structures are formed around the old ones after the latter have separated from the epidermis. The remains of the old spiracle and tracheal lining are then pulled out through what becomes the new aperture. This method of moulting is the most common one, but where the aperture or atrium of the new spiracle is so obstructed by cuticular processes that withdrawal is mechanically impossible, one or other of the following methods is adopted. In the Mecoptera and some Diptera the spiracle opens by numerous small apertures ar-

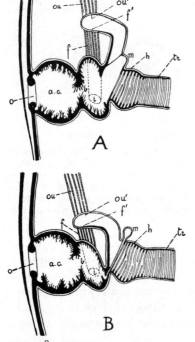

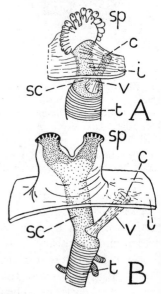

FIG. 108
Longitudinal sections of the last abdominal spiracle of an ant

A, open; B, closed; *o*, spiracular opening; *a.c*, anterior chamber; *b*, occluding chamber; *f*, closing muscle and *f'* mobile insertion of same; *h*, thickened portion of trachea; *i*, fixed insertion of closing muscle; *m*, flexible membrane; *o*, spiracular opening; *ou*, opening muscle; *ou'*, fixed insertion of same; *tr*, trachea. *After* Janet.

FIG. 109
Spiracles of *Rhagoletis* (Trypetidae)

A, anterior spiracle of 3rd instar larva. B, right pronotal spiracle of pupa. *c*, cicatrix or closed end of remains of passage *v* through which spiracle of 2nd instar was cast off during ecdysis; *i*, integument; *sc*, spiracular chamber; *sp*, spiracle, *t*, trachea. Adapted from Snodgrass, *J. agric. Res.*, 1924.

ranged around a central solid area. Before moulting occurs a new spiracle forms around the old one so that the former possesses a central ecdysial aperture through which the old structures are drawn out and which then closes by hardening and contraction of the cuticle to form the solid central part. In other Diptera and many Coleoptera there develops around the old spiracle a simple cuticular tube (the ecdysial tube) connected with which, and situated to one side of the old structures, is the new spiracle. The old spiracle and tracheae are then pulled out through the unobstructed ecdysial tube at moulting and the tube later shrivels, its former opening giving rise to the stigmatic scar (Fig. 109).

In the apneutic and hemipneustic systems of immature forms all or some of the spiracles are non-functional. They consist of a surface scar from the inside of which a more or less solid, cuticular stigmatic cord runs to an adjacent part of the tracheal system. Before moulting, an ecdysial tube, continuous with the new tracheal system, forms around the stigmatic cord and the appropriate part of the old tracheal lining is later pulled out through the ecdysial tube by means of the stigmatic cord. The tube then shrivels to form the stigmatic cord and scar of the new instar. If however, this new instar is to have a functional spiracle in that position, the necessary structures form around the old stigmatic cord and remain functional after moulting. The functions of the stigmatic cords are therefore: (a) to anchor the tracheal system to the cuticle, (b) to draw out the old tracheal lining at ecdysis and (c) to form a structure around which an ecdysial tube or new functional spiracle can develop.

The Tracheae and Tracheoles

The tracheae are elastic, cuticular tubes which, when filled with air, have a silvery appearance in dissections. Their functional morphology and ultra-structure has been reviewed by Whitten (1972). The cuticular lining or *intima* is continuous with that of the external body surface and secreted by tracheal matrix cells derived from the epidermis. It comprises an epicuticle of several layers, an outer cuticular layer in which the chitin micelles are axially oriented and an inner cuticular layer containing tangentially arranged chitin micelles (Beaulaton, 1968; Edwards *et al.*, 1958; Locke, 1957). The last-mentioned layer is thickened locally to form the characteristic and readily visible helical thread or *taenidium* that runs round the trachea, enabling its walls to resist compression. The taenidium probably arises through buckling of the trachea under the influence of simple compression and shearing stresses in the cuticle as it is being laid down (Locke, 1958). In some insects several taenidia exist side by side and in teased preparations a ribbon-like band uncoils which is formed of several parallel thickenings. Taenidia are usually absent from the large tracheae close to the spiracles, the intima in such positions presenting a tesselated or other type of thickening. In some insects (*Zaitha*, *Lampyris*, *Luciola*, etc.) cuticular piliform processes arise from the taenidia and project into the cavity of the trachea. The epithelial layer outside the intima is composed of pavement cells with relatively large nuclei. The larger tracheae of some insects are faintly coloured with reddish-brown or violet pigment which is lodged in the cells of the epithelial layer. A delicate basement membrane forms the outermost coat of the tracheae.

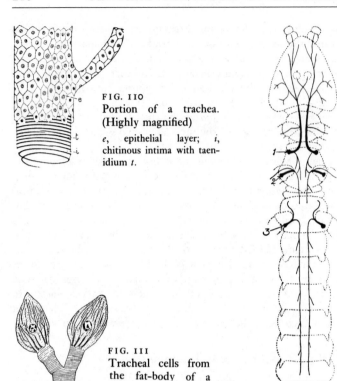

FIG. 110
Portion of a trachea.
(Highly magnified)

e, epithelial layer; i, chitinous intima with taenidium t.

FIG. 111
Tracheal cells from the fat-body of a *Gasterophilus* larva

FIG. 112
Tracheal system of *Campodea*

1, 2 and 3, Spiracles. *After* Grassi.

The ultimate branches of the tracheal system are termed *tracheoles* and are canals with a diameter of $0 \cdot 2–0 \cdot 3$ μm whose thin walls often bear helical or circular taenidia visible only under the electron microscope. They may contain liquid or air and end blindly or in anastomosis with each other. Their lining is not shed at ecdysis and their walls are freely permeable to water. The tracheoles are intracellular structures, developing in almost all cases from large, stellate end-cells (tracheoblasts) (Fig. 113) and later becoming joined to a developing trachea by a ring of cement. The tracheoblasts, though ectodermal, develop independently of the tracheal epithelium and their processes commonly anastomose to form a fenestrated membrane over the surface of various viscera—the tracheated 'peritoneal layer'. Tracheoles also ramify between the cells of the insect tissues and may penetrate the cells of muscle, and perhaps other tissues, to end intracellularly (Beinbrech, 1970; Weis-Fogh, 1964b); they may even penetrate mitochondria (Afzelius and Gonnert, 1972). In the fat-body of the larvae of *Gasterophilus*, the tracheoles lie wholly within the cytoplasm of large tracheal end-cells of a special type (Fig. 111) which contain haemoglobin and act as an oxygen store (Dinulescu, 1932). In the highly tracheated luminous organs of the firefly *Photuris* (Coleoptera), the tracheoles are partially enclosed by tracheal end-cells, but are also enveloped by tracheolar cells comparable to those secreting the tracheae (Smith, 1963). In *Rhodnius* air-filled tracheoles appear to migrate into regions deficient in a supply of oxygen; in fact they are pulled into these regions by contractile filaments that are emitted by the epidermal cells there and become attached to the tracheoles (Wigglesworth, 1959).

The general arrangement and distribution of tracheae present important differences among the various groups of insects (see, for example, Whitten, 1955, 1960; Tonapi, 1959–60). In *Campodea* (Fig. 112), the Sminthuridae and those Protura with a tracheal system, the tracheae arising from each spiracle remain unconnected with those from the others, but the Thysanura

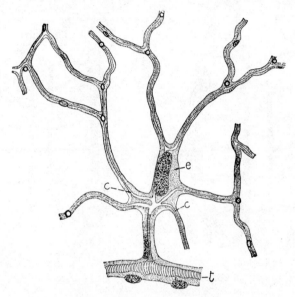

FIG. 113 Tracheal end-cell and tracheoles from the silk gland of the larva of *Phalera bucephala* (Lepidoptera)

e, end-cell; *c*, tracheoles; *t*, trachea. *After* Holmgren.

and Japygidae resemble the majority of Pterygotes in having a system developed from the union of a series of tracheospiracular metameres by transverse and longitudinal trunks (Stobbart, 1956; see also Fig. 114). Clear indications of such a metameric arrangement are present from the earliest postembryonic stages, though growth is accompanied by an increase in the complexity of the branches present (Fuller, 1919; Keister, 1948). A metameric basis is evident not only in holopneustic forms but also in the hemipneustic and apneustic systems owing to their retention of non-functional spiracles. The most constant features of well-developed tracheal systems are the presence of lateral longitudinal (spiracular) trunks (rarely absent, as in *Cimex*), of dorsal longitudinal trunks connected with the lateral trunks by palisade tracheae, and, less frequently, of ventral longitudinal trunks. Transverse dorsal or ventral commissures connect the systems of each side.

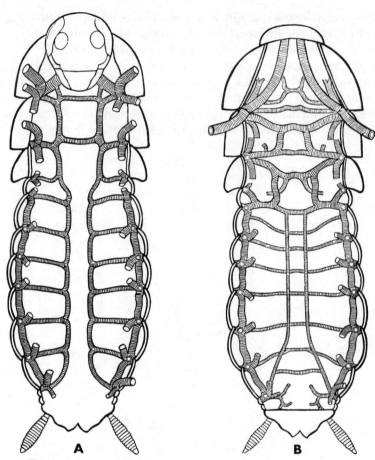

FIG. 114   Tracheal system of *Periplaneta*

A, with the ventral integument and viscera removed showing dorsal tracheae; B, with dorsal integument and viscera removed showing ventral tracheae. *After* Miall and Denny.

The dorsal longitudinal trunks give off segmental branches which pass to the heart and dorsal musculature. Visceral branches, which supply the digestive canal and reproductive organs, take their origin from the palisade tracheae or directly from the spiracular tracheae. The nerve cord and ventral musculature are supplied by branches derived from the ventral transverse commissures. The tracheae supplying the legs arise from the spiracular (or, in Odonata, the dorsal longitudinal) trunks in the thoracic region, and the basal tracheae of the developing wings usually take their origin in close association with those of the leg tracheae of the meso- and metathorax (Comstock, 1918). The head and mouthparts are principally supplied by branches derived from the anteriormost spiracle and the dorsal longitudinal trunk.

Hypopneustic tracheal systems deviate to varying degrees from the segmental arrangement, longitudinal trunks and transverse commissures tend to be reduced or disappear and each of the few spiracles gives rise directly to a greater or lesser number of branches which supply different parts of the body. The hypopneustic system may be markedly reduced, as in some Coccoidea.

The Air-Sacs

In many winged insects the tracheae are dilated in various parts of the body to form thin-walled vesicles or *air-sacs*. Their cuticular intima may show taenidia, as in the abdominal air-sacs of the Blattaria and Diptera, or it may exhibit various transitional forms leading to the irregular or punctate thickenings that predominate in the thoracic air-sacs (Faucheux, 1972; Faucheux and Sellier, 1971). The sacs are readily distensible and are easily seen as glistening white vesicles when inflated. They are well developed in the abdomen of large Scarabaeid and Buprestid beetles (Miller, 1966*a*) and in *Melolontha*, for example, where they are dilations of the secondary tracheae, the sacs are relatively small but very numerous. In *Melanoplus* there is a pair of large thoracic air-sacs and five pairs in the abdomen which are likewise dilations of the secondary tracheae; there are also many smaller vesicles among the muscles. The air-sacs attain great development in *Volucella*, *Musca* and other Cyclorrhapha and in *Apis* and *Bombus* among Hymenoptera (Fig. 115). In these instances the abdominal air-sacs are especially large and are dilations of the main longitudinal tracheal trunks. Air-sacs are also met with among Lepidoptera and Odonata.

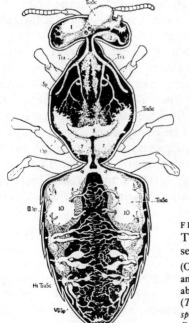

FIG. 115
Tracheal system of worker honey bee seen from above

(One pair of abdominal air-sacs removed and transverse ventral commissures of abdomen not shown.) The air-sacs (*TraSc*) are indicated in arabic numerals: *sp*, spiracles. *After* Snodgrass, *U.S. Bur. Entom. Tech. Ser.* No. 18.

The air-sacs may have one or more of the following functions (Wigglesworth, 1963): (a) they assist flight by reducing the specific gravity of an insect of given size; (b) they increase the volume of 'tidal air' which is changed when respiratory movements occur (p. 223); (c) the periodic compression and expansion of thoracic air-sacs during flight causes air to be pumped to the very actively respiring thoracic muscles (Weis-Fogh, 1964a); (d) the presence of air-sacs reduces the mechanical damping of wing-movements by the blood; (e) the large abdominal air-sacs of some female Diptera Cyclorrhapha ensure that the body shape remains constant during sexual maturation by providing room for the developing ovaries to expand; (f) in some aquatic insects, such as the larvae of *Chaoborus* and *Mochlonyx*, the specialized air-sacs have a hydrostatic function (Damant, 1924); (g) in the more highly developed auditory organs an air-sac allows the free vibration of the tympanic membrane; and (h) by restricting the volume of circulating blood, as they do in the newly-emerged *Drosophila*, the air-sacs increase its concentration of sugar, which can therefore be utilized more rapidly by the flight muscles.

Physiology of Respiration

Atracheate insects and those stages of the life-cycle during which the tracheal system is filled with liquid obtain their oxygen by direct diffusion from the environment into the body fluids. When a functional tracheal system is present, however, the carriage of oxygen to the tissues requires that the gas (a) enters the tracheal system, (b) is transported to its finest branches and (c) passes from them into the cells concerned. Various aspects of the subject are reviewed by Buck (1962), Miller (1974) and Keister and Buck (1974).

The entry of oxygen into the tracheal system of terrestrial forms occurs for the greater part through the spiracles in all except a few apneustic insects like the larvae of *Forcipomyia*. In this, as in the apneustic aquatic and endoparasitic forms mentioned later, there is, over part or most of the body, a rich subcutaneous plexus of fine tracheae which provides a large surface across which diffusion can occur. The transport of oxygen to the internal tracheal endings of insects was long thought to depend simply on diffusion of oxygen along a gradient of partial pressure, but only with the work of Krogh (1920) was an accurate study of this undertaken. The relation between the oxygen consumption of an insect and its tracheal dimensions is given by the formula:

$$S = \frac{k(p - p')A}{L}$$

where S = ml of oxygen used per second; p = partial pressure of oxygen in the atmosphere (c. 0·2 of an atmosphere); p' = partial pressure of oxygen at the ends of the tracheae; A = mean cross-sectional area of the tracheae in sq. cm, L = mean tracheal length in cm and k = the diffusion constant for oxygen (i.e. 0·18). From this formula it may be shown that diffusion alone is quite sufficient to account for transport of oxygen through the entire tracheal system of small or inactive insects – the observed oxygen uptake of *Cossus* larvae, for instance, could result from a difference of only 11 mm Hg in oxygen tension between the atmosphere and the beginning of the tracheoles. Despite inaccuracies in some of the assumptions involved (Buck, 1962; Nunome, 1944–51), the principle is valid; diffusion is probably the only factor involved in most larvae, pupae and small adults and is the sole mode of transfer in the terminal parts of the tracheal system. It may, however, be supplemented in other cases by two forms of mass transfer of respiratory gases: *passive suction ventilation* and active, skeletomuscular *mechanical ventilation*. Suction ventilation has been studied most fully in the diapausing pupae of *Hyalophora cecropia*, where it is associated with discontinuous respiration (e.g., Schneiderman and Schechter, 1966; Levy and Schneiderman, 1966; Brockway and Schneiderman, 1967). Carbon dioxide is retained and released in bursts during short periods of spiracular activity; there is an accompanying cycle of intratracheal pressure, during part of which air is sucked in through the spiracles. In active ventilation, compression of the abdomen and thorax results in the expulsion through the spiracles of air from the air-sacs and those tracheae whose walls are less resistant to collapse. The ensuing expansion of these parts causes a fresh supply of air to be drawn in, as much as two-thirds of the tracheal volume being changed in some cases. The ventilatory movements may take the form of dorsoventral compression – in which the terga or sterna or both are moved – or, in the Orthoptera and Aculeate Hymenoptera of alternate telescoping and protrusion of the abdominal segments. Reference to the diffusion formula cited above will show that in non-ventilating insects the maximum possible body size is limited, since the oxygen consumption is affected more by a linear increase in size than is the rate of diffusion. The same effect also operates, though to a lesser extent, in ventilating forms since oxygen passes along the finer branches only by diffusion.

The regulation of respiratory activity results from *spiracular control*, due to the opening and closing of spiracles, from *ventilation control* caused by variations in the frequency and intensity of respiratory movements, and from the coordinated activity of both processes (Miller, 1966*b*, 1974). In the flea *Xenopsylla cheopis*, for example, spiracular movements are regulated by the oxygen concentration of the tracheal air and the accumulation of carbon dioxide in the tissues, the spiracles being open for a greater length of time when the oxygen is depleted and the concentration of carbon dioxide is high (Wigglesworth, 1935). The effective stimulus in some such cases is probably the direct, local action of carbon dioxide (Case, 1957; Hoyle, 1960), but in others there is evidence that carbon dioxide and oxygen lack induce ganglia of the ventral nerve cord to modify their motor discharge to the spiracular muscles. Starvation, changes in water balance and temperature, and injury can also affect spiracular control, which differs in detail from one species to another or even between different spiracles in the same insect. The ventilating movements are likewise stimulated by lack of oxygen and accumulation of carbon dioxide and are subject to nervous control by local, segmental (primary) centres and by secondary thoracic centres. In *Schistocerca* these complicated neural mechanisms seem to depend on a metathoracic pacemaker whose activity is distributed by a coordinating interneuron that runs along the abdominal nerve cord to segmental centres, each comprising further interneurons and motor neurons (Lewis *et al.*, 1973). By combining ventilation movements with controlled opening and closing of selected spiracles a directed flow of air through the tracheal system is made possible, as in some Acridids in which air enters at times through the two thoracic and two anterior abdominal pairs of spiracles and leaves by the more posterior abdominal pairs (Fraenkel, 1932; Weis-Fogh, 1967).

Respiratory exchange between the tracheoles and the tissues depends on diffusion. This process is very much slower if the oxygen is diffusing through liquids than it is in the gas phase, but the tissue diffusion path is normally very short (10 μm or less) so that no other physical process need be invoked (Weis-Fogh, 1964*b*). The tracheoles may be entirely filled with gas or contain liquid in their terminal parts, the amount of liquid in the tracheole being affected by the osmotic pressure of the surrounding fluid. Thus, in active muscle with an inadequate oxygen supply the accumulation of metabolites raises the osmotic pressure of the tissue fluids, liquid is withdrawn from the tracheoles through colloid imbibition and is replaced by air, so improving the supply of oxygen to these tissues (Wigglesworth, 1930–31; 1953).

The diffusion processes which account for the inward transport of oxygen also permit the outward movement of carbon dioxide, but the tissues and cuticle are more permeable to carbon dioxide than to oxygen so that an appreciable proportion of the former can escape through the tracheal walls and the general surface of the body. The blood plays a relatively minor role in respiration (p. 239).

Water loss from the general body surface is restricted by the epicuticular wax layer (p. 12) but the permeability of the tracheoles makes loss of water through the spiracles a serious matter in terrestrial species. Excessive loss is prevented by reduction of ventilatory movements (Loveridge, 1968) or by the use of the spiracular closing mechanism or the development of a felt-chamber which reduces the diffusion of water-vapour across the spiracular opening. Xerophilous species such as the Tenebrionid beetles of deserts may show strongly developed mechanisms of this sort (Ahearn, 1970; see also Bergold, 1935) while, on the other hand, for example, aquatic Heteroptera lack a closing mechanism (Mammen, 1912).

Respiration of Aquatic and Endoparasitic Insects

These modes of life are similar in that the insects are surrounded by a liquid or semi-liquid medium and must either extract dissolved oxygen from it by diffusion or retain a connexion with an atmospheric supply (Wesenberg-Lund, 1943; Keilin, 1944; Clausen, 1950; Mill, 1974).

(a) **Aquatic Insects** – The least highly modified to an aquatic habit are those with an open respiratory system, air entering the tracheae through one or more pairs of spiracles. They include several different types:

(i) There are first the oligopneustic forms without external air-stores (e.g. larvae of many Diptera and Dytiscidae). They are commonly metapneustic and possess a hydrofuge spiracular area with which they penetrate the surface film and so secure atmospheric oxygen. In many cases the spiracles are situated on the end of a siphon which may be short (e.g. some Stratiomyid and Culicid larvae) or long and retractile as in the 'rat-tailed' larvae of Ptychopteridae and most Eristaline Syrphidae. Comparable open prothoracic 'respiratory horns' occur in the pupae of many aquatic Nematoceran Diptera.

(ii) An essentially similar respiratory adaptation occurs in those forms where the spiracles are situated on sharply pointed processes which penetrate the air-containing cavities of the submerged parts of aquatic plants. Such a method of obtaining oxygen has been evolved in the larvae and pupae of several widely distinct genera of Diptera and Coleoptera including *Donacia, Lissorhoptrus, Noterus, Taeniorhynchus, Notiphila* (Varley, 1937; Hartley, 1958; Houlihan, 1969–70).

(iii) Many aquatic Coleoptera and Cryptocerate Heteroptera are provided with renewable external air-bubbles which lie beneath the elytra (e.g. *Dytiscus*) or are trapped in a hydrofuge hair-pile on other parts of the body. Such bubbles are in contact with the spiracles and act not only as hydrostatic organs and stores of oxygen but also as 'physical gills' (Ege, 1915; Popham, 1960, 1962; Rahn and Paganelli, 1968; Vlasblom, 1970; Wolvekamp, 1955). As oxygen is removed from the bubble by the insect, equilibrium is restored by the diffusion of oxygen into the bubble from the water more rapidly than the nitrogen of the bubble diffuses out. The insect can thus use the bubble to extract from the water far more oxygen than was originally present, though the bubble gradually decreases in size as the nitrogen diffuses away and requires periodic replacement from the atmosphere. This may be brought about by the insect pushing the hind end of the body through the surface film or by the use of a more specialized respiratory siphon (Nepidae, Belostomatidae) or, in the Hydrophilidae (Hrbáček, 1950), by hydrofuge antennal hairs forming an air-channel along part of the antennae to the ventral surface of the insect.

(iv) A considerable number of insects such as the Heteropteran *Aphelocheirus* and representatives of several families of beetles (*Haemonia, Phytobius*, some Dryopidae, Elmidae and others) have adopted 'plastron respiration' (Thorpe, 1950; Hinton, 1969). The plastron is a special type of air-store in the form of a thin film communicating with the spiracles and so held by a system of hydrofuge hairs, scales or other cuticular processes that its volume remains constant. Provided there is adequate oxygen dissolved in the water, the plastron can act as a *permanent* physical gill which needs no renewal and insects so equipped can remain submerged continually. A plastron of various kinds is also found in almost all spiracular gills (see below) and may form part of the respiratory system of insect eggs (p. 300). A rather different form of 'permanent physical gill' occurs in the African beetle *Potamodytes tuberosus*, where the bubble holds gas below atmospheric pressure and can therefore draw in oxygen from the saturated water that flows quickly past it (Stride, 1955).

Other aquatic insects are usually provided with a closed (apneustic) tracheal system into which oxygen passes from the surrounding medium (Koch, 1938). Such an arrangement is found in the immature stages of the

Ephemeroptera, Odonata, Plecoptera, Trichoptera, Sialoidea and larvae of many Dipteran families. Respiratory exchange occurs over the general body surface and across the thin cuticle of special respiratory organs – the gills or *branchiae*. The latter name was given in the past to almost any thin-walled integumentary outgrowth with a relatively large surface area and it was usual to distinguish between tracheal gills and blood-gills. Physiological studies (Thorpe, 1933, etc.) have shown that the blood-gills have little or no respiratory function while in some cases the tracheal gills may take up less oxygen than the remainder of the body surface. Three kinds of gills or gill-like structures have been distinguished.

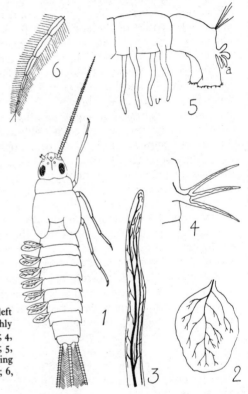

FIG. 116

Gills of aquatic insects

1, Nymph of *Cloeon* showing tracheal gills of left side; 2, 7th tracheal gill of *Cloeon* more highly magnified; 3, tracheal gill of a *Phryganea* larva; 4, tracheal gill of a larva of *Nymphula stratiotata*; 5, hind extremity of a larva of *Chironomus* showing anal blood-gills (*a*) and ventral blood-gills (*v*); 6, jointed tracheal gill of a larva of *Sialis*.

Tracheal gills are filiform or more or less lamellate structures that are well supplied with tracheae and tracheoles, the latter often arranged regularly at optimal distances apart and very close to the cuticle (Wichard, 1973; Wichard and Komnick, 1971, 1974). They are present in the majority of aquatic larvae and in some aquatic pupae. In many cases they are the only organs of respiration but in others (larval Culicidae for example) they are accessory in function and co-exist with open spiracles. Tracheal gills are usually borne on the abdomen: they are less frequently present on the thorax, and are only very rarely found on the head, as in some Plecoptera and in *Jolia* and *Oligoneuria* among the Ephemeroptera. Experimental proof of their respiratory role – at least when oxygen is scarce – has been obtained in several insects, such as *Ephemera* (Eriksen, 1963), *Agrion* (Harnisch, 1958) and the stonefly *Paragnetina* (Kapoor, 1974). The immature stages of Trichoptera, *Cloeon* (Eastham, 1958) and other species carry out movements that cause a water current to flow over the body surface and gills. In a few instances the gills of the larvae persist thoughout life in the imago; this is best exhibited in *Pteronarcys* whose imagines possess thirteen pairs of gill-tufts on the ventral surface of the thoracic and first two abdominal segments. Tracheal gills similarly persist in other Plecoptera and in *Hydropsyche* among Trichoptera but they are usually retained in a more or less shrivelled condition.

In the Ephemeroptera tracheal gills are usually borne on the first seven abdominal segments and may be either lamellate or filamentous. When lamellate each gill may consist of a simple leaf-like expansion (*Cloeon*) or the lamella may form a cover which protects a tuft of filamentous gills beneath (*Heptagenia*). In *Caenis* the upper lamellae of the 2nd pair of gills form opercula which conceal and protect the gills behind. In *Prosopistoma* the gills are entirely hidden within a special branchial chamber.

In the Plecoptera primitive abdominal gills occur in the Eustheniidae, but in the nymphs of other forms they are replaced by secondary tufts of filaments which are variable in position.

Tracheal gills are universally present in the nymphs of Odonata. In the Anisoptera they form an elaborate system of folds in the wall of the rectum, the latter chamber being modified to form what is termed the branchial basket. In most Zygoptera there are three external caudal gills and no rectal gills; in a few rare cases lateral filamentous abdominal gills are also present.

Among Neuroptera gills are present in the larvae of the Sialoidea and in *Sisyra* among the Plannipennia. They consist of seven or eight pairs of filaments, usually jointed, borne segmentally on the abdomen.

Filamentous abdominal gills are present in most larval Trichoptera and frequently persist in the pupae. In some genera although the larvae are without gills the pupae are provided with well-developed branchial organs.

Among Lepidoptera tracheal gills have long been known in the larva of *Nymphula stratiotata* and a few other species; they consist of a series of delicate filaments arising from the sides of the trunk segments.

Among Coleopteran larvae tracheal gills are filamentous in character and are only present in a few families. In *Hygrobia* they are ventral and are located near the bases of each of the pairs of legs and on the first three abdominal segments. In the Gyrinidae there are 10 pairs of hair-fringed lateral abdominal gills; somewhat similar organs are also found in *Hydrocharis* and *Berosus* among the Hydrophilidae. In *Peltodytes* they take the form of numerous elongate jointed filaments which arise from the dorsal surface of the thorax and abdomen.

Among Dipteran larvae there are four lamellate anal gills in the Culicidae; in *Phalacrocera* the tracheal gills are in the form of numerous elongate filamentous processes which arise from almost all parts of the body segments; in *Simulium* and *Eristalis* rectal gills are present.

Spiracular Gills. In some aquatic pupae the peritreme and atrial regions of one or more pairs of spiracles are drawn out to form long processes, the spiracular gills (Hinton, 1968). They have been evolved on several separate occasions in over 1400 species among the Coleopteran families Psephenidae and Torridincolidae and several Dipteran families such as the Tipulidae, Blephariceridae, Deuterophlebiidae, Simuliidae and Empididae. Spiracular gills are also found in the larvae of other Myxophagan beetles (Sphaeriidae and Hydroscaphidae). In all cases except the Chironomidae they are provided with a plastron which is held by a variety of different forms of cuticular sculpturing. Insects with spiracular gills are characteristically found in well-oxygenated aquatic habitats subject to alternate flooding and drying out. Such pupae as have been tested are able to live in damp air and to complete their development out of water, so that the spiracular gills are adapted both for aquatic and aerial respiration.

The so-called **blood-gills** are commonly tubular or digitiform and are sometimes eversible. They derive their name from the fact that they contain blood but not as a rule tracheae, although occasional tracheoles may be present. In some instances there is little real distinction between these organs and tracheal gills. Blood-gills are of infrequent occurrence and are not exclusively confined to aquatic insects. They are found among many larval Trichoptera which have 4 to 6 finger-like tubes at the anal extremity. Among Diptera they are well developed in the larvae of *Chironomus*, some species of which bear two pairs of ventral blood-gills on the penultimate segment, and a group of four shorter anal gills. Small anal blood-gills are also met with among aquatic Tipulid larvae. In Culicids and Chironomids, at least, the function of these structures is the absorption of water and inorganic ions (Wigglesworth, 1933; Koch, 1938) rather than respiration and it is worth noting that there is also some ultrastructural evidence for chloride absorption in the tracheal gills of the Plecopteran *Paragnetina media* (Kapoor and Zachariah, 1973).

(*b*) **Endoparasitic Insects** – In most endoparasitic Dipteran larvae and a few parasitic Hymenoptera, a connection is established between the atmosphere and the open tracheal system of the parasite. In 1st-instar larvae of the Encyrtidae (Maple, 1947) this is brought about by the larva retaining a close association with the egg-shell which is connected with the surface of the host by a stalk-like prolongation and bears the so-called aeroscopic plate – a strip of modified chorion acting as an air-channel. Larvae of the Conopidae become attached to a tracheal trunk of their host though they appear not to penetrate it. Others perforate the body-wall or trachea so as to put their spiracles in contact with the atmosphere (e.g. the larvae of Nemestrinidae (Prescott, 1961; Leonide, 1963) and old 3rd instar larvae of *Cryptochaetum*, Thorpe, 1934) and the host is sometimes stimulated to surround the parasite, except for the spiracular region, with a cuticular sheath (cf. Salt, 1968). In the remaining endoparasitic forms with a closed tracheal system, respiratory exchange occurs through the thin cuticle, beneath which lies a rich tracheal supply. Other early instars of parasitic species are either atracheate or have a system filled with liquid and in them the oxygen must diffuse directly into the haemolymph. The larvae of many Ichneumonidae, Braconidae and Chalcidoidea possess tail-like appendages which are probably not respiratory structures, but the paired, richly tracheated tail filaments of *Cryptochaetum* larvae and the blood-filled caudal vesicle of Braconid larvae are responsible for some oxygen uptake (Thorpe, 1932–41).

References on the Respiratory System

AFZELIUS, B. A. AND GONNERT, N. (1972), Intramitochondrial tracheoles in flight-muscle from the hornet, *Vespa crabro* (Hym., Vespidae, *J. Submicr. Cytol.*, **4**, 1–6.

AHEARN, G. A. (1970), The control of water loss in desert Tenebrionid beetles, *J. exp. Biol.*, **53**, 573–595.

BEAULATON, J. (1968), Modifications ultrastructurales des trachées et genèse de petites trachées et trachéoles chez les vers à soie en periode de mue. *J. Microscopie*, **7**, 621–646.

BEINBRECH, G. (1970), Zur Flugmuskelentwicklung von *Phormia regina*: Beziehungen zwischen dem Sarkotubulären und dem Trachealsystem, *Zool. Anz.*, Suppl., **33**, 401–407.

BERGOLD, G. (1935), Die Ausbildung der Stigmen bei Coleopteren verschiedener Biotope, *Z. Morph. Ökol. Tiere*, **29**, 511–526.

BHATNAGAR, B. S. (1972), Spiracles in certain terrestrial Heteroptera, *Int. J. Insect Morphol & Embryol.*, **1**, 207–217.

BROCKWAY, A. P. AND SCHNEIDERMAN, H. A. (1967), Strain-gauge transducer studies on intratracheal pressure and pupal length during discontinuous respiration in diapausing silkworm pupae, *J. Insect Physiol.*, **13**, 1413–1451.

BUCK, J. (1962), Some physical aspects of insect respiration, *A. Rev. Ent.*, **7**, 27–56.

BURSELL, E. (1955), The polypneustic lobes of the tsetse larva (*Glossina*, Diptera), *Proc. R. Soc* (B), **144**, 275–286.

CASE, J. F. (1957), Differentiation of the effects of pH and CO_2 on spiracular function of insects. *J. cell. comp. Physiol.*, **49**, 103–113.

CLAUSEN, C. P. (1950), Respiratory adaptations in the immature stages of parasitic insects, *Arthropoda*, **1**, 197–224.

COMSTOCK, J. H. (1918), *The Wings of Insects*, Comstock Publ. Co., Ithaca, New York, 430 pp.

DAMANT, G. C. C. (1924), The adjustment of the buoyancy of the larva of *Corethra plumicornis*, *J. Physiol.*, **59**, 345–356.

DINULESCU, G. (1932), Recherches sur la biologie des Gastrophiles. Anatomie, physiologie, cycle évolutif, *Ann. Sci. nat. Paris*, (10), **15**, 1–183.

EASTHAM, L. E. S. (1958), The abdominal musculature of nymphal *Chloeon dipterum* L. (Insecta: Ephemeroptera) in relation to gill movements and swimming, *Proc. zool. Soc. Lond.*, **131**, 279–291.

EDWARDS, G. A., RUSKA, H. AND DE HARVEN, E. (1958), The fine structure of insect tracheoblasts, tracheae and tracheoles, *Arch. Biol.*, **69**, 351–369.

EGE, R. (1915), On the respiratory function of the air stores carried by some aquatic insects, *Z. allg. Physiol.*, **17**, 81–124.

ERIKSEN, C. H. (1963), Respiratory regulation in *Ephemera simulans* Walker and *Hexagenia limbata* (Serville) (Ephemeroptera), *J. exp. Biol.*, **40**, 455–467.

FAUCHEUX, M. J. (1972), Relations entre l'ultrastructure de l'intima cuticulaire et les fonctions des sacs aeriens chez les insectes, *C. r. Acad. Sci., Paris*, D, **274**, 1518–1521.

FAUCHEUX, M. J. AND SELLIER, R. (1971), L'ultrastructure de l'intima cuticulaire des sacs aeriens chez les insectes, *C.r. Acad. Sci., Paris*, D, **272**, 2197–2200.

FRAENKEL, G. (1932), Beiträge zur Physiologie der Atmung der Insekten. *Arch. Zool., Torino*, **16**, 905–921.

FULLER, C. (1919), The wing venation and respiratory system of certain South African termites, *Ann. Natal Mus.*, **4**, 19–102.

GRYSE, J. J. DE (1926), The morphogeny of certain types of respiratory systems in insect larvae, *Trans. R. Soc. Canada*, **20**, 483–503.

HARNISCH, O. (1958), Untersuchungen an den Analkiemen der Larve von *Agrion*, *Biol. Zbl.*, **77**, 300–310.

HARTLEY, J. C. (1958), The root-piercing spiracles of the larva of *Chrysogaster hirtella* Loew (Diptera: Syrphidae) *Proc. R. ent. Soc. Lond.* (A), **33**, 81–87.

HASSAN, A. A. G. (1944), The structure and mechanism of the spiracular regulatory mechanism in adult Diptera and certain other groups of insects, *Trans. R. ent. Soc. Lond.*, **94**, 105–153.

HINTON, H. E. (1947), On the reduction of functional spiracles in the aquatic larvae of the Holometabola, with notes on the moulting process of spiracles, *Trans. R. ent. Soc. Lond.*, **98**, 449–473.

—— (1967a), Structure and ecdysial process of the larval spiracles of the Scarabaeoidea, with special reference to those of *Lepidoderma*, *Aust. J. Zool.*, **15**, 947–953.

—— (1967b), On the spiracles of the larvae of the suborder Myxophaga (Coleoptera), *Aust. J. Zool.*, **15**, 955–959.

—— (1968), Spiracular gills, *Adv. Insect Physiol.*, **5**, 65–162.

—— (1969), Plastron respiration in adult beetles of the suborder Myxophaga, *J. Zool. Lond.*, **159**, 131–137.

HOULIHAN, D. F. (1969a), Respiratory physiology of the larva of *Donacia simplex*, a root-piercing beetle, *J. Insect Physiol.*, **15**, 1517–1536.

—— (1969b), The structure and behaviour of *Notiphila riparia* and *Erioptera squalida*, two root-piercing insects, *J. Zool. Lond.*, **159**, 249–267.

—— (1970), Respiration in low oxygen partial pressures: the adults of *Donacia simplex* that respire from the roots of aquatic plants, *J. Insect Physiol.*, **16**, 1607–1622.

HOYLE, G. (1960), The action of carbon dioxide gas on an insect spiracular muscle, *J. Insect Physiol.*, **4**, 63–79.

HRBÁČEK, J. (1950), On the morphology and function of the antennae of the Central European Hydrophilidae (Coleoptera), *Trans. R. ent. Soc. Lond.*, **101**, 239–256.

KAPOOR, N. N. (1974), Some studies on the respiration of stonefly nymph, *Paragnetina media* (Walker), *Hydrobiologia*, **44**, 37–41.

KAPOOR, N. N. AND ZACHARIAH, K. (1973), A study of specialized cells of the tracheal gills of *Paragnetina media* (Plecoptera), *Can. J. Zool.*, **51**, 983–986.

KEILIN, D. (1944), Respiratory systems and respiratory adaptations in larvae and pupae of Diptera, *Parasitology*, **36**, 1–66.

KEISTER, M. L. (1948), The morphogenesis of the tracheal system of *Sciara*, *J. Morph.*, **83**, 373–423.

KEISTER, M. AND BUCK, J. (1974), Respiration: some exogenous and endogenous effects on rate of respiration, In: Rockstein, M. (ed.), *The Physiology of Insecta*, 2nd edn, Academic Press, New York, **6**, 469–509.

KOCH, H. J. A. (1938), The absorption of chloride ions by the anal papillae of Diptera larvae, *J. exp. Biol.*, **15**, 152–160.

KROGH, A. (1920), Studien über Tracheenrespiration. II, III, *Pflügers Arch. ges. Physiol.*, **179**, 95–112.

LÉONIDE, J.-C. (1963), Formation du pore respiratoire et de la partie proximale du tube respiratoire de la larve de *Symmictus costatus* Loew. (Diptera, Nemestrinidae) selon des diverses régions du corps de l'hôte, *Bull. Soc. zool. France*, **87** (1962), 550–558.

LEVY, R. I. AND SCHNEIDERMAN, H. A. (1966), Discontinuous respiration in insects. II–IV, *J. Insect Physiol.*, **12**, 83–104; 105–121; 465–492.

LEWIS, G. W., MILLER, P. L. AND MILLS, P. S. (1973), Neuro-muscular mechanisms of abdominal pumping in the locust, *J. exp. Biol.*, **59**, 149–168.

LOCKE, M. (1957), The structure of insect tracheae, *Q. Jl microsc. Sci.*, **98**, 487–492.

—— (1958), The formation of tracheae and tracheoles in *Rhodnius prolixus*, *Q. Jl microsc. Sci.*, **99**, 29–46.

LOTZ, G. (1961), Vergleichend morphologische und histologische Untersuchungen an den Stigmen der Lamellicornier-Larven mit Beiträgen zur Entwicklungsgeschichte, *Z. Morph Ökol. Tiere*, **50**, 726–784.

LOVERIDGE, J. P. (1968), The control of water loss in *Locusta migratoria migratorioides* R. & F. II. Water loss through the spiracles, *J. exp. Biol.*, **49**, 15–29.

MAMMEN, H. (1912), Über die Morphologie der Heteropteren- und Homopterenstigmen, *Zool. Jb.* (*Anat.*), **34**, 121–178.

MAPLE, J. D. (1947), The eggs and first instar larvae of Encyrtidae and their morphological adaptations for respiration, *Univ. Calif. Publs Ent.*, **8**, 25–122.

MILL, P. J. (1974), Respiration – aquatic insects, In: Rockstein, M. (ed.), *The Physiology of Insecta*, 2nd edn, Academic Press, New York, **6**, 403–467.

MILLER, P. L. (1966a), The supply of oxygen to the active flight muscles of some large beetles, *J. exp. Biol.*, **45**, 285–304.

—— (1966b), The regulation of breathing in insects, *Adv. Insect Physiol.*, **3**, 279–354.

—— (1974), Respiration – aerial gas transport, In: Rockstein, M. (ed.), *The Physiology of Insecta*, 2nd edn, Academic Press, New York, **6**, 345–402.

NUNOME, Z. (1944–51), Studies on the respiration of the silkworm, I–III, *Bull. Sericult. Exp. Sta.*, **12**, 17–39; 41–90; *J. Sericult. Sci. Japan*, **20**, 111–127.

POPHAM, E. J. (1960), On the respiration of aquatic Hemiptera Heteroptera with special reference to the Corixidae, *Proc. zool. Soc. Lond.*, **135**, 209–242.

—— (1962), A repetition of Ege's experiments and a note on the efficiency of the physical gill of *Notonecta* (Hemiptera–Heteroptera), *Proc. R. ent. Soc. Lond.* (A), **37**, 154–160.

PRESCOTT, H. W. (1961), Respiratory pore construction in the host by the nemestrinid parasite *Neorhynchocephalus sackenii* (Diptera), with notes on respiratory tube characters, *Ann. ent. Soc. Am.*, **54**, 557–566.

RAHN, H. AND PAGANELLI, C. V. (1968), Gas exchange in gas gills of diving insects, *Respiration Physiology*, **5**, 145–164.

RITCHER, P. O. (1969a), Spiracles of adult Scarabaeoidea (Coleoptera) and their phylogenetic significance. I. The abdominal spiracles, *Ann. ent. Soc. Am.*, **62**, 869–880.

—— (1969b), Spiracles of adult Scarabaeoidea (Col.) and their phylogenetic significance. II. Thoracic spiracles and adjacent sclerites, *Ann. ent. Soc. Am.*, **62**, 1388–1389.

SALT, G. (1968), The resistance of insect parasitoids to the defence reactions of their hosts, *Biol. Rev.*, **43**, 200–232.

—— (1970), *The Cellular Defence Reactions of Insects*, Cambridge Univ. Press, Cambridge, 117 pp.

SCHNEIDERMAN, H. A. AND SCHECHTER, A. N. (1966), Discontinuous respiration in insects – V, *J. Insect Physiol.*, **12**, 1143–1170.

SMITH, D. S. (1963), The organization and innervation of the luminescent organ in a firefly, *Photuris pennsylvanicus* (Coleoptera), *J. Cell Biol.*, **16**, 323–359.

STOBBART, R. H. (1956), A note on the tracheal system of the Machilidae, *Proc. R. ent. Soc. Lond.* (A), **31**, 34–36.

STRIDE, G. O. (1955), On the respiration of an aquatic African beetle, *Potamodytes tuberosus* Hinton, *Ann. ent. Soc. Am.*, **48**, 344–351.

THORPE, W. H. (1932), Experiments upon respiration in the larvae of certain parasitic Hymenoptera, *Proc. R. Soc.* (B), **109**, 450–471.

—— (1933), Experiments on the respiration of aquatic and parasitic insect larvae, *Trans. 5 Congr. int. Ent.*, **2**, 345–351.

—— (1934), The biology and development of *Cryptochaetum grandicorne* (Diptera), an internal parasite of *Guerinia serratulae* (Coccidae), *Q. Jl microsc. Sci.*, **77**, 273–304.

—— (1941), The biology of *Cryptochaetum* (Diptera) and *Eupelmus* (Hymenoptera), parasites of *Aspidoproctus* (Coccidae) in East Africa, *Parasitology*, **33**, 149–168.

—— (1950), Plastron respiration in aquatic insects, *Biol. Rev.*, **25**, 344–390.

TONAPI, G. T. (1957), A comparative study of spiracular structure and mechanism in some Hymenoptera, *Trans. R. ent. Soc., Lond.*, **110**, 489–520.

—— (1959–60), A comparative study of the respiratory system of some Hymenoptera, *Indian J. Ent.*, **20**, 108–120; 203–220; 245–269.

VARLEY, G. C. (1937), Aquatic insect larvae which obtain oxygen from the roots of plants, *Proc. R. ent. Soc. Lond.* (A), **12**, 55–60.

VLASBLOM, A. G. (1970), The respiratory significance of the physical gill in some adult insects, *Comp. Biochem. Physiol.*, **36**, 377–385.

WEBB, J. E. (1946), Spiracle structure as a guide to the phylogenetic relationships of the Anoplura (biting and sucking lice) with notes on the affinities of the mammalian hosts, *Proc. zool. Soc. Lond.*, **116**, 49–119.

WEIS-FOGH, T. (1964a), Functional design of the tracheal system of flying insects as compared with the avian lung, *J. exp. Biol.*, **41**, 207–227.

—— (1964b), Diffusion in insect wing muscle, the most active tissue known, *J. exp. Biol.*, **41**, 229–256.

—— (1967), Respiration and tracheal ventilation in locusts and other flying insects, *J. exp. Biol.*, **47**, 561–587.

WESENBERG-LUND, C., (1943), *Biologie der Süsswasserinsekten*, Copenhagen, 682 pp.

WHITTEN, J. M. (1955), A comparative morphological study of the tracheal system in larval Diptera. Part 1, *Q. Jl microsc. Sci.*, **96**, 257–278.

—— (1960), The tracheal patterns in selected Diptera Nematocera, *J. Morph.*, **107**, 233–257.

—— (1972), Comparative anatomy of the tracheal system, *A. Rev. Ent.*, **17**, 373–402.

WICHARD, W. (1973), Zur Morphogenese des respiratorischen Epithels der Tracheenkiemen bei Larven der Limnephilini Kol. (Insecta, Trichoptera), *Z. Zellforsch. mikrosk. Anat.*, **144**, 585–592.

WICHARD, W., AND KOMNICK, H. (1971), Zur Feinstruktur der Tracheenkiemen von *Glyphotaelius pellucidus* (Insecta, Trichoptera), *Cytobiologie*, **3**, 106–110. ·

—— (1974), Structure and function of the respiratory epithelium in the tracheal gills of stonefly larvae, *J. Insect. Physiol.*, **20**, 2397–2406.

WIGGLESWORTH, V. B. (1930), A theory of tracheal respiration in insects, *Proc. R. Soc.*, (B), **106**, 229–250.

—— (1931), The extent of air in the tracheoles of some terrestrial insects, *Proc. R. Soc.* (B), **109**, 354–369.

—— (1933), The function of the anal gills of the mosquito larva, *J. exp. Biol.*, **10**, 16–26.

—— (1935), The regulation of respiration in the flea, *Xenopsylla cheopis* Roths. (Pulicidae), *Proc. R. Soc.* (B), **118**, 397–419.

—— (1953), Surface forces in the tracheal system of insects, *Q. Jl microsc. Sci.*, **94**, 507–522.

—— (1959), The role of the epidermal cells in the 'migration' of tracheoles in *Rhodnius prolixus* (Hemiptera), *J. exp. Biol.*, **36**, 632–640.

—— (1963), A further function of the air sacs in some insects, *Nature*, **198**, 106.

WOLVEKAMP, H. P. (1955), Die physikalische Kieme der Wasserinsekten, *Experientia*, **11**, 294–301.

Chapter 14

THE CIRCULATORY SYSTEM

Among insects the circulatory system is usually open, with only a single closed dorsal vessel. The greater part of the circulation takes place in the cavities of the body and its appendages, the blood occupying the spaces not appropriated by the internal organs. The larger spaces may be enclosed by special membranes and form definite sinuses but except for the aorta, and the segmental blood-vessels of the Dictyoptera, there are usually no definite veins or arteries such as are found in many other Arthropoda. In the appendages and wing-veins, however, the blood flows in streams along defined channels analogous to blood-vessels and in the nymph of *Cloeon* the hindmost chamber of the heart gives off three caudal arteries which enter the tail appendages.

The organs and tissues belonging to the circulatory system are separately dealt with below; for a review see Gouin (1970).

The Diaphragms and Sinuses

When the diaphragms are completely developed the general body-cavity or haemocoele is divided into three sinuses by means of two fibro-muscular septa (Fig. 117). The *dorsal diaphragm* – which is partly double in *Grylloblatta* and some Tettigonioidea – is the principal septum and the one most generally prevalent. It extends across the abdominal cavity above the alimentary canal and the blood-space thus enclosed is known as the *dorsal* or *pericardial sinus*. This is situated beneath the abdominal terga and within it is located the heart. The *ventral diaphragm*, when present, stretches across the abdominal cavity just above the ganglia of the ventral nerve-cord, and the space limited by it is the *ventral* or *perineural sinus* (Richards, A. G., 1963; Dierichs, 1972). Between the dorsal and ventral sinuses is the large central cavity or *visceral sinus* containing the principal internal organs.

Pairs of *aliform* or *alary muscles*, composed of striated fibres (Sanger & McCann, 1968*b*) arise from the terga and spread out fanwise over the surface of the dorsal diaphragm. The fibres of one alary muscle attach to the heart or meet, beneath the heart, those of the corresponding muscle of the opposite

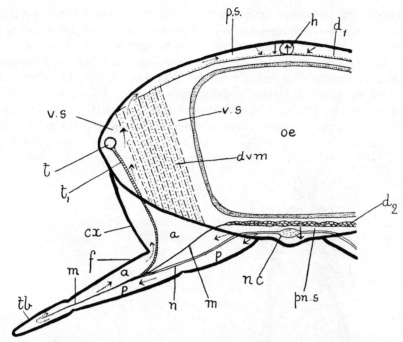

FIG. 117   Schematic transverse section of the thorax of *Periplaneta* showing the diaphragms and sinuses

The plain arrows indicate the course of the circulation towards the head and the dotted arrows signify transverse currents more or less parallel with the plane of the paper. *h*, dorsal vessel; d_1, dorsal diaphragm; d_2, ventral diaphragm; *p.s*, pericardial sinus; *oe*, oesophagus; *v.s*, visceral sinus; *dvm*, dorsoventral muscles; *t*, lateral tracheal trunk; t_1, leg trachea; *pn.s*, perineural sinus; *nc*, nerve-cord; *n*, nerve to leg. The cavity of the leg is divided into an anterior sinus *a* and a posterior sinus *p* either by muscles or by a membrane *m*; in the femur the trachea and nerve are attached to the membrane; *cx*, coxa; *tb*, tibia and tarsus. Adapted from Brocher, *Ann. Soc. ent. Fr.*, 1922.

side of the body. These muscles vary in number; in *Periplaneta*, for example, there are 12 pairs of alary muscles (Fig. 118), in the honey bee 4 pairs, in *Haematopinus* 3 pairs and in the larva of *Chironomus* 2 pairs (see also Nutting, 1951; Hinks, 1966).

The Dorsal Vessel

The dorsal vessel extends from near the caudal end of the body, through the thorax, and terminates in the head. It lies along the median dorsal line just beneath the integument and is protected by the dorsal diaphragm below. Morphologically it is a continuous tube, usually closed posteriorly, and always open at its cephalic extremity. It comprises two regions, the *heart* or pumping organ and a conducting vessel or *aorta*.

The *heart* (Gerould, 1938; Nutting, 1951; Hinks, 1966; Richter, 1973) is maintained in position within the pericardial sinus by suspensory filaments attached to the abdominal terga and frequently to the dorsal diaphragm also. It may be divided by successive constrictions into a series of chambers but is otherwise a uniform tube, and its segmentation is only shown by the presence of paired incurrent ostia (described below) and the alary muscles. In the most primitive condition there are thus signs that the heart occupies the three thoracic and first nine abdominal segments. Among most insects, however, it is restricted to the abdomen and is variously shortened from both extremities so that fewer segments participate. Thus, in the

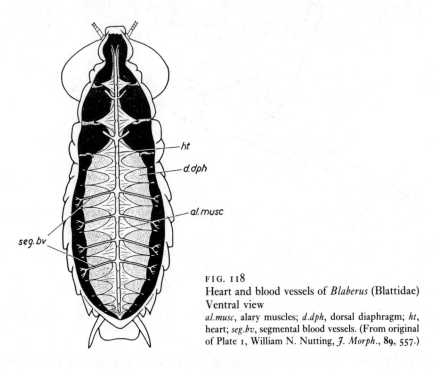

FIG. 118
Heart and blood vessels of *Blaberus* (Blattidae) Ventral view
al.musc, alary muscles; *d.dph*, dorsal diaphragm; *ht*, heart; *seg.bv*, segmental blood vessels. (From original of Plate 1, William N. Nutting, *J. Morph.*, **89**, 557.)

Dictyoptera, *Grylloblatta* and some primitive Tettigonioidea the generalized number of 12 segments is involved; in *Japyx* there are 10, in *Lucanus cervus* 7, among Aculeate Hymenoptera and in *Musca* there are 5 divisions; in a few insects the heart is reduced to only one chamber. Histologically the heart is composed of a single layer of striated muscle cells (Edwards and Challice, 1960; Sanger and McCann, 1968a; Myklebust, 1975). Longitudinally the cells are linked by simple apposition of the plasma membranes; transversely they join by modifications of the cell border which resemble the classical intercalated disks of vertebrate cardiac muscle and can be resolved into interfibrillar junctions and septate desmosomes. Transverse tubules and sarcoplasmic reticular vesicles are irregularly spaced throughout the cells, and the heart is lined externally and internally by connective tissue sheaths known respectively as the *adventitia* and *intima*. The blood enters the heart through lateral inlets or *incurrent ostia*, a pair of which is situated at each constriction. The wall of the heart is reflected inwards and forwards at each ostium to form an *auricular valve*, which precludes the return flow of the blood into the dorsal sinus. In many insects each pair of auricular valves also prevents the backward flow of the blood in the heart itself (Fig. 122). In the larva of *Aeshna* the ventricular valves are separately developed and situated some distance in front of each pair of ostia.

In many Orthopteroid insects Nutting (1951) has described various types of excurrent apertures through which blood may leave the heart. These include (i) the *lateral segmental blood-vessels*, of which 2 thoracic and 4 abdominal pairs occur in the cockroaches and the latter 4 pairs also in some Mantids; (ii) the valved *excurrent ostia* of which *Grylloblatta* has as many as 2

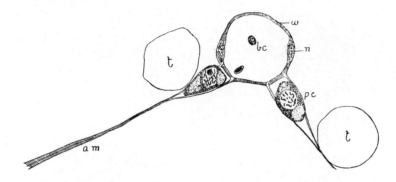

FIG. 119   Transverse section of the heart of a Tachinid larva

bc, blood cells; *w*, wall of heart; *n*, nucleus; *am*, alary muscle; *pc*, pericardial nephrocyte; *t*, trachea. *after* Pantel, *La Cellule*, 1898.

thoracic and 8 abdominal pairs and which, in different insects, may open above or below the dorsal diaphragm or between its two layers in a few cases or which, in the Grylloidea and some Tettigonioidea, lead into blind diverticula through whose phagocytic walls the blood plasma is thought to diffuse; (iii) 1 to 6 *unpaired ventral excurrent ostia* opening into the undivided haemocoele of a few species (*Pteronarcys*, *Oligotoma* and *Thermobia*).

The *aorta* is the anterior prolongation of the dorsal vessel and it functions as the principal artery of the body (Meyer, 1958, Hessel, 1966). Its junction with the heart is frequently marked by the presence of *aortic valves*. The aorta extends forwards through the thorax to terminate in the head near the brain. In some insects its anterior extremity is an open funnel-like mouth but it may instead divide into two or more *cephalic arteries*, each of which may subdivide into smaller vessels.

Accessory Pulsatory Organs

In addition to the heart, accessory pulsatory organs have been described in many insects. They are sac-like structures situated in various regions of the body and pulsate independently of the heart, ensuring an adequate circulation of blood through the appendages. Brocher (1919) has observed thoracic pulsatile organs in *Agrius* (= *Herse*) and *Dytiscus* where they are present just beneath the meso- and metathoracic terga. In *Agrius* the mesotergal pulsatile organ is well developed and is directly connected with a special diverticulum from the loop of the aorta (Fig. 120 and see Hessel, 1966): the metathoracic organ, on the other hand, is very small. In *Dytiscus* the opposite is true, the metathoracic organ being the best developed. Comparable pterothoracic pulsatile organs also occur in the Odonata (Whedon, 1938) and other orders. Among Hemiptera special pulsatile organs are present in the legs (Weber,

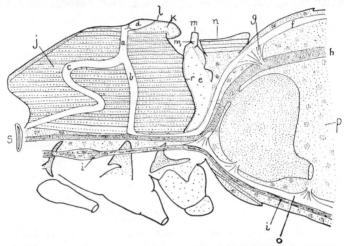

FIG. 120   Section through the thorax and base of the abdomen of *Agrius* (= *Herse*) *convolvuli* showing the circulatory system (Diagrammatic)

a, branch of aorta to mesothoracic pulsatile organ *d*; *b*, *c*, loop of aorta; *e* and *k*, air-sacs; *f*, heart; *g*, alary muscle; *h*, gut; *i*, ventral nerve-cord; *j*, mesotergal muscles; *l*, scutellum; *m*, metatergum and its pulsatile organ *m₁*; *n*, 1st abdominal tergum; *o*, ventral diaphragm; *p*, blood space; *r*, mesophragma; *s*, spiracle. Adapted from Brocher, *Arch. Zool. Exp.*, 1919.

1930) and in many insects there is an accessory heart at the base of each antenna (Clements, 1953; Dönges, 1954; Pinet, 1964; Schneider and Kaissling, 1959). A pair of accessory hearts occurs in the mesoscutellum of several Cyclorrhaphan Diptera and four pulsatile regions are found in the veins at the base of the wing (Thomsen, 1938; Perttunen, 1955).

The Blood

The blood or *haemolymph* of insects is contained in the general body-cavity, where it bathes the various internal organs and also enters the appendages and the tubular cavities of the wing-veins. Blood is the only extracellular fluid in the insect body and makes up 15–75 per cent of the volume of the insect, the amount and composition varying with the species and its physiological condition. It consists of the liquid *plasma* and numerous blood-cells or *haemocytes*.

The plasma, which contains about 85 per cent water, is usually slightly acid and includes inorganic ions, amino acids, proteins, fats, sugars, organic acids and other substances in variable amounts (Wyatt, 1961; Jeuniaux, 1971; Florkin and Jeuniaux, 1974). In most Exopterygota, as Sutcliffe (1963) has shown, sodium and chloride ions account for major fractions of the total osmolar concentration whereas potassium, calcium and magnesium make up only minor proportions. This kind of ionic constitution perhaps represents the basic type of blood in the winged insects and resembles that found in the few Apterygotes and Myriapods studied. In the Endopterygota the basic type of haemolymph is modified by a reduction of chloride and an increase in organic acids, including free amino acids. In Lepidoptera and Hymenoptera the composition of the blood is more specialized, with the inorganic ions contributing only a small fraction of the total osmolar concentration and high levels of free amino acids. The ratio of sodium to potassium varies considerably from one species to another, but the generalization that herbivorous insects have a low ratio and carnivorous species a high one (Boné, 1944) has only limited validity. The carbohydrates, fats and proteins of insect blood have been less fully studied. The major blood sugar is α-trehalose, but in *Apis* this is replaced by a reducing sugar, apparently glucose. The use of electrophoretic and immunological methods has demonstrated a large number of different proteins in the blood of many insects (e.g. Laufer, 1960; Mjeni and Morrison, 1973; Le Bras *et al.*, 1973; Ramade and Le Bras, 1973; Whitmore and Gilbert, 1974). They vary between and within species and change qualitatively and quantitatively during metamorphosis and the reproductive cycle of the female (Schmidt, 1974); some are enzymes and others are conjugated with carbohydrates and lipids. The free amino acids also vary considerably from one species to another and in different developmental stages; in *Rhodnius*, for example, the concentration of tyrosine increases more than tenfold during the moulting cycle, presumably in preparation for cuticular sclerotization (Barrett, 1974). The plasma is often pigmented, though little is known of the chemical nature of the coloured materials; haemoglobins have long been known from the plasma of some Chironomid larvae (see below) and Chefurka and Williams (1951) have identified α-carotene, chlorophylls a and b, riboflavin and taraxanthin from the blood of *Hyalophora* pupae. The plasma has numerous functions. It provides a store of water on which the tissues can draw during desiccation and in some cases acts as an appreciable organ of food-storage (e.g. the blood-sugar of *Apis*, Beutler, 1936). It also transports food materials and hormones and exerts a mechanical function in the eversion of protrusible structures such as the ptilinum of Cyclorrhaphan Diptera or the penis of many male insects and in the dilation of the wings at the emergence of the adult. Because of the well-developed tracheal system of most insects, the blood has a relatively small respiratory role and is less concerned with oxygen transfer than with the carriage of carbon dioxide, from 30–80 per cent of the total carbon dioxide of the blood being bound as bicarbonate

(Levenbook, 1950). In the Chironomid larvae with haemoglobin, the latter is of some respiratory significance at low oxygen tensions (e.g. Walshe, 1947, 1950; Redmond, 1971). Reflex bleeding – i.e. the emission of blood through pores or slits in the cuticle – occurs when some insects are disturbed (e.g. Blum, 1974); it presumably has a protective function though it has not always been distinguished from a similar secretion by epidermal glands (q.v.).

The blood-cells or *haemocytes* arise in the embryo from the median walls of the coelomic sacs and are thought to increase during postembryonic development partly by the division of existing haemocytes and partly through the activity of more or less discrete haemopoietic organs (Hoffman, 1970). The number of cells present in the circulating blood of various adult species varies from about 1000 per cubic mm to about 100 times this number (Tauber and Yeager, 1935–36). Females tend to have higher counts than males and Endopterygote larvae have relatively more cells than the corresponding adults, though the reverse is true of Exopterygote nymphs. Particularly high counts are recorded from insects which are infected with bacteria or other parasites and at the time of ecdysis and oviposition. Such blood-cell counts do not record all the haemocytes present since large numbers are found attached to the surfaces of the viscera and some insects have masses of 'fixed haemocytes'. Very many attempts have been made to classify the haemocytes histologically but they are not entirely satisfactory since, apart from differences between species, the cells vary greatly in shape according to whether they are free or attached and according to the techniques used for observing them. It is also probable that many, if not all, cell-types are merely stages in the development of a single type and that others are arbitrary selections from a continuous range of variation. Among those who have proposed or discussed classifications of insect haemocytes may be mentioned Jones (1962), Gupta (1969) and Price and Ratcliffe (1974); see also the general reviews by Wigglesworth (1959), Arnold (1974) and Crossley (1975) and the many accounts of haemocytes in individual species, listed among the references on p. 244. After a light-microscope study of the living haemocytes from fifteen orders of insects, Price and Ratcliffe (1974) have simplified the classification of Jones (1962) to recognize six cell types, all of which they regard as developmental or functional stages of a single basic form. These six types, of which the first three occur in all insects studied, are as follows: (*a*) the *prohaemocytes* are rounded or oval cells, 6–13 μm in diameter with relatively large nuclei and a few large cytoplasmic granules. (*b*) The *plasmatocytes* are rather larger, variable, round, oval, spindle-shaped or amoeboid cells whose nuclei occupy about 40 per cent of the cell volume and which have more or less finely granular or vacuolated cytoplasm. (*c*) The *granular cells* are round or oval, 10–17 μm in diameter and with nuclei which occupy about half the cell volume; their cytoplasm is packed with granules and can form fine extensions which attach the cells to their substrate. They are sometimes indistinguishable from plasmatocytes. (*d*) The *spherule* cells are oval or spindle-shaped, 8–16 μm in diameter and usually full of large granules or spherules 1·5–2 μm across. Like the granular cells they exude fine attachment processes. (*e*) *Cystocytes* are round, 9–14 μm in diameter, with a relatively large nucleus; in undiluted blood the plasma membrane of the cystocytes becomes ill-defined *in vitro* and the cell swells and gives rise to granular material associated with the clotting of the blood (Grégoire, 1974). (*f*) The *oenocytoids* are relatively large cells, up to 19 μm in diameter with a relatively smaller nucleus and few cytoplasmic granules; they are most often

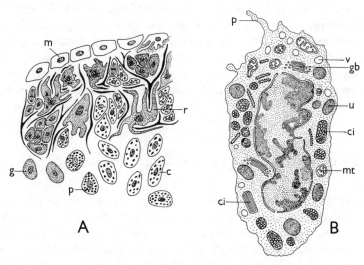

FIG. 121   A, Histology of haemopoietic organ of *Gryllus bimaculatus* (*after* Hoffmann, 1970). B, Ultrastructure of haemocyte of *Periplaneta americana* (*after* Baerwald and Boush, 1970)

c, coagulocyte; ci, cylinder inclusion; g, granulocyte; gb, Golgi body; m, mesothelial cells, enclosing the haemopoietic organ; mt, mitochondrion; p, pseudopodium; r, reticulocyte; u, 'unstructured inclusion'; v, vesicle.

seen in Lepidoptera and can be confused with large spherule cells. Ultra-structural studies, such as those by Akai and Sato (1973), Lai-Fook (1973), Lea and Gilbert (1966) and Neuwirth (1973), as well as histochemical obser-vations, yield categories similar to those indicated above though the *adipohaemocytes* recognized by many authors seem to be indistinguishable from fat-body cells.

The principal function of the haemocytes is *phagocytosis*, i.e. the ingestion of small solid particles. This is demonstrated by the fact that most of the blood-cells other than the prohaemocytes, cystocytes and oenocytoids readily take up injected particles of Indian ink, but they also ingest bacteria and the tissue fragments which result from histolysis during pupation (Crossley, 1964). Accumulations of phagocytic blood-cells may form discrete phagocy-tic organs near the excurrent ostia of some insect hearts (Nutting, 1951) while fully laden phagocytes may agglomerate and become encapsulated by other blood-cells in various tissues or in the haemocoele (Ermin, 1939). In the larvae of some species of *Chironomus* there are no free haemocytes, but a network of fixed phagocytes occurs in the posterior part of the abdomen, while in other Chironomid larvae there are free haemocytes with or without fixed cells (Lange, 1932; Maier, 1969). Haemocytes also form a sheath around the bodies of parasites (Salt, 1968, 1970) while the non-nucleate connective tissue membrane around such viscera as the ovaries, the basement membrane of the epidermis and the fibrous non-cellular capsule which surrounds experimentally introduced foreign bodies may all be derived from plasmatocytes which degenerate after forming a fibrillar, argentophil membrane (Clark and Harvey, 1965; Grimstone *et al.*, 1967; Wigglesworth, 1973). Blood-cells collect at the site of wounds, ingesting debris and forming a plug. Finally, some haemocytes fulfil the function of food storage since they contain inclusions of glycogen and fat which decrease on starvation of the insect (Yeager and Munson, 1941; Munson and Yeager, 1944).

The Circulation of the Blood (Beard, 1953; Selman, 1965; Bayer, 1968; McCann, 1970)

The heart is the principal pulsatory organ and undergoes rhythmical contractions which are brought about by the muscle fibrillae of its walls. The rhythm of the insect heart appears to be purely myogenic (Miller, 1974) though the frequency and amplitude of the beat are influenced by nervous stimuli. Contraction of the heart takes the form of a wave of peristalsis which runs forward from the posterior end. In some species this wave travels so quickly that the whole heart seems to contract at once; in other cases it is so slow that two or three waves can be seen travelling forwards at the same time. Diastole results from the relaxation of the heart muscle aided, in some insects, by contraction of the alary muscles or the elastic tension of their fibres (De Wilde, 1948). During diastole the blood enters the heart by the incurrent ostia, which may exclude some or all of the haemocytes. On contraction the expanded lips of these ostia act as valves so that the blood is prevented from returning through them to the pericardial sinus and is propelled forwards (Fig. 122). In its forward passage, some blood leaves the heart by the excurrent ostia and lateral segmental vessels when these are present. The remainder enters the cephalic haemocoele, from which a part is pumped into the antennae by accessory pulsatile organs. The circulation of blood through the wings of many insects occurs in definite channels between the inner walls of the veins and the enclosed trachea and takes the form of a distally directed flow in the front half of the wing and a return flow along the posterior margin (Clare and Tauber, 1940–42; Arnold, 1964). It is brought about by the thoracic pulsatile organs which aspirate blood from the posterior part of the wing-base. Circulation through the legs may also be achieved through accessory hearts at their bases, helped by movements of the legs and the presence of diaphragms within them. In *Sialis*, with a complicated and rapid circulation, the appendages contain sheath-like vessels formed from connective tissue associated with the tracheae and nerves (Selman, 1965). From the head and thorax the blood flows backwards in the haemocoele of many insects and is probably assisted in this direction by the undulatory movements of the ventral diaphragm. There are few measurements of the blood pressures of insects, but those of *Locusta* are very low (Bayer, 1968), with mean systolic pressures of about 8·6 cm of water in the aorta and small negative pressures in the pericardial and perineural sinuses. The pressure in the general haemocoele varies but positive pressures only occur there intermittently during respiratory and other body movements.

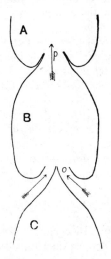

FIG. 122
Valves of the heart

A, B, C, chambers of the heart; AB, at the moment of systole; BC, at the moment of diastole. *p*, interventricular passage; *o*, ostium.

The rate at which the heart beats is greatly affected by temperature and the activity and physiological state of the insect and also varies in different species, there being 14 beats per minute in *Lucanus* larvae at 18° C, and over 150 in *Campodea* at 20° C. In *Sphinx ligustri* and *Bombyx mori* the rate declines steadily throughout larval and pupal development and slower rates are recorded from some parasitized Lepidopterous larvae. A great many other factors affect the heart-beat of different insects under experimental conditions (Jones, 1974); they include light, the secretions of the corpora allata and corpora cardiaca, various tissue extracts, and such pharmacologically active substances as acetylcholine, adrenaline and 5-hydroxytryptamine. Their role in regulating normal cardiac activity, as well as the part played by the cardiac nerves and neurons (Miller, 1974), is still under active investigation. Reversal of the direction in which peristalsis is propagated may occur as a temporary phenomenon in many insects and in *Bombyx mori* it occurs regularly in the late larva and pupa and intermittently in the adult (Gerould, 1938).

Literature on the Circulatory System

AKAI, H. AND SATO, S. (1973), Ultrastructure of the larval haemocytes of the silkworm, *Bombyx mori* L. (Lepidoptera: Bombycidae), *Int. J. Insect Morph. Embryol.*, **2**, 207–231.

ARNOLD, J. W. (1964), Blood circulation in insect wings, *Mem. ent. Soc. Canada*, **38**, 1–48.

—— (1972), A comparative study of the haemocytes (blood cells) of cockroaches (Insecta, Dictyoptera, Blattaria), with a view of their significance in taxonomy, *Can. Ent.*, **104**, 309–348.

—— (1974), The haemocytes of insects, In: Rockstein, M. (ed.), *The Physiology of Insecta*, **5**, 202–254.

BARRETT, F. M. (1974), Changes in the concentration of free amino acids in the haemolymph of *Rhodnius prolixus* during the fifth instar, *Comp. Biochem. Physiol.*, **48** B, 241–250.

BAYER, R. (1968), Untersuchungen am Kreislaufsystem der Wanderheuschrecke (*Locusta migratoria migratorioides* R. & F., Orthopteroidea) mit besonderer Berücksichtigung des Blutdrucks, *Z. vergl. Physiol.*, **58**, 76–135.

BEARD, R. L. (1953), Circulation, In : Roeder, K. D. (ed.), *Insect Physiology*, pp. 232–272.

BEUTLER, R. (1936), Ueber den Blutzucker der Bienen, *Z. vergl. Physiol.*, **24**, 71–115.

BLUM, M. S. (1974), Reflex bleeding in the Lampyrid *Photinus pyralis*: defensive function, *J. Insect Physiol.*, **20**, 451–460.

BONÉ, J. G. (1944), Le rapport sodium-potassium dans le liquide coelomique des insects. I. Ses relations avec le régime alimentaire, *Ann. Soc. zool. Belg.*, **75**, 123–132.

BROCHER, F. (1919), Les organes pulsatiles méso- et métatergaux des Lépidoptères, *Arch. Zool. exp. gén.*, **58**, 149–171.

CHEFURKA, W. AND WILLIAMS, C. M. (1951), Biochemical changes accompanying the metamorphosis of the blood of the Cecropia silkworm, *Anat. Rec.*, **111**, 516–517.

CLARE, S. AND TAUBER, O. E. (1940–42), Circulation of the haemolymph in the wings of the cockroach, *Blattella germanica* L. I–III, *Iowa St. Coll. J. Sci.*, **14**, 107–127; **16**, 349–356; *Ann. ent. Soc. Am.*, **35**, 57–67.

CLARK, R. M. AND HARVEY, W. R. (1965), Cellular membrane formation by plasmatocytes of diapausing Cecropia pupae, *J. Insect Physiol.*, **11**, 161–175.

CLEMENTS, A. N. (1953), The antennal pulsatile organs of mosquitos and other Diptera, *Q. Jl microsc. Sci.*, **97**, 429–433.

CROSSLEY, A. C. S. (1964), An experimental analysis of the origins and physiology of haemocytes in the blue blowfly *Calliphora erythrocephala* (Meig.), *J. exp. Zool.*, **157**, 375–398.

—— (1975), The cytophysiology of insect blood, *Adv. Insect Physiol.*, **11**, 117–221.

DIERICHS, R. (1972), Elektronenmikroskopische Untersuchungen des ventralen Diaphragmas von *Locusta migratoria* und der langsamen Kontraktionswelle nach Fixation durch Gefriersubstitution, *Z. Zellforsch.*, **126**, 402–420.

DÖNGES, G. (1954), Der Kopf von *Cionus scrophulariae*, *Zool. Jb. (Anat.)*, **74**, 1–76.

EDWARDS, G. A. AND CHALLICE, C. E. (1960), The ultrastructure of the heart of the cockroach, *Blattella germanica*, *Ann. ent. Soc. Am.*, **53**, 369–383.

ERMIN, R. (1939), Ueber Bau und Funktion der Lymphocyten bei Insekten (*Periplaneta americana* L.), *Z. Zellforsch.*, **29**, 613–669.

FLORKIN, M. AND JEUNIAUX, C. (1974), Haemolymph composition, In: Rockstein, M. (ed.), *The Physiology of Insecta*, **5**, 256–307.

GEROULD, J. H. (1938), Structure and action of the heart of *Bombyx mori* and other insects, *Acta Zool.*, **19**, 297–352.

GILLIAM, M. AND SHIMANUKI, H. (1971), Blood cells of the worker honeybee, *J. apicult. Res.*, **10**, 79–85.

GOUIN, J. F. (1970), Morphologie, Histologie und Entwicklungsgeschichte der Insekten und der Myriapoden. V. Angiologie und Hämatologie, *Fortschr. Zool.*, **20**, 269–299.

GRÉGOIRE, C. (1974), Haemolymph coagulation, In : Rockstein, M. (ed.), *The Physiology of Insecta*, **5**, 309–360.

GRIMSTONE, A. V., ROTHERHAM, S. AND SALT, G. (1967), An electron-microscope study of capsule formation by insect blood cells, *J. Cell Sci.*, **2**, 281–292.

GUPTA, A. P. (1969), Studies of the blood of Meloidae (Coleoptera). I. The haemocytes of *Epicauta cinerea* (Forster) and a synonymy of haemocyte terminologies, *Cytologia*, **34**, 300–344.

HESSEL, J. H. (1966), A preliminary comparative anatomical study of the mesothoracic aorta of the Lepidoptera, *Ann. ent. Soc. Am.*, **59**, 1217–1227.

HINKS, C. F. (1966), The dorsal vessel and associated structures in some Heteroptera, *Trans. R. ent. Soc. Lond.*, **118**, 375–392.

HOFFMAN, J. A. (1970), Les organes hématopoiétiques de deux insectes orthoptères: *Locusta migratoria* et *Gryllus bimaculatus*, *Z. Zellforsch.*, **106**, 451–472.

JEUNIAUX, C. (1971), Hemolymph–Arthropoda, In: Florkin, M. and Scheer, B. T. (eds), *Chemical Zoology*, **6 B**, 63–118.

JONES, J. C. (1962), Current concepts concerning insect haemocytes, *Am. Zool.*, **2**, 209–246.

—— (1965), The haemocytes of *Rhodnius prolixus* Stål, *Biol. Bull.*, **129**, 282–294.

—— (1974), Factors affecting heart rates in insects, In: Rockstein, M. (ed.), *The Physiology of Insecta*, **5**, 119–167.

LAI-FOOK, J. (1973), The structure of the haemocytes of *Calpodes ethlius* (Lepidoptera), *J. Morph.*, **139**, 79–103.

LANGE, H. H. (1932), Die Phagocytose bei Chironomiden, *Z. Zellforsch.*, **16**, 753–805.

LAUFER, H. (1960), Blood proteins in insect development, *Ann. N.Y. Acad. Sci.*, **89**, 490–515.

LEA, M. S. AND GILBERT, L. J. (1966), The haemocytes of *Hyalophora cecropia* (Lepidoptera), *J. Morph.*, **118**, 197–216.

LE BRAS, S., ECHAUBARD, M. AND RAMADE, F. (1973), Variations de la protéinhémie chez *Musca domestica* au cours d'un cycle gonotrophique dans les conditions normales et après action d'agents toxiques, *Bull. Soc. zool. Fr.*, **98**, 385–403.

LEVENBOOK, L. (1950), The physiology of carbon dioxide transport in insect blood. I–III, *J. exp. Biol.*, **27**, 158–174; 175–183; 184–191.

MAIER, W. (1969), Die Hämocyten der Larve von *Chironomus thummi* (Dipt.), *Z. Zellforsch.*, **99**, 55–63.

MARSCHAL, L. J. (1966), Bau und Funktion der Blutzellen des Mehlkäfers *Tenebrio molitor* L., *Z. Morph. Ökol. Tiere*, **58**, 182–446.

MCCANN, F. V. (1970), Physiology of insect hearts, *A. Rev. Ent.*, 15, 173–200.

MEYER, G. F. (1958), Der feinere Bau der Aorta im Thorax der Honigbiene. *Z. Zellforsch.*, 48, 635–638.

MILLER, T. A. (1974), Electrophysiology of the insect heart, In: Rockstein, M. (ed.), *The Physiology of Insecta*, 5, 169–200.

MJENI, A. M. AND MORRISON, P. E. (1973), Changes in haemolymph proteins in the normal and allatectomized blowfly *Phormia regina* Meig., during the first reproductive cycle, *Can. J. Zool.*, 51, 1069–1079.

MORAN, D. T. (1971), The fine structure of cockroach blood cells, *Tissue and Cell*, 3, 413–422.

MUNSON, S. C. AND YEAGER, J. F. (1944), Fat inclusions in blood cells of the southern armyworm, *Prodenia eridania* (Cram.), *Ann. ent. Soc. Am.*, 37, 396–400.

MURRAY, V. I. E. (1974), The haemocytes of *Hypoderma* (Diptera: Oestridae), *Proc. ent. Soc. Ontario*, 102, 46–63.

MYKLEBUST, R. (1975), The ultrastructure of the myocardial cell in the dragonfly *Aeschna juncea* L., *Norwegian J. Zool.*, 23, 17–26.

NEUWIRTH, M. (1973), The structure of the hemocytes of *Galleria mellonella* (Lepidoptera), *J. Morph.*, 139, 105–124.

NUTTING, W. L. (1951), A comparative anatomical study of the heart and accessory structures of the Orthopteroid insects, *J. Morph.*, 87, 501–597.

PERTTUNEN, V. (1955), The blood circulation and the accessory pulsatile organs in the wings of *Drosophila funebris* and *D. melanogaster* (Dipt., Drosophilidae), *Ann. ent. fennici*, 21, 78–88.

PINET, J. M. (1964), Les cœurs accessoires antennaires de *Rhodnius prolixus* Stål (Heteroptera. Reduviidae), *Bull. Soc. zool. Fr.*, 89, 443–449.

PRICE, C. D. AND RATCLIFFE, N. A. (1974), A reappraisal of insect haemocyte classification by the examination of blood from fifteen insect orders, *Z. Zellforsch.*, 147, 537–549.

RAMADE, F. AND LE BRAS, S. (1973), Études sur les variations de la protéinhémie au cours de la vie imaginale chez *Musca domestica*, *Ann. Soc. ent. Fr.*, 9 (1), 211–217.

RATCLIFFE, N. A. AND PRICE, C. D. (1974), Correlation of light and electron microscopic haemocyte structure in the Dictyoptera, *J. Morph.*, 144, 485–498.

REDMOND, J. R. (1971), Blood respiratory pigments – Arthropoda, In: Florkin, M. and Sheer, B. T. (eds), *Chemical Zoology*, 6 B, 119–144.

RICHARDS, A. G. (1963), The ventral diaphragm of insects, *J. Morph.*, 113, 17–34.

RICHTER, K. (1973), Struktur und Funktion der Herzen wirbelloser Tiere, *Zool. Jb. (Allg. Zool.)*, 77, 477–668.

SALT, G. (1968), The resistance of insect parasitoids to the defence reactions of their hosts, *Biol. Rev.*, 43, 200–232.

—— (1970), *The Cellular Defence Reactions of Insects*, Cambridge Univ. Press, Cambridge, 117 pp.

SANGER, J. W. AND MCCANN, F. V. (1968a), Ultrastructure of the myocardium of the moth, *Hyalophora cecropia*, *J. Insect. Physiol.*, 14, 1105–1111.

—— (1968b), Ultrastructure of moth alary muscles and their attachment to the heart wall, *J. Insect Physiol.*, 14, 1539–1544.

SCHMIDT, G. H. (1974), Disk-elektrophoretische Fraktionierung der Hämolympheproteine während der Metamorphose verschiedener Kasten und Geschlechter von *Formica pratensis, J. Insect Physiol.*, 20, 1421–1466.

SCHNEIDER, D. AND KAISSLING, K. E. (1959), Der Bau der Antenne des Seidenspinners *Bombyx mori*: Das Bindegewebe und das Blutgefass, *Zool. Jb.* (*Anat.*), 77, 111–132.

SELMAN, B. J. (1965), The circulatory system of the alder fly *Sialis lutaria, Proc. zool. Soc. Lond.*, 144, 487–535.

SUTCLIFFE, D. W. (1963), The chemical composition of haemolymph in insects and some other arthropods, in relation to their phylogeny, *Comp. Biochem. Physiol.*, 9, 121–135.

TAUBER, O. E. AND YEAGER, J. F. (1935–36), On the total haemolymph (blood) cell count of insects, *Ann. ent. Soc. Am.*, 28, 229–240; 29, 112–118.

THOMSEN, E. (1938), Ueber den Kreislauf im Flügel der Musciden, mit besonderer Berücksichtigung der akzessorischen pulsierenden Organe, *Z. Morph. Ökol. Tiere*, 34, 416–438.

WALSHE, B. M. (1947), The function of haemoglobin in *Tanytarsus* (Chironomidae), *J. exp. Biol.*, 24, 343–351.

—— (1950), The function of haemoglobin in *Chironomus plumosus* under natural conditions, *J. exp. Biol.*, 27, 73–95.

WEBER, H. (1930), *Biologie der Hemipteren*, Springer, Berlin, 543 pp.

WHEDON, A. D. (1938), The aortic diverticula of the Odonata, *J. Morph.*, 63, 229–262.

WHITMORE, E. AND GILBERT, L. I. (1974), Haemolymph proteins and lipoproteins in Lepidoptera – a comparative electrophoretic study, *Comp. Biochem. Physiol.*, 47 B, 63–78.

WIGGLESWORTH, V. B. (1959), Insect blood cells, *A. Rev. Ent.*, 4, 1–16.

—— (1973), Haemocytes and basement membrane formation in *Rhodnius, J. Insect. Physiol.*, 19, 831–844.

WILDE, J. DE (1948), Contribution to the physiology of the heart of insects, with special reference to the alary muscles, *Arch. néerl. Sci.*, (III C), 28, 530–542.

WYATT, G. R. (1961), The biochemistry of insect hemolymph, *A. Rev. Ent.*, 6, 75–102.

YEAGER, J. F. AND MUNSON, S. C. (1941), Histochemical detection of glycogen in blood cells of the southern armyworm (*Prodenia eridania*) and in other tissues, especially midgut epithelium, *J. agric. Res.*, 63, 257–294.

ZACHARY, D. AND HOFFMAN, J. A. (1973), The hemocytes of *Calliphora erythrocephala* (Meig.) (Diptera), *Z. Zellforsch.*, 141, 55–73.

Chapter 15

THE EXCRETORY ORGANS, FAT-BODY AND OTHER HAEMOCOELIC STRUCTURES

The function of an excretory system is the maintenance of a relatively constant internal environment for the tissues of the body; among the regulatory processes involved are the elimination of the nitrogenous waste-products of protein breakdown and the control of the ionic composition of the haemolymph. The principal excretory organs of insects are the Malpighian tubules, working in conjunction with the hind gut (p. 252). Other haemocoelic structures, such as the nephrocytes, fat-body and some cephalic glands, have also been regarded in the past as excretory organs. This interpretation now needs to be treated with reserve, but the structures concerned, together with the oenocytes and mycetomes, are dealt with in this chapter.

The Malpighian Tubules (Fig. 123)

First discovered by the Italian anatomist Malpighi, these organs occur in almost all insects. They are long, slender, blind tubes lying in the haemocoele where they are freely bathed by the blood. They open at their proximal extremities into the intestine, near the junction of the hind gut with the mid gut. They have generally been regarded as derivatives of the ectodermal proctodaeum (e.g. Srivastava and Khare, 1966), but others believe them to originate from an undifferentiated zone of cells between mid and hind gut and treat them as outgrowths of the mesenteron (Henson, 1946; Savage, 1956, 1962). Distally they are usually free but in some insects the distal ends of the tubules are closely applied to the hind gut. This 'cryptonephridial' condition occurs in the larvae and adults of many Coleoptera (Saini, 1964; Grimstone et al., 1968; Ramsay, 1971), in some Symphytan larvae (Poll, 1937), in Myrmeleontid larvae (Poll, 1936) and in almost all Lepidopteran larvae (Poll, 1939). The tubules may be held against the rectum by a thin cylindrical sheath or lateral sac of peritoneal cells (Coleoptera and Symphytan larvae) or lie below the muscle layer of the hind gut, as in Lepidopteran larvae (Poll, 1939). Cryptonephry enables the insect

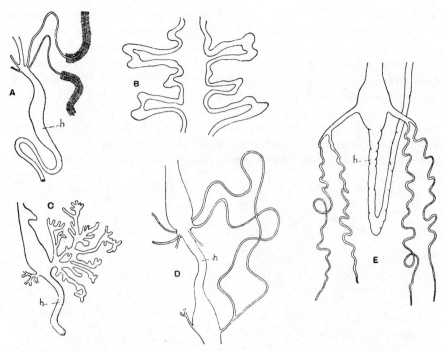

FIG. 123 Malpighian tubules

A, *Melolontha melolontha*. B, portion with diverticula more highly magnified. C, *Galleria mellonella*. D, *Timarcha tenebricosa*. E, *Calliphora* (larva). *h*, hind gut. A–D *after* Veneziani, *Redia*, 1904.

to conserve water by withdrawing it from the faeces. The exterior of the Malpighian tubules is richly supplied with a reticulum of fine tracheae whose larger branches serve to maintain these organs in position. The number of Malpighian tubules is very variable but tolerably constant within most of the orders (Veneziani, 1905). They usually occur in twos, or multiples of two, and their primitive number according to Wheeler (1893) is six. It is only very exceptionally that more than six are present in the embryo and they are often reduced to four. Specialization either by addition or reduction is frequent; their number is often less than 6 and may exceed 100 through postembryonic differentiation of so-called secondary tubules (Henson, 1944; Savage, 1956, 1962).

The typical number of tubules present in the various orders is given below.

Anoplura, Thysanoptera, Hemiptera, Diptera and Siphonaptera 4.
Psocoptera, Coleoptera 4–6. Isoptera 2–8. Thysanura 4–16.
Mecoptera, Trichoptera and Lepidoptera 6. Neuroptera 6–8. Dermaptera 8–20.
Ephemeroptera 8–100. Plecoptera 50–60. Odonata 50–200. Orthoptera 30–200.
Hymenoptera, 6–20 in ants and over 100 in many Aculeata.

The Coccoidea and larval parasitic Hymenoptera are exceptional in having only two Malpighian tubes and the Culicidae have the unusual number of five. In certain Diplura, the Protura, and Strepsiptera these vessels are doubtfully represented by papillae; in the Collembola, *Japyx*, and the Aphidoidea they are wanting altogether. Although usually simple tubes they are sometimes arborescent, as in *Galleria mellonella*, or they may give off short closely-packed diverticula as in *Melolontha* (Fig. 123). Very frequently the Malpighian tubules unite in groups of two or three and they may open into a common ampulla or bladder, which discharges into the gut. When very numerous they may be grouped in bunches, each bunch discharging by

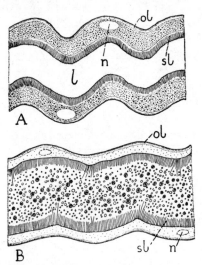

FIG. 124
Malpighian tubules of *Rhodnius* (Reduviidae) seen in optical sections of living material. A, upper region of tubule with excretory granules in the cells. B, lower region of tubule with excretory spheres in the lumen

l, lumen; *n*, nucleus of excretory cell; *ol*, outer striated zone; *sl*, inner striated margin (filamentous in B). Adapted from Wigglesworth, *J. exp. Biol.*, 1931.

a separate duct or ureter; in the Gryllidae all the tubules converge to open into a common ureter of considerable length. Not infrequently the Malpighian tubules exhibit morphological and physiological differences. Thus, in *Haltica* and *Donacia* four of them discharge into a common ampulla while the remaining two shorter vessels have isolated insertions. In many species some or all of the tubules show a division into two or three zones (Fig. 124) which differ in structure and function (see also Smith and Littau, 1960).

When viewed in transverse section a Malpighian tubule is seen to be composed of about three to eight large and variably-shaped epithelial cells with prominent nuclei. The latter increase in size during development and may become palmate or form giant endopolyploid nuclei; in some species the cells are binucleate. Where each cell borders the lumen of the tubule its surface is produced into closely packed microvilli while in the opposite, basal region of the cell the plasma membrane is thrown into deep, elaborate, ultramicroscopic infoldings which divide the cytoplasm into a series of compartments (Berkaloff, 1960; Mazzi and Baccetti, 1963; Taylor, 1971; Wessing, 1962). Mitochondria are associated with both the microvillar and basal regions, Golgi bodies occur, and the tubule is enclosed by a basement membrane apparently permeable to large molecules (Kessel, 1970). In some species there are two or more ultrastructurally distinct kinds of cells inter- mingled in the same tubule, as in *Calliphora* where the primary secretory cells are accompanied by a smaller number of stellate cells perhaps con- cerned with resorption (Berridge and Oschman, 1969). In others, such as *Rhodnius* (Wigglesworth and Salpeter, 1962), each tubule is differentiated into two regions, the proximal segment having cells with a reduced microvillar surface. In still other cases, there are two or more distinct types of tubule, as in *Carausius* (Savage, 1962; Taylor, 1971). In many insects the tubules are provided with an investment of muscle fibres. These fibres, which have practically all the usual ultrastructural features of striated muscle (Crowder and Shankland, 1972), may run in bands or as a reticulum over the whole tube or be restricted to the proximal region; they appear not to be innervated but are responsible for peristaltic movements of the tubules. In the Thysanura, Dermaptera and Thysanoptera there are no muscles and peris- talsis does not occur (Palm, 1946). In some of the species with a cryptoneph- ridial arrangement, the distal parts of the tubules have small thin areas (*leptophragmata*) adjoining the haemocoele and giving a strong reaction for chlorides.

Physiology of Excretion

The physiology of the associated processes of nitrogenous excretion and osmoregulation have been much studied in recent years (see reviews by Craig, 1960; Bursell, 1967; Razet, 1966; Berridge, 1970; Maddrell, 1971*a*, 1971*b*; Riegel, 1971; Schoffeniels and Gilles, 1970; and Stobbart and Shaw, 1974). The predominant excretory end-product of protein breakdown in insects is uric acid, but its derivatives allantoin and allantoic acid occur as major excretory materials in some species among the Lepidoptera (especially larvae), the Heteroptera and some other groups. All three substances are relatively very insoluble in water, contain similar proportions of nitrogen and are appropriate excretory materials for terrestrial animals that need generally to conserve water. On the other hand, many insects excrete much of their nitrogen in the form of ammonia, as in the aquatic larvae of *Sialis* and *Aeshna* (Staddon, 1955, 1959) as well as the aphids and the larvae of meat- eating Cyclorrhaphan Diptera such as *Calliphora* and *Lucilia* (Brown, 1938). Urea and amino acids occur in insect excreta less often and in smaller amounts, but it is difficult to make simple generalizations on the major excretory materials. In many insects they change at metamorphosis or vary with the diet and even in a well-known insect like *Periplaneta* there is now evidence that ammonia rather than uric acid is the main product (Mullins and Cochran, 1973). In some insects, much of the uric acid is permanently segregated in special urate cells of the fat-body or in the epidermis or elsewhere. Otherwise, however, it is excreted through the Malpighian tub- ules. In *Rhodnius* the distal two-thirds of each tubule secretes a solution of sodium or potassium acid urate while in the proximal segment the resorption of water and bases results in the precipitation of spherical masses of free uric acid (Wigglesworth, 1931). In *Carausius* (Ramsay, 1955–58) the process is comparable but the tubules are uniformly secretory and resorption of water and bases is deferred until their secretion reaches the rectum, where precipitation occurs.

The role of the Malpighian tubules in regulating water balance and ionic concentrations presents complex biophysical problems still not fully elucidated (e.g. Maddrell, 1971; Berridge, 1970). Essentially, however, the secretory region of typical Malpighian tubules actively transports potassium ions into the lumen (Ramsay, 1953–58), followed by passive movement of other ions and water. In this connection the basal infoldings and microvilli of the tubules are perhaps the sites of standing osmotic gradients (Berridge and Oschman, 1969). The fluid produced by the tubule is then acted on by the hind gut which selectively absorbs water and ions (and probably also some metabolically useful substances like sugars and amino acids which have been secreted into the lumen of the tubules). Such processes are, in many insects, supplemented by water and ionic regulation through the activity of other tissues. Thus, in Saturniid larvae, which feed on leaves containing large amounts of potassium, the mid gut epithelium is able to secrete these ions actively into the lumen of the alimentary canal and so prevent their accumulation in the haemolymph (Harvey and Nedergaard, 1964). Again, many aquatic insects are able to take up sodium, chloride and other ions, either through the anal papillae in Culicid larvae (Wigglesworth, 1938: Copeland, 1964) or at other localized sites in the integument of Ephemeropteran, Odonate and Trichopteran larvae (Wichard and Komnick, 1972–74). Active absorption of atmospheric water is also possible in some insects, notably in *Thermobia* and in larvae of *Tenebrio*, where it apparently occurs through the rectum (Noble-Nesbitt, 1969, 1970). In a wide sense, therefore, insect excretory mechanisms involve many interrelated processes occurring in a variety of organs and tissues, including some others dealt with below; they are certainly not confined to the Malpighian tubules and they must, in any case, take place in species where these are absent or vestigial.

The Fat-body

The fat-body (Pardi, 1939; Kilby, 1963; Price, 1973) is composed of irregular masses or lobes of rounded or polyhedral cells (trophocytes) which are usually vacuolated and contain inclusions of various kinds. In many insects the fat-body is built up of tightly compacted cells; in others it is a more or less laminate tissue with numerous lacunae, or it may take the form of loose strands. In colour it may be either white, yellow, orange or greenish. The lobes of the fat-body are enclosed in an extracellular connective tissue lamella and its cytoplasm usually contains abundant granular endoplasmic reticulum (Evans, 1967; Gaudecker, 1963; Huddart, 1972; Walker, 1965). In well-fed insects fat is formed and stored in the cisternae of the endoplasmic reticulum, ultimately giving rise to numerous large fat-vacuoles. Protein is stored as granules or in multivesicular bodies (Locke and Collins, 1965, 1968) and small glycogen granules are present. Like the vertebrate liver, the insect fat-body synthesizes, stores and mobilizes lipids, proteins and glycogen; and in the fat-body cells of *Calpodes* larvae the resemblance between the two tissues is further emphasized by the presence of characteristic liver organelles – the membrane-bound vesicles or microbodies, containing catalase and uric oxidase (Locke and McMahon, 1971).

The fat-body is derived from the mesoderm by differentiation of the walls of the coelomic cavities and it consequently has a primitive metameric disposition. With the breaking down of the embryonic coelom and the development of a haemocoele, the fat-body forms the irregular boundaries of the permanent body-cavity. In many insects it is possible to distinguish an outer or *parietal layer*, beneath the body-wall, and an inner or *visceral layer*, which surrounds and lies between the various organs (Fig. 125). In some

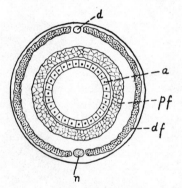

FIG. 125
Schematic transverse section of an insect larva showing distribution of the fat-body

a, alimentary canal; *d*, dorsal vessel; *n*, ventral nerve-cord; *pf*, *df*, visceral and parietal layers of fat-body.

larvae the parietal layer is interrupted at each segment and thus retains a segmental arrangement; the visceral layer, on the other hand, forms a continuous sheet passing from one segment to another. In *Carausius* and some other insects a sheath of fat-body encloses the ventral nerve-cord, but it seems permeable to relatively large molecules and does not – as was once thought – play a part in regulating the sodium concentration of the blood in contact with the nervous system (Treherne, 1972). The fat-body alters considerably in its histological structure during the life of an insect (e.g. Lartchenko, 1937; Locke and McMahon, 1971). In the earlier instars its nuclei are rounded or oval (Fig. 126, 3) but they often later alter in character, becoming stellate or ribbon-like (Fig. 126, 2 and 4). In many cases the cellular structure is no longer evident and the fat-body has the appearance of a syncytium (Fig. 126, 1) though cell-boundaries may reappear on starvation. The cells of the fat-body seem to be closely related to the haemocytes. They may be indistinguishable in young stages, or haemocytes may become laden with fat and incorporated into the fat-body and fat-body cells are said in some species to be capable of phagocytosis.

In some insects deposits of uric acid or urates appear in the fat-body during life. In the Collembola, *Blatta*, and parasitic Hymenopteran larvae, these excretory granules are formed in specialized urate cells of the fat-body, the Malpighian tubules being either absent or non-functional or not secreting urates. In other cases, even though the tubules secrete urates, some may also be deposited in the trophocytes. In the larvae of the Apocrita, Lepidoptera and Cyclorrhaphan Diptera, the urates of the fat-body are transferred to the Malpighian tubules at pupation. Calcium salts, in the form of calcospherites, accumulate in the fat-body of phytophagous Dipteran larvae (Keilin, 1921).

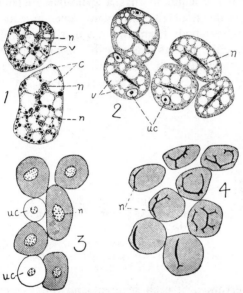

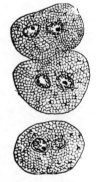

FIG. 126   Fat-body of various insects

1, Adult Termite (soldier); 2, Adult larva of *Caliroa limacina*; 3, Young larva of *Formica rufa*; 4, Adult larva of same; *c*, urate concretions; *n*, nucleus; *uc*, urate cells; *v*, vacuoles filled with fat globules. Nos. 3 and 4 adapted from Pérez, *Bull. Sci. Fr. et Belg.*, 37.

FIG. 127
Three ventral nephrocytes from the larva of *Limnophora riparia*, highly magnified
After Keilin, *Parasitology*, 1917.

Variations in the food reserves of the fat-body depend on the state of nutrition and other factors. Larvae of *Aedes aegypti* (Wigglesworth, 1942) accumulate glycogen in the fat-body when fed on starch and some sugars, while fat is deposited when fed on olive oil. Feeding on casein causes an accumulation of protein together with glycogen and some fat. The reserves of the insect fat-body are depleted during starvation (Walker, 1965) and are also mobilized during periods of great activity, such as prolonged flight (p. 65), in moulting (Wigglesworth, 1947) or during hibernation, maturation of eggs and at pupation (Ishizaki, 1965; Larsen, 1970); during the last-mentioned phase the larval fat-body disintegrates in some species.

Other Haemocoelic Tissues

The labial glands of the Collembola, Diplura and Thysanura discharge at the base of the labium (Fahlander, 1940; Altner, 1968). In the Collembola, according to Feustel (1958) these organs consist of a terminal coelomic sac, which accumulates injected trypan blue and lithium carmine, followed by a short cellular duct and a longer, coiled syncytial duct. There are also rudimentary coelomic sacs in the syncytial organs at the base of each antenna and at each side of the retinaculum and ventral tube. The precise physiological role of the labial glands in these Apterygote insects is not clear, but salt and water regulation may be one of the functions of the labial salivary glands of higher insects (see p. 268).

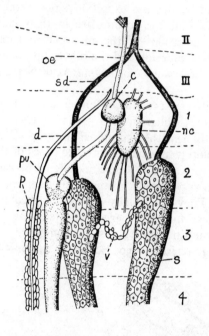

FIG. 128
Dissection of the anterior region of the larva of *Phaonia cincta* (Anthomyidae), showing the pericardial cells *p*, and the ventral nephrocytes *v*

II, III, 2nd and 3rd thoracic segments; 1–4, abdominal segments; *c*, cerebral ganglion; *d*, dorsal vessel; *nc*, ventral ganglionic centre; *oe*, oesophagus; *pv*, proventriculus; *s*, salivary gland; *sd*, salivary duct. Adapted from Keilin, *Parasitology*, 1917.

The Nephrocytes (Fig. 127) – The nephrocytes consist of scattered or localized groups of cells or syncytial aggregates often with more than one nucleus per cell. When localized they occur in two principal groups: (1) the dorsal nephrocytes or pericardial cells and (2) the ventral nephrocytes. The pericardial cells consist of two chains or segmental groups of cells arranged in a linear series one on either side of the heart in the pericardial sinus; such cells are present in the immature stages and adults of most insects. The ventral nephrocytes occur in Dipterous larvae where they constitute the 'garland-like cell-chain' of Weismann. In these insects they usually form a chain of cells which is suspended in the body-cavity below the fore gut and attached by its two extremities to the salivary glands. The nephrocytes are able to absorb colloidal particles, chiefly of 1·6–2·0 nm radius, from the blood; proteins, chlorophyll and such dyestuffs as trypan blue and ammonia carmine appear in them after haemocoelic injection. For this reason they were regarded as organs of storage excretion but except in a few cases where pigments or other materials accumulate in them, their contents do not seem to increase with age. They are now considered to be analogous to the reticulo-endothelial system of vertebrates and take up proteins and other particles of colloidal dimensions from the blood by pinocytosis whereas the haemocytes take up larger bodies by phagocytosis. Electron microscopy has revealed the presence of abundant endoplasmic reticulum and a profusion of coated vesicles and peripheral infolding of the plasma membrane. For further information see Kessel (1962), Mills and King (1965), Sanger and McCann (1968) and Wigglesworth (1970).

The Gut perhaps plays some part in deposit excretion since certain injected dye-stuffs are secreted by its walls or accumulate in their cells (Feustel, 1958) while the periodic 'renovation' of the mid gut in some Collembola (p. 467) suggests an excretory function and the biliverdin resulting from haemoglobin breakdown in the tissues of *Rhodnius* is discharged into its lumen (Wigglesworth, 1943).

The Oenocytes – These large cells (15–150 μm or more across) have been recorded from most orders of insects (Richards, A. G., 1951). They arise from the ectoderm or epidermis, usually near the abdominal spiracles (Figs. 129 and 130) and sometimes remain closely associated with the bases of the epidermal cells. In other cases they project into the haemocoele or become separated from the epidermis to form segmentally arranged clusters or are even dispersed among the fat-body. They are usually amber-coloured but may be brown, red, green or colourless and either arise repeatedly during postembryonic development or, in some Endopterygotes, are composed of a larval and an imaginal generation (e.g. Kaiser, 1950; Schmidt, 1961). The

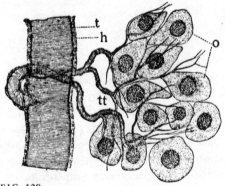

FIG. 129
Cluster of oenocytes from a nearly mature Phryganeid larva

o, oenocytes; *t*, trachea; *tt*, small tracheal branches; *h*, tracheal matrix cells.

ultrastructure and cytochemistry of oenocytes has been studied in many species (e.g. Lhoste, 1950; Huber, 1958; Schmidt, 1961; Gnatzy, 1970; and Romer, 1974). Their cytoplasm contains abundant smooth tubular endoplasmic reticulum and the plasma membrane is sometimes infolded to form a network of tubules; Golgi regions, mitochondria and peroxidase-containing microbodies have also been reported, as well as glycogen, lipid vesicles and other metabolic inclusions. In the permanent oenocytes of the larvae of *Calpodes* (Locke, 1969) there is some destruction of cytoplasmic structures at

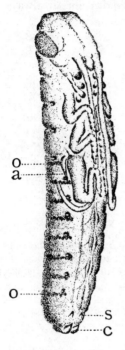

FIG. 130
A nearly mature
embryo of *Xiphidium*

o, oenocyte cluster; *a*, appendage of 1st abdominal segment; *s*, style; *c*, cercus. This and Fig. 129 after Wheeler, *Psyche*, 1892.

all times and it reaches massive proportions just before the last two larval moults, though at these the oenocytes show pinocytosis of blood proteins, nuclear replication, RNA synthesis and the regeneration of endoplasmic reticulum. The functions of the oenocytes are still problematic; they have been thought to secrete cuticular lipids but their ultrastructure resembles that of vertebrate cells concerned with steroid synthesis and it has even been suggested that they may be the site of ecdysone production.

Mycetomes – Many insects contain specific, symbiotic micro-organisms, usually yeasts, *Actinomyces* or other bacteria, which are often transmitted from one generation to the next through the egg and which are known in several cases to supply their hosts with water-soluble vitamins or essential amino acids. In some species of insects the symbionts live in the lumen or wall of the gut or in compact caeca evaginated from it. The latter case is exemplified by *Stegobium* and other Anobiids in which, at metamorphosis, some symbionts are transferred to newly-formed adult caeca while others pass to tubular pouches that open near the base of the penultimate segment of the oviscapt. From these organs they are smeared over the surface of the egg while it is being laid and are eventually eaten by the newly-emerged larva. In other insects the symbionts are segregated in specialized cells, the mycetocytes. These in turn may be scattered singly in the fat-body, as in the cockroaches or *Mastotermes*, or they may be grouped into more or less compact organs, the mycetomes. Such organs are well developed in many Coleoptera, for example *Oryzaephilus*, *Lyctus*, *Sitophilus* and *Rhizopertha*, as well as in many Hemiptera. The Aphidoidea almost always have a pair of

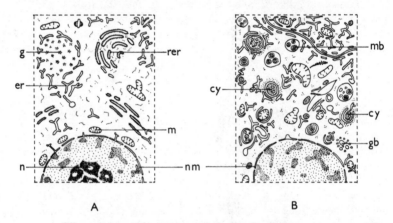

A   B

FIG. 131   Ultrastructure of oenocytes of *Gryllus bimaculatus* at two stages in the moulting cycle (*after* Romer, 1974). A, 30 hours after moult; B, just before the next moult, at culmination of cytolytic activity

cy, stages in cytolysis, the cytolysing structure enclosed in an isolation membrane; *er*, endoplasmic reticulum; *g*, glycogen deposits; *gb*, Golgi bodies; *m*, mitochondria; *mb*, myelin bodies (final stages in cytolysis); *n*, nucleoli; *nm*, nuclear membrane.

strand-like mycetomes extending over several abdominal segments and sometimes containing two or three different forms of bacterial symbionts. Aleyrodids have a pair of rounded mycetomes containing rod-like bacteria and Psyllids a single organ with two types of symbionts. In the Coccoidea there is greater variation: most families have bacterial or yeast-like symbionts housed in scattered mycetocytes in the fat-body, but Pseudococcids generally have a single, large, characteristic mycetome, the Apiomorphidae lack symbionts, and *Stictococcus* has them only in the female. The Auchenorrhyncha show many different di- and polysymbiotic mycetomes of various shapes and discrete mycetomes may also occur in some Heteroptera, such as *Ischnodemus* and *Cimex*, though far more Heteroptera have bacterial symbionts in the lumen of tubular or crypt-like evaginations of the mid gut. The great diversity of symbiotic organisms and the devices for accommodating them and ensuring their transmission have been extensively studied and authoritatively reviewed by Buchner (1965) but see also p. 700.

Literature on Excretory Organs, Fat-body, Oenocytes, etc.

ALTNER, H. (1968), Die Ultrastruktur Labialnephridien von *Onychiurus quadriocellatus* (Collembola), *J. Ultrastruct. Res.*, **24**, 349–366.

BERKALOFF, A. (1960), Contribution à l'étude des tubes de Malpighi et de l'excrétion chez les insectes. Observation au microscope électronique, *Annls Sci. nat.* (*Zool.*) *Paris* (Sér. 12), **2**, 869–947.

BERRIDGE, M. J. (1970), Osmoregulation in terrestrial arthropods, In: Florkin, M. and Scheer, B. T. (eds), *Chemical Zoology*, V(A), 287–319.

BERRIDGE, M. J. AND OSCHMAN, J. L. (1969), A structural basis for fluid secretion by Malpighian tubules, *Tissue and Cell*, **1**, 247–272.

BROWN, A. W. A. (1938), The nitrogen metabolism of an insect (*Lucilia sericata* Mg.). I–II, *Biochem. J.*, **32**, 895–902; 903–912.

BUCHNER, P. (1965), *Endosymbiosis of Animals with Plant Microorganisms*, Interscience, New York, 909 pp.

BURSELL, E. (1967), The excretion of nitrogen in insects, *Adv. Insect Physiol.*, **4**, 33–67.

COPELAND, E. (1964), A mitochondrial pump in the cells of the anal papillae of mosquito larvae, *J. Cell Biol.*, **23**, 253–264.

CRAIG, R. (1960), The physiology of excretion in the insect, *A. Rev. Ent.*, **5**, 53–68.

CROWDER, L. AND SHANKLAND, D. L. (1972), Structure of the Malpighian tubule muscle of the American cockroach, *Periplaneta americana*, *Ann. ent. Soc. Am.*, **65**, 614–619.

EVANS, J. T. (1967), Development and ultrastructure of the fat body cells and oenocytes of the Queensland fruit fly *Dacus tryoni* (Frogg.), *Z. Zellforsch.*, **81**, 49–61.

FÄHLANDER, K. (1940), Die Segmentalorgane der Diplopoda, Symphyla und Insecta Apterygota, *Zool. Bidr. Upps.*, **18**, 243–251.

FEUSTEL, H. (1958), Untersuchungen über die Exkretion bei Collembolen. (Ein Beitrag zur Exkretion bei Arthropoden), *Z. wiss. Zool.*, **161**, 209–238.

GAUDECKER, B. VON (1963), Über den Formwechsel einiger Zellorganelle bei der Bildung der Reservestoffe im Fettkörper von *Drosophila*-Larven, *Z. Zellforsch.*, **61**, 56–95.

GNATZY, W. (1970), Struktur und Entwicklung des Integuments und der Oenocyten von *Culex pipiens* L. (Dipt.), *Z. Zellforsch.*, **110**, 401–443.

GRIMSTONE, A. V., MULLINGER, A. M. AND RAMSAY, J. A. (1968), Further studies on the rectal complex of the mealworm *Tenebrio molitor* L. (Coleoptera, Tenebrionidae), *Phil. Trans. R. Soc.*, Ser. B, **253**, 343–382.

HARVEY, W. R. AND NEDERGAARD, S. (1964), Sodium-independent active transport of potassium in the isolated midgut of the cecropia silkworm, *Proc. natn. Acad. Sci. U.S.A.*, **51**, 757–765.

HENSON, H. (1944), The development of the Malpighian tubules of *Blatta orientalis* (Orthoptera), *Proc. R. ent. Soc. Lond.* (A), **19**, 73–91.

—— (1946), The theoretical aspects of insect metamorphosis, *Biol. Rev.*, **21**, 1–14.

HUBER, M. (1958), Histologische und experimentelle Untersuchungen über die Oenocyten der Larve von *Sialis lutaria* L., *Z. Zellforsch.*, **49**, 661–697.

HUDDART, H. (1972), Fine structure of the neural fat-body sheath in the stick insect and its physiological significance, *J. exp. Zool.*, **179**, 145–156.

ISHIZAKI, H. (1965), Electron microscopic study of changes in the subcellular organization during metamorphosis of the fat body cells of *Philosamia ricini* (Lepidoptera), *J. Insect Physiol.*, **11**, 845–855.

KAISER, P. (1950), Zur Kenntnis der Oenocyten. Histologische Untersuchungen am Kohlweissling, *Zool. Anz.*, **145**, 364–372.

KEILIN, D. (1921), On the calcium carbonate and the calcospherites in the Malpighian tubes and fat-body of Dipterous larvae and the ecdysial elimination of these products of secretion, *Q. Jl microsc. Sci.*, **65**, 611–625.

KESSEL, R. G. (1962), Light and electron microscope studies on the pericardial cells of nymphal and adult grasshoppers, *Melanoplus differentialis differentialis* (Thomas), *J. Morph.*, **110**, 79–88.

—— (1970), The permeability of dragon-fly Malpighian tubule cells to protein using horseradish peroxidase as a tracer, *J. Cell Biol.*, **47**, 299–303.

KILBY, B. A. (1963), The biochemistry of the insect fat body, *Adv. Insect Physiol.*, **1**, 111–174.

LARSEN, W. (1970), Genesis of mitochondria in insect fat body, *J. Cell Biol.*, **47**, 373–383.

LARTCHENKO, K. I. (1937), Le cycle évolutif du corps adipeux de *Loxostege sticticalis* L. et *Feltia segetum* Schiff. et sa relation avec la fécondité, *Rev. ent. U.R.S.S.*, **27**, 29–75.

LHOSTE, J. (1950), Sur quelques aspects cytologiques et histochimiques des oenocytes de *Forficula auricularia* L. imago, *Bull. Soc. zool. Fr.*, **75**, 162–163.

LOCKE, M. (1969), The ultrastructure of oenocytes in the molt/intermolt cycle of an insect, *Tissue and Cell*, **1**, 103–154.

LOCKE, M. AND COLLINS, J. V. (1965), The structure and formation of protein granules in the fat body of an insect, *J. Cell Biol.*, **26**, 857–884.

—— (1968), Protein uptake into multivesicular bodies and storage granules in the fat body of an insect, *J. Cell Biol.*, **36**, 453–483.

LOCKE, M. AND MCMAHON, J. T. (1971), The origin and fate of microbodies in the fat body of an insect, *J. Cell Biol.*, **48**, 61–78.

MADDRELL, S. H. P. (1971*a*), Fluid secretion by the Malpighian tubules of insects, *Phil. Trans. R. Soc.*, Ser. B, **262**, 197–207.

—— (1971*b*), The mechanisms of insect excretory systems. *Adv. Insect Physiol.*, **8**, 199–331.

MAZZI, V. AND BACCETTI, B. (1963), Ricerche istochimice e al microscopio elettronico sui tubi malpighiani di *Dacus oleae* Gmel. I. La larva, *Z. Zellforsch.*, **59**, 47–70.

MILLS, R. P. AND KING, R. C. (1965), The pericardial cells of *Drosophila melanogaster*, *Q. Jl microsc. Sci.*, **106**, 261–268.

MULLINS, D. E. AND COCHRAN, D. G. (1973), Nitrogenous excretory materials from the American cockroach, *J. Insect Physiol.*, **19**, 1007–1018.

NOBLE-NESBITT, J. (1969, 1970), Water balance in the fire brat, *Thermobia domestica*, *J. exp. Biol.*, **50**, 745–769; **52**, 193–200.

PALM, N. B. (1946), Studies on the peristalsis of Malpighian tubules of insects, *Acta Univ. lund.*, **42**, 1–39.

PARDI, L. (1939), I corpi grassi degli insetti, *Redia*, **25**, 87–288.

POLL, M. (1937), Contribution à l'étude de histophysiologie de l'appareil urinaire des larves de Myrméléontides, *Mém. Mus. Hist. nat. Belg.* (2), **3**, 635–666.

—— (1937), Note sur les tubes de Malpighi des larves de Tenthredinoïdes, *Bull. Ann. Soc. ent. Belg.*, **77**, 433–442.

—— (1939), Contribution à l'étude de l'appareil urinaire des chenilles de Lépidoptères, *Ann. Soc. zool. Belg.*, **69**, 9–52.

PRICE, G. M. (1973), Protein and nucleic acid metabolism in insect fat body, *Biol. Rev.*, **48**, 333–375.

RAMSAY, J. A. (1955), The excretion of sodium, potassium and water by the Malpighian tubules of the stick insect, *Dixippus morosus* (Orthoptera, Phasmidae), *J. exp. Biol.*, **32**, 200–216.

—— (1956), Excretion by the Malpighian tubules of the stick insect, *Dixippus morosus* (Orthoptera, Phasmidae): calcium, magnesium, chloride, phosphate and hydrogen ions, *J. exp. Biol.*, **33**, 697–708.

—— (1958), Excretion by the Malpighian tubules of the stick insect, *Dixippus morosus* (Orthoptera, Phasmidae): amino acids, sugars and urea, *J. exp. Biol.*, **35**, 871–891.

—— (1971), Insect rectum, *Phil. Trans. R. Soc.*, *Ser. B*, **262**, 251–260.

RAZET, P. (1966), Les éléments terminaux du catabolisme azoté chez les insectes, *Année biol.*, **5**, 43–73.

RICHARDS, A. G. (1951), *The Integument of Arthropods*, University of Minnesota Press, Minneapolis, 411 pp.

RIEGEL, J. (1971), Excretion–Arthropoda, In: Florkin, M. and Scheer, B. T. (eds), *Chemical Zoology*, **6B**, 249–277.

ROMER, F. (1974), Ultrastructural changes of the oenocytes of *Gryllus bimaculatus* Deg (Saltatoria, Insecta) during the moulting cycle, *Cell and Tissue Res.*, **151**, 27–46.

SAINI, R. S. (1964), Histology and physiology of the cryptonephridial system of insects, *Trans. R. ent. Soc. Lond.*, **116**, 347–392.

SANGER, J. W. AND MCCANN, F. V. (1968), Fine structure of the pericardial cells of the moth, *Hyalophora cecropia*, and their role in protein uptake, *J. Insect Physiol.*, **14**, 1839–1845.

SAVAGE, A. A. (1956), The development of the Malpighian tubules of *Schistocerca gregaria* (Orthoptera), *Q. Jl microsc. Sci.*, **97**, 599–615.

—— (1962), The development of the Malpighian tubules of *Carausius morosus* (Orthoptera), *Q. Jl microsc. Sci.*, **103**, 417–437.

SCHMIDT, G. (1961), Sekretionsphasen und cytologische Beobachtungen zur Funktion der Oenocyten während der Puppenphase verschiedener Kasten und Geschlechter von *Formica polyctena* Foerst. (Ins., Hym., Form.), *Z. Zellforsch.*, **55**, 707–723.

SCHOFFENIELS, E. AND GILLES, R. (1970), Osmoregulation in aquatic arthropods, In: Florkin, M. and Scheer, B. T. (eds), *Chemical Zoology* **5A**, 255–286.

SMITH, D. S. AND LITTAU, V. C. (1960), Cellular specialization in the excretory epithelia of an insect, *Macrosteles fasciifrons* Stål (Homoptera), *J. biophys. biochem. Cytol.*, **8**, 103–133.

SRIVASTAVA, U. S. AND KHARE, M. K. (1966), The development of Malpighian tubules and associated structures in *Philosamia ricini* (Lepidoptera, Saturniidae), *J. Zool., Lond.*, **150**, 145–163.

STADDON, B. W. (1955), The excretion and storage of ammonia by aquatic larvae of *Sialis lutaria* (Neuroptera), *J. exp. Biol.*, **32**, 84–94.

—— (1959), Nitrogen excretion in nymphs of *Aeshna cyanea* (Müll.) (Odonata, Anisoptera), *J. exp. Biol.*, **36**, 566–574.

STOBBART, R. H. AND SHAW, J. (1974), Salt and water balance; excretion, In: Rockstein, M. (ed.), *The Physiology of Insecta*, **5**, 361–446.

TAYLOR, H. H. (1971), Water and solute transport by the Malpighian tubules of the stick insect, *Carausius morosus*. The normal ultrastructure of the Type I cells, *Z. Zellforsch.*, **118**, 333–368.

TREHERNE, J. E. (1972), A study of the function of the neural fat-body sheath in the stick insect, *Carausius morosus*, *J. exp. Biol.*, **56**, 129–137.

VENEZIANI, A. (1905), Valore morfologico e fisiologico dei tubi malpighiani, *Redia*, **2**, 177–230.

WALKER, P. A. (1965), The structure of the fat body in normal and starved cockroaches as seen with the electron microscope, *J. Insect Physiol.*, **11**, 1625–1631.

WESSING, A. (1962), Elektronenmikroskopische Studien zur Funktion der Malpighischen Gefässe von *Drosophila melanogaster*. I. Die Gefässe der Larve und Imago, *Protoplasma*, **55**, 264–293.

WHEELER, W. M. (1893), The primitive number of Malpighian vessels in insects. *Psyche*, **6**, 457–460 ; 485–486; 497–498; 509–510; 539–541; 545–547; 561–564.

WICHARD, W. AND KOMNICK, H. (1973), Fine structure and function of the abdominal chloride epithelia in caddisfly larvae, *Z. Zellforsch.*, **136**, 579–590.

—— (1974), Fine structure and function of the rectal chloride epithelia of damselfly larvae, *J. Insect Physiol.*, **20**, 1611–1621.

WICHARD, W. KOMNICK, H. AND ABEL, J. H. (1972), Typology of ephemerid chloride cells, *Z. Zellforsch.*, **132**, 533–551.

WIGGLESWORTH, V. B. (1931), The physiology of excretion in a blood-sucking insect, *Rhodnius prolixus* (Hemiptera, Reduviidae). I–III, *J. exp. Biol.*, **8**, 411–451.

—— (1938), The regulation of osmotic pressure and chloride concentrations in the haemolymph of mosquito larvae, *J. exp. Biol.*, **15**, 235–247.

—— (1942), The storage of protein, fat, glycogen and uric acid in the fat body and other tissues of mosquito larvae, *J. exp. Biol.*, **19**, 56–77.

—— (1943), The fate of haemoglobin in *Rhodnius prolixus* (Hemiptera) and other blood-sucking Arthropods, *Proc. R. Soc.* (B), **131**, 313–339.

—— (1947), The epicuticle in an insect, *Rhodnius prolixus* (Hemiptera), *Proc. R. Soc.* (B), **134**, 163–181.

—— (1970), The pericardial cells of insects: analogue of the reticuloendothelial system, *J. Reticuloendothelial Soc.*, **7**, 208–216.

WIGGLESWORTH, V. B. AND SALPETER, M. M. (1962), Histology of the Malpighian tubules in *Rhodnius prolixus* Stål (Hemiptera), *J. Insect Physiol.*, **8**, 299–307.

Chapter 16

THE GLANDS OR ORGANS
OF SECRETION

The glands of insects are composed of one or more cells which secrete substances used in or eliminated from the body. Two main types of secretory structures occur. The *exocrine glands* are provided with a more or less well-defined aperture or duct and discharge their secretions outside the body or into the lumen of one or other of the viscera, while the *endocrine glands* have no specialized ducts, their products (hormones) usually diffusing into the blood which transports them to all parts of the body.

Exocrine Glands

The normal epidermal cells, the secretory cells of the mid gut and those of the Malpighian tubules all form glandular epithelia but, for convenience, are dealt with on pp. 15, 195 and 248 respectively. The numerous other exocrine glands considered here are almost all derived from ectoderm and may retain a thin or perforated cuticular layer. They are sometimes unicellular or simple aggregations of single cells (Figs. 132 and 133) or form more highly

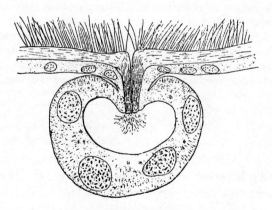

FIG. 132   Section of the integument of the larva of
Gnophomyia tripudians, showing simple gland
After Keilin, 1913.

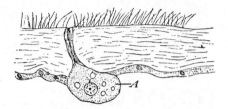

FIG. 133
Section of the integument of the larva of *Ula macroptera* (Tipulidae) showing unicellular epidermal gland (A)
After Keilin, *Arch. Zool. exp.*, 1913.

organized multicellular structures. The latter may be tubular, in the form of a simple alveolus (Fig. 132) or, when the central cavity or duct is branched or divided, the gland is said to be compound. There are consequently compound tubular and compound alveolar (or racemose) glands. The secretory cells line the subdivisions of a tubular gland and the ultimate alveoli, or acini, of the other type. Such elaborate glands are usually provided with a non-secretory duct, opening by a specialized aperture, and may also possess a reservoir for temporary storage of the secretion (Fig. 137).

Histologically, a gland is composed mainly of the secretory epithelial cells which are often very large and elaborate in structure (Noirot and Quennedey, 1974). Their nuclei may be ovoidal or much branched and may contain polytene chromosomes as in the salivary glands of Dipteran larvae (Metz, 1939; Painter, 1939; see also Laufer and Nakase, 1965; Ashburner, 1970; Pearson, 1974). Apically the plasma membrane is often thrown into

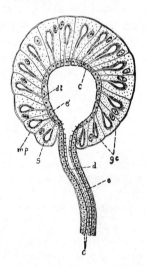

FIG. 134
Semi-diagrammatic section of an acinus of the pygidial gland of a Carabid, *Pterostichus*

c, cuticular lining; *d*, lumen of duct; *dt*, ducteole; *e*, epithelial lining of duct and *e'* of acinus; *gc*, gland cells; *mp*, membrana propria; *s*, striated zone. Based on Dierckx, *La Cellule*, 16.

microvilli where it is surmounted by cuticle. Tubular or vesicular agranular endoplasmic reticulum is sometimes well developed and mitochondria are often very numerous, but granular endoplasmic reticulum is only prominent when the secretion is a protein. The Golgi regions are variably developed but when conspicuous they seem usually to be associated with the formation of secretory granules or vesicles. Electron-dense inclusions of various shapes and sizes are often encountered; their functional significance is not clear though some are probably lysosomes. Basally the plasma membrane may be infolded and beneath it is a basement membrane of variable thickness. The relationship between ultrastructure and secretory activity is still not clear in many cases and in relatively few epidermal glands has a nerve supply been identified. Many exocrine glands that secrete toxic materials have, associated with each gland cell, an 'intracellular ducteole' or end-apparatus. This comprises an invagination bounded by a microvillar plasma membrane, lined by fibrous material, and connected to the surface by a cuticle-lined duct that allows the secretion to pass externally (Crossley and Waterhouse, 1969a).

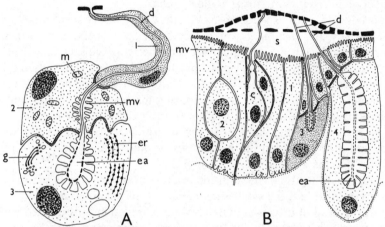

FIG. 135   A, Epidermal gland of *Tenebrio*, comprising a secretory cell (1), intercalary cell (2) and canal cell (3). B, Sternal gland of an Isopteran (Rhinotermitidae), comprising three types of secretory cells: cell 1 discharges its secretion directly through its apical border and across the integument; cell 2 passes its secretion on to an intercalary cell similar to 1; cells 3 and 4 both discharge via an end-apparatus and ductule, the latter associated in each case with a separate canal cell

c, campaniform sensillum; d, intracellular ductule; ea, end-apparatus; er, rough endoplasmic reticulum; g, Golgi apparatus; m, mitochondria; mv, microvilli; s, sub-cuticular space. (After Noirot and Quennedey, 1974).

Some of the major kinds of exocrine glands are dealt with separately below.

Wax Glands – Glands which secrete wax are more especially characteristic of Homoptera where they are uni- or multicellular structures distributed in various parts of the integument. They are particularly evident in the Coccoidea (Bénassy, 1962). The wax is secreted in the form of a powdery covering, as a clothing of threads, or as thin lamellae. Chinese white wax, which was formerly a commercial product, is secreted by the Coccid *Ericerus pela*. Wax glands are also frequent among Aphidoidea and in *Eriosoma lanigerum* the secretion (which is, strictly speaking, a fat rather than a wax) is exuded both in a powdery and a filamentous condition. In the latter case it is discharged through plates composed of a ring or an aggregation of several large cells, each cell containing a central excavation, or wax chamber, within which the secretion accumulates. In the Flatid *Phromnia marginella* Šulc (1929) has shown that the dense clothing of waxy filaments which covers the nymphs is secreted by groups of unicellular glands composed of greatly elongated epidermal cells. Overlying each group of cells is a cuticular plate studded with small pores which are the openings of the separate gland cells. The larvae of some Coccinellidae and of a species of *Selandria* (Tenthredinidae) are invested with a mass of flocculent material believed to be of wax. The wax glands of the honey bee are alluded to under Hymenoptera (Reimann, 1952; Callow, 1963).

Lac Glands – Lac is secreted by certain Coccoidea (Lacciferidae) and, in particular, by *Laccifer lacca* which yields the lac of commerce, a resinous substance produced in large quantities by the female insect as a protective covering (Glover, 1937). The lac is a product of gland cells distributed in the integument (Mahdihassan, 1938, 1961; Haque, 1975). Chemically it consists very largely of resin together with colouring matter, wax, proteins and small amounts of other substances. It is noteworthy that *Laccifer lacca* flourishes best on trees containing gums or resins, or which are rich in certain kinds of latex, and the food-plant influences the colour and quantity of the lac produced (Imms and Chatterjee, 1915).

Cephalic Glands – In addition to the peculiar frontal gland of the Isoptera (p. 612) and the small antennal glands of unknown function in some species, the head of insects contains several paired glands associated with the mouthparts. These may be, at least in part, homologous with the segmentally arranged coelomoducts of Annelids and lower Arthropods (Fahlander, 1940) and appear generally to be of ectodermal origin, though Pflugfelder (1934) claims that the labial glands of *Pontania* larvae are largely derived from the mesodermal coelom-sac walls. Three types of cephalic glands occur widely:

(1) *Mandibular Glands*. Small glands open near the base of the mandibles in Apterygota, Isoptera, Dictyoptera, Coleoptera, Trichoptera and Hymenoptera. In the larvae of the Lepidoptera they attain a considerable size (Fig. 136) and secrete saliva, the normal salivary (labial) glands having become specialized for silk-production in this group. Mandibular glands show varying degrees of elaboration within the Apoidea (Nedel, 1960), where they secrete attractant or alarm pheromones in some species and produce the queen substance in *Apis mellifera* (p. 143).

(2) *Maxillary Glands*. Glands belonging to the maxillary segment are sometimes present. They are found for example in Collembola, Protura, in the larvae of Neuroptera Planipennia and of certain Trichoptera. Part of the salivary gland complex of *Icerya* is possibly a maxillary gland (Pesson, 1944) and they occur in the larvae and adults of some Coleoptera (Srivastava, 1959).

(3) *Labial Glands*. These organs are commonly known as salivary glands and are paired structures, generally situated in the thorax, on either side of the fore intestine. Their ducts combine to form a median salivary duct which normally opens into the pre-oral food cavity near the base of the hypopharynx. In many insects the

ducts of the labial glands possess taenidia in their cuticular lining, and bear a close resemblance to tracheae. Although these glands have been detected in very few Coleoptera (Srivastava, 1959), they are present in most other insects and assume a great variety of form and structure. Among Orthoptera and Dictyoptera they are commonly very large and composed of a number of lobes, each consisting of groups of glandular acini; a salivary reservoir is also present in many species in relation with each gland (Fig. 137; see also Bland and House, 1971, and Kendall, 1969). In

FIG. 136
Right mandibular gland (*g*) of the larva of *Acherontia atropos*

m, mandible and its adductor muscle *a*; *o*, external aperture of gland. *After* Bordas, *Ann. Sci. nat., Zool.*, 1910.

Hemiptera the salivary glands are often differentiated into several lobes and a reservoir ('accessory gland') is sometimes also present (Baptist, 1941; Balasubramanian and Davies, 1968); in phytophagous forms their secretions are not only digestive but may be toxic to the plant and also form the stylet sheath that encloses the mandibles and maxillae after they have penetrated the plant tissues (Miles, 1968). In adult Lepidoptera the labial glands form filamentous tubes whose secretion, in newly emerged Saturniidae and Bombycidae, forms a solvent for an enzyme, cocoonase. This is produced in a semi-solid form by epidermal glands at the bases of the reduced galeae and the solution softens the silken cocoon and so allows the adult moth to escape (Kafatos *et al.*, 1967; Kafatos, 1972). Among Diptera the labial glands are likewise tubular organs which, in the Muscidae, may considerably exceed the total length of the body. Among Hymenoptera they are extremely well developed and assume great complexity. In the honey bee they consist of two pairs of racemose organs, one pair being cephalic and the other thoracic in position and their four ducts unite to form a common canal (Simpson, 1960, and see p. 1191). In the Psocoptera there are also two pairs of labial glands which differ in structure and function (Weber, 1938) and very similar glands occur in the few investigated Mallophaga (e.g. Risler, 1951). In *Panorpa*, the labial glands are greatly enlarged in the male and their secretion is eaten by the female in copulation (Grell, 1938).

The primary function of insect saliva is to assist the digestion of food, either

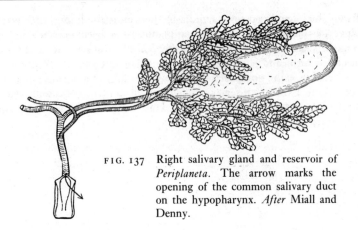

FIG. 137 Right salivary gland and reservoir of *Periplaneta*. The arrow marks the opening of the common salivary duct on the hypopharynx. *After* Miall and Denny.

externally or in the fore gut. Amylases and invertases are most often present but proteases and lipases also occur, the different enzymes corresponding to the predominant dietary constituents and sometimes changing at metamorphosis (e.g. Rodems *et al.*, 1969). Several ultrastructurally and histochemically distinct salivary constituents may be recognized among the secretion granules or the cell organelles with which they are first associated (e.g. Orr *et al.*, 1961; McGregor and Mackie, 1967; Wright, 1969). In many cases there are a few distinct cell types, some of which secrete enzymes while others produce non-digestive mucoproteins or transport water and inorganic ions (Bland and House, 1971; Oschman and Berridge, 1970).

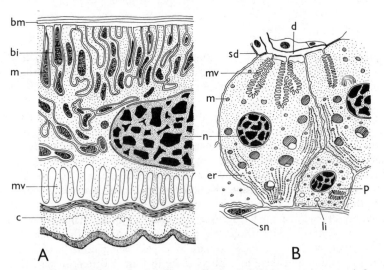

A

B

FIG. 138 Ultrastructure of salivary gland of *Schistocerca gregaria* (*after* Kendall, 1969). A, Salivary duct. B, Acinar cells. N.B. Diagram A is shown at a much higher magnification than B

bi, basal infolding of duct cell; *bm*, basement membrane; *c*, cuticular intima of duct; *d*, duct; *er*, rough endoplasmic reticulum; *li*, lipid globule; *m*, mitochondrion; *mv*, microvilli; *n*, nucleus; *p*, parietal cell; *sd*, septate desmosome; *sn*, nucleus of sheath cell.

The saliva of blood-sucking Diptera may contain anticoagulants and toxic materials (e.g. Yorke and Macfie, 1924; Kahan, 1964; Wright, 1969). The labial glands of *Sciara* larvae apparently produce the slime with which the body is coated as well as the structural mucoproteins of the pupal cocoon (Phillips and Swift, 1965) while those of older *Drosophila* larvae secrete a mucoprotein used to cement the puparium to its substrate (Harrod and Kastritsis, 1972). A subsidiary role of the large salivary reservoir of cockroaches is to store water which is drawn on when the insect is desiccated (Sutherland and Chillseyzn, 1968).

Silk Glands – The silks produced by insects are usually fibrous proteins which vary somewhat in chemical composition and molecular configuration. The glycine content of the protein may vary from over sixty to less than ten per cent and some silks may resemble closely the material of the Chrysopid

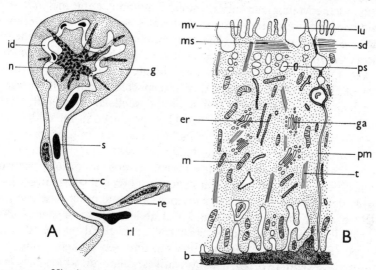

FIG. 139　Histology of silk glands. A, Secretory cell from silk gland of a Symphytan larva (*after* Rudall and Kenchington, 1971). B, Cytoplasmic ultrastructure of anterior portion of silk gland of *Bombyx mori* larva (*after* Voigt, 1965)

b, basement membrane; *c*, canal from secretory cell to reservoir; *er*, granular endoplasmic reticulum; *g*, gland cell; *ga*, Golgi apparatus; *id*, intracellular duct; *lu*, lumen of gland; *m*, mitochondria; *ms*, apical sheath of microtubules; *mv*, microvilli; *n*, nucleus of gland cell; *pm*, plasma membrane; *ps*, prosecretion globule; *re*, reservoir epithelium; *rl*, reservoir lumen; *s*, silk globule; *sd*, septate desmosome; *t*, microtubule.

egg-stalk or the Mantid ootheca (Lucas and Rudall, 1968; Rudall and Kenchington, 1971). Silk is usually produced as fine threads which are more or less closely woven to form larval shelters or cocoons, but ribbons or even parchment-like material may also be formed. The classical fibroin silks of Lepidopteran larvae, notably of the Bombycidae and Saturniidae, have a characteristic molecular arrangement of pleated protein sheets in the extended β-configuration. Others, however, such as the silks produced by the larvae of *Nematus* and allied Symphytan genera are structurally more like collagen, while a few other sawfly larvae form polyglycine silks, as in *Blennocampa*, and the Aculeata and Siphonaptera produce distinctive silks with an α-helical structure. Unusual silks occur in the cocoons of Chrysopid larvae, which are composed of a cuticulin-like material, and of *Ptinus tectus* larvae which are formed from an unbroken chitinous thread of peritrophic membrane extruded from the anus.

In the Lepidoptera silk is secreted by the labial glands in the form of fibroinogen which undergoes denaturation on extrusion to form the tough, elastic protein, fibroin, and is surrounded by an outer layer of a water-soluble, gelatinous protein known as sericin. In the larvae of the Carabid *Lebia scapularis*, and of the Neuroptera Planipennia, silk is produced as a secretion of the Malpighian tubes; among Embioptera and male *Hilara* spp. (Empididae) it is secreted by dermal glands situated in the anterior tarsi, while the egg-cocoon of *Hydrophilus* is formed from secretions of the female accessory reproductive glands (Laabs, 1939). In the silk glands of *Bombyx mori* (Voigt, 1965) fibroin is secreted proximally, appearing first in the Golgi regions of the secretory cells and then transported peripherally in vacuoles, from which it is released by fusion of the vacuolar membrane with the plasmalemma. Sericin, on the other hand, is produced in the middle section of the gland from ultrastructurally distinct cells and its precursor granules have no direct connection with the Golgi vacuoles.

Repugnatorial Glands – Many dermal glands, in various parts of the body of different insects, produce a variety of chemical substances which are probably defensive as they have a nauseous smell and other repellent properties (Roth and Eisner, 1962). The dorsal abdominal scent glands of many nymphal Heteroptera and the metepisternal scent glands of the adults are of this type (Remold, 1962; Stein, 1966–69; and see p. 696). They secrete mixtures of several hydrocarbon derivatives, some of which may also be alarm pheromones (Calam and Youdeowei, 1968); the secretions of Coreids usually include hexanal, hexanol, acetic acid and hexyl acetate, while Pentatomid secretions include hexenal, octenal and decenal. Among the Coleoptera, especially the Adephaga, there are often complex pygidial glands which open near the anus and secrete pungent or corrosive material (Forsyth, 1970, 1972; Kendall, 1969; and see p. 829). In the bombardier beetles (*Brachinus* spp., see p. 829), the pygidial glands produce a mixture of quinones and hydrogen peroxide which accumulate in a reservoir where they react explosively under the influence of an accessory secretion to produce a defensive spray (Schildknecht and Holoubek, 1961). In the Isoptera the enlarged frontal gland of the nasute soldiers secretes a sticky defensive mixture of terpenes and resins (Moore, 1964), while among Lepidopteran larvae there are eversible repugnatorial glands (osmeteria) on the 6th and 7th abdominal segments of Lymantriidae and elsewhere in other groups (p. 1092; Crossley and Waterhouse, 1969*b*).

Attractant Glands – Many insects possess glands that secrete pher-

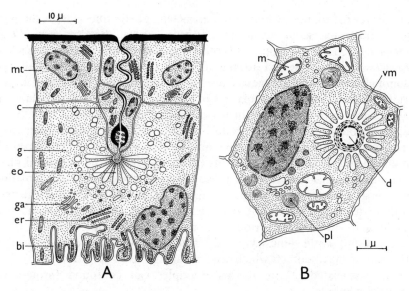

FIG. 140   Ultrastructure of repugnatorial glands. A, Dorsal abdominal gland of
Pyrrhocoris apterus nymph (*after* Stein, 1967). B, Secretory cell from
gland of *Pterostichus madidus* (*after* Forsyth, 1970)

bi, basal infolding of plasma membrane; c, canal cell; d, vesicular duct; eo, end-
organ; er, endoplasmic reticulum; g, gland cell; ga, Golgi apparatus; m, mitochondrion; mt, matrix cell; mv, microvilli; pl, phospholipid body; vm, vesicular membrane.

omones responsible for attracting individuals to one another (see p. 142). In
social species this attraction promotes the cohesion of the colony; in normally solitary forms it facilitates mating by bringing the sexes together or it
leads to the formation of temporary aggregations. The chemical stimuli may
operate over distances of several kilometres or they may be confined to
short-range attraction and stimulation during the later stages of courtship
and mating (Shorey, 1973). Such pheromones play a major part in controlling insect behaviour and are the subject of active investigation (Beroza,
1970; Wood et al., 1970; Wilson, 1971; Jacobson, 1972; Birch, 1974). They
are chemically diverse and the glands producing them are equally varied in
their taxonomic distribution, their position and structure, and in the detailed
biological roles played by their secretions. Many Lepidoptera, for example,
possess scent glands that differ widely according to species and sex. Scattered or clustered together on the wings of males there are often characteristically shaped scales known as *androconia* with gland cells at their bases.
Males of *Hepialus hectus* have swollen hind tibiae that bear long clavate or
cylindrical scales associated with secretory cells. In other male Lepidoptera
the males may have eversible abdominal hair-pencils (modified glandular
scales) which produce an aphrodisiac secretion, as in the queen butterfly
Danaus gilippus (Brower et al., 1965). In many Lepidoptera it is the female

which bears the glands secreting the attractant pheromones, as in the well-known abdominal scent organs of *Porthetria dispar*, *Bombyx mori*, *Plodia interpunctella*, *Ephestia* spp. and others. Females of various species of *Trogoderma* produce sexual attractants from secretory cells lining the 7th abdominal sternum (Hammack *et al.*, 1973) and another is secreted by the pygidial glands of some female Coccoidea (Moreno, 1972). Males of *Harpobittacus* have two orange-brown eversible vesicles between the 6th, 7th and 8th abdominal terga; they produce a sex pheromone and are made up of secretory and other cells (Crossley and Waterhouse, 1969). The dorsal abdominal glands of some male Blattids (Roth, 1969) and the metanotal gland of *Oecanthus* males secrete a substance licked by the females at copulation. The aromatic secretions of various symphiline Coleoptera, living in the nests of ants or termites, are produced by dermal glands situated at the bases of tufts of hairs located in various regions of the integument (Mou, 1938; Pasteels, 1969) while some Lycaenid larvae have eversible spinose tubercles at the end of the abdomen which are similarly licked by the ants with which they are associated. Many further examples can be found through the reviews cited above.

Poison Glands – These are best developed in Apocritan Hymenoptera where they are modified accessory reproductive glands associated with the ovipositor or sting (Beard, 1963; Robertson, 1968; Roth and Eisner, 1962; Bücherl and Buckley, 1972). The secretions are generally a complex mixture of several substances, among which may be mentioned the protein melittin of honey-bee venom and the polypeptide kinins of *Vespula* venom (Prado *et al.*, 1966; Kreil, 1973). The venoms of Braconidae have been studied in some detail and have been shown to differ in their toxicity from one host species to another; *Microbracon* venom perhaps acts by blocking neuromuscular transmission at a presynaptic site (Piek and Engels, 1969). Several Lepidopteran larvae, including representatives of the Megalopygidae, Eucleidae, Saturniidae, Arctiidae and Lymantriidae, are provided with epidermal poison glands associated with setae or spines which, when broken, allow the discharge of a secretion causing urticaria in man (Bücherl and Buckley, 1972). Many insects are also able to induce an allergic response in man; this may be due to powdery dust or scales emanating incidentally from the body of living or dead insects or to the secretion of salivary or poison glands (Shulman, 1967; Frazier, 1969; Feingold *et al.*, 1968).

Accessory Reproductive Glands – See pp. 291 and 298.

Endocrine Glands

The anterior part of the insect body contains a number of endocrine glands, many of which, from their position in the head, are known as the retrocerebral glands. Functionally they form a co-ordinated and balanced system whose secretions exercise a major influence on many important biological functions such as growth, postembryonic development, metamorphosis and diapause, as well as certain aspects of reproduction, water balance and excretion, and such metabolic activities as the regulation of blood sugar and lipid levels. Various modes of behaviour, the deposition, hardening and darkening of cuticle, digestive enzyme secretion, and perhaps even the control of caste and morph differentiation in certain polymorphic species are also among those processes that are subject to hormonal influences. Insect endocrinology is discussed in many reviews and general works, including those by Engelmann (1968), Burdette (1974), Highnam and Hill (1969), Joly (1968), Novak (1975), Sehnal (1971), Sláma et al. (1974), Truman and Riddiford (1974), Wigglesworth (1964, 1970) and Willis (1974).

Neurosecretory Cells – Relatively large cells, containing inclusions that usually stain selectively with paraldehyde-fuchsin, occur in the ganglia of the nervous system and in the corpora cardiaca (q.v.). These neurosecretory cells are specialized neurons, showing electrical signs of nervous activity (e.g. Cook and Milligan, 1972), and also possessing endocrine functions. They are best known from the brain, where they are typically arranged in clusters in the pars intercerebralis of the protocerebrum and in a more lateral group on each side. Stainable secretory material, thought to comprise the hormones along with carrier proteins, leaves the perikarya of these cells and passes in granular form along their axons, from the ends of which it is ultimately discharged. In the brain, most of the axons from the paired groups of pars intercerebralis cells cross to the opposite side and run back, as do the uncrossed axons and those from the lateral cell groups, eventually to end in the corpora cardiaca (Fig. 141). Variations on this general plan are described by Cazal (1948) and Juberthie and Cassagnau (1971), and in many other descriptions of individual species, including those listed on pp. 282–286. Neurosecretory cells, with somewhat varied staining reactions, are also distributed among the ganglia of the ventral nerve-cord (e.g. Fraser, 1959; Geldiay, 1959; Delphin, 1965), they occur in the frontal ganglion, and they form the so-called intrinsic cells of the corpora cardiaca, discussed below. The ultrastructure of insect neurosecretory cells is now known in some detail (see references on pp. 282–286); their cytoplasm commonly contains abundant granular endoplasmic reticulum, free ribosomes and well-developed Golgi areas, in association with which the secretion granules are formed. Neurosecretory granules vary in size, with diameters from about 60–600 nm and they show different degrees of electron density. These and other characteristics allow the cells to be arranged in a number of categories, more or less similar to those distinguished earlier by their staining reactions.

The hormones produced by the neurosecretory cells have a number of physiological effects (Goldsworthy and Mordue, 1974; Maddrell, 1974; Miller, 1975) but they have not yet been fully characterized chemically nor can their sites of origin and release always be identified exactly. The following are among those which can be recognized. The brain hormone is probably a protein or mixture of proteins with molecular weights up to about 40 000 (Ishizaki and Ichikawa, 1967; Yamazaki and Kobayashi, 1969; Gersch and Stürzebecher, 1968). It initiates the production of ecdysone (moulting hormone) by the prothoracic glands (q.v.) and is therefore indirectly responsible for moulting and, in some cases, for the termination of diapause (p. 384). The neurosecretory cells of the pars intercerebralis of *Calliphora* and some other insects exert a more direct effect in promoting the development of the ovary by stimulating the growth of the oocytes. Further, the hardening and darkening of the cuticle of *Calliphora*, *Locusta*, *Periplaneta* and other species depends on *bursicon*, a peptide or protein hormone produced by neurosecretory cells in the brain and released into the blood from the thoracic or abdominal ganglia (Cottrell, 1964; Fraenkel and Hsaio, 1965; Vincent, 1972). Neurosecretory cells in the compound ventral ganglion of *Rhodnius* discharge a diuretic hormone (Maddrell, 1963–64) and the control of excretion and water balance by neurosecretory mechanisms has now been established in several species (Pilcher, 1970; Mordue, 1972). Metabolic effects of neurosecretion are also known, as with the hyperglycaemic peptides that raise the trehalose content of the blood and the adipokinetic hormones that induce changes in the lipids of the blood and fat-body. For these and still further effects, both physiological and behavioural, see the reviews cited above. In many cases the release of neurosecretory material from the swollen axon terminals occurs in discrete *neurohaemal organs*, of which the corpora cardiaca are the best developed though less conspicuous ones make up the perisympathetic system associated with the ventral nerve-cord (Brady and Maddrell, 1967; Raabe, 1971; Provensal and Grillot, 1972). Neurosecretory granules of a kind may also be found in peripheral axons supplying many viscera such as the heart, gut, muscles, epidermis, spermatheca and even the corpora allata and prothoracic glands. Such nervous structures are perhaps best regarded as neurosecretomotor rather than neurosecretory in the classical sense, since they release their secretions close to the effector cells rather than into the haemolymph (Miller, 1975). Related to such systems are a variety of 'neurohumours' including the catecholamines and serotonin (5-hydroxytryptamine) as well as extractable materials whose action on the heart and other organs is likely to be pharmacological rather than physiological.

Corpora Cardiaca (Fig. 141) – Situated in close association with the aorta behind the brain, in all Pterygotes, some Thysanura and Japygidae, is a

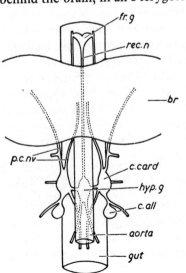

FIG. 141
Stomatogastric nervous system and associated endocrine glands. Diagram of typical condition (*after* Cazal, 1948)

br, brain; *c.all*, corpus allatum; *c.card*, corpus cardiacum; *fr.g*, frontal ganglion; *hyp.g*, hypocerebral ganglion; *p.c.nv*, external and internal paracardiac nerves; *rec.n*, recurrent nerve.

pair of small bodies, the corpora cardiaca (corpora paracardiaca, postcerebral or pharyngeal 'ganglia'). Each is connected to the protocerebrum by a pair of nerves and to the hypocerebral ganglion of the stomatogastric system by a single nerve and the two glands often show some degree of median fusion. Ultrastructural and experimental evidence shows that the corpora cardiaca are neurohaemal organs containing the terminations of axons from the neurosecretory cells of the brain, whose hormones they store and release into the haemolymph. A simple condition of this sort occurs in many Apterygota (Juberthie and Cassagnau, 1971), where there are no true corpora cardiaca and the neurosecretory axons end in the wall of the aorta. The Pterygote corpora cardiaca, however, also contain intrinsic neurosecretory cells (parenchymal cells) and interstitial, glia-like tissue, so that they not only store and release material from the cerebral neurosecretory cells but are also endocrine glands in their own right. In some genera such as *Locusta* there is a clear distinction into a storage lobe and a glandular lobe (e.g. Cazal *et al.*, 1971); in *Calliphora* the intrinsic cells are peripherally arranged (Normann, 1965); and in *Carausius* the cellular components form a loosely-knit tissue that incorporates ramifying extracellular spaces (Smith and Smith, 1966). The secretory granules of the intrinsic cells of *Leucophaea* first become visible in the Golgi zones as electron-opaque granules which gradually become less dense as they accumulate in the cellular processes (Scharrer, 1963). While some of the axons entering the corpora cardiaca from the brain end blindly there, others pass through to the corpora allata and other viscera (Aggarwal and King, 1971; Bowers and Johnson, 1966). The hormones of the corpora cardiaca are those produced by its intrinsic cells and by the neurosecretory cells of the brain. Their general properties have therefore been discussed above though it is worth mentioning again the extractable substances that are thought to increase the rate of heart-beat but whose exact physiological role is not understood (Davey, 1964; Miller, 1975; Goldsworthy and Mordue, 1974).

Prothoracic Glands – These paired structures, also known as thoracic glands, pericardial glands, ecdysial glands or ventral glands, apparently arise as epithelial invaginations of the labial segment of the head. They have been found in the cephalic or anterior thoracic region in the immature stages of all but a few orders of insects (Hermann, 1967). They vary in appearance from more or less compact structures to very diffuse organs and are composed of large cells, often with polyploid nuclei. The glands are well supplied with tracheae and in most cases they are provided with nerves, the axons of which contain neurosecretory granules. Ultrastructurally the prothoracic gland cells show a close resemblance to the steroid-secreting cells of vertebrates; their cytoplasm contains variable, often sparse, amounts of agranular endoplasmic reticulum, inconspicuous Golgi areas, free ribosomes, numerous microtubules, mitochondria (perhaps involved in secretory activity) and cyclically prominent lysosome-like inclusions. (See, for example, Scharrer, 1964; Beaulaton, 1968; Yin and Chippendale, 1974.) The

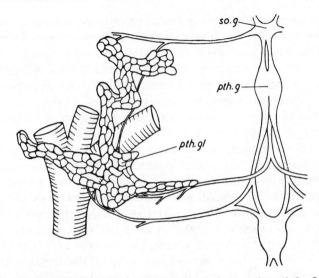

FIG. 142 Prothoracic gland and its nerve supply in *Saturnia* larva (*after* Lee, 1948)
pth.g, prothoracic ganglion; *pth.gl*, prothoracic gland; *so.g*, suboesophageal ganglion.

plasmalemma is thrown into strongly developed invaginations with intracellular channels and peripheral lacunae. Changes in the nuclei, mitochondria and lysosomes seem to be associated with phases of the moulting cycle and the glands atrophy completely in adult Pterygota. Only in the Thysanura, where moulting continues after sexual maturity, are the prothoracic glands retained throughout life.

Experimental evidence shows that the secretions of the prothoracic glands exert a number of effects. In particular, they initiate events in the epidermis (and associated muscles) which induce moulting and cause new cuticle to be laid down. They also stimulate the growth of imaginal discs and their activity leads to the termination of postembryonic diapause, as was shown in the classical experiments of Williams (1946–53). Many of their effects require increased protein synthesis and this seems to be due to the prothoracic gland hormone activating specific sites of messenger RNA synthesis in the chromosomes, as shown by the production of characteristic 'puffs' in the polytene chromosomes of Dipteran salivary glands (Clever, 1963; Ilan and Ilan, 1973). The first prothoracic gland hormone to be isolated was α-ecdysone, a steroid which is now known to originate from cholesterol by a sequence of enzymatically controlled steps. α-ecdysone is hydroxylated to ecdysterone (20-hydroxyecdysone) and both hormones can be further modified chemically. There is thus a 'family' of interconvertible ecdysones, each of which seems to differ somewhat, according to the species or stage in which it predominates, in biological activity and in the pattern of chromosomal puffing that it induces. It is uncertain whether one of these is the 'real' moulting hormone or if each has its own function (Karlson and Koolman, 1974; Svoboda *et al.*, 1975).

Corpora Allata – These endocrine glands are situated laterally behind the brain, usually near the corpora cardiaca, and are present in all insects.

They arise from ectoderm in most cases (though derivation from other germ-layers has been reported) and though generally paired, globular bodies they are fused into a median mass beneath the aorta in some insects, such as the Hemiptera and Dermaptera. In most Apterygota and in the Ephemeroptera the corpora allata are innervated only from the suboesophageal ganglia but in other insects they are also supplied by cerebral nerves that run through the corpora cardiaca, and the original suboesophageal innervation may be lost, as in the Diptera (Juberthie and Cassagnau, 1971); some of the axons supplying the corpora allata contain neurosecretory granules. The glands increase in size throughout development but may decline somewhat in older adults (e.g. Pflugfelder, 1948). Two main histological types are found: the vesicular type is composed of an epithelial wall with a central, more or less vacuolar region, while the other type is solid and may be differentiated into cortex and medulla. The cells are often irregularly shaped, but their homogeneous cytoplasm shows few striking features under the light-microscope. Ultrastructurally it exhibits numerous mitochondria, frequent microtubules and abundant free ribosomes but there is little granular endoplasmic reticulum nor are the Golgi areas very conspicuous. Actively secreting corpora allata contain abundant tubular agranular endoplasmic reticulum, from which are derived lipid droplets that may perhaps represent the endocrine secretion. For further details see Waku and Gilbert (1964), Odhiambo (1966), Thomsen and Thomsen (1970), Scharrer (1971) and others.

The corpora allata are known from experimental work to produce secretions whose most important function is to inhibit the realization of imaginal characters during postembryonic development. These juvenile hormones, sometimes denoted as JH I, JH II and JH III, have been isolated and are known to be terpenoids, in fact chemically related esters of tridecadienoic acid (Röller and Dahm, 1968; Menn and Beroza, 1972; Judy et al., 1973). In the immature stages they are actively liberated and, acting in conjunction with the ecdysones, they ensure that the normal sequence of pre-imaginal instars is produced. During the last larval or nymphal instar the corpora allata become inactive and the resulting change in hormonal balance leads to metamorphosis. In addition to the many morphogenetic effects embraced by this inhibitory action (Willis, 1974), the juvenile hormones also act independently of ecdysones to exert a gonadotropic influence in the adult females of many species. Experimental removal of the adult corpora allata, which are once more active after metamorphosis has been completed, prevents the deposition of yolk in the developing oocytes (Engelmann, 1970). In males there is no gonadotropic effect but the secretions of the corpora allata may be needed in some species for full development of the accessory reproductive glands and for normal sexual behaviour (Barth and Lester, 1973). Juvenile hormones also have a number of other, sometimes more specific, effects: they ensure the maintenance of the prothoracic glands of immature insects, they break the adult reproductive diapause of several species, and they modify epigamic behaviour or cocoon construction in others. Their action seems also to underlie some striking cases of polymorphism. Experimental application of juvenile hormone to young worker larvae of *Apis mellifera* induces the development of queen-like individuals in the worker cells (Wirtz and Beetsma, 1972;

but cf. Rembold *et al.*, 1974). In *Kalotermes*, caste differentiation appears to depend on the balance between ecdysones and juvenile hormones (Lüscher and Springhetti, 1960), while in *Locusta* treatment with juvenile hormone or the implantation of corpora allata leads to the development of a *solitaria* facies (Staal, 1975). The effects of the natural juvenile hormones – and also those of the ecdysones – are mimicked by a large number of more or less closely related substances obtained synthetically or from plants. Many of these are being actively investigated as alternatives to the traditional insecticides used in pest control (Jacobson and Crosby, 1971; Sláma, 1971; Sláma *et al.*, 1974; Staal, 1975).

Weismann's Ring (Ring Gland) – The larvae of Cyclorrhaphan Diptera do not display the usual arrangement of retrocerebral endocrine glands. Instead, there is present behind the brain and around the aorta a small ring-like structure supported by tracheae (Fig. 143). In addition to the tracheal matrix cells and the hypocerebral ganglion, Weismann's ring contains three types of glandular cells which are respectively homologous with

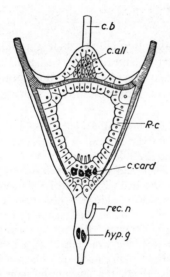

FIG. 143
Diagram of Weismann's ring from *Calliphora* larva (*after* Thomsen, 1951)

c.b, cephalo-pharyngeal band; *R-c*, R-cells (probably homologous with prothoracic glands of other insects). Other lettering as in Fig. 141.

the corpora allata, corpora cardiaca and pericardial (or prothoracic) glands (Thomsen, 1951; Aggarwal and King, 1969). Conditions in some Brachyceran larvae are intermediate between the above and those in the more normal Nematocera. Physiological studies involving the extirpation and implantation of all or parts of the ring (Possompès, 1949; 1950) show that its various portions control metamorphosis in a manner similar to the corresponding glands of other insects.

Literature on the Glands

(a) Exocrine Glands and their Secretions:

ASHBURNER, M. (1970), Function and structure of polytene chromosomes during insect development, *Adv. Insect Physiol.*, **7**, 1–95.

BALASUBRAMANIAN, A. AND DAVIES, R. G. (1968), The histology of the labial glands of some Delphacidae (Hemiptera: Homoptera), *Trans. R. ent. Soc. Lond.*, **120**, 239–251.

BAPTIST, B. A. (1941), The morphology and physiology of the salivary glands of Hemiptera Heteroptera, *Q. Jl microsc. Sci.*, **83**, 91–139.

BEARD, R. L. (1963), Insect toxins and venoms, *A. Rev. Ent.*, **8**, 1–18.

BENASSY, C. (1962), Les sécrétions tegumentaires chez les Coccides, *Année Biol.* (3ᵉ serie) 37 (1961), 321–341.

BEROZA, M. (ed.) (1970), *Chemicals Controlling Insect Behavior*, Academic Press, New York, 170 pp.

BIRCH, M. (ed.) (1974), *Pheromones*, North-Holland, Amsterdam, 495 pp.

BLAND, K. P. AND HOUSE, C. R. (1971), Functions of the salivary glands of the cockroach, *Nauphoeta cinerea*, *J. Insect Physiol.*, **17**, 2069–2084.

BROWER, L. P., BROWER, J. V. Z. AND CRANSTON, F. P. (1965), Courtship behavior of the queen butterfly, *Danaus gilippus berenice* (Cramer), *Zoologica, N.Y.*, **50**, 1–39.

BÜCHERL, W. AND BUCKLEY, E. E. (eds) (1972), *Venomous Animals and their Venoms, Vol. 3: Venomous Invertebrates.* New York, 560 pp.

CALAM, D. H. AND YOUDEOWEI, A. (1968), Identification and functions of secretion from the posterior scent gland of the fifth-instar larvae of the bug *Dysdercus intermedius*, *J. Insect Physiol.*, **14**, 1147–1158.

CALLOW, R. K. (1963), Chemical and biochemical problems of beeswax, *Bee World*, **44**, 95–101.

CROSSLEY, A. C. AND WATERHOUSE, D. F. (1969a), The ultrastructure of a pheromone-secreting gland in the male scorpion-fly *Harpobittacus australis* (Bittacidae: Mecoptera), *Tissue and Cell*, **1**, 273–294.

—— (1969b), The ultrastructure of the osmeterium and the nature of its secretion in *Papilio* larvae (Lepidoptera), *Tissue and Cell*, **1**, 525–554.

FAHLANDER, K. (1940), Die Segmentalorgane der Diplopoda, Symphyla und Insecta Apterygota, *Zool. Bidr. Upps.*, **18**, 243–251.

FEINGOLD, B. F., BENJAMINI, E. AND MICHAELI, D. (1968), The allergic responses to insect bites, *A. Rev. Ent.*, **13**, 137–158.

FORSYTH, D. J. (1970), The ultrastructure of the pygidial defence glands of the Carabid *Pterostichus madidus* F., *J. Morph.*, **131**, 397–416.

—— (1972), The structure of the pygidial defence glands of Carabidae (Coleoptera), *Trans. zool. Soc. Lond.*, **32**, 249–309.

FRAZIER, C. A. (1969), *Insect Allergy: Allergic and toxic reactions to insects and other Arthropods*, Warren H. Green, St. Louis, Missouri, 493 pp.

GLOVER, P. M. (1937), *Lac Cultivation in India*, Indian Lac Institute, Nankum, Ranchi, 147 pp.

GRELL, K. G. (1938), Der Darmtraktus von *Panorpa communis* L. und seine Anhänge bei Larve und Imago (Ein Beitrag zur Anatomie und Histologie der Mecopteren), *Zool. Jb. (Anat.)*, **64**, 1–86.

HAMMACK, L., BURKHOLDER, W. E. AND MA, M. (1973), Sex pheromone localization in females of six *Trogoderma* species (Coleoptera: Dermestidae), *Ann. ent. Soc. Am.*, **66**, 545–550.

HAQUE, M. S. (1975), Cells secreting non-resinous substances in the female lac insect *Kermia laçca* (Homoptera–Coccoidea), *J. Zool.*, **176**, 27–38.

HARROD, M. J. E. AND KASTRITSIS, C. D. (1972), Developmental studies in *Drosophila*. II, VI, *J. Ultrastruct. Res.*, **38**, 482–499; **40**, 292–312.

IMMS, A. D. AND CHATTERJEE, N. C. (1915), On the structure and biology of *Tachardia lacca*, Kerr, with observations on certain insects predaceous or parasitic upon it, *Indian For. Mem., For. Zool. Series*, **3** (1), 42 pp.

JACOBSON, M. (1972), *Insect Sex Pheromones*, Academic Press, New York, 382 pp.

KAFATOS, F. C. (1972), The cocoonase zymogen cells of silk moths: a model of terminal cell differentiation for specific protein synthesis, In: *Current Topics in Developmental Biology*, **7**, 125–191.

KAFATOS, F. C., TARTAKOFF, A. M. AND LAW, J. H. (1967), Cocoonase. I. Preliminary characterization of a proteolytic enzyme from silkmoths, *J. biol. Chem.*, **242**, 1477–1487.

KAHAN, D. (1964), The toxic effect of the bite and the proteolytic activity of the saliva and stomach contents of the robber flies (Diptera Asilidae), *Israel J. Zool.*, **13**, 47–57.

KENDALL, D. A. (1974), The structure of defence glands in some Tenebrionidae and Nilionidae (Coleoptera), *Trans. R. ent. Soc. Lond.*, **125**, 437–487.

KENDALL, M. D. (1969), The fine structure of the salivary glands of the Desert Locust *Schistocerca gregaria* Forskål, *Z. Zellforsch.*, **98**, 399–420.

KREIL, G. (1973), Biosynthesis of melittin, a toxic peptide from bee venom. Aminoacid sequence of the precursor, *European J. Biochem.*, **33**, 558–566.

LAABS, A. (1939), Brutfürsorge und Brutpflege einiger Hydrophiliden mit Berücksichtigung des Spinnapparates, seines äusseren Baues und seiner Tätigkeit, *Z. Morph. Ökol. Tiere*, **36**, 123–178.

LAUFER, H. AND NAKASE, Y. (1965), Salivary gland secretion and its relation to chromosomal puffing in the dipteran, *Chironomus thummi*, *Proc. natn. Acad. Sci. U.S.A.*, **53**, 511–516.

LUCAS, F. AND RUDALL, K. M. (1968), Extracellular fibrous proteins: the silks. In: Florkin, M. and Stotz, E. H. (eds), *Comprehensive Biochemistry*, **26 B**, 475–558.

MCGREGOR, H. C. AND MACKIE, J. B. (1967), Fine structure of the cytoplasm in salivary glands of *Simulium*, *J. Cell Sci.*, **2**, 137–144.

MAHDIHASSAN, S. (1938), Die Struktur des Stocklacks und der Bau der Lackzelle, *Z. Morph. Ökol. Tiere*, **33**, 527–554.

—— (1961), Waxes of the lac insects and their glands, *Z. angew. Ent.*, **48**, 433–444.

METZ, C. W. (1939), The visible organization of the giant salivary gland chromosomes in Diptera, *Am. Nat.*, **73**, 457–466.

MILES, P. W. (1968), Insect secretions in plants, *A. Rev. Phytopath.*, **6**, 137–164.

MOORE, B. P. (1964), Volatile terpenes from *Nasutitermes* soldiers. (Isoptera, Termitidae), *J. Insect Physiol.*, **10**, 371–375.

MORENO, D. S. (1972), Location of the site of production of the sex pheromone in the yellow scale and the California red scale, *Ann. ent. Soc. Am.*, **65**, 1283–1286.

MOU, Y. C. (1938), Morphologische und histologische Studien über Paussidendrüsen, *Zool. Jb. (Anat.)*, **64**, 287–346.

NEDEL, J. O. (1960), Morphologie und Physiologie der Mandibeldrüse einiger Bienen-Arten (Apidae), *Z. Morph. Ökol. Tiere*, **49**, 139–183.

NOIROT, C. AND QUENNEDEY, A. (1974), Fine structure of insect epidermal glands, *A. Rev. Ent.*, **19**, 61–80.

ORR, C. W. M., HUDSON, A. AND WEST, A. S. (1961), The salivary glands of *Aedes aegypti*: histological-histochemical studies, *Can. J. Zool.*, **39**, 265–272.

OSCHMAN, J. L. AND BERRIDGE, M. J. (1970), Structural and functional aspects of salivary fluid secretion in *Calliphora*, *Tissue and Cell*, **2**, 281–310.

PAINTER, T. S. (1939), The structure of salivary gland chromosomes, *Am. Nat.*, **73**, 315–330.

PASTEELS, J. M. (1969), Les glandes tégumentaires des staphylins termitophiles. III, *Insectes soc.*, **16**, 1–26.

PEARSON, M. J. (1974), Polyteny and the functional significance of the polytene cell cycle, *J. Cell Sci.*, **15**, 457–479.

PESSON, P. (1944), Contribution à l'étude morphologique et fonctionelle de la tête, de l'appareil buccal et du tube digestif des femelles de Coccides, *Monogr. Sta. Lab. Rech. agron., Paris*, **1944**, 266 pp.

PFLUGFELDER, O. (1934), Bau und Entwicklung der Spinndrüse der Blattwespen, *Z. wiss. Zool.*, **145**, 261–282.

PHILLIPS, D. M. AND SWIFT, H. (1965), Cytoplasmic fine structure of *Sciara* salivary glands. I. Secretion, *J. Cell Biol.*, **27**, 395–409.

PIEK, T. AND ENGELS, E. (1969), Action of the venom of *Microbracon hebetor* Say on larvae and adults of *Philosamia cynthia* Hübn., *Comp. Biochem. Physiol.*, **28**, 603–618.

PRADO, J. L., TAMURA, Z., FURANO, E., PISANO J. J. AND UDENFRIEND, S. (1966), Characterisation of kinins in wasp venom, In: Erdös, E. G., Back, N., Sicuteri, F. and Wilde, A. F. (eds), *Hypotensive peptides*, Springer, Berlin and New York.

REIMANN, K. (1952), Neue Untersuchungen über die Wachsdrüsen der Honigbiene, *Zool. Jb. (Anat.)*, **72**, 251–272.

REMOLD, H. (1962), Über die biologische Bedeutung der Duftdrüsen bei den Landwanzen (Geocorisae), *Z. vergl. Physiol.*, **45**, 636–694.

RISLER, H. (1951), Der Kopf von *Bovicola caprae* (Gurlt.) (Mallophaga), *Zool. Jb. (Anat.)*, **71**, 325–374.

ROBERTSON, P. L. (1968), A morphological and functional study of the venom apparatus in representatives of some major groups of Hymenoptera, *Aust. J. Zool.*, **16**, 133–166.

RODEMS, A. E., HENDRIKSON, P. A. AND CLEVER, U. (1969), Proteolytic enzymes in the salivary gland of *Chironomus tentans*, *Experientia*, **25**, 686–687.

ROTH, L. M. (1969), The evolution of male tergal glands in the Blattaria, *Ann. ent. Soc. Am.*, **62**, 176–208.

ROTH, L. M. AND EISNER, T. (1962), Chemical defenses of Arthropods, *A. Rev. Ent.*, **7**, 107–136.

RUDALL, K. M. AND KENCHINGTON, W. (1971), Arthropod silks: the problem of fibrous proteins in animal tissue, *A. Rev. Ent.*, **16**, 73–96.

SCHILDKNECHT, H. AND HOLOUBEK, K. (1961), Die Bombardierkäfer und ihre Explosionschemie, *Angew. Chem.*, **73**, 1–7.

SHOREY, H. H. (1973), Behavioral responses to insect pheromones, *A. Rev. Ent.*, **18**, 349–380.

SHULMAN, S. (1967), Allergic responses to insects, *A. Rev. Ent.*, **12**, 323–346.

SIMPSON, J. (1960), The functions of the salivary glands of *Apis mellifera*, *J. Insect Physiol.*, **4**, 107–121.

SRIVASTAVA, U. S. (1959), The maxillary glands of some Coleoptera, *Proc. R. ent. Soc. Lond.* (A), **34**, 57–62.

STEIN, G. (1966–67), Über den Feinbau der Duftdrüsen von Feuerwanzen (*Pyrrhocoris apterus* L., Geocorisae), *Z. Zellforsch.*, **74**, 271–290; **75**, 501–516; **79**, 49–63.

—— (1969), Über den Feinbau der Duftdrüsen von Heteropteren. Die hintere larvale Abdominaldrüse der Baumwollwanze *Dysdercus intermedius* Dist. (Insecta, Heteroptera), *Z. Morph. Tiere*, **65**, 374–391.

ŠULC, K. (1929), Die Wachsdrüsen und ihre Produkte bei den Larven von *Flata* (*Phromnia*) *marginella* d'Olivier, *Biolog. Spisy.*, **8**, (2), 1–23.

SUTHERLAND, D. J. AND CHILLSEYZN, J. M. (1968), Function and operation of the cockroach salivary reservoir, *J. Insect Physiol.*, **14**, 21–31.

VOIGT, W. H. (1965), Zur funktionellen Morphologie der Fibroin- und Sericinsekretion der Seidendrüse von *Bombyx mori* L. I, II, *Z. Zellforsch.*, **66**, 548–570; 571–582.

WEBER, H. (1938), Beiträge zur Kenntnis der Ueberordnung Psocoidea. I. Die Labialdrüsen der Copeognathen, *Zool. Jb.* (*Anat.*), **64**, 243–286.

WILSON, E. O. (1971), *The Insect Societies*, Cambridge, Mass., 548 pp.

WOOD, D. L., SILVERSTEIN, R. M. AND NAKAJIMA, M. (eds) (1970), *Control of Insect Behavior by Natural Products*, Academic Press, New York, 346 pp.

WRIGHT, K. A. (1969), The anatomy of salivary glands of *Anopheles stephensi* Liston, *Can. J. Zool.*, **47**, 579–587.

YORKE, E. AND MCFIE, J. W. S. (1924), The action of the salivary secretion of mosquitoes and of *Glossina tachinoides* on human blood, *Ann. trop. Med. Parasitol.*, **18**, 103–108.

(*b*) Endocrine Glands and their Secretions:

AGGARWAL, S. K. AND KING, R. C. (1969), A comparative study of the ring glands from wild type and 1(2)gl mutant *Drosophila melanogaster*, *J. Morph.*, **129**, 171–200.

—— (1971), An electron microscopic study of the corpus cardiacum of adult *Drosophila melanogaster* and its afferent nerves, *J. Morph.*, **134**, 437–446.

BARTH, R. H. AND LESTER, L. J. (1973), Neuro-hormonal control of sexual behavior in insects, *A. Rev. Ent.*, **18**, 445–472.

BEATTIE, T. M. (1971), Histology, histochemistry, and ultrastructure of the neurosecretory cells in the optic lobe of the cockroach, *Periplaneta americana*, *J. Insect Physiol.*, **17**, 1843–1855.

BEAULATON, J. (1968), Étude ultrastructurale et cytochimique des glandes prothoraciques de vers à soie aux quatrième et cinquième âges larvaires. I, *J. Ultrastruct. Res.*, **23**, 474–498.

BLOCH, B., THOMSEN, E. AND THOMSEN, M. (1966), The neurosecretory system of the adult *Calliphora erythrocephala*. III. Electron microscopy of the medial neurosecretory cells of the brain and some adjacent cells, *Z. Zellforsch.*, **70**, 185–208.

BORG, T. K. AND MARKS, E. P. (1973), Ultrastructure of the median neurosecretory cells of *Manduca sexta in vivo* and *in vitro*, *J. Insect Physiol.*, **19**, 1913–1920.

BOWERS, B. AND JOHNSON, B. (1966), An electron microscope study of the corpora cardiaca and secretory neurons in the aphid *Myzus persicae* (Sulz.), *Gen. Comp. Endocr.*, **6** (2), 213–230.

BRADY, J. AND MADDRELL, S. H. P. (1967), Neurohaemal organs in the medial nervous system of insects, *Z. Zellforsch.*, **76**, 389–404.

BRANDENBURG, J. (1956), Das endokrine System des Kopfes von *Andrena vaga* Pz. (Ins. Hymenopt.) und Wirkung der Stylopisation (*Stylops* Ins. Strepsiptera), *Z. Morph. Ökol. Tiere*, **45**, 343–364.

BURDETTE, W. J. (1974), *Invertebrate Endocrinology and Hormone Heterophylly*, Springer, Berlin, 437 pp.

CASSIER, P. AND FAIN-MAUREL, M. A. (1970), Contribution à l'étude infrastructurale du système neurosécréteur rétrocérébral chez *Locusta migratoria migratorioides* (R. et F.). II. Le transit des neurosécrétions, *Z. Zellforsch.*, **111**, 483–492.

CAZAL, P. (1948), Les glandes endocrines rétro-cérébrales des insectes (étude morphologique), *Bull. biol. Fr. Belg.*, **32**, 227 pp.; **33**, 9–18.

CAZAL, M., JOLY, L. AND PORTE, A. (1971), Etude ultrastructurale des corpora cardiaca et de quelques formations annexes chez *Locusta migratoria* L., *Z. Zellforsch.*, **114**, 61–72.

CLEVER, U. (1963), Von der Ecdysonkonzentration-abhangige Genaktivitätsmuster in den Speicheldrüsenchromosomen von *Chironomus tentans*, *Develop. Biol.*, **6**, 73–98.

COOK, D. J. AND MILLIGAN, J. V. (1972), Electrophysiology and histology of the medial neurosecretory cells in adult male cockroaches, *Periplaneta americana*, *J. Insect Physiol.*, **18**, 1197–1214.

COTTRELL, C. B. (1964), Insect ecdysis with particular emphasis on cuticular hardening and darkening, *Adv. Insect Physiol.*, **2**, 175–218.

DAVEY, K. G. (1964), The control of visceral muscles in insects, *Adv. Insect Physiol.*, **2**, 219–245.

DELPHIN, F. (1965), The histology and possible functions of neurosecretory cells in the ventral ganglia of *Schistocerca gregaria* Forskål (Orthoptera: Acrididae), *Trans. R. ent. Soc. Lond.*, **117**, 167–214.

DOGRA, G. S. (1967), Studies on the neurosecretory system of the female mole cricket *Gryllotalpa africana* (Orthoptera: Gryllotalpidae), *J. Zool., Lond.*, **152**, 163–178.

DORN, A. (1972), Die endokrinen Drüsen im Embryo von *Oncopeltus fasciatus* Dallas (Insecta, Heteroptera), *Z. Morph. Tiere*, **71**, 52–104.

ENGELMANN, F. (1968), Endocrine control of reproduction in insects, *A. Rev. Ent.*, **13**, 1–26.

—— (1970), *The Physiology of Insect Reproduction*, Pergamon Press, Oxford, 307 pp.

EWEN, A. B. (1962), Histophysiology of the neurosecretory system and retrocerebral endocrine glands of the Alfalfa plant bug, *Adelphocoris lineatus* (Goeze) (Hemiptera: Miridae), *J. Morph.*, **111**, 255–269.

FRAENKEL, G. AND HSIAO, C. (1965), Bursicon, a hormone which mediates tanning of the cuticle in the adult fly and other insects, *J. Insect Physiol.*, **11**, 513–556.

FRASER, A. (1959), Neurosecretory cells in the abdominal ganglia of larvae of *Lucilia caesar* (Diptera), *Q. Jl microsc. Sci.*, **100**, 395–399.

GELDIAY, S. (1959), Neurosecretory cells in ganglia of the roach *Blaberus craniifer*, *Biol. Bull.*, **117**, 267–274.

GERSCH, M. AND STÜRZEBECHER, J. (1968), Weitere Untersuchungen zur Kennzeichnung des Aktivationshormons der Insektenhäutung, *J. Insect Physiol.*, **14**, 87–96.

GIRARDIE, A. AND GIRARDIE, J. (1967), Étude histologique, histochimique et ultrastructurale de la pars intercerebralis chez *Locusta migratoria* L. (Orthoptère), *Z. Zellforsch.*, **78**, 54–78.

GOLDSWORTHY, G. J. AND MORDUE, W. (1974), Neurosecretory hormones in insects, *J. Endocr.*, **60**, 529–558.

HERMAN, W. S. (1967), The ecdysial glands of arthropods, *Int. Rev. Cytol.*, **22**, 269–347.

HIGHNAM, K. C. (1961), The histology of the neurosecretory system of the adult female desert locust, *Schistocerca gregaria*, *Q. Jl microsc. Sci.*, **102**, 27–38.

HIGHNAM, K. C. AND HILL, L. (1969), *Comparative Endocrinology of Invertebrates*, Arnold, London, 270 pp.

HIGHNAM, K. C. AND WEST, M. W. (1971), The neuropilar neurosecretory reservoir of *Locusta migratoria migratorioides*, *Gen. comp. Endocr.*, **16**, 574–585.

HINKS, C. F. (1971), A comparative survey of the neurosecretory cells occurring in the adult brain of several species of Lepidoptera, *J. Ent.* (A), **46**, 13–26.

ISHIZAKI, M. AND ICHIKAWA, M. (1967), Purification of the brain hormone of the silkworm *Bombyx mori*, *Biol. Bull.*, **133**, 355–368.

ILAN, J. AND ILAN, J. (1973), Protein synthesis and insect morphogenesis, *A. Rev. Ent.*, **18**, 167–182.

JACOBSON, M. AND CROSBY, D. G. (eds) (1971), *Naturally occurring Insecticides*, Dekker, New York, 548 pp.

JOHANSSON, A. S. (1958), Relation of nutrition to endocrine-reproductive functions in the milkweed bug *Oncopeltus fasciatus* (Dallas) (Heteroptera: Lygaeidae), *Nytt Mag. Zool.*, **7**, 1–132.

JOLY, P. (1968), *Endocrinologie des Insectes*, Masson et Cie., Paris, 344 pp.

JUBERTHIE, C. AND CASSAGNAU, P. (1971), L'évolution du système neurosécréteur chez les Insectes; l'importance des Collemboles et des autres Aptérygotes, *Rev. Ecol. Biol. Sol.*, **8**, 59–80.

JUDY, K. J., SCHOOLEY, D. A., DUNHAM, L. L., *et al.* (1973), Isolation, structure, and absolute configuration of a new natural insect juvenile hormone from *Manduca sexta*, *Proc. natn. Acad. Sci. U.S.A.*, **70**, 1509–1513.

KARLSON, P. AND KOOLMAN, J. (1974), Zur physiologischen Bedeutung des Ecdysonstoffwechsels der Insekten, *Fortschr. Zool.*, **22**, 23–33.

LANGLEY, P. A. (1965), The neuroendocrine system and stomatogastric nervous system of the adult tsetse fly *Glossina morsitans*, *Proc. zool. Soc. Lond.*, **144**, 415–423.

LÜSCHER, M. AND SPRINGHETTI, A. (1960), Untersuchungen über die Bedeutung der Corpora allata für die Differenzierung der Kasten bei der Termite *Kalotermes flavicollis* F., *J. Insect Physiol.*, **5**, 190–212.

MADDRELL, S. H. P. (1963–4), Excretion in the blood-sucking bug *Rhodnius prolixus* Stål. I–III., *J. exp. Biol.*, **40**, 247–256; **41**, 163–176; **41**, 459–472.

—— (1974), Neurosecretion, In: Treherne, J. E. (ed.), *Insect Neurobiology*, pp. 307–358.

MCLEOD, D. G. R. AND BECK, S. D. (1963), The anatomy of the neuroendocrine complex of the European Corn Borer, *Ostrinia nubilalis*, and its relation to diapause, *Ann. ent. Soc. Am.*, **56**, 723–727.

MENN, J. J. AND BEROZA, M. (eds) (1972), *Insect juvenile Hormones: Chemistry and Action*, Academic Press, New York, 341 pp.

MILLER, T. A. (1975), Neurosecretion and the control of visceral organs in insects, *A. Rev. Ent.*, **20**, 133–149.

MORDUE, W. (1972), Hormones and excretion in locusts, *Gen. comp. Endocr. Suppl.*, **3**, 289–298.

NAYAR, K. K. (1955), Studies on the neurosecretory system of *Iphita libata* Stål. I. Distribution and structure of the neurosecretory cells of the nerve ring, *Biol. Bull.*, **108**, 296–307.

NORMANN, T. C. (1965), The neurosecretory system of the adult *Calliphora erythrocephala*. I. The fine structure of the corpus cardiacum, with some observations on adjacent organs, *Z. Zellforsch.*, **67**, 461–501.

NOVAK, V. J. A. (1975), *Insect Hormones*, 2nd edn, Chapman & Hall, London.

ODHIAMBO, T. R. (1966), The fine structure of the corpus allatum of the sexually mature male of the desert locust, *J. Insect Physiol.*, **12**, 819–828.

PFLUGFELDER, O. (1948), Volumetrische Untersuchungen an der Corpora allata der Honigbiene *Apis mellifica* L., *Biol. Zbl.*, **67**, 223–241.

PILCHER, D. E. M. (1970a), Hormonal control of the Malpighian tubules of the stick insect, *Carausius morosus*, *J. exp. Biol.*, **52**, 653–665.

—— (1970b), The influence of the diuretic hormone on the process of urine secretion by the Malpighian tubules of *Carausius morosus*, *J. exp. Biol.*, **53**, 465–484.

PROVANSAL, A. AND GRILLOT, J. P. (1972), Les organes périsympathiques des insectes holométaboles. I. Coléoptères, *Annls Soc. ent. Fr.*, **8**, 863–913.

RAABE, M. (1971), Neurosécrétion dans la chaine nerveuse ventrale des insectes et organes neurohémaux métamériques, *Archs Zool. exp. gén.*, **112**, 679–694.

REMBOLD, H., CZOPPELT, C. AND RAO, P. J. (1974), Effect of juvenile hormone treatment on caste differentiation in the honeybee, *Apis mellifera*, *J. Insect Physiol.*, **20**, 1193–1202.

RÖLLER, H. AND DAHM, K. H. (1968), The chemistry and biology of juvenile hormone, *Recent Prog. Hormone Res.*, **24**, 651–680.

SCHARRER, B. (1963), Neurosecretion, XIII. The ultrastructure of the corpus cardiacum of the insect *Leucophaea maderae*, *Z. Zellforsch.*, **60**, 761–796.

—— (1964), The fine structure of blattarian prothoracic glands, *Z. Zellforsch.*, **64**, 301–326.

—— (1971), Histophysiological studies on the corpus allatum of *Leucophaea maderae*. V. Ultrastructure of sites of origin and release of a distinctive cellular product, *Z. Zellforsch.*, **120**, 1–16.

SCHOONEVELD, H. (1974), Ultrastructure of the neurosecretory system of the Colorado potato beetle, *Leptinotarsa decemlineata* (Say). I, II, *Cell and Tissue Res.*, **154**, 275–288; 289–301.

SEHNAL, F. (1971), Endocrines of Arthropods, In: Florkin, M. and Scheer, B. T. (eds), *Chemical Zoology*, **6B**, 307–345.

SLÁMA, K. (1971), Insect juvenile hormone analogues, *A. Rev. Biochem.*, **40**, 1079–1102.

SLÁMA, K., ROMAŇUK, M. AND ŠORM, F. (1974), *Insect Hormones and Bioanalogues*, Springer, Berlin & London, 477 pp.

SMITH, U. AND SMITH, D. S. (1966), Observations on the secretory processes in the corpus cardiacum of the stick insect, *Carausius morosus*, *J. Cell Sci.*, **1**, 59–66.

STAAL, G. B. (1975), Insect growth regulators with juvenile hormone activity, *A. Rev. Ent.*, **20**, 417–460.

SVOBODA, J. A., KAPLANIS, J. N., ROBBINS, W. E. AND THOMPSON, M. J. (1975), Recent developments in insect steroid metabolism, *A. Rev. Ent.*, **20**, 205–220.

THOMSEN, E. AND THOMSEN, M. (1970), Fine structure of the corpus allatum of the female blowfly *Calliphora erythrocephala*, *Z. Zellforsch.*, **110**, 40–60.

THOMSEN, M. (1951), Weismann's ring and related organs in larvae of Diptera, *Dan. Biol. Skr.*, **6** (5), 32 pp.

—— (1965), The neurosecretory system of the adult *Calliphora erythrocephala*. II. Histology of the neurosecretory cells of the brain and some related structures, *Z. Zellforsch.*, **67**, 693–717.

TOMBES, A. S. AND SMITH, D. S. (1970), Ultrastructural studies on the corpora cardiaca-allata complex of the adult alfalfa weevil, *Hypera postica*, *J. Morph.*, **132**, 137–147.

TRUMAN, J. W. AND RIDDIFORD, L. M. (1974), Hormonal mechanisms underlying insect behaviour, *Adv. Insect Physiol.*, **10**, 297–352.

VINCENT, J. F. V. (1972), The dynamics of release and the possible identity of bursicon in *Locusta migratoria migratorioides*, *J. Insect Physiol.*, **18**, 757–780.

WAKU, Y. AND GILBERT, L. I. (1964), The corpora allata of the silkmoth, *Hyalophora cecropia*: an ultrastructural study, *J. Morph.*, **115**, 69–96.

WIGGLESWORTH, V. B. (1964), The hormonal regulation of growth and reproduction in insects, *Adv. Insect Physiol.*, **2**, 247–336.

—— (1970), *Insect Hormones*, Oliver & Boyd, Edinburgh, 159 pp.

WILLIAMS, C. M. (1946–53), Physiology of insect diapause. I–IV, *Biol. Bull.*, **90**, 234–243; **93**, 90–98; **94**, 60–65; **103**, 120–138.

WILLIS, J. (1974), Morphogenetic action of insect hormones, *A. Rev. Ent.*, **19**, 97–115.

WIRTZ, P. AND BEETSMA, J. (1972), Induction of caste differentiation in the honeybee (*Apis mellifera*) by juvenile hormone, *Entomologia exp. appl.*, **15**, 517–520.

YAMAZAKI, M. AND KOBAYASHI, M. (1969), Purification of the proteinic brain hormone of the silkworm, *Bombyx mori*, *J. Insect Physiol.*, **15**, 1981–1990.

YIN, C.-M. AND CHIPPENDALE, G. M. (1974), Insect prothoracic glands: function and ultrastructure in diapause and non-diapause larvae of *Diatraea grandiosella*, *Can. J. Zool.*, **53**, 124–131.

Chapter 17

THE REPRODUCTIVE SYSTEM

The reproductive organs present a very wide range of variation in different insects. In their embryonic condition they are at first essentially similar in the male and female, becoming differentiated later in development. Among the more primitive orders (Fig. 144) much of this similarity is still evident but an increasing divergence in structure becomes noticeable in the higher groups. The gonads are a pair of mesodermal structures which probably

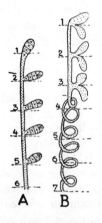

FIG. 144
Gonads of a young *Lepisma* (mesodermal portions only)

A, female; B, male. The numerals refer to the abdominal segments. Adapted from Grassi, 1887.

evolved from coelom-sacs whose paired ducts go to form all or most of the lateral gonoducts of the adult insect (lateral oviducts or vasa deferentia according to sex). Primitively, this pair of ducts opened separately and it is not improbable that the paired reproductive openings of the Ephemeroptera represent a retention of the generalized condition. Among the vast majority of insects, however, the mesodermal gonoducts do not open directly to the exterior but join a median passage (the common oviduct and vagina or the ejaculatory duct) which develops from one or more invaginations of the ventral body-wall and is lined by cuticle. The extent of this unpaired portion and its precise mode of development vary in the sexes and in different groups but the result is that the definitive system of adult efferent ducts includes both mesodermal and ectodermal parts. The situation is complicated by the development of accessory glands as outpocketings of either part of the genital tract and the occurrence in the female of one or more spermathecae and a pouch-like bursa copulatrix.

In males, the genital opening usually lies behind the 9th abdominal sternum at the end of the aedeagus and in females it is on or behind the 8th or 9th sternum with, in the Ditrysian Lepidoptera and some others, separate openings leading respectively into the vagina and bursa copulatrix. Fuller details of the less common arrangements are given in the sections dealing with the orders concerned, especially the Protura, Collembola, Ephemeroptera, Dermaptera, Raphidioidea and Lepidoptera. For general reviews of the morphology of the reproductive system of insects see Heberdey (1931), Snodgrass (1935) and Weber (1939–43; 1952). General, mainly physiological, accounts of reproductive biology are provided by Highnam (1964), Davey (1965), Engelmann (1970) and Adiyodi and Adiyodi (1974).

The sexual organs, and their counterparts in the male and female, are tabulated below.

MALE REPRODUCTIVE ORGANS	FEMALE REPRODUCTIVE ORGANS
1. Paired testes composed of follicles (testicular tubes)	Paired ovaries composed of ovarioles (ovarian tubes)
2. Paired vasa deferentia	Paired oviducts
3. Vesiculae seminales	Egg-calyces
4. Median ejaculatory duct	Common oviduct and vagina
5. Accessory glands:	Accessory glands:
(a) Mesadenia	(a) ——
(b) Ectadenia	(b) Colleterial glands
6. ——	Spermatheca
	Bursa copulatrix
7. Genitalia	Ovipositor

The sexes of insects are almost always separate but mention must be made of the abnormal individuals known as gynandromorphs and intersexes as well as the very few established cases of functional hermaphroditism referred to on p. 307. *Gynandromorphs* or sex-mosaics are teratological forms in which some parts of the body show female characteristics while the remaining parts are male. They therefore have a striking appearance when secondary sexual differences in colour-pattern or structure occur, as in the many cases recorded from the Lepidoptera. Frequently one side of the insect is male and the other is female but anteroposterior gynandromorphs and forms with an irregular mosaic-like distribution of sexual characters are also known. In *Drosophila* and perhaps also in *Carausius morosus*, gynandromorphs can arise through the loss of a sex-chromosome in one of the early cleavage nuclei of the embryo, so that deficient cells form male tissues while those with a full complement of sex-chromosomes yield female tissues (Morgan *et al.*, 1919; Wilbert, 1953). In other species, gynandromorphs can result from 'double fertilization' of abnormal eggs possessing two nuclei, one of which gives rise to male tissue, the other to female (Cockayne, 1935; White, 1968). Some parasitic Hymenoptera that normally reproduce by thelytokous parthenogenesis (p. 303) yield many gynandromorphs at unusually high temperatures through what is really a mosaic-like incidence of male-producing parthenogenesis (Flanders, 1945; Wilson, 1962; Bowen and Stern, 1966). *Intersexes* are forms in which, owing to a disturbance of the normal balance or

epistatic relationship between male- and female-determining genes during development, an adult is produced which is more or less intermediate between the two
sexes. Several cases have been intensively studied because of the light they shed on
the genetic and environmental control of sexual differentiation. Among these are
species of *Aedes* and also *Carausius morosus*, where intersexuality follows exposure to
high temperatures (e.g. Anderson, 1967; Brust, 1966; Bergerard, 1961, 1972). In
Drosophila and the Psychid moth *Solenobia triquetrella* intersexuality is associated
with triploidy (Laugé, 1969, Nüesch, 1947; Seiler *et al.*, 1958; Seiler, 1969) and in
the classical example of *Porthetria* (= *Lymantria*) *dispar* it may follow exposure of
the pupae to high temperatures or the crossing of genetically distinct strains
(Goldschmidt, 1938, 1955). Further examples of intersexuality due to parasitism are
discussed on p. 308. It should also be noted that superficial examination of a
sexually abnormal specimen may not enable one to decide whether it is a gynandromorph or an intersex; physiological, genetic or cytological study may be needed.

The many cases of gynandromorphism seem to rule out the possibility that the
gonads of insects secrete sex-hormones like those that control secondary sexual
differentiation in the vertebrates. Experimental transplantation of gonads has
confirmed this view for many species. In *Lampyris notiluca*, however, the presence of
active neuroendocrine cells in the brain leads to the appearance of special secretory
tissue in the apex of the testis rudiment of male larvae. This in turn controls the
development of male structures and has a masculinizing effect when transplanted into
female larvae (Naisse, 1966, 1969).

The Male Reproductive System (Figs. 145–147)

The Testes – In most Apterygotes the testes resemble the ovaries in form
and size, but in the majority of insects they are much smaller. They are
variably situated in relation to the gut, lying dorsal, lateral or even ventral to
it. The testes are maintained in position by tracheae and fat-body and lack the
suspensory filaments found in the ovary. As a rule each testis is a more or
less ovoid body, partly or completely divided into a variable number of lobes
or *follicles* that show very many variations in form and arrangement among
the different insect species. In *Lepisma* there are 3 or 4 segmentally ar-

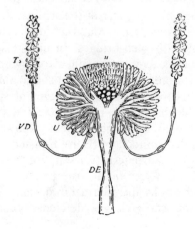

FIG. 145
Male reproductive organs of
Periplaneta, ventral view

Ts, testis; *VD*, vas deferens; *u*, *U*, accessory
glands; *DE*, ejaculatory duct. *After* Miall
and Denny.

ranged, bilobed follicles but in most Apterygotes each testis is a simple sac or greatly enlarged follicle. Among the Neuroptera, Adephagan Coleoptera and Diptera the testes are small and unifollicular; in *Pediculus* and *Pthirus* they are bifollicular, and in the Orthoptera and related orders the follicles are very numerous. They may be short and globular as in *Periplaneta* and *Tetrix*, or elongate and tubular as in *Oedipoda*. In the Mallophaga, Siphunculata and certain Coleoptera, each follicle is connected with the vas deferens by a relatively well developed slender tube or *vas efferens*. In many insects the peritoneal investment of the follicles is developed to the extent of enveloping the testis as a whole in a common coat or *scrotum* which is frequently pigmented. In most Lepidoptera, *Gryllotalpa* and certain Hymenoptera, the testes are in close contact medially and are enclosed in a single scrotum.

The Structure of a Testicular Follicle – The testicular follicles are lined with a layer of epithelium, whose cells rest externally upon a basement membrane, outside which there is a peritoneal coat of connective tissue. Each follicle is divided into a series of zones characterized by the presence of the sex cells in different stages of development, corresponding to the successive generations of these cells (Depdolla, 1928; Bairati, 1967–68; Phillips, 1970). These zones are as follows:

(1) The *germarium* is the region containing the primordial germ cells or spermatogonia which undergo multiplication.

(2) The *zone of growth* is where the spermatogonia increase in size, undergo repeated mitosis and develop into spermatocytes (Krishan and Buck, 1965).

(3) The *zone of division and reduction* where the spermatocytes undergo meiosis and give rise to spermatids.

(4) The *zone of transformation* where the spermatids become transformed into spermatozoa.

Masses of developing spermatozoa from the spermatocyte stage onwards are enclosed in, and perhaps nourished by, the *testicular cyst-cells* (Anderson, 1950b; Phillips, 1970), from which they are released in the vas deferens, the abandoned cyst-cells finally degenerating after transfer of the seminal liquid to the female. While in the testis each cyst contains around 2^5 to 2^8 synchronously differentiating spermatids, linked by cell-bridges. In most insects the spermatogonial and meiotic divisions occur in the pupal or nymphal instars so that the imaginal testis contains only spermatids and spermatozoa, but in long-lived insects the adult testis may show earlier stages of spermatogenesis. In addition, the testis of the Orthoptera, Dictyoptera and some Lepidoptera, Diptera and Homoptera contains large elements known as *Verson's cells* or apical cells whose abundant mitochondria are transferred to the spermatogonial cytoplasm during spermatogenesis (Carson, 1945).

The Male Genital Ducts – The *vasa deferentia* are the paired canals leading from the testes and are partly or wholly mesodermal in origin. They vary greatly in length and in most insects each vas deferens becomes enlarged along its course to form a sac or *vesicula seminalis* in which the spermatozoa congregate and which is sometimes a large, complex structure; in some Diptera the vasa deferentia open into a common vesicula seminalis. Histologically the vas deferens consists of an outer peritoneal coat, a middle coat of muscle fibres, and an inner coat of epithelial cells. Posteriorly, the vasa deferentia unite to form a short common canal which is continuous with a

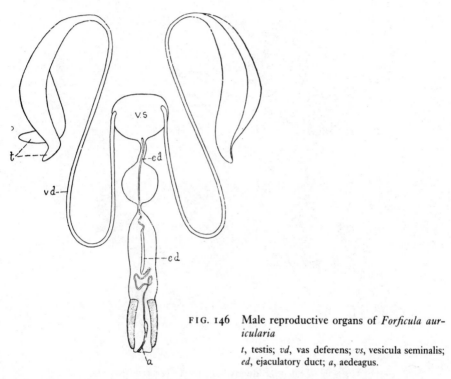

FIG. 146 Male reproductive organs of *Forficula auricularia*

t, testis; *vd*, vas deferens; *vs*, vesicula seminalis; *ed*, ejaculatory duct; *a*, aedeagus.

median ectodermal tube or ejaculatory duct. The latter is provided with a powerful muscular coat consisting of an outer layer of circular fibres and an inner layer of longitudinal fibres. Within the muscle layers is a stratum of epithelial cells which secrete a cuticular lining to the lumen of the ejaculatory duct. In *Antheraea pernyi* a polypeptide secreted by the ductus ejaculatorius activates the sperms which, prior to ejaculation, are motionless in the seminal vesicles (Shepherd, 1974, 1975).

The Aedeagus – The terminal section of the ejaculatory duct is enclosed in a finger-like evagination of the ventral body-wall which forms the male intromittent organ or *aedeagus* which, with its associated structures, assumes a wide variety of forms and is dealt with on p. 77 and in the sections dealing with the separate orders of insects.

The Male Accessory Glands – From one to three pairs of tubular or sac-like accessory glands are usually present (Leopold, 1976). In most cases their secretions mix with the spermatozoa (e.g. Anderson, 1950a) and in some insects they are directly concerned with the production of the proteinaceous spermatophore (Davey, 1959; Gregory, 1965; Odhiambo, 1969). Peptide secretions of the male accessory glands may also have effects on the female, reducing her receptivity to further mating or stimulating oviposition (Leahy, 1967; Hinton, 1974). Following Escherich (1894) they may be divided into two categories: (i) *mesadenia*, derived from the mesoderm and formed as

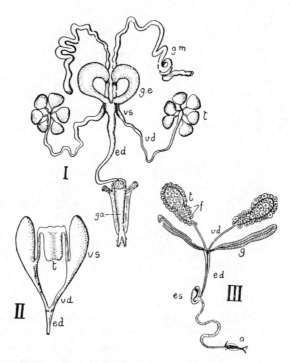

FIG. 147   Male reproductive organs. I, *Tenebrio obscurus*, *after* Bordas, 1900. II, *Sphecodes fuscipennis, after* Bordas, 1895. III, *Calliphora, after* Lowne

a, aedeagus; *ed*, ejaculatory duct; *es*, ejaculatory sac; *f*, fat cells; *ga*, external genitalia; *g*, accessory gland; *ge*, do., ectadenes; *gm*, do., mesadenes; *t*, testis; *vd*, vas deferens; *vs*, vesicula seminalis.

evaginations of the vasa deferentia, and (ii) *ectadenia*, derived from ectoderm and formed as evaginations of the ejaculatory duct (Fig. 147). In the Adephagan Coleoptera the ectadenia are the only accessory glands but in the Polyphagan beetles two or more pairs of mesadenia are also present. Among Orthoptera and Dictyoptera the accessory glands are very greatly developed, forming dense bunches of tubuli which, in *Periplaneta*, form the 'mushroom-shaped gland' of Huxley; an additional 'conglobate gland' is also present in cockroaches. The accessory glands are wanting in some insects, including the Apterygota, and *Musca, Tabanus* and other Diptera.

The Female Reproductive System (Figs. 148–151)

The Ovaries (Beams *et al.*, 1962) – The ovaries are typically more or less compact bodies lying in the body-cavity of the abdomen on either side of the alimentary canal. Each organ is composed of a variable number of separate egg-tubes or *ovarioles* which open into the oviduct. The primitive number of ovarioles composing an ovary is uncertain and probably does not exceed eight, the latter number being retained in *Periplaneta* for example. In some Thysanura and Diplura (*Japyx, Campodea* and *Lepisma*) there are 5–7 ovarioles on each side, opening one behind the other into an elongate oviduct. A comparable longitudinal repetition of ovarioles also occurs in the Ephemeroptera, Acridoidea and some Dermaptera but elsewhere the ovarioles radiate from the inner end of the lateral oviducts. Specialization either by the reduction or the multiplication of the ovarioles is very frequent. In insects which produce a small number of relatively large eggs, such as *Glossina*, there are two ovarioles to each ovary and in *Termitoxenia* only one. In the sexual females of some aphids there is a single ovary with one ovariole, the other ovary having atrophied. Two ovarioles are present in each ovary of *Melophagus, Hippobosca*, and certain Coleoptera and Hymenoptera; among Lepidoptera there are commonly four. Examples of specialization by multiplication are much more frequent. Thus in *Calliphora* and *Hypoderma* there are 100 or more ovarioles to an ovary; in some ants there are over 200; in *Meloe* they are even more numerous while the maximum number is attained in the Isoptera where among species of *Eutermes* it exceeds 2400. In a few apparently anomalous instances ovarioles are wanting and the ovaries are more or less sac-like without any serial arrangement of the developing eggs. Such ovaries are well exhibited among Collembola. In some Braconidae (*Aphidius*) although there is an evident differentiation into follicles ovarioles are wanting: this is a secondary and highly modified condition.

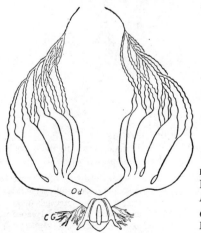

FIG. 148
Female   reproductive   organs   of *Periplaneta*

Od, oviduct; *CG*, colleterial gland. *After* Miall and Denny.

The Ovarioles – A typical ovariole is an elongate tube in which the developing eggs are disposed one behind the other in a single chain, the oldest oocytes being situated nearest the union with the oviduct. The wall of an ovariole is a delicate transparent membrane; its inner coat is a layer of epithelium whose cells rest on a basement membrane or tunica propria, outside which is a peritoneal coat of connective tissue that may contain muscle fibres and is covered by a reticulate layer of tracheal end-cells (e.g. Bier and Meyer, 1952; Bonhag and Arnold, 1961; Baccetti, 1963; Cruickshank, 1973).

Three zones or regions are recognizable in an ovariole (Fig. 151) – (1) The *terminal filament*. This is the slender thread-like apical prolongation of the

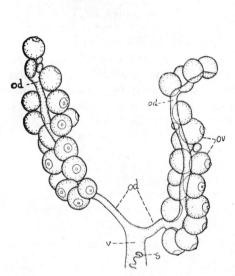

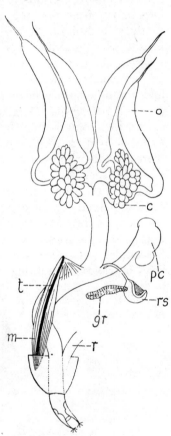

FIG. 149

Female reproductive organs of *Forficula auricularia*

od, oviduct; *ov*, ovarioles; *v*, vagina; *s*, spermatheca.

FIG. 150

Female reproductive system of *Anthonomus pomorum*

c, egg-calyx; *m*, vaginal muscle and spiculum *t*; *o*, ovariole; *pc*, bursa copulatrix; *r*, rectum; *rs*, receptaculum seminis and gland *gr*. *After* Henneguy, *Les Insectes*.

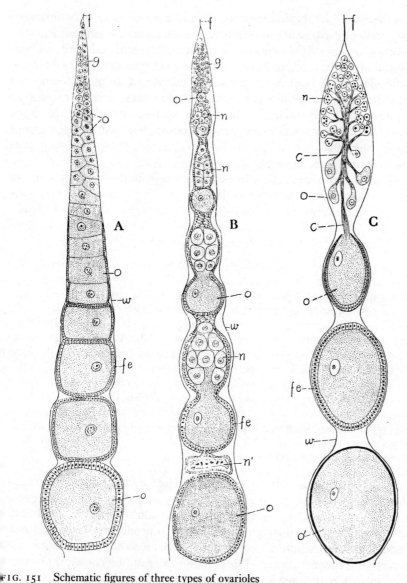

FIG. 151   Schematic figures of three types of ovarioles

A, panoistic. B, polytrophic. C, acrotrophic. *f*, terminal filament; *g*, germarium; *o*, oocytes; *o′*, mature oocyte with chorion; *n*, nurse-cells; *n′*, remains of same; *w*, wall of ovariole; *fe*, follicular epithelium; *c*, trophic cord.

peritoneal layer. The filaments of the ovarioles of one ovary combine to form a common thread which unites with that from the ovary of the opposite side to form a median ligament. The latter helps to maintain the ovaries in position and is attached to the body-wall, the fat-body or the pericardial diaphragm. In some insects the ovarian ligament is wanting and the filaments end free in the body-cavity. (2) The *germarium*. This forms the apex of an ovariole, below the terminal filament, and consists of a mass of cells from which are differentiated the primordial germ cells and, in many insects, the nutritive cells, all at first linked by cell-bridges. (3) The *vitellarium*. The vitellarium constitutes the major portion of an ovariole and contains the developing eggs and the nutritive cells (nurse-cells or *trophocytes*) when present. The epithelial layer of the wall of the vitellarium grows inwards in such a manner as to enclose each oocyte in a definite egg-chamber or follicle. The cells of the follicle secrete the chorion of the egg and in some cases serve to nourish the oocytes.

Three principal types of ovarioles are recognized depending on the presence or absence of nutritive cells and on their location when present (Fig. 151). For histological details see Bonhag (1955, 1958, 1959), Schlottmann and Bonhag (1956), King (1970), Huebner and Anderson (1972), Mahowald (1972), and Wightmann (1973).

(*a*) The *panoistic type*. Specially differentiated nutritive cells are wanting. This type of ovariole is primitive and is found in the Thysanura, Orthopteroid orders, Ephemeroptera, Odonata, Siphonaptera and others.

(*b*) The *polytrophic type*. Nurse-cells are present and alternate with the oocytes. In many cases (Neuroptera, Coleoptera-Adephaga and Hymenoptera) the nurse-cells are grouped together so as to lie in chambers, each chamber being separated from that containing an oocyte by a well-marked constriction; in others (Lepidoptera, Diptera) these constrictions are wanting.

(*c*) The *acrotrophic type*. Nurse-cells are present and situated at the apices of the ovarioles (Coleoptera-Polyphaga and Hemiptera). In certain Heteroptera the nurse-cells are connected with the oocytes by cytoplasmic strands, the trophic cords.

Types (*b*) and (*c*) are often grouped as one type – *meroistic*, which is characterized by the presence of nurse-cells. Other atypical arrangements are occasionally found, e.g. Martoja (1964), Matsuzaki (1973).

Ovarial Maturation – In some insects a large number of eggs mature in the pupa or last nymphal stage and are ready to be laid soon after emergence and mating. In others, the ovary of the young adult female is small and a period of feeding and maturation is necessary to produce fully formed eggs. This entails the deposition of yolk and the formation of the membranes that are to surround the mature egg. The physiological and biochemical processes involved in yolk deposition have been studied in considerable detail though much still remains to be ascertained. There are reviews by Bonhag (1958), Telfer (1965, 1975), Telfer and Smith (1970), Engelmann (1970) and King (1970); a selection of further references is also included in the list on p. 309. The constituents of insect yolk include proteins (especially mucoproteins and lipoproteins), neutral lipids, carbohydrates such as glycogen, and the ribonucleic acids. Their mode of synthesis and deposition seems to vary somewhat among the different species. In some panoistic ovarioles the oocyte itself can synthesize proteins, glycogen and RNA, an ability that is apparently related to the presence of characteristic 'lamp-brush chromosomes' in its nucleus. In meroistic ovarioles the nurse-cells are mainly responsible for providing RNA (and sometimes also extranuclear DNA) which pass along the trophic cords of the Hemipteran ovary, together with glycogen and lipids. Protein yolk is mainly derived from proteins that are synthesized in the cells of the fat-body and mid gut, accumulate in the blood, pass selectively by an intercellular route through the follicular epithelium and are then absorbed by pinocytosis at the oocyte surface. Adequate nutrition is necessary for normal vitellogenesis which in many cases also requires the presence of the median cerebral neurosecretory cells and of the juvenile hormone secreted by the corpora allata (Wigglesworth, 1964). It seems possible that the neurosecretory cells control the synthesis of the proteins and that the juvenile hormone is needed for their uptake by the oocyte. The mature eggs are discharged from the ovariole – an insufficiently studied process known as ovulation (Curtin and Jones, 1961) – and are usually stored temporarily in the efferent ducts before being laid. After ovulation the cells of the empty follicle undergo condensation and disorganization to form the *corpus luteum*, a body of unknown function which is sometimes yellow or orange in colour and which finally degenerates before the next egg is discharged. In some species, if conditions are unfavourable to oviposition, the mature egg is resorbed before ovulation (e.g. Flanders, 1942; Singh, 1958; Lusis, 1963).

The Female Genital Ducts – The lateral oviducts are paired canals leading from the ovaries and are usually formed from mesoderm at the hinder extremities of the embryonic gonads. The two lateral oviducts join the common oviduct, which is initially developed from an invagination of the body-wall behind the 7th abdominal sternum but which generally becomes extended through the 8th segment to join the vagina, which arises from an infolding behind segment 8 (Fig. 152). In many adult insects there is no obvious distinction between these parts of the median reproductive duct but in some viviparous insects such as *Glossina* and *Melophagus* the vagina is greatly enlarged to form a chamber or uterus for the reception of the developing larva. In the more generalized insects it forms little more than a shallow genital chamber into which the common oviduct opens. If present, the bursa copulatrix is a pouch-like development of the vaginal region and when, as often happens, the latter becomes extended into the 9th segment to open there by the definitive genital aperture, the bursa loses direct connection with the exterior. In the Ditrysian Lepidoptera, however, the opening behind the 8th segment is retained as a copulatory aperture leading into the bursa while the opening of the vagina proper behind the 9th segment is the pore through which the eggs are laid (p. 1087). In the Anobiidae a pair of posteriorly placed vaginal pouches contain symbiotic yeasts which are smeared over the surface of the egg and pass to the young larva when it eats the egg-shell after hatching. In other insects the transmission of symbionts to the offspring occurs in different ways, including passage into the ovarial eggs (Buchner, 1965). Structurally the oviducts and vagina are composed of an epithelial layer, the cells of the ectodermal part secreting a cuticular lining which is continuous with the cuticle of the body-wall. The epithelial layer rests upon a basement membrane and outside the latter is a coat of powerful, mainly circular, muscle fibres.

The Spermatheca (*receptaculum seminis*) – This is a pouch or sac for the reception and storage of the spermatozoa and is rarely absent. It varies greatly in form and usually opens by a duct (often reduced to a mere neck) into the dorsal wall of the vagina or genital cavity; in some Coleoptera it or its duct is also connected by a fecundation canal to the vagina. In many insects pairing only takes place once and, since the maturation of the eggs

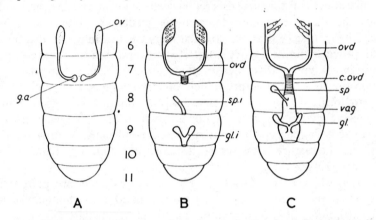

FIG. 152   A, B. Stages in development of female reproductive ducts. C. Adult condition

c.ovd, common oviduct; *g.a*, genital ampulla; *gl*, accessory glands; *gl.i*, invagination of developing accessory glands; *ov*, developing ovary; *ovd*, lateral oviduct; *sp*, spermatheca; *sp.i*, invagination of developing spermatheca; *vag*, vagina. 6–11, Abdominal segmentation.

may extend over a prolonged period, the provision of a spermatheca allows for their fertilization from time to time. Although commonly ovoid or spherical it is tubular in some Coleoptera, or even branched as in *Paederus*. As a rule the spermatheca is a single organ but in *Blaps*, *Phlebotomus* and *Dacus* there are two and in *Culex*, the Tabanidae and most Calyptratae three. Since it is derived from the ectoderm the spermatheca is lined with cuticle which is sometimes dark or brightly coloured. A stratum of columnar epithelium rests upon a basement membrane which is followed by a muscular coat. In some cases glandular cells are present in the wall of the spermatheca (e.g. *Periplaneta*; Gupta and Smith, 1969) and in others a special *spermathecal gland* opens into the duct of the spermatheca, or near the junction of the latter with the vagina.

The Female Accessory Glands – One or two pairs of accessory (colleterial) glands are present in most female insects and usually open into the distal portion of the vagina. They are large and important organs in many Orthoptera and Dictyoptera where they provide the material for the formation of the egg-pod or ootheca (Brunet, 1951–52; Mercer and Brunet, 1959).

In *Chironomus* they secrete a mucus-like substance which forms the gelatinous investment of the eggs, and in other insects such as *Hyalophora* (Berry, 1968) they provide a cement-like secretion which serves to fasten the eggs down to the substratum upon which they are laid. The poison glands of Hymenoptera are modified accessory glands (p. 1188), but the nutritive glands associated with the uterus of some viviparous insects are perhaps separately evolved structures.

The Sex Cells and Sperm Transfer

The Eggs – The cytoplasm of the mature eggs usually forms a reticulum in whose meshes occurs a large, centrally placed nucleus and abundant globules or granules of yolk. The egg is invested by a delicate homogeneous vitelline membrane and a shell or chorion which is secreted by the follicle

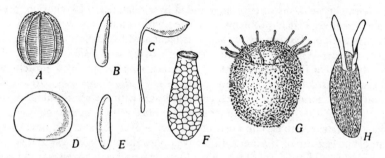

FIG. 153   Eggs of various insects

A, butterfly, *Polygonia interrogationis*; B, *Musca domestica*; C. chalcid, *Bruchophagus funebris*; D, *Papilio troilus*; E, midge, *Dasyneura trifolii*; F hemipteron, *Orius insidiosus*; G, hemipteron, *Podisus maculiventris*; H, *Drosophila melanogaster*. Greatly magnified. *After* Folsom, 1923.

cells of the ovary (Beams and Kessel, 1969; Cummings, 1972; Hamon, 1972). The chorion may be very thin, membranous and elastic, as in many parasitic Hymenoptera, or even absent (some viviparous insects), but more usually it is composed of two main layers (endo- and exochorion) and is relatively tough and rigid, protecting the contents and giving the egg its characteristic form. In the well investigated egg of *Rhodnius* the exo- and endochorion consist respectively of two and five distinct layers, some formed of relatively soft protein or lipoprotein and others rich in polyphenols and harder tanned protein. Beneath this chorion is the primary wax-layer, about 0·5 μm thick, and a fertilization membrane to which material is later added by the serosal cells of the developing embryo (Beament, 1946a; 1946b; 1948–49). Other complex types of egg-membranes are described by Slifer (1937–38), Davies, L. (1948), Moscona (1950), Matthée (1952), Smith *et al.* (1971) and Kawasaki *et al.* (1974).

In the greater number of insects the chorion exhibits some form of external sculpture: very commonly it is marked out into hexagonal areas which correspond with the overlying follicular cells. In the Lepidoptera the eggs of many species are conspicuously ribbed (Döring, 1955) and in some Ephemeroptera they are covered with fine processes resembling pile. The form assumed by the eggs presents innumerable variations (Fig. 153): one of the commonest types is the elongate-ovoid and slightly curved egg prevalent among Orthoptera, and in many Diptera and Aculeate Hymenoptera. Among Lepidoptera the eggs may be almost spherical, cake-like or somewhat cylindrical and flattened at one end: in many parasitic Hymenoptera there is a tubular prolongation or pedicel. In some insects an operculum is formed as a special differentiation of the chorion at the anterior extremity. This structure is lifted up at eclosion and is well seen in the Embioptera, the Cimicimorph Heteroptera, and the Mallophaga and Siphunculata. In the Oestridae the eggs are provided with flanges which attach them to the hairs of the host upon which they are laid.

The normal type of insect egg, surrounded by a relatively thick chorion, presents a number of physiological problems associated with respiration, water relations and sperm penetration. Respiratory exchange is facilitated by various devices described by Tuft (1950), Wigglesworth and Beament (1950), and Hinton (1969). These frequently involve the development of a porous, spongy, air-filled protein which may be distributed throughout the chorion (e.g. *Psylla, Calliphora, Drosophila*) or forms an inner chorionic layer which communicates with the atmosphere by special channels (aeropyles). These are sometimes also filled with a similar porous protein and may be restricted to certain areas of the shell – over 180 such channels occur in the rim of the egg-cap of *Rhodnius*. In many species the chorionic respiratory system forms a plastron (Hinton, 1969) which enables terrestrial eggs to withstand flooding. Many different forms of plastron have evolved independently; they may occur over the whole surface or are restricted to horn-like processes of the egg, as in some Diptera. The loss of water from the egg in dry surroundings is restricted by the primary wax-layer (Beament, 1946) but many eggs absorb water actively from their environment as a normal part of their development. In Aleyrodid eggs this occurs through the stalk by which they are attached to the host-plant (Weber, 1931) and in Acridids the serosal cuticle of the developing egg is modified over a small area – the hydropyle – through which uptake or loss of water occurs (Slifer, 1938, 1949; Matthée, 1952). The hydropic eggs of some parasitic Hymenoptera (Flanders, 1942) are thinly chorionated, small and deficient in yolk when laid but increase enormously during development by absorption of food and water through their specialized embryonic membranes. Except where fertilization occurs in the ovary before an impenetrable chorion is laid down, some provision is necessary to enable the spermatozoa to gain admittance. One or more specialized pores or canals known as *micropyles* are present for this purpose, usually at the anterior or cephalic pole of the egg. Unfortunately, older workers sometimes described as micropyles what are know known to be respiratory channels. In *Rhodnius* the respiratory channels of the cap-rim are accompanied by about 15 rather similar true micropyles while in *Oncopeltus* the 'sperm cups' at the anterior end of the egg function both as micropyles and respiratory channels (Wigglesworth and Beament, 1950).

The Spermatozoa – As in other animals these consist of a head largely made up of chromatin, a middle piece and a vibratile tail of variable and often complex structure (Depdolla, 1928; Phillips, 1970; Baccetti, 1972). Mature sperm lack ribosomes, a Golgi apparatus and some other organelles, but they generally possess giant mitochondria, a flagellum (usually with a 9 + 9 + 2 pattern of tubules), microtubules and an acrosome. The sperms of some species are atypical: those of the Lepidoptera have longitudinal, lamellate, paddle-like processes; Membracid sperms have 4-branched tails (Phillips, 1969), while *Eosentomon transitorium* (Protura) has disk-shaped, aflagellate sperms (Baccetti *et al.*, 1973), as has the termite *Macrotermes natalensis* (Geyer, 1951).

Sperm Transfer – In the normal methods of copulation the spermatozoa are transferred to the genital tract of the female after more or less elaborate forms of mating behaviour (Richards, O. W., 1927; Alexander, 1964; Manning, 1966; Fowler, 1973; Spieth, 1974). In some species free spermatozoa are deposited directly in the spermatheca (e.g. Downes, 1968) and the males may have a long extension of the aedeagus (flagellum) which traverses the spermathecal duct, as in some Heteroptera (Bonhag & Wick, 1953). In other cases, free spermatozoa are deposited in the vagina or bursa copulatrix or the sperms are enclosed in a proteinaceous sac or *spermatophore* which may be formed in various ways (Khalifa, 1949a; Gerber, 1970). In Dictyoptera, Orthoptera and Neuroptera it develops in a mould in the ejaculatory duct or aedeagus and is transferred to the female only when it is fully formed, often protruding from the female genital aperture after mating (e.g. Khalifa, 1949b; Gregory, 1965). In other insects such as *Rhodnius* (Davey, 1959), the Dytiscidae, Scarabaeidae and Ceratopogonidae, it develops in the spermatophore sac of the male while the aedeagus is everted into the bursa of the female. Again, in the Trichoptera, Lepidoptera, Simuliidae, Coccinellidae and *Glossina* (Pollock, 1974), secretions of the male accessory glands are ejected into the bursa or vagina and encapsulate the sperm there, so that the form of the spermatophore depends on the shape of the female ducts. Finally the spermatophore may be less constant in form or reduced to little more than a mating plug, as in the Culicidae (Giglioli & Mason, 1966). The sperms which escape from the spermatophore or those deposited free in the bursa or vagina eventually reach the spermatheca, which they later leave to fertilize the egg. Multiple mating may be followed by 'competition' between sperms from two or more males (Parker, 1970).

Several unusual methods of sperm transfer have been described. In the Thysanura, Campodeidae and Collembola there is indirect transfer, without copulation (Schaller, 1971). The sperm is deposited on the substrate in droplets, either at the end of short stalks, as in the Campodeidae and Collembola, or associated with threads produced from the aedeagus. The female then locates the sperm – sometimes assisted by the male – and takes it up into her genital tract. The Odonata also show an unusual method of mating, before which the male transfers sperm to a secondary copulatory organ at the base of the abdomen. From here it is passed to the female genital opening during the mating flight and the whole process may perhaps be derived from an ancestral, indirect method of transfer (Brinck, 1962). In the Cimicidae and in *Xylocoris* the sperms are deposited in Ribaga's organ (p. 699) and from there pass through the haemocoele to the ovaries (Carayon, 1966). Haemocoelic migration of sperms also occurs in *Hesperoctenes* (Hagan, 1931), *Lyctocoris* and in the Nabids *Prostemma* and *Alloeorhynchus*; in the last three the migration follows the puncture of the wall of the genital chamber by the male at copulation (Carayon, 1952).

Types of Reproduction

Insect reproduction usually involves the meeting of the two sexes and the fertilization of the ovum by the spermatozoon. Most insects are oviparous, i.e. they lay eggs which hatch after deposition. Exceptions to these generalizations, however, are somewhat numerous and are separately dealt with below.

Viviparity – Species in which embryonic development is completed within the body of the female parent and which therefore produce larvae or nymphs instead of laying eggs are said to be viviparous. The phenomenon occurs in scattered representatives of many orders but is particularly characteristic of the parthenogenetic Aphidoidea, the Strepsiptera and the 'Pupiparan' Diptera and in a few striking cases it is associated with paedogenesis (q.v.). Viviparity may mean little more than the retention of the eggs in the reproductive tract and the expulsion of the young when they rupture the chorion, but in other cases the structure and physiology of the parent and the mode of development of the embryo show elaborate adaptations to the habit and Hagan (1951), who reviews the whole subject, distinguishes four main types of viviparity:

1. *Ovoviviparity*. Here the eggs contain enough yolk to nourish the developing embryos which are deposited by the mother soon after hatching. No special nutritive structures have therefore been evolved though the chorion may be thin and the female may have a reduced number of ovarioles and a saccular vagina while fertilization sometimes occurs in the ovarioles. This type is found in various representatives of the Thysanoptera, Blattidae, Muscidae, Tachinidae, Coleoptera, etc. It may be noted that the term ovoviviparity is sometimes also used to denote a different condition, i.e. the laying of eggs containing embryos in an advanced state of development.

2. *Adenotrophic viviparity*. The thinly chorionated eggs are ovulated singly and the embryos develop at the expense of the yolk which they contain. After hatching, however, the somewhat degenerate larva – whose gut is closed posteriorly – is retained in the large, muscular vagina (uterus) of the mother where it feeds *per os* on the secretion of her hypertrophied uterine glands and moults twice, being deposited as a mature larva which soon pupates. This condition occurs only in the 'Pupiparan' Diptera and in *Glossina*.

3. *Haemocoelous viviparity*. This highly specialized type occurs in the Strepsiptera and the paedogenetic larvae of the Cecidomyiids *Miastor*, *Mycophila* and *Heteropeza*. In all these there are no oviducts and the ovaries lie free among the fat-body, breaking up readily when mature so that the eggs are dispersed in the

haemocoele. The eggs have no chorion but become surrounded from an early stage by a trophic membrane through which nutrient materials are supplied from the maternal tissues. When development is complete the young larvae of the Strepsiptera escape through the brood canal, but those of the Cecidomyiids first devour the tissues of the maternal larva before escaping through its integument.

4. *Pseudoplacental viviparity*. Here the embryo develops in an enlarged part of the maternal vagina from a practically yolkless egg which is almost always devoid of a chorion and, when not parthenogenetic, is fertilized in the ovariole. The embryo is nourished through placenta-like structures which are formed from maternal and/or embryonic tissues and which are either in close contact with or actually fused to the maternal tissues. Oral feeding does not occur. This type of viviparity is found in the Aphidoidea, Polyctenidae and a few other Heteroptera (Carayon, 1962), *Arixenia*, *Hemimerus*, *Diploptera* (Blattidae) and *Archipsocus* (Psocoptera).

Parthenogenesis – This phenomenon, in which eggs undergo full development without having been fertilized, is well shown by various insects though among the better-known orders it is not known with certainty in the Odonata and Heteroptera. It occurs occasionally in what are otherwise normal, bisexually reproducing species, many examples occurring in the Lepidoptera (Cockayne, 1938). As a regular feature of the life-cycle it is particularly characteristic of certain groups and as it may be associated with various unusual life-cycles, with a sex-determining mechanism and with atypical gametogenesis, various types of parthenogenesis have been recognized (White, 1973; Suomalainen, 1962). Thus, it may be facultative, when it co-exists with bisexual reproduction, or obligatory when males are absent or extremely rare and perhaps functionless. Again, the parthenogenetically developing eggs may have either a haploid or a diploid set of chromosomes and may give rise to both sexes (amphitoky) or exclusively to males (arrhenotoky) or females (thelytoky). Furthermore, parthenogenesis may be combined in the life-cycle with paedogenesis and viviparity or may occur in alternation with a bisexual generation. It may be found throughout a relatively large systematic group, in one or a few species of otherwise bisexual groups or may even occur in only part of the geographical range of a single species. In some species a parthenogenetic 'race' or 'races' is known, but some of these cases probably involve sibling species. Some of the better investigated examples of parthenogenesis are cited below; further details are given in the systematic section of this book.

1. *Haploid facultative arrhenotoky*. The significance of this type of parthenogenesis is that it constitutes a sex-determining mechanism which, moreover, can be more flexible than the usual chromosomal type. Females lay fertilized (diploid) eggs which give rise to females and unfertilized (haploid) eggs which develop parthenogenetically into males. It is characteristic of all Hymenoptera in which males are at all frequent, of some Aleyrodids, the Iceryine Coccoidea, some Thysanoptera, some species of *Xyleborus* and the peculiar beetle *Micromalthus debilis* (see below).

2. *Facultative thelytoky*. This is well seen in one race of *Coccus hesperidum*. Oogenesis is accompanied by meiosis and if the eggs are fertilized they produce both males and females. The nuclei of those eggs which are not fertilized return to the

diploid condition by fusion with the second polar body and develop only into females (Thomsen, 1927; 1929). Comparable processes occur in some Tetrigidae (Nabours, 1929, etc.), some Phasmids and some Symphytan Hymenoptera.

3. *Obligate thelytoky.* This is a 'striking and frequent type of parthenogenesis, males being absent or extremely rare and, at least sometimes, non-functional. The eggs are often formed without meiosis (apomictic parthenogenesis) or, when this occurs, there is a later doubling of the chromosome number, e.g. by fusion of cleavage nuclei or oocyte and polar body (automictic parthenogenesis); some species with this type of parthenogenesis have achieved polyploidy in the germ-line. The main significance of obligate thelytoky is that it permits more rapid reproduction by allowing all the activity of the female to be concentrated on feeding and the production of young and by eliminating any competition for food which might otherwise have resulted from the presence of males. These apparent advantages are, however, offset by the absence of the genetic variability which results from bisexual reproduction. Among other cases, it occurs in some Curculionidae (Suomalainen, 1969), many Psychids (e.g. *Solenobia triquetrella*, Seiler and Schaeffer, 1939–41) and such Phasmids as *Carausius*, *Clonopsis* and *Bacillus*. See also below under cyclical parthenogenesis.

4. *Cyclical parthenogenesis in Aphids and Cynipids.* Some Aphids and Cynipids are exclusively parthenogenetic and provide striking examples of obligate thelytoky. In the remainder, however, a bisexual generation is interposed in the colder part of the year between one or more parthenogenetic generations. In such Aphidoidea (Cognetti, 1962) there are normally several parthenogenetic generations with obligate thelytoky but the sexupara generation immediately preceding the bisexual one either consists of one type of female which gives rise parthenogenetically to both males and females (obligate amphitoky, e.g. *Tetraneura*, Schwartz, 1932) or, as in *Phylloxera* (Morgan, 1912), is made up of two sorts of females which produce respectively males and females by parthenogenesis. In the Cynipid *Neuroterus quercusbaccarum* (Doncaster, 1910–16; Dodds, 1939) there is one parthenogenetic and one sexual generation, the former consisting of two types of females which are respectively male- or female-producers. In some other Cynipids each parthenogenetic female produces both sexes or gives rise to families in which one or other sex predominates (Patterson, 1928). Cyclical parthenogenesis can combine the genetic advantages of bisexual reproduction with the greater reproductive rate of thelytoky.

Akin to parthenogenesis is the rare phenomenon of gynogenesis (pseudogamy), in which sperms are needed only to activate the egg; their nuclei do not fuse with the egg nuclei and they contribute no genetic material to the progeny. The bisexual moth *Luffia lapidella* has such a gynogenetic form, in which the diploid number is restored in the egg nucleus after activation (Narbel-Hoffstetter, 1963). The beetle *Ptinus clavipes* is bisexual but also has a female triploid gynogenetic form (*mobilis*) which produces fertile eggs after mating with a *clavipes* male (Sanderson, 1960).

Paedogenesis – A few immature insects possess functional ovaries, the eggs of which develop parthenogenetically so that reproduction is effected by the immature organism, a condition known as paedogenesis. This is most clearly shown in *Micromalthus debilis* (Coleoptera) and in some Cecidomyiids, of which the best known are *Miastor*, *Heteropeza* (= *Oligarces*) and *Mycophila*. In the N. American form of *Micromalthus* Scott (1936–41) distinguishes adult males and females – which are not known with certainty to be capable of reproduction – and two types of paedogenetic larvae. One of these is the principal reproductive form and produces viviparously 4–20 young larvae, these consuming their parent before escaping to adopt a phytophagous life. They later develop either into adult females or into paedogenetic larvae like their parents or into a second type of paedogenetic larva. The latter lays a single egg from which a male eventually develops, the young male larva normally devouring its parent. If, for some reason, this male larva is not produced or if the parent escapes being eaten she then continues her reproductive life as a paedogenetic larva of the first type. In the S. African form of the species the males and male-producing larvae are absent (Pringle, 1938). In *Heteropeza pygmaea* there are normal males and females (the latter laying fertilized or parthenogenetic eggs) as well as three types of viviparous, paedogenetic larvae (Ulrich, 1936–43; Camenzind, 1962; Hauschtek, 1962; Kaiser, 1970). The first type produces daughter-larvae which can develop, according to conditions, into any of the following: (*a*) adult females, (*b*) paedogenetic larvae like their parent, (*c*) purely male-producing paedogenetic larvae and (*d*) the third type of paedogenetic larvae whose offspring develop either into adult males or paedogenetic larvae of the first type. For other paedogenetic Cecidomyiids see Nikolei (1961), Nicklas (1960) and Ulrich *et al.* (1972). The parthenogenetic viviparous aphids may also be described as paedogenetic in a sense since their earliest offspring begin embryonic development long before the mother is mature, some, in fact, before she has even been born (Uichanco, 1924). Similarly in *Hesperoctenes* (Polyctenidae) fertilization and the early stages of embryonic development take place in the mother before her last moult is carried out (Hagan, 1931). The cases of oviparous pupal paedogenesis reported among Chironomids are probably better regarded as the laying of parthenogenetic eggs by a fully-formed adult which has not yet escaped from the pupal integument (Hinton, 1948).

Polyembryony – This term denotes the production of two or more (often very many) embryos from a single egg. It has been reported as a rare abnormality in, for example, the Acrididae (Slifer and Shulow, 1947) and is described as a normal occurrence in the Strepsipteran *Halictoxenos* (Noskiewicz and Poluszyński, 1935) but is otherwise known only in certain parasitic Hymenoptera. Here it is found in several Encyrtidae (*Ageniaspis, Copidosoma, Litomastix*), in *Platygaster* (Platygasteridae), in the Braconids *Amicroplus* and *Macrocentrus*, in *Aphelopus theliae* (Dryinidae) and some others. Silvestri (1937), Clausen (1940) and Ivanova-Kasas (1972) review the subject and indicate the many similarities which exist in the development of polyembryonic species (see also Doutt, 1947; and Koscielska, 1963). One or more of the poorly yolked parasite eggs, which may be fertilized or develop parthenogenetically, are laid in the egg of the host species and the polar bodies which arise in the maturation of the parasite egg are not thrown off. Instead, with some cytoplasm from the egg, their nuclei give rise to a membrane – the trophamnion (Fig. 154) – which invests and helps to transmit food to the developing embryonic part of the egg (but see Ivanova-Kasas, *l.c.*, who regards the trophamnion as a precociously developed serosa). The developing egg now divides to form a number of morulae each of which forms an embryo and in some cases the whole structure also becomes surrounded by an adventitious sheath of cells derived from the developing host. When numerous, the embryos, surrounded by the trophamnion, form

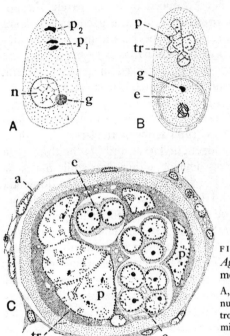

FIG. 154

Ageniaspis fuscicollis, polyembryonic development

A, egg with 1st and 2nd polar bodies p_1, p_2; n, oocyte nucleus. B, division of egg into embryonic area e and trophamnion tr with paranucleus p; g, germ cell determinant. C, transverse section of polyembryonic mass; a, adventitious sheath; e, e, embryos. From Martin.

an elongate, irregularly shaped, sometimes branching mass known as an embryo-chain (Fig. 155). This later breaks up and the separate embryos give rise to first-stage larvae. Abortive embryos and larvae may occur and degenerate or, together with trophamniotic fragments, are eaten by the remaining larvae which otherwise feed on the tissues of the developing host and reach maturity in its late larval, pupal or adult stage. In this way a single host may produce a brood of from 2 to about 3000 parasites. The larger recorded broods are probably all due to the development of more than one parasite egg though each egg of *Copidosoma* and *Litomastix* may give rise to a thousand larvae. The brood from a single host may, as a result of superparasitism, consist of both sexes (as is usually the case in *Platygaster felti*) or be unisexual (e.g. most instances of *Copidosoma gelechiae*). Polyembryony clearly enables the species to achieve a high reproductive potential but its effects are partly offset by the fact that females of polyembryonic species tend to produce fewer eggs than do those of related monembryonic ones and

Clausen (1940) says that field records do not indicate polyembryonic species to be particularly effective parasites.

Hermaphroditism – Functional hermaphroditism is an extremely rare phenomenon in insects. In the scale-insect *Icerya purchasi* Hughes-Schrader (1927, 1930) found that in addition to the infrequent but normal males there are forms which are externally similar to the females of related species but which are actually functional hermaphrodites, true females being absent. The hermaphrodite reproductive system is not unlike the usual female condition but the gonad – which is formed by the anterior fusion of two originally separate organs – produces both spermatozoa and eggs. The outer cells of the gonad form ovarioles while the inner ones give rise to sperms. The eggs sometimes develop parthenogenetically into haploid males but are

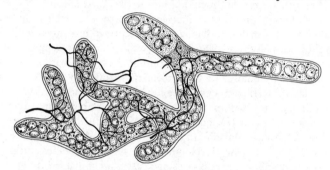

FIG. 155   Ramifying chain of embryos of *Ageniaspis fuscicollis* (*after* Marchal)

usually fertilized in the gonad by sperms from the same individual or, rarely, by sperms from the males with which the hermaphrodites mate. The hermaphrodites do not copulate with each other. A similar situation has now been reported in another scale insect, *Icerya bimaculata*, in which cytological evidence suggests that the hermaphrodites are self-fertilizing (Hughes-Schrader, 1963). In the Phorid *Termitostroma* Mergelsberg (1935) found each individual to have a pair of ovaries and a testis, each gonad with its own efferent duct and separate male and female gonopores. Hermaphroditism is probable in other Termitoxeniine Phorids; it is likely that the testis is functional before the ovaries ripen and that though cross-fertilization is the rule self-fertilization can also occur.

Non-functional, accessory hermaphroditism is also very rare but is found in the stone-fly *Perla marginata*. Schoenemund (1912) has shown that all the males of this species possess a well-developed ovary connecting the anterior extremities of the testes. The eggs in this ovary show the male chromosome number (22), and although they develop up to a late period they neither mature nor function (Junker, 1923). What may be a similar case occurs in the termite *Neotermes zuluensis*, in which a high proportion of male individuals (nymphs, soldiers and young alates) have gonads that contain oocytes (Geyer, 1951).

Castration – Castration, in the broad sense, implies any process which inhibits completely or to a considerable extent the production of mature gametes by the organism, whose gonads may be greatly atrophied or well developed but not normally functional. Two types may be distinguished:

1. *Physiological castration.* This occurs as a normal phenomenon in social insects with one or more sterile castes. The processes leading to the production of sterile forms have not been fully investigated but probably differ in the different groups concerned. In the Isoptera, where the workers and soldiers are sterile representatives of both sexes, it is probable that pheromones exuded by the sexual forms inhibit the development of the gonads in those nymphs which receive them (p. 623). In many Hymenoptera (p. 1176) it is not unlikely that differences in the amount or quality of the food received by the developing larvae help to decide whether the females to which they give rise are fertile (queen) or sterile (workers) and in *Apis mellifera* the diet of future workers is known to contain less 'royal jelly' though adult workers will develop ovaries and lay male-producing eggs if deprived of contact with the queen, from whom they normally receive an inhibitory substance (Hess, 1942; Butler, 1954; Rembold *et al.*, 1974; and see p. 143). In *Polistes* (Vespidae) within 24 hours of the death or experimental removal of the queen, the previously sterile females in the nest start to lay unfertilized eggs, suggesting that their sterility was due to behavioural relationships within the normal colony rather than to more deeply seated physiological causes (Deleurance, 1946). For further information see the reviews by Weaver (1966), Wilson (1971), Richards, O. W. (1971) and Kerr (1974).

2. *Parasitic castration.* The development of various parasites within an adult insect may induce sterility which is, in some cases, accompanied by changes in the secondary sexual characteristics of the host. The phenomenon is well shown in Aculeate Hymenoptera and Auchenorrhynchan Homoptera parasitized by Strepsiptera where it is known as 'stylopization' (p. 927). The adult stylopized host often differs in many features from normal specimens (Salt, 1927, 1931; Clausen, 1940; Lindberg, 1949; Baumert-Behrisch, 1960): There is a general loss of vitality and the rate of development may be accelerated or retarded; the ovaries are atrophied and though the testes are usually little affected, stylopized males may be sterile; in some host species, various changes also occur in the external genitalia and such secondary sexual characters as coloration and the structure of the antennae and pollen-collecting apparatus of the Hymenoptera. The stylopized host shows inter-sexual features, females tending towards maleness and vice versa. A somewhat similar phenomenon is shown by female Chironomids infected with Mermithid Nematodes (Rempel, 1940; Götz, 1964; Wülker, 1961) – these show, to varying degrees, a number of male features in the external genitalia and reproductive system, even extending in extreme cases to the replacement of the ovaries by testes in which spermatogenesis takes place. Other Nematodes cause castration without affecting secondary sexual characters. For example, the frit fly (*Oscinella frit*) is attacked by females of *Tylenchinema oscinellae* (Goodey, 1930) while *Sphaerularia bombi* causes female sterility in *Bombus* through the production of a toxic substance which probably affects the ovaries through its action on the gonadotropic function of the corpora allata (Palm, 1948). Those insect hosts which live as adults following attack by Hymenopteran or Dipteran parasites are also partially or completely castrated. Clausen (1940) summarizes many examples attacked by representatives of the following parasitic families: Tachinidae, Braconidae (Euphorinae and Aphidiinae),

Aphelinidae, Encyrtidae and Dryinidae. Male hosts attacked by the last-mentioned family tend to exhibit female external characters but the equivalent changes do not occur in castrated female hosts.

Literature on the Reproductive System and Reproduction

ADIYODI, K. G. AND ADIYODI, R. G. (1974), Comparative physiology of reproduction in arthropods, *Adv. comp. Physiol. Biochem.*, **5**, 37–107.

ALEXANDER, R. D. (1964), The evolution of mating behaviour in arthropods, In: Highman, K. C. (ed.), Insect Reproduction, *Symp. R. ent. Soc. Lond.*, **2**, 78–94.

ANDERSON, J. F. (1967), Histopathology of intersexuality in mosquitoes, *J. exp. Zool.*, **165**, 475–495.

ANDERSON, J. M. (1950a), A cytological and cytochemical study of the male accessory glands in the Japanese beetle, *Popillia japonica* Newman, *Biol. Bull.*, **99**, 49–64.

—— (1950b), A cytological and histological study of the testicular cyst-cells in the Japanese beetle, *Physiol. Zool.*, **23**, 308–316.

BACCETTI, B. (1963), Indagini ultramicroscopiche sul problema della costituzione delle tuniche peritoneali ovariche e dell'esistenza negli insetti di fibre muscolari liscie, *Redia*, **48**, 1–13.

—— (1972), Insect sperm cells, *Adv. Insect Physiol.*, **9**, 315–397.

BACCETTI, B., DALLAI, R. AND FRATELLO, B. (1973), The spermatozoon of Arthropoda. XXII. The '12+0, '14+0' or aflagellate sperm of Protura, *J. Cell Sci.*, **13**, 321–335.

BAIRATI, A. (1967–68), Struttura ed ultrastruttura dell'apparato genitale maschilo di *Drosophila melanogaster* Meig. I, II., *Z. Zellforsch.*, **76**, 56–99. *Monit. Zool.*, **2**, 105–182.

BAUMERT-BEHRISCH, A. (1960), Ein Einfluss des Strepsipteren-Parasitismus auf die Geschlechtsorgane einiger Homoptera. I, II., *Zool. Beitr.*, **6**, 85–126; 291–332.

BEAMENT, J. W. L. (1946a), The waterproofing process in eggs of *Rhodnius prolixus* Stål., *Proc. R. Soc.*, *Ser.* B, **133**, 407–418.

—— (1946b), The formation and structure of the chorion in an Hemipteran, *Rhodnius prolixus*, *Q. Jl microsc. Sci.*, **87**, 393–439.

—— (1947), The formation and structure of the micropylar complex in the eggshell of *Rhodnius prolixus* Stål (Heteroptera, Reduviidae), *J. exp. Biol.*, **23**, 213–233.

—— (1948–49), The penetration of insect eggshells. I, II, *Bull. ent. Res.*, **39**, 359–383; **39**, 467–488.

BEAMS, H. W., ANDERSON, E. AND KESSEL, R. (1962), Electron microscope observations on the phallic (conglobate) gland of the cockroach, *Periplaneta americana*, *Jl R. microsc. Soc.*, **81**, 85–89.

BEAMS, H. W. AND KESSEL, R. G. (1969), Synthesis and deposition of oocyte envelopes (vitelline membrane, chorion) and the uptake of yolk in the dragonfly (Odonata: Aeschnidae), *J. Cell Sci.*, **4**, 241–264.

BELL, W. J. AND SAMS, G. R. (1974), Factors promoting vitellogenic competence and yolk deposition in the cockroach ovary: the post ecdysis female, *J. Insect Physiol.*, **20**, 2475–2485.

BERGERARD, J. (1961), Intersexualité expérimentale chez *Carausius morosus* Br. (Phasmidae), *Bull. biol. Fr. Belg.*, **95**, 273–300.

BERGERARD, J. (1972), Experimental and physiological control of sex determination and differentiation, *A. Rev. Ent.*, **17**, 57–74.

BERRY, S. J. (1968), The fine structure of the collaterial glands of *Hyalophora cecropia* (Lepidoptera), *J. Morph.*, **125**, 259–280.

BIER, K. (1963a), Synthese, interzellulärer Transport, und Abbau von Ribonukleinsäure im Ovar der Stubenfliege *Musca domestica*, *J. Cell Biol.*, **16**, 436–440.

—— (1963b), Autoradiographische Untersuchungen über die Leistungen des Follikelepithels und der Nährzellen bei der Dotterbildung und Eiweisssynthese im Fliegenovar, *Arch. EntwMech. Org.*, **154**, 552–575.

BIER, K. AND MEYER, G. F. (1952), Ueber die Struktur der peritonealen Hülle des Formicidenovars, *Zool. Anz.*, **148**, 317–324.

BONHAG, P. F. (1955), Histochemical studies of the ovarian nurse tissues and oocytes of the milkweed bug, *Oncopeltus fasciatus* (Dallas), *J. Morph.*, **96**, 381–439.

—— (1958), Ovarian structure and vitellogenesis in insects, *A. Rev. Ent.*, **3**, 137–160.

—— (1959), Histological and histochemical studies on the ovary of the American cockroach *Periplaneta americana* (L.), *Univ. Calif. Publs Ent.*, **16**, 81–124.

BONHAG, P. F. AND ARNOLD, W. J. (1961), Histology, histochemistry and tracheation of the ovariole sheaths in the American Cockroach *Periplaneta americana* (L.), *J. Morph.*, **108**, 107–129.

BONHAG, P. F. AND WICK, J. R. (1953), The functional anatomy of the male and female reproductive systems of the milkweed bug, *Oncopeltus fasciatus* (Dallas) (Heteroptera: Lygaeidae), *J. Morph.*, **93**, 177–284.

BOWEN, W. R. AND STERN, V. M. (1966), Effect of temperature on the production of males and sexual mosaics in an uniparental race of *Trichogramma semifumatum* (Hymenoptera: Trichogrammatidae), *Ann. ent. Soc. Am.*, **59**, 823–834.

BRINCK, P. (1962), Die Entwicklung der Spermaübertragung der Odonaten, *Verh. XI. int. Kongr. Ent.*, **1**, 715–718.

BRUNET, P. C. J. (1951–52), The formation of the ootheca by *Periplaneta americana*. I, II, *Q. Jl microsc. Sci.*, **92**, 113–127; **93**, 47–69.

BRUST, R. A. (1966), Gynandromorphs and intersexes in mosquitoes (Diptera: Culicidae), *Can. J. Zool.*, **44**, 911–921.

BUCHNER, P. (1965), *Endosymbiosis of Animals with Plant Microorganisms*, Interscience, New York, 909 pp.

BUTLER, C. G. (1954), The method and importance of the recognition by a colony of honeybees (*A. mellifera*) of the presence of its queen, *Trans. R. ent. Soc. Lond.*, **105**, 11–29.

CAMENZIND, R. (1962), Untersuchungen über die bisexuelle Fortpflanzung einer pädogenetischen Gallmücke, *Revue suisse Zool.*, **69**, 377–384.

CARAYON, J. (1952), Existence chez certains Hémiptères Anthocoridae d'un organe analogue à l'organe de Ribaga, *Bull. Mus. Hist. nat.*, Paris (2), **24**, 89–97.

—— (1962), La viviparité chez les Héteroptères, *Verh. XI int. Kongr. Ent.*, **1**, 711–714.

—— (1966), Traumatic insemination and the paragenital system, In: Usinger, R. L. (ed.), *Monograph of Cimicidae (Hemiptera–Heteroptera)*, pp. 81–166.

CARSON, H. L. (1945), A comparative study of the apical cell of the insect testis, *J. Morph.*, **77**, 141–162.

CLAUSEN, C. P. (1940), *Entomophagous Insects*, McGraw-Hill, New York, 688 pp.

COCKAYNE, A. E. (1935), The origin of gynandromorphs in the Lepidoptera from binucleate ova, *Trans. R. ent. Soc. Lond.*, **83**, 509–521.

—— (1938), The genetics of sex in Lepidoptera, *Biol. Rev.*, **13**, 107–132.

COGNETTI, G. (1962), Citogenetica della partenogenesi negli afidi, *Archo zool. ital.*, **46** (1961), 89–122.

CRUICKSHANK, W. J. (1973), The ultrastructure and functions of the ovariole sheath and tunica propria in the flour moth, *J. Insect Physiol.*, **19**, 577–592.

CUMMINGS, M. R. (1972), Formation of the vitelline membrane and chorion in developing oocytes of *Ephestia kuehniella*, *Z. Zellforsch.*, **127**, 175–178.

CURTIN, T. J. AND JONES, J. C. (1961), The mechanism of ovulation and oviposition in *Aedes aegypti*, *Ann. ent. Soc. Am.*, **54**, 298–313.

DAVEY, K. G. (1959), Spermatophore production in *Rhodnius prolixus*, *Q. Jl microsc. Sci.*, **100**, 221–230.

—— (1965), *Reproduction in the Insects*, Oliver & Boyd, Edinburgh and London, 96 pp.

DAVIES, L. (1948), Laboratory studies on the egg of the blowfly *Lucilia sericata* (Mg.), *J. exp. Biol.*, **25**, 71–85.

DELEURANCE, E. (1946), Une régulation à base sensorielle périphérique. L'inhibition de la ponte des ouvrières par la présence de la fondatrice des *Polistes* (Hyménoptère–Vespidae), *C.r. Acad. Sci.*, Paris, **223**, 871–872.

DEPDOLLA, P. (1928), Die Keimzellenbildung und die Befruchtung bei den Insekten, In: Schröder's *Handb. d. Entomol.*, **1**, 825–1116.

DODDS, K. S. (1939), Oogenesis in *Neuroterus baccarum*, *Genetica*, **21**, 177–190.

DONCASTER, L. (1910–16), Gemetogenesis of the gall fly, *Neuroterus lenticularis* (*Spathegaster baccarum*), I–III, *Proc. R. Soc. Lond.*, Ser. B, **82**, 88–113; **83**, 476–489; **89**, 183–200.

DÖRING, E. (1955), *Zur Morphologie der Schmetterlingseier*, Akademie-Verlag, Berlin, 154 pp.

DOUTT, R. L. (1947), Polyembryony in *Copidosoma koehleri* Blanchard, *Am. Nat.*, **81**, 435–453.

DOWNES, J. A. (1968), Notes on the organs and processes of sperm-transfer in the lower Diptera, *Can. Ent.*, **100**, 608–617.

ENGELMANN, F. (1968), Endocrine control of reproduction in insects, *A. Rev. Ent.*, **13**, 1–26.

—— (1970), *The Physiology of Insect Reproduction*, Pergamon Press, Oxford, 307 pp.

ENGELS, W. (1973), Das zeitliche und räumliche Muster der Dottereinlagerung in der Oocyte von *Apis mellifica*, *Z. Zellforsch.*, **142**, 409–430.

ESCHERICH, K. (1894), Anatomische Studien über das männliche Genitalsystem der Coleopteren, *Z. wiss. Zool.*, **57**, 620–641.

FLANDERS, S. E. (1942), Oosorption and ovulation in relation to oviposition in the parasitic Hymenoptera, *Ann. ent. Soc. Am.*, **35**, 251–266.

—— (1945), The bisexuality of uniparental Hymenoptera, a function of the environment, *Am. Nat.*, **79**, 122–141.

FOWLER, G. L. (1973), Some aspects of the reproductive biology of *Drosophila*: sperm transfer, sperm storage and sperm utilisation, *Adv. Genet.*, **17**, 293–360.

GERBER, G. H. (1970), Evolution of the methods of spermatophore formation in pterygotan insects, *Can. Ent.*, **102**, 358–362.

GEYER, J. W. C. (1951), The reproductive organs of certain termites, with notes on the hermaphrodites of *Neotermes, Ent. Mem. Dept. Agric. S. Africa,* **2** (9), 233–325.

GIGLIOLI, M. E. C. AND MASON, G. F. (1966), The mating plug in anopheline mosquitoes, *Proc. R. ent. Soc. Lond.* (A), **41**, 123–129.

GOLDSCHMIDT, R. (1938), Intersexuality and development, *Am. Nat.,* **72**, 228–242.

—— (1955), *Theoretical Genetics,* Univ. Calif. Press, Berkeley, 573 pp.

GOODEY, T. (1930), On a remarkable new Nematode, *Tylenchinema oscinellae* gen. et sp. n., parasitic on the Frit-fly, *Oscinella frit* L., attacking oats. *Phil. Trans. R. Soc. Ser. B,* **218**, 315–343.

GÖTZ, P. (1964), Der Einfluss unterschiedlicher Befallsbedingungen auf die mermithogene Intersexualität von *Chironomus, Z. Parasitenk.,* **24**, 484–545.

GREGORY, G. E. (1965), The formation and fate of the spermatophore in the African migratory locust, *Locusta migratoria migratorioides* Reiche and Fairmaire, *Trans. R. ent. Soc. Lond.,* **117**, 33–36.

GUPTA, B. L. AND SMITH, D. S. (1969), Fine structural organization of the spermatheca in the cockroach, *Periplaneta americana, Tissue and Cell,* **1**, 295–324.

HAGAN, H. R. (1931), The embryogeny of the polyctenid *Hesperoctenes fumarius* Westwood, with reference to viviparity in insects, *J. Morph.,* **51**, 1–118.

—— (1951), *Embryology of the Viviparous Insects,* The Ronald Press Co., New York, 472 pp.

HAMON, C. (1972), Formation du chorion et pénétration des symbiontes dans l'oeuf d'un insecte Homoptère Auchénorrhynque *Ulopa reticulata* Fab, *Z. Zellforsch.,* **123**, 112–120.

HAUSCHTECK, E. (1962), Die Cytologie der Pädogenese und der Geschlechtsbestimmung einer heterogonen Gallmücke, *Chromosoma,* **13**, 163–182.

HEBERDEY, R. F. (1931), Zur Entwicklungsgeschichte, vergleichenden Anatomie und Physiologie der weiblichen Geschlechtsausführwege der Insekten, *Z. Morph. Ökol. Tiere,* **22**, 416–586.

HESS, G. (1942), Ueber den Einfluss der Weisellosigkeit und des Fruchtbarkeitsvitamins E auf die Ovarien der Bienenarbeiterin, *Schweiz. Bienenztg,* **2**, 33–110.

HIGHNAM, K. C. (1964) (ed.), Insect Reproduction, *Symp. R. ent. Soc. Lond.,* **2**, 120 pp.

HINTON, H. E. (1948), On the origin and function of the pupal stage, *Trans. R. ent. Soc. Lond.,* **99**, 395–409.

—— (1969), Respiratory systems of insect egg-shells, *A. Rev. Ent.,* **14**, 343–368.

—— (1974), Accessory functions of seminal fluid, *J. med. Ent.,* **11**, 19–25.

HOPKINS, C. R. AND KING, P. E. (1966), An electron-microscopical and histochemical study of the oocyte periphery in *Bombus terrestris* during vitellogenesis, *J. Cell Sci.,* **1**, 201–216.

HUEBNER, E. AND ANDERSON, E. (1972), A cytological study of the ovary of *Rhodnius prolixus.* I–III, *J. Morph.,* **136**, 459–494; **137**, 385–415; **138**, 1–40.

HUGHES-SCHRADER, S. (1927), Origin and differentiation of the male and female germ-cells in the hermaphrodite of *Icerya purchasi* (Coccidae), *Z. Zellforsch.,* **6**, 509–540.

—— (1930), Contribution to the life-history of the iceryine coccids, with special reference to parthenogenesis and haploidy, *Ann. ent. Soc. Am.,* **23**, 359–380.

—— (1963), Hermaphroditism in an African Coccid, with notes on other

Margarodids (Coccoidea–Homoptera), *J. Morph.*, **113**, 173–184.

IVANOVA-KASAS, O. M. (1972), Polyembryony in insects, In: Counce, S. J. and Waddington, C. H. (eds), *Developmental Systems: Insects*, **1**, 243–271.

JOHANSSON, A. S. (1958), Relation of nutrition to endocrine-reproductive functions in the milkweed bug *Oncopeltus fasciatus* (Dallas), *Nytt Mag. Zool.*, **7**, 3–132.

JUNKER, H. (1923), Cytologische Untersuchungen an den Geschlechtsorganen der halbzwittrigen Steinfliege *Perla marginata*, *Arch. Zellforsch.*, **17**, 185–259.

KAISER, P. (1970), Hormonalorgane, Zentralnervensystem und Imaginalanlagen von *Heteropeza pygmaea*, *Zool. Jb.* (*Anat.*), **87**, 377–385.

KAWASAKI, H., SATO, H. AND SUZUKI, M. (1974), Structural proteins in the egg envelopes of dragonflies, *Sympetrum infuscatum* and *S. frequens*, *Insect Biochem.*, **4**, 99–111.

KERR, W. E. (1974), Advances in cytology and genetics of bees, *A. Rev. Ent.*, **19**, 253–268.

KHALIFA, A. (1949*a*), Spermatophore production in Trichoptera and some other insects, *Trans. R. ent. Soc. Lond.*, **100**, 449–479.

—— (1949*b*), The mechanism of insemination and the mode of action of the spermatophore in *Gryllus domesticus*, *Q. Jl microsc. Sci.*, **90**, 281–292.

KING, P. E., RATCLIFFE, N. A. AND FORDY, M. R. (1971), Oögenesis in a Braconid, *Apanteles glomeratus* (L.) possessing an hydropic type of egg, *Z. Zellforsch.*, **119**, 43–57.

KING, P. E. AND RICHARDS, J. G. (1969), Oögenesis in *Nasonia vitripennis* (Walker) (Hymenoptera: Pteromalidae), *Proc. R. ent. Soc. Lond.* (A), **44**, 143–157.

KING, R. C. (1970), *Ovarian Development in* Drosophila melanogaster, Academic Press, New York, 227 pp.

KING, R. C. AND AGGARWAL, S. K. (1965), Oögenesis in *Hyalophora cecropia*, *Growth*, **29**, 17–83.

KOSCIELSKA, M. K. (1963), Investigations on polyembryony in *Ageniaspis fuscicollis* Dalm. (Chalcidoidea, Hymenoptera), *Zool. Polon.*, **13** (3–4), 255–276.

LAUGÉ, G. (1969), Recherches expérimentales sur la détermination et la différenciation des caractères morphologiques et histologiques des intersexués triploïdes de *Drosophila melanogaster*. I, II, *Ann. Embryol. Morphog.*, **2**, 245–270; 273–299.

LEAHY, M. G. (1967), Non-specificity of the male factor enhancing egg-laying in Diptera, *J. Insect Physiol.*, **13**, 1283–1292.

LEOPOLD, R. A. (1976), 'The role of male accessory glands in insects, *A. Rev. Ent.*, **21**, 199–221.

LINDBERG, H. (1949), On stylopization of Areopids, *Acta zool. fenn.*, **57**, 40 pp.

LUSIS, O. (1963), The histology and histochemistry of development and resorption in the terminal oocytes of the desert locust, *Schistocerca gregaria*, *Q. Jl microsc. Sci.*, **104**, 57–68.

MAHOWALD, A. P. (1972), Oogenesis, In: Counce, S. J. and Waddington, C. H. (eds), *Developmental Systems: Insects*, **1**, 1–47.

MANNING, A. (1966), Sexual behaviour, *Symp. R. ent. Soc. Lond.*, **3**, 59–68.

MARTIN, J. S. (1969), Lipid composition of the fat body and its contribution to the maturing oocytes in *Pyrrhocoris apterus*, *J. Insect Physiol.*, **15**, 1025–1045.

MARTOJA, R. (1964), Un type particulier d'appareil génital femelle chez les insectes: les ovarioles adénomorphes du Coléoptère *Steraspis speciosa* (Heterogastra, Buprestidae), *Bull. Soc. zool. Fr.*, **89**, 614–641.

MATSUZAKI, M. (1973), Oogenesis in the springtail, *Tomocerus minutus* Tullberg

(Collembola: Tomoceridae), *Int. J. Insect Morphol. and Embryol.*, **2**, 335–349.

MATTHÉE, J. J. (1952), The structure and physiology of the egg of *Locustana pardalina* (Walk.), *Scient. Bull. Dep. Agric. S. Afr.*, **316**, 83 pp.

MAYS, U. (1972), Stofftransport im Ovar von *Pyrrhocoris apterus* L. Autoradiographische Untersuchungen zum Stofftransport von den Nährzellen zur Oocyte der Feuerwanze *Pyrrhocoris apterus* L. (Heteroptera), *Z. Zellforsch.*, **123**, 395–410.

MELIUS, M. E. AND TELFER, W. H. (1969), An autoradiographic analysis of yolk deposition in the cortex of the Cecropia moth oocyte, *J. Morph.*, **129**, 1–16.

MERCER, E. H. AND BRUNET, P. C. J. (1959), The electron microscopy of the left colleterial gland of the cockroach, *J. biophys. biochem. Cytol.*, **5**, 257–261.

MERGELSBERG, O. (1935), Ueber die postimaginale Entwicklung (Physogastrie) und den Hermaphroditismus bei afrikanischen Termitoxenien (Dipt.), *Zool. Jb. (Anat.)*, **60**, 345–398.

MORDUE, W. (1965), Neuro-endocrine factors in the control of oocyte production in *Tenebrio molitor* L., *J. Insect Physiol.*, **11**, 617–629.

MORGAN, T. H. (1912), The elimination of the sex chromosomes from the male-producing eggs of Phylloxerans, *J. exp. Zool.*, **12**, 479–498.

MORGAN, T. H., BRIDGES, C. B. AND STURTEVANT, A. H. (1919), Contributions to the genetics of *Drosophila melanogaster*, *Carnegie Inst. Publ.*, **278**, 388 pp.

MOSCONA, A. (1950), Studies of the egg of *Bacillus libanicus* (Orthoptera, Phasmidae). I, II, *Q. Jl microsc. Sci.*, **91**, 183–193; 195–203.

NABOURS, R. K. (1929), The genetics of the Tettigidae, *Bibliogr. genet.*, **5**, 27–104.

NAISSE, J. (1966), Controle endocrine de la différenciation sexuelle chez l'insecte *Lampyris noctiluca*. I–III, *Archs Biol., Liège*, **77**, 139–201. *Gen. comp. Endocr.*, **7**, 85–104; 105–110.

—— (1969), Rôle des neurhormones dans la différenciation sexuelle de *Lampyris noctiluca*, *J. Insect Physiol.*, **15**, 877–892.

NARBEL-HOFSTETTER, M. (1963), Cytologie de la pseudogamie chez *Luffia lapidella* Goeze (Lepidoptera, Psychidae), *Chromosoma*, **13**, 623–645.

NICKLAS, R. B. (1960), The chromosome cycle of a primitive Cecidomyid – *Mycophila speyeri*, *Chromosoma*, **11**, 402–418.

NIKOLEI, E. (1961), Vergleichende Untersuchungen zur Fortpflanzung heterogoner Gallmücken unter experimentellen Bedingungen, *Z. Morph. Ökol. Tiere*, **50**, 281–329.

NOSKIEWICZ, J. AND POLUSZYŃSKI, G. (1935), Embryologische Untersuchungen an Strepsipteren. II. Teil. Polyembryonie, *Zool. Polon.*, **1**, 53–94.

NÜESCH, H. (1947), Entwicklungsgeschichtliche Untersuchungen über die Flügelreduktion bei *Fumea casta* und *Solenobia triquetrella* (Lep.) und Deutung der *Solenobia*-Intersexen, *Arch. Klaus.-Stift. Vererbforsch.*, **22**, 221–293.

ODHIAMBO, T. R. (1969), The architecture of the accessory reproductive glands of the male desert locust. 1: Types of glands and their secretions; 2: Microtubular structures, *Tissue and Cell*, **1**, 155–182; 325–340.

PALM, N. B. (1948), Normal and pathological histology of the ovaries in *Bombus* Latr. (Hymenoptera), *Opusc. ent., Suppl.*, **7**, 101 pp.

PATTERSON, J. T. (1928), Sex in the Cynipidae, and male-producing and female-producing lines, *Biol. Bull.*, **54**, 201–211.

PHILLIPS, D. M. (1969), Exceptions to the prevailing pattern of tubules $(9 + 9 + 2)$ in the sperm flagella of certain insect species, *J. Cell Biol.*, **40**, 28–43.

—— (1970), Insect sperm: their structure and morphogenesis, *J. Cell Biol.*, **44**, 243–277.

POLLOCK, J. N. (1974), Anatomical relations during sperm transfer in *Glossina austeni* Newstead (Glossinidae, Diptera), *Trans. R. ent. Soc. Lond.*, **125**, 489–501.

PARKER, G. A. (1970), Sperm competition and its evolutionary consequences in the insects, *Biol. Rev.*, **45**, 525–567.

PRINGLE, J. A. (1938), A contribution to the knowledge of *Micromalthus debilis* LeC. (Coleoptera), *Trans. R. ent. Soc. Lond.*, **87**, 271–286.

REMBOLD, H., LACKNER, B. AND GEISTBECK, J. (1974), The chemical basis of honeybee, *Apis mellifera*, caste formation. Partial purification of queen bee determinator from royal jelly, *J. Insect Physiol.*, **20**, 307–314.

REMPEL, J. G. (1940), Intersexuality in Chironomidae induced by Nematode parasitism, *J. exp. Zool.*, **84**, 261–289.

RICHARDS, O. W. (1927), Sexual selection and allied problems in the insects, *Biol. Rev.*, **2**, 298–364.

—— (1971), The biology of the social wasps (Hymenoptera, Vespidae), *Biol. Rev.*, **46**, 483–528.

SALT, G. (1927), The effect of stylopization on Aculeate Hymenoptera, *J. exp. Zool.*, **48**, 223–331.

—— (1931), A further study of the effect of stylopization on wasps, *J. exp. Zool.*, **59**, 133–166.

SANDERSON, A. R. (1960), The cytology of a diploid bisexual spider beetle, *Ptinus clavipes* Panzer and its triploid gynogenetic form *mobilis* Moore, *Proc. R. Soc. Edinb.*, **67**, 333–350.

SCHALLER, F. (1971), Indirect sperm transfer by soil arthropods, *A. Rev. Ent.*, **16**, 407–446.

SCHLOTTMAN, L. L. AND BONHAG, P. F. (1956), Histology of the ovary of the adult mealworm *Tenebrio molitor* L. (Coleoptera, Tenebrionidae), *Univ. Calif. Publs Ent.*, **11**, 351–394.

SCHOENEMUND, E. (1912), Zur Biologie und Morphologie einiger *Perla*-Arten, *Zool. Jb. (Anat.)*, **34**, 1–56.

SCHWARTZ, H. (1932), Der Chromosomenzyklus von *Tetraneura ulmi* DeGeer, *Z. Zellforsch.*, **15**, 645–687.

SCOTT, A. C. (1936), Haploidy and aberrant spermatogenesis in a Coleopteran, *Micromalthus debilis* LeConte, *J. Morph.*, **59**, 485–509.

—— (1938), Paedogenesis in the Coleoptera, *Z. Morph. Ökol. Tiere*, **33**, 633–653.

—— (1941), Reversal of sex production in *Micromalthus*, *Biol. Bull.*, **81**, 420–431.

SEILER, J. (1969), Intersexuality in *Solenobia triquetrella* F. R. and *Lymantria dispar* L. (Lepid.). Questions of determination, *Monit. Zool. Ital.* (N.S.), **3**, 185–212.

SEILER, J., PUCHTA, O., BRUNOLD, E. AND RAINER, M. (1958), Die Entwicklung des Genitalapparates bei triploiden Intersexen von *Solenobia triquetrella* F. R.; Deutung des Intersexualitätsphänomens, *Arch. EntwMech. Org.*, **150**, 199–372.

SEILER, J. AND SCHAEFFER, K. (1939–41), Der Chromosomenzyklus einer diploid parthenogenetischen *Solenobia triquetrella*, *Arch. exp. Zellforsch.*, **22**, 215–216; *Revue suisse Zool.*, **48**, 537–540.

SHEPHERD, J. G. (1974), Sperm activation in Saturniid moths: some aspects of the mechanism of activation, *J. Insect Physiol.*, **20**, 2321–2328.

—— (1975), A polypeptide sperm activator from male Saturniid moths, *J. Insect Physiol.*, **21**, 9–22.

SILVESTRI, F. (1937), Insect polyembryony and its general biological aspects, *Bull. Mus. comp. Zool. Harv.*, **81**, 469–498.

SINGH, T. (1958), Ovulation and corpus luteum formation in *Locusta migratoria* and *Schistocerca gregaria*, *Trans. R. ent. Soc. Lond.*, **110**, 1–20.

SLIFER, E. H. (1937), The origin and fate of the membranes surrounding the grasshopper egg; together with some experiments on the source of the hatching enzyme, *Q. Jl microsc. Sci.*, **79**, 493–506.

—— (1938), The formation and structure of a special water-absorbing area in the membranes covering the grasshopper egg, *Q. Jl microsc. Sci.*, **80**, 437–458.

—— (1949), Changes in certain of the grasshopper egg coverings during development as indicated by fast green and other dyes, *J. exp. Zool.*, **110**, 183–203.

SLIFER, E. H. AND SHULOW, A. (1947), Sporadic polyembryony in grasshopper eggs, *Ann. ent. Soc. Am.*, **40**, 652–655.

SMITH, D. S., TELFER, W. H. AND NEVILLE, A. C. (1971), Fine structure of the chorion of a moth, *Hyalophora cecropia*, *Tissue and Cell*, **3**, 477–498.

SNODGRASS, R. E. (1935), *Principles of Insect Morphology*, McGraw-Hill, New York, 667 pp.

SPIETH, H. T. (1974), Courtship behaviour in *Drosophila*, *A. Rev. Ent.*, **19**, 385–405.

SUOMALAINEN, E. (1962), Significance of parthenogenesis in the evolution of insects, *A. Rev. Ent.*, **7**, 349–366.

—— (1969), Evolution in parthenogenetic Curculionidae, In: Dobzhansky, T., Hecht, H. K. and Steere, W. C. (eds) (1969), *Evolutionary Biology* 3, 261–296.

TELFER, W. H. (1965), The mechanism and control of yolk formation, *A. Rev. Ent.*, **10**, 161–184.

TELFER, W. H. AND SMITH, D. S. (1970), Aspects of egg formation, In: Neville, A. C. (ed.), Insect Ultrastructure, *Symp. R. ent. Soc. Lond.*, **5**, 117–134.

THOMSEN, M. (1927), Studien über die Parthenogenese bei einigen Cocciden und Aleurodinen, *Z. Zellforsch.*, **5**, 1–116.

—— (1975), Development and physiology of the oocyte-nurse cell syncytium, *Adv. Insect Physiol.*, **11**, 223–319.

—— (1929), Sex determination in *Lecanium*, *Trans. IVth int. Congr. Ent., Ithaca, N.Y.*, 18–24.

TUFT, P. H. (1950), The structure of the insect eggshell in relation to the respiration of the embryo, *J. exp. Biol.*, **26**, 327–334.

UICHANCO, L. B. (1924), Studies on the embryogeny and post-natal development of the Aphididae with special reference to the history of the 'symbiotic organ' or 'mycetom', *Phillip. J. Sci.*, **24**, 143–247.

ULLMANN, S. L. (1973), Oogenesis in *Tenebrio molitor*: histological and autoradiographical observations on pupal and adult ovaries, *J. Embryol. exp. Morph.*, **30**, 179–217.

ULRICH, H. (1936), Experimentelle Untersuchungen über den Generationswechsel der heterogonen Cecidomyide *Oligarces paradoxus*, *Z. indukt. Abstamm.- u. VererbLehre*, **71**, 1–60.

—— (1939), Untersuchungen über die Morphologie und Physiologie des Generationswechsels von *Oligarces paradoxus* Mein., einer Cecidomyide mit lebendgebärenden Larven, *Verhandl. VII int. Kongr. Ent.*, **2**, 955–974.

—— (1943), Ueber den Einfluss verschiedener, in Ernährungsgrad bestimmender Kulturbedingungen auf Entwicklungsgeschwindigkeit, Wachstum und Nachkommenzahl der lebendgebärenden Larven von *Oligarces paradoxus* (Cecidom., Dipt.), *Biol. Zbl.*, 63, 109–142.

ULRICH, H., PETALAS, A. AND CAMENZIND, R. (1972), Der Generationswechsel von *Mycophila speyeri* Barnes, einer Gallmücke mit paedogenetischer Fortpflanzung, *Revue suisse Zool.*, 79 (Suppl.), 75–83.

WEAVER, N. (1966), Physiology of caste determination, *A. Rev. Ent.*, 11, 79–102.

WEBER, H. (1931), Lebensweise und Umweltbeziehungen von *Trialeurodes vaporariorum* (Westwood) (Homoptera–Aleurodina), *Z. Morph. Ökol. Tiere*, 23, 575–753.

—— (1939–43), Morphologie und Entwicklungsgeschichte der Arthropoden. *Fortschr. Zool.*, 4, 95–136; 5, 51–96; 7, 104–165.

—— (1952), Morphologie, Histologie und Entwicklungsgeschichte der Articulaten, *Z. Morph. Ökol. Tiere*, 9, 18–231.

WHITE, M. J. D. (1968), A gynandromorphic grasshopper produced by double fertilization, *Aust. J. Zool.*, 16, 101–109.

—— (1973), *Animal Cytology and Evolution*, Cambridge, 3rd edn, 961 pp.

WIGGLESWORTH, V. B. (1964), The hormonal regulation of growth and reproduction in insects, *Adv. Insect Physiol.*, 2, 247–336.

WIGGLESWORTH, V. B. AND BEAMENT, J. W. L. (1950), The respiratory mechanisms of some insect eggs, *Q. Jl microsc. Sci.*, 91, 429–452.

WIGHTMAN, J. A. (1973), Ovariole microstructure and vitellogenesis in *Lygocoris pabulinus* (L.) and other Mirids (Hemiptera: Miridae), *J. Ent.* (A), 48, 103–115.

WILBERT, H. (1953), Normales und experimental beeinflusstes Auftreten von Männchen und Gynandromorphen der Stabheuschrecke, *Zool. Jb.* (*Allg. Zool.*), 64, 470–495.

WILSON, E. O. (1971), *The Insect Societies*, Harvard Univ. Press, Cambridge, Mass.

WILSON, F. (1962), Sex determination and gynandromorph production in aberrant and normal strains of *Ooencyrtus submetallicus*, *Aust. J. Zool.*, 10, 349–359.

WÜLKER, W. (1961), Untersuchungen über die Intersexualität der Chironomiden nach *Paramermis*-Infektion, *Arch. Hydrobiol.*, *Suppl.*, 25, 127–181.

Part II
DEVELOPMENT AND METAMORPHOSIS

Oviposition or egg-laying takes place among insects in various ways: the eggs are often protected and are generally deposited in places suitable for the immediate needs of the subsequent offspring. In some cases the female simply drops the eggs at random while flying low, as in a few Lepidoptera whose larvae feed on grasses or their roots. Often they are laid singly or in clusters on the leaves of the future larval food-plants (Heteroptera, Lepidoptera, certain Coleoptera), or they may be superficially inserted into plant tissues (many Tettigoniidae and Homoptera). When inserted more deeply, excrescences (galls) of the plant may arise (Tenthredinidae, Cynipidae). In the Chrysopidae the eggs are laid at the apices of stiff pedicels made of a hardened secretion. In other cases they may be glued to some surface or are laid beneath a web or a cottony covering. There are also insects which enclose their eggs in a firm capsule or *ootheca* (Blattidae, Mantidae) or a less compact pod (Acrididae). Many aquatic species surround the eggs with a gelatinous secretion which swells in water forming a jelly-like spawn (Trichoptera, *Chironomus*). A considerable number of insects lay their eggs beneath the soil (Gryllidae, many Coleoptera). Parasitic species usually oviposit on or within the bodies of the hosts which support their future offspring (Tachinidae, parasitic Hymenoptera); when the host is a vertebrate, the eggs are often fastened to the hair or feathers (Mallophaga, Siphunculata, Oestridae).

Embryonic development may take place entirely after oviposition, or partly while the eggs are still within the parental body, or the whole phase may be passed within the latter in viviparous species (p. 302). Almost every transition between these conditions may be found, notably among the higher Diptera. The duration of the egg stage after oviposition is very variable. In some Sarcophaginae it is only momentary, the larvae emerging immediately; in *Musca domestica* it lasts about 8–12 hours, according to temperature. At the other extreme are certain Lepidoptera which pass about nine months in the egg, and among the Phasmida this stage may last nearly two years.

Development may proceed without interruption from egg to adult or the onset of unfavourable climatic conditions – especially temperature – may result in hibernation or aestivation, when little or no development takes place. Growth may also be interrupted in some insects by the animal entering a condition of diapause. During this state – which may arise at any stage of the life-cycle – development is suspended and cannot be resumed, even in the presence of apparently favourable conditions, unless the diapause is first 'broken' by an appropriate environmental change (see p. 384).

In discussing the development of insects the subject falls naturally into two divisions: (*a*) embryology and (*b*) postembryonic development.

Chapter 18

EMBRYOLOGY

Detailed accounts of embryonic development in the Insecta are given by Johannsen and Butt (1941), Jura (1972) and Anderson (1972). The evolutionary significance of insect development in relation to that of other arthropods is discussed in detail by Anderson (1973); physiological aspects of embryology are reviewed by Pflugfelder (1958), Krause (1957, 1958), Seidel (1961), Counce (1961, 1973) and Kühn (1971).

In the eggs of most insects there is a distinction between the anterior and posterior poles which bears a definite relation to the position of the future embryo. The eggs are located in the ovarioles so that the cephalic pole of each is directed towards the head of the parent while the dorsal and ventral aspects of the egg correspond with those of the parent and future embryo (Hallez, 1886). Exceptions to this rule occur in *Pyrrhocoris* (Seidel, 1924), *Pteronarcys* (Miller, 1939) and others. The ultrastructural and biochemical basis of polarity is not understood (Mahowald, 1972).

The contents of the egg are made up of two portions, cytoplasm and deutoplasm or yolk. The cytoplasm forms a reticulum which pervades the substance of the egg and usually also forms a bounding layer or *periplasm*, which lies just beneath the vitelline membrane and completely surrounds the egg (Fig. 156). The yolk is contained within the meshes of the cytoplasm and consists of spheres and globules of reserve material – proteins, lipids, glycogen and other polysaccharides – laid down during vitellogenesis (p. 296). In addition to these constituents, many eggs contain symbiotic micro-organisms which they receive from the mother and which may come to occupy a special organ, the mycetome, in later development (Buchner, 1965).

In the unfertilized egg the nucleus is situated in the central part of the yolk, enclosed in an island of cytoplasm. During maturation the nucleus migrates towards the periphery of the egg where it undergoes division and the polar bodies are formed and later resorbed (Fig. 156). After fertilization the zygote nucleus passes inwards and there begins to divide into daughter nuclei.

Cleavage and Blastoderm Formation – The products of the division of the zygote nucleus are the cleavage nuclei, each of which becomes enveloped by a stellate mass of protoplasm. When a considerable number of cleavage cells have been formed, the majority migrate to the periphery of the egg, where they become merged with the periplasm to form a continuous

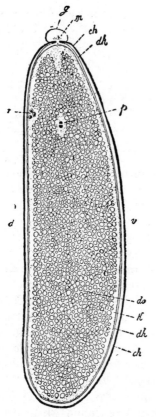

FIG. 156
Longitudinal section of the egg
of *Musca* at time of fertilization

ch, chorion; *d*, dorsal; *v*, ventral side;
dh, vitelline membrane; *do*, yolk; *g*,
gelatinous cap over micropyle (*m*); *k*,
periplasm; *p*, male and female pronu-
clei; *r*, polar bodies. From
Korscheldt and Heider *after*
Henking and Blochmann.

cellular layer or *blastoderm* surrounding the yolk (Fig. 158). At or about the time of blastoderm formation one or more of the cleavage nuclei of some insects become segregated at the posterior pole of the egg to form the future germ-cells (Fig. 158, *g*) but in other species this segregation is postponed somewhat (see p. 242).

As it forms, or at a slightly later stage, the blastoderm consists of a layer of columnar cells on the ventral side of the egg and a flattened epithelial stratum over the remainder. Those cleavage cells which remain in the yolk form the *primary yolk cells* or vitellophages, which become augmented by *secondary yolk cells* derived by the immigration of cells from the blastoderm. In some cases it appears that the yolk cells are only derived from the latter source. Among several orders of insects, notably Orthoptera, Lepidoptera and Coleoptera, the yolk undergoes secondary cleavage, becoming thereby divided into polyhedral masses each of which contains one or more yolk nuclei (Fig. 174). The function of the yolk cells is to liquefy the yolk and bring about its assimilation.

In a few insects cleavage is total rather than peripheral as described above. This feature is exhibited among Collembola (Jura, 1965) and certain endoparasitic Hymenoptera (e.g. *Ageniaspis*, Martin, 1914). In the Collembola cleavage is slightly

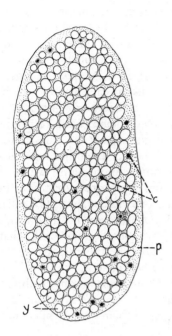

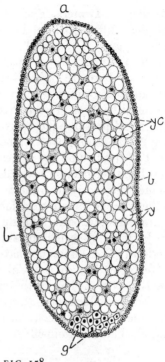

FIG. 157
Longitudinal section of the egg
of *Clytra laeviuscula*, 24 hours
old

The cleavage cells (*c*) are seen
migrating towards the periphery of
the egg. *p*, periplasm; *y*, yolk spheres.
After Lecaillon.

FIG. 158
Median longitudinal section of the
egg of *Clytra laeviuscula* at the
time of completion of segmentation

a, anterior pole; *b*, blastoderm; *g*, genital cells; *y*, yolk spheres; *yc*, yolk cells.
After Lecaillon.

unequal and subsequently becomes peripheral; among some parasitic Hymenoptera
it is total and complete, presumably as a secondary condition following the loss of
yolk.

Differentiation of the Blastoderm – Irrespective of whether the newly
formed blastoderm is quite uniform or already shows signs of precocious
differentiation, the next major stage of development leads to a condition in
which some of the blastoderm forms a thin layer of extra-embryonic
ectoderm covering part of the yolk surface, while the remainder of the
blastoderm develops into a thicker, though still one-layered, *embryonic
primordium*. It is the embryonic primordium that gives rise to most of the
tissues and organs of the later embryo and it varies considerably in size and
shape. In the Diplura and Collembola it occupies much of the surface of the
yolk mass; in the Thysanura and lower Pterygota it is a thin layer restricted
to a relatively small posteroventral region, while in the Hemipteroid orders
and the Endopterygota it is thicker and more extensive, especially in the

Cyclorrhaphan Diptera. The cells of the embryonic primordium show little histological differentiation, but the subsequent developmental fate of its various regions is remarkably constant within the Pterygota (Anderson, 1972, 1973). The various 'presumptive areas' of the differentiated blastoderm can therefore be marked out on 'fate maps' which show that the embryonic primordium comprises five main regions (Fig. 159): (i) a narrow mid-ventral band of presumptive mesoderm cells, at each end of which lie (ii) the small presumptive areas of the anterior and posterior mid gut; (iii) and (iv) the presumptive stomodaeum and proctodaeum, lying respectively in front of and behind the corresponding mid gut areas; (v) a pair of latero-ventral bands of presumptive ectoderm which border the other regions and are joined anteriorly in front of the presumptive stomodaeum, where they take the form of broad head-lobes. In the few Apterygota that have been adequately studied, the mid gut is formed from vitellophages (Woodland, 1957; Jura, 1966) so that presumptive mid gut areas are not found in the blastoderm. The Thysanura have a presumptive mesoderm area like that of the Pterygota, but in the Collembola and Diplura the mesoderm arises diffusely over the whole of the presumptive ectoderm area.

As the blastoderm of *Campodea* and the Collembola begins to differentiate, some of its dorsal cells form a conspicuous *primary dorsal organ* (Tiegs, 1942*a*, *b*). This is a fluid-filled, spherical or mushroom-shaped invagination

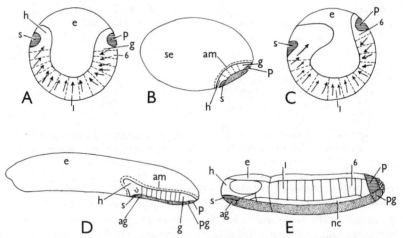

FIG. 159   Fate maps of the insect blastoderm (*after* Anderson, 1973). A, Diplura; B, Thysanura; C, Collembola; D, Odonata (*Platycnemis*); E, Diptera (*Dacus*). All in lateral view, with cephalic area to the left. Presumptive mesoderm stippled in B, D and E; presumptive stomodaeum and proctodaeum horizontally cross-hatched. The arrows in A and C denote immigration of diffusely distributed presumptive mesoderm.

ag, anterior mid gut; *am*, amnion; *e*, dorsal extra-embryonic ectoderm; *g*, posterior growth zone; *h*, head ectoderm; *nc*, nerve cord; *p*, proctodaeum; *pg*, posterior mid gut; *s*, stomodaeum; *se*, serosa; *1*, first thoracic segment ectoderm; *6*, sixth abdominal segment ectoderm.

that secretes radiating threads and subsequently degenerates; it may assist elongation and change of shape in the Collembolan embryo (Jura, 1967). A vestigial primary dorsal organ occurs in *Thermobia* (Woodland, 1957) but not in the Pterygotes, though these possess a so-called *secondary dorsal organ* derived from the serosa (q.v.).

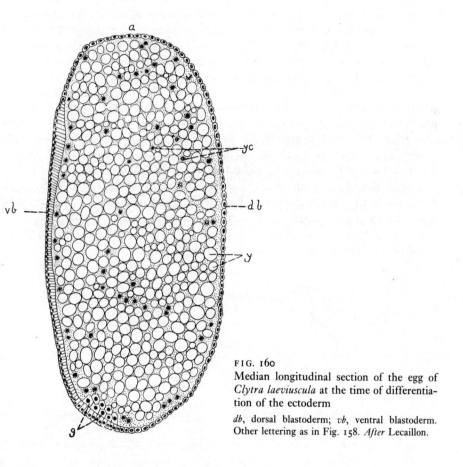

FIG. 160

Median longitudinal section of the egg of *Clytra laeviuscula* at the time of differentiation of the ectoderm

db, dorsal blastoderm; *vb*, ventral blastoderm. Other lettering as in Fig. 158. *After* Lecaillon.

The Germ Band and Gastrulation – The germ band arises by growth and differentiation of the embryonic primordium. It forms an elongate or oval area that may increase considerably in length during further development, undergoing segmentation and gastrulation. The latter process results in the conversion of the single-layered germ band into a two-layered structure and is most clearly exhibited in the large germ band of Endopterygote insects. In the Diplura and Collembola gastrulation is represented by little more than inward migration of presumptive mesoderm cells over the length

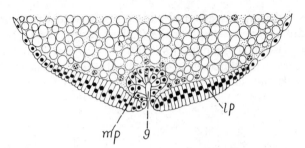

FIG. 161   Transverse section of the germ-band of *Clytra*
laeviuscula at the time of formation of the gastral
groove (*g*)
lp, lateral plate; *mp*, median plate. *After* Lecaillon.

of the germ band. In the Thysanura and Pterygota, however, the process is more clearly defined and occurs through the formation of a more or less short-lived mid-ventral groove, the *gastral groove* (Fig. 161), which varies in length and depth (e.g. Görg, 1959; Striebel, 1960; Roonwal, 1936–37; Louvet, 1964; Goss, 1952–53; Ullmann, 1964; Amy, 1961). The cells of the groove proliferate and sink inwards, while those bordering it eventually meet externally and close the groove (Fig. 162). There is thus formed an *inner layer* that will give rise to mesodermal structures and mid gut rudiments, while the external cells of the germ band (*outer layer*) will form the *embryonic ectoderm* and the stomodaeal and proctodaeal rudiments (Fig. 164).

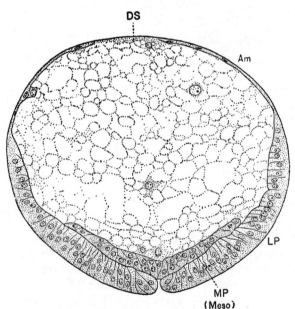

FIG. 162   Transverse section of egg of the honey bee
MP, middle plate; *LP*, lateral plate; *Am*, amnion; *DS*,
dorsal strip of blastoderm. *After* Nelson, 1915.

Extra-embryonic Membranes – It is characteristic of most insects that the germ band does not remain freely exposed on the surface of the yolk, but becomes covered by *extra-embryonic membranes*. These are not formed in the Diplura and Collembola (nor in other Arthropods) but they are seen in a simple form in the Thysanura (Fig. 163). Here, as the germ band grows longer it sinks into the yolk, its margins proliferate to form the walls of the

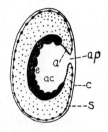

FIG. 163
Diagram of the embryo and embryonic membranes in *Lepisma* according to Heymons

a, amnion; *ac*, amniotic cavity; *ap*, amniotic pore; *c*, chorion; *e*, embryo; *s*, serosa.

cavity into which it sinks, and the opening of the cavity eventually closes (*Petrobius, Ctenolepisma*) or is restricted to a small pore, as in *Lepisma* (Sharov, 1953; Woodland, 1957; Larink, 1969). The cavity is the *amniotic cavity*, its enclosing cellular membrane is the *amnion*, and the outer extra-embryonic ectoderm now constitutes the *serosa*. In most Pterygote insects the same extra-embryonic membranes arise through the formation of *amniotic folds* at the edges of the germ band. These grow towards each other, meeting and fusing, so that the germ band is covered by a double cellular envelope made up of amnion and serosa (Fig. 164; and see, for example, Görg, 1959; Leuzinger *et al.*, 1926; Roonwal, 1936–37; and Mahr, 1960). Variations on this mode of formation of the extra-embryonic membranes occur in several orders of insects and are summarized and discussed by Anderson (1972). The modifications may be related to the greater extent to which the germ band becomes immersed in the yolk, as in the Orthoptera, Plecoptera and Hemipteroid orders. In the Lepidoptera, where the germ band is also immersed, the extra-embryonic membranes are formed in a

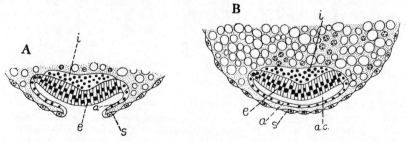

FIG. 164   A, Transverse section of the germ-band of *Clytra* at the time of formation of the amniotic folds. B, at the time of fusion of the amniotic folds

a, amnion; *ac*, amniotic cavity; *e*, ectoderm; *i*, inner layer (mesoderm); *s*, serosa. Based on Lecaillon.

rather different way (Christensen, 1943; Anderson and Wood, 1968). Extra-embryonic ectoderm cells first cover the embryonic primordium, forming the serosa, then the edges of the germ band become reflexed downwards and proliferate to form the amnion, and finally yolk invades the amnioserosal space. In the Hymenoptera and Cyclorrhaphan Diptera the large embryonic primordium and relatively small area of extra-embryonic ectoderm lead to a reduction of the membranes: the Apocrita usually lack the amnion and the Cyclorrhapha have neither amnion nor serosa. It is perhaps surprising that the functions of the extra-embryonic membranes are not properly under-

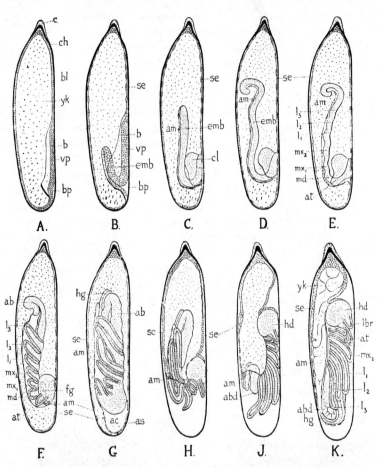

FIG. 165   Stages in the embryonic development of *Agrion*

Lateral view, ventral surface to the right. A, Formation of ventral plate. B–D, Invagination of embryo in yolk. E–G, Formation of appendages. H–K, Rupture of amnion and reversion of embryo. *ab*, abdomen; *ac*, amniotic cavity; *am*, amnion; *as*, union of amnion and serosa; *at*, antenna; *b*, lateral border of ventral plate; *bl*, blastoderm; *bp*, blastopore; *c*, cap or pedicel; *ch*, chorion; *cl*, cephalic lobe; *emb*, embryo; *fg*, stomodaeum; *hd*, head; *hg*, proctodaeum; l_1–l_3, legs; *lbr*, labrum; *md*, mandible; mx_1, first maxilla; mx_2, labium; *se*, serosa; *vp*, ventral plate; *yk*, yolk. From Tillyard, *Biology of Dragonflies, after* Brandt.

stood, though the serosa of many insects secretes an external cuticle beneath the vitelline membrane.

Blastokinesis – During its development the germ band of many insects undergoes a sequence of movements known collectively as *blastokinesis* (Wheeler, 1893). In the first stage or *anatrepsis* the germ band becomes immersed in the yolk. The second stage, *katatrepsis*, follows elongation and segmentation of the germ band and the accompanying development of limb-rudiments (see below). It is preceded by rupture of the extra-embryonic membranes and entails active displacement of the embryo into a more normal position over the ventral surface and anterior pole of the egg (Fig. 165). The extent and details of these movements vary somewhat in the different groups of insects. The Dictyoptera show the simplest pattern since their germ band remains on the surface of the yolk mass and does not move much as the membranes rupture and roll back. In other Exopterygote insects there is more or less extensive immersion in the yolk as growth and anatrepsis leads to dorsal migration of the germ band, with its posterior end moving though the yolk towards the anterior pole of the egg, followed by a considerable migration in the opposite direction during katatrepsis (Fig. 166). In most Endopterygota there is little movement before or after the rupture of

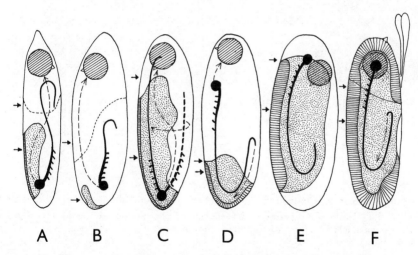

A   B   C   D   E   F

FIG. 166   Diagram summarizing growth, development and blastokinetic movements in the embryos of various insects (*after* Weber, 1953). A, *Platycnemis*; B, *Tachycines*; C, *Notonecta*; D, *Tenebrio*; E, *Ephestia*; F, *Drosophila*

All diagrams are drawn with the ventral surface to the left. The stippled area is the embryonic primordium, its thickness denoted by cross-hatching. The large cross-hatched circle represents the position of the head, with the ventral side indicated, at the end of development. The earlier position of the germ band is indicated by the heavy lines and the extent and direction of movements at katatrepsis by the broken-line arrows. Small arrows to the left of each diagram indicate the level of the differentiation centre (above) and the cleavage centre (below).

the extra-embryonic membranes (Fig. 166). The embryos of some Nematoceran Diptera and Hymenoptera, with elongate eggs, rotate through 180° about the long axis of the egg, but it is only the Lepidoptera that otherwise show major movements. These are usually referred to as blastokinesis but are actually displacements of a unique kind (Reed and Day, 1966; Anderson and Wood, 1968). In *Epiphyas postvittana*, where they have been studied in detail, the embryonic primordium undergoes immersion in the yolk and after provisional dorsal closure (*q.v.*) the embryo lies free in the amniotic cavity, curled almost into a circle with its ventral side outwards. Then, by taking in amniotic fluid through a suctorial action of the stomodaeum, the embryo increases in length and reverses its curvature. A subsequent quiescent phase is eventually followed by the more fully developed Lepidopteran embryo breaking through the amnion and eating the remains of the yolk and the extra-embryonic membranes.

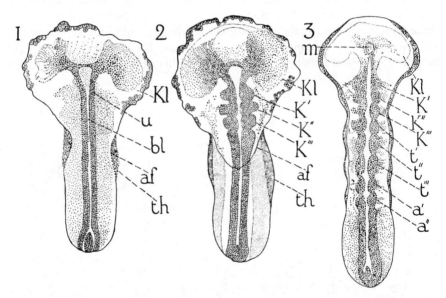

FIG. 167　Three stages in the segmentation of the germ band of a leaf-beetle (*Chrysomela*) *a'a''*, abdominal segments; *af*, amniotic fold; *bl*, blastopore; *K'–K'''*, gnathal segments; *Kl*, head-lobe; *t'–t'''*, thoracic segments. *After* Graber.

Segmentation of the Embryo – At an early stage in development the germ band becomes divided by transverse furrows into a series of segments and in this condition it may be referred to as the embryo (Figs. 167–169). The segmentation may even occur contemporaneously with the formation of the gastral groove, as in *Hydrophilus* and *Chalicodoma*, but as a rule it does not become apparent until after the separation of the inner layer. The embryo is at first divisible into a *protocephalic* or *primary head region*, and a *protocormic* or *primary trunk region*. The protocephalic region is conspicuous on account of its large lateral lobes, which give rise to the *pre-antennal segment* which bears evanescent *pre-antennal appendages* in a few cases (cf. p. 37). Immediately in front of the preantennal segment is a median, often bilobed swelling which is the future *labrum*. It is not generally regarded as a true segment but rather as a pre-oral outgrowth – the acron of Heymons. The 2nd primary head segment is the *deutocerebral* or *antennal segment* which bears a pair of outgrowths representing the future antennae. The 3rd segment is the *tritocerebral* or *intercalary segment*; in some cases it bears a pair of evanescent rudimentary appendages (Fig. 170) homologous with the Crustacean 2nd antennae. Immediately behind the labrum is a pit-like invagination of the ectoderm which is the beginning of the future stomodaeum. The first three of the primary trunk segments subsequently combine with the protocephalic region to form the future head. These segments bear the developing rudiments of the mandibles, maxillae and labium. The *premandibular* and *mandibular segments* give rise to the hypopharynx and, in *Anurida* and *Campodea*, a pair of small protuberances situated near the mid line are the rudiments of the future *superlinguae* (Fig. 170, B). The appendages of the labial segment ultimately fuse to form the *labium*. The next three protocormic segments bear the rudiments of the future three pairs of *thoracic legs* and eventually form the thorax, while the remaining segments constitute the abdomen. In most insect embryos this consists of ten segments, together with a non-segmental terminal region or *telson*, the latter bearing a median invagination which is the beginning of the proctodaeum. There is good reason to believe, however, that the primitive number of abdominal segments is 11 which, with the telson, make up a total of 12 divisions; this number has been recognized by Heymons (1895–96) in the embryos of Dermaptera, Orthoptera and Odonata, and by Nelson (1915) in that of the hive bee. All the abdominal segments may carry a pair of embryonic appendages and in some orders the first pair is often more pronounced than those on the remaining segments and may later take on very different appearances in different groups. They are then known as pleuropodia and appear to be fitted for a variety of functions. In some insects they secrete an enzyme which helps to dissolve part of the egg-shell before hatching (Slifer, 1937; Miller, 1940; Jones, 1956; Ibrahim, 1958) while in some viviparous species they form a sheath (*Hesperoctenes*) or a pair of elongate filaments (*Diploptera*) and perhaps subserve respiration, excretion or the assimilation of food (Hagan, 1951; Polivanova, 1965).

According to the degree of development of the segmentation and appendages of the abdomen, it is possible to distinguish in some insects three embryonic stages which succeed each other more or less clearly (Berlese, 1913). In the earliest or *protopod* phase (Fig. 167) segmentation is absent or indistinct and there are rudimentary appendages only on the head and thorax. The *polypod* phase (Fig. 169) has a clearly segmented abdomen and each segment of the body bears a pair of appendages, while the *oligopod* embryo is clearly segmented but lacks abdominal appendages (Fig. 168).

There are, however, many deviations from such a pattern of development –
e.g. the protopod phase is the only one in some parasitic Hymenoptera but is
less well defined in many other insects while the polypod phase is absent
from many Endopterygotes and the oligopod stage is apparently missing in
others.

In general, some or all of the abdominal appendages degenerate and are
lost before the end of embryonic development. The Apterygota retain at least
one and usually more than one pair of pregenital abdominal appendages but

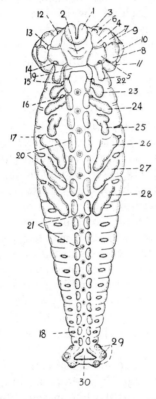

FIG. 168

Embryo of *Leptinotarsa*

1, labrum; 2, stomodaeum; 3–5,
brain segments; 6–8, segments of
optic ganglion; 9–11, segments of
optic plate; 12–16, tentorial
invaginations, etc.; 17, 18, first and
last spiracles; 19, tritocerebral commissure; 20, neuromeres; 21, mid-
dle-cord thickenings; 22, antenna;
23, mandible; 24, maxilla; 25, labium; 26–28, legs; 29, rudiments of
Malpighian tubes; 30, proctodaeum. *After* Wheeler, *J. Morph.*,
3.

the Pterygotes lose them unless it is considered that they are represented by
the prolegs of Lepidopteran and Symphytan larvae or the gills of the
Ephemeroptera, *Sialis*, etc. The appendages of the 8th to 10th segments
form the external genitalia or disappear according to sex and species (p. 77)
while those of the 11th segment are retained in many orders as the cerci.

**Dorsal Closure of the Embryo and Degeneration of the Embryonic
Envelopes** – As the embryo develops it grows round the yolk and the dorsal
or non-embryonic portion of the former blastoderm becomes more and more
restricted. The final closure of the embryo and the fate of the extra-
embryonic membranes exhibit important differences among various insects
which are classified below into four main types. It should be noted, however,

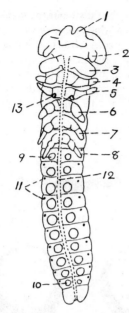

FIG. 169
Embryo of silkworm

1, labrum; 2–5, head-
appendages; 6–8, legs; 9, 10,
first and last abdominal appendages; 11, spiracles; 12,
neural furrow; 13, opening
of silk gland on labium.
After Toyama, *Bull. agric.
Coll., Tokyo*, 5.

hat the processes are somewhat different in insects with an anomalous
rrangement of extra-embryonic membranes and that in all cases it is neces-
ary to distinguish between the final or *definitive dorsal closure* of the embryo,
vhich is accomplished by mid-dorsal junction of the upwardly growing
ctoderm of each side and the *provisional dorsal closure* which may precede
his and is brought about by the extra-embryonic membranes.

1. *Involution through the Formation of a Dorsal Amnioserosal Sac.* – This process
ccurs in the more generalized orders of Pterygota but exhibits various modifica-
ions. The two envelopes rupture and, with the upward growth of the embryo, their

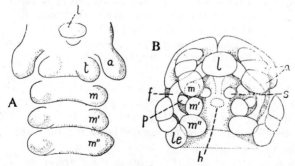

FIG. 170   Cephalic region of embryo of *Anurida* showing
developing appendages

A, at an early stage; B, later. *a*, antenna; *f*, oral fold; *h*,
hypopharynx; *l*, labrum; *le*, leg; *m*, mandible; *m′*, maxilla;
m″, labium; *p*, maxillary palp; *s*, superlingua; *t*, tritocerebral appendage. *After* Folsom, *Bull. Mus. Zool. Harvard*,
36 (redrawn).

contracted remains become carried on to the dorsal side of the yolk. Here they sink into the latter, forming a tubular sac known as the *secondary dorsal organ* (not to be confused with the primary dorsal organ, p. 326). Ultimately the secondary dorsal organ undergoes dissolution and the embryonic ectoderm completes the dorsal closure. In *Hydrophilus* the two flaps formed by the rupture of the amnion and serosa become carried to the upper side of the yolk, with a small contracted area of the original dorsal serosa between them. The flaps then overgrow the latter until

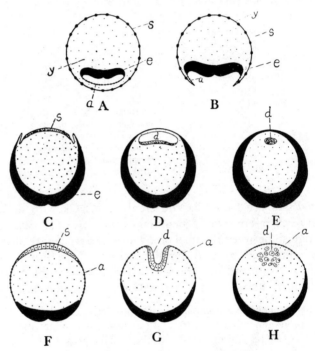

FIG. 171   Diagrams illustrating the dorsal closure of the embryo and the fate of the embryonic membranes. A, B, General. C–E, *Hydrophilus*. F–H, *Oecanthus*

a, amnion; *d*, dorsal organ; *e*, embryo; *s*, serosa; *y*, yolk. Based on Ayers, Graber and Kowalevsky.

their edges unite. By this means a tubular dorsal organ is formed, which sinks into the yolk and becomes enclosed by the developing mesenteron, while the embryonic ectoderm completes the dorsal closure (Fig. 171, A–E). In *Oecanthus* the contracted serosa alone forms the dorsal organ, the amnion persisting, for a while, as a covering of the yolk (Fig. 171, A, B, F–H).

2. *Involution of the Amnion with the Retention of the Serosa.* In *Leptinotarsa* and other Chrysomelids, the amnion ruptures ventrally and grows round the yolk so as to enclose it dorsally, becoming at the same time separated from the serosa. With the upward growth of the embryo the amnion becomes compressed into a small dorsal tract – the *dorsal organ*. The latter disintegrates in the yolk with the dorsal closure of the embryo. The serosa persists, until a late stage, as a complete membrane applied to the inner aspect of the chorion (Fig. 172, A–C).

3. *Involution of the Serosa with Retention of the Amnion*. In *Chironomus* the serosa alone ruptures and contracts to form the dorsal organ, which becomes absorbed into the yolk. The amnion afterwards grows over this area, so as to entirely enclose the egg, and persists until the time of hatching (Fig. 172, D–F).

4. *Retention of both the Amnion and Serosa*. In Lepidoptera and Tenthredinidae the amnion ultimately grows entirely round the yolk and becomes separated from the serosa. The egg is now enclosed by two complete envelopes up to the time of hatching, when they are ruptured. In Lepidoptera a quantity of yolk is retained between these two envelopes, and serves as the first food of the young larva (Fig. 172, G, H).

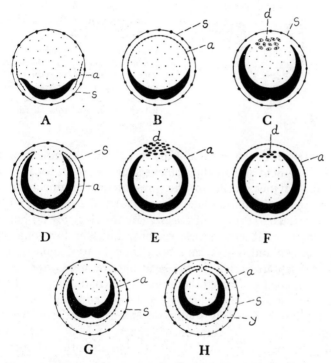

FIG. 172 Diagrams illustrating the dorsal closure of the embryo and the fate of the embryonic membranes in A–C, *Leptinotarsa*; D–F, *Chironomus*; G, H, Lepidoptera
Based on Wheeler, Graber and Tichomiroff.

The Mesoderm – The inner layer gives rise to the mesoderm which becomes arranged into two longitudinal bands, connected across the mid line by a single layer of cells. These bands usually come to be constricted transversely, and consequently the mesoderm becomes divided into segments. These mesoblastic somites appear before or, less often, after the corresponding divisions of the ectoderm. In most insects some or all of the mesodermal somites form coelom sacs by acquiring paired cavities, the coelomic cavities, which form either as clefts in the solid mesoderm or through a folding over of the lateral margins of the somites (Fig. 174). In most insects there are

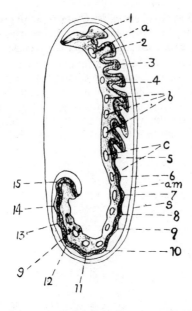

FIG. 173

Diagrammatic sagittal section (a little to one side of the median line) of the embryo of *Donacia crassipes* showing coelom sacs

1–4, cephalic appendages; 5–15, abdominal segments; *a*, coelomic sac of intercalary segment; *b*, coelom sacs of thoracic segments; *c*, coelom sacs of abdomen; *am*, amnion; *s*, serosa; *g*, genital cells. Adapted from Hirschler, *Z. wiss. Zool.*, 1909.

coelomic cavities in the antennal and the three gnathal segments of the head, each thoracic segment and all abdominal segments except usually for the last one or two. Cavities occur in the pre-antennal segment of *Carausiu.* (Wiesmann, 1926), *Rhodnius* (Mellanby, 1936) and a few other species and also in the intercalary segment of *Locusta* (Roonwal, 1936–37) but are otherwise rudimentary or absent in these regions. The labral coelomic cavities of *Carausius, Locusta* and others have been thought to demonstrate the existence of a true labral segment in the head but are more often regarded as a secondary phenomenon. Some other specialized conditions may be noted. In the Nematocera and some Cyclorrhapha the paired trunk somites are solid while other Diptera may lack somite-development entirely, the mesoderm developing directly into its component organs. *Habrobracon* also shows this mode of development (Amy, 1961), but in *Apis* and other Hymenoptera the coelomic cavities on each side of the body are confluent from their earliest appearance, thus forming a pair of longitudinal tubes, while in some Hemiptera the sacs remain open towards the yolk.

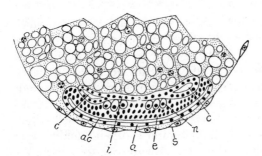

FIG. 174

Transverse section of the germ band of *Clytra* at the time of appearance of the neuroblasts (*n*) and coelomic cavities (*c*)

Other lettering as in Fig. 164. *After* Lecaillon.

The outer or somatic layer of the mesodermal somites gives rise to the
body muscles, dorsal diaphragm, and pericardial cells; from the inner or
splanchnic layer the visceral muscles, genital ridges, and the greater part of
the fat-body are produced. At the upper angles, where the somatic and
splanchnic layers meet, are peculiar cells termed *cardioblasts* (Fig. 175)
which take part in the formation of the heart. The middle layer of
mesoderm, which unites the somites of the two sides of the body, appears in
some insects to dissociate and form blood-cells.

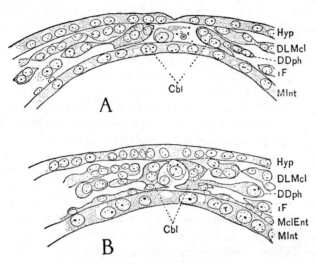

FIG. 175   Transverse sections of dorsal region of 4th trunk segment of late embryo of hive bee × 600

Cb, cardioblasts; *DDph*, dorsal diaphragm; *DLMcl*, dorsal
longitudinal muscles; *Hyp*, epidermis; *IF*, fat-body; *MInt*,
mid intestine and its muscles *MclEnt*. *After* Nelson, 1915.

Alimentary Canal – There has been much dispute over the interpretation of gut development in the insects, especially in connection with the
formation of the mid gut epithelium and its relation to the endoderm of the
germ layer theory (see, for example, Johannsen and Butt, 1941; Eastham,
1930; Henson, 1946). The modern tendency (Anderson, 1972) is to lay less
emphasis on germ layers and to stress instead the differences between the
cells of the various presumptive areas of the blastoderm. In the Pterygotes,
as indicated on p. 326, the blastoderm includes anterior and posterior mid
gut rudiments and, associated with them, the rudiments of the stomodaeum
and proctodaeum. From each mid gut rudiment cells grow towards each
other in various ways. Often they first appear as paired ribbons, but other
methods are known, and eventually the rudiments from each outgrowth
meet and surround the yolk as an epithelial sac, being joined in some cases
by cells liberated from the inner layer. Within the mid gut so formed lies the

yolk and the vitellophages, both of which are gradually absorbed. Only in the Apterygotes (Jura, 1966; Woodland, 1957) does the mid gut arise in a quite different way. Presumptive mid gut areas cannot be recognized in the blastoderm of these insects and the mid gut epithelium differentiates instead from vitellophages which move peripherally in the yolk and then proliferate to enclose it. In all insects the fore and hind gut arise by invagination and subsequent differentiation of the stomodaeal and proctodaeal rudiments. The Malpighian tubules are formed by further invagination at the inner ends of the proctodaeum and this has led to a controversy on their germ layer origin (see p. 248). As the stomodaeum grows in it forms a tube that gives rise dorsally to the frontal, hypocerebral and ventricular ganglia of the stomatogastric nervous system while its inner portion may differentiate into the proventriculus. Eventually the apposed walls of the mid gut and the stomodaeal and proctodaeal invaginations break down to establish a continuous lumen throughout the alimentary canal.

The Nervous System – The beginnings of the central nervous system appear as a pair of longitudinal *neural ridges* of the ectoderm of the germ band, about the time when the latter becomes segmented. They begin at the sides of the stomodaeum, and continue backwards until they unite behind the proctodaeum. These ridges are separated by a median furrow – the *neural groove* (Figs. 169, 176). A chain of cells forming the *median cord* is separated from the ectoderm lining the neural groove. The ectoderm cells forming the neural ridges become segregated into two layers – an outer thin layer of *dermatoblasts* which forms the ventral body-wall and an inner layer of *neuroblasts* which forms the nervous tissue (Fig. 174). When the embryonic appendages start to appear the neural ridges become segmented into definite swellings at their bases, and each pair of swellings constitutes a neuromere. The intrasegmental portions of the median cord and neural ridge give rise to the definitive ganglia, while the intersegmental portions of the ridges form the connectives.

In the cephalic region the neural ridges expand into broad *procephalic lobes*, forming the future brain and they become divided into three neuromeres corresponding with the three primary cephalic segments. These neuromeres are known respectively as the *proto-*, *deuto-* and *tritocerebrum* (see also p. 103 ff.); since the first two lie in front of the stomodaeum, they are pre-oral in position, while the tritocerebrum is postoral and the commissure uniting its two halves (ganglia) passes below the stomodaeum. The optic lobes are formed separately be delamination (Exopterygota), or invagination (Endopterygota) from the cephalic ectoderm and neuroblasts do not appear to participate in their formation.

The neuromeres of the first three protocormic segments fuse to form the suboesophageal ganglion, while the remainder form the ganglia of the ventral nerve-cord. These latter are nine to eleven in number and subsequently undergo varying degrees of fusion in different insects.

The Tracheal System – Shortly after the appearance of the neuromeres the tracheae appear as ectodermal invaginations lying just outside the bases of the appendages (Figs. 168, 178). As a rule, ten pairs are developed, and they occur on the last two thoracic and first eight abdominal segments. In a few species embryonic tracheal invaginations develop on the prothorax or 9th and 10th abdominal segments but close before hatching. In *Apis* there is a pair of invaginations on the labial segment of the head which forms the anterior prolongations of the main tracheal trunks and subsequently closes. Each invagination gives rise to a T-shaped horizontal outgrowth, which extends longitudinally until it meets and fuses with those of the segment in front and behind, thus forming the main longitudinal trunks. The mouths of the original invaginations contract and form the spiracles. After the main tracheal trunks are laid down branches from them extend inwards and, at the

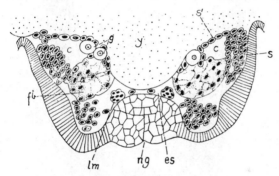

FIG. 176　Transverse section of the abdomen of the
embryo of *Blattella germanica*
For lettering, see Fig. 178.

ends of the finer vessels, certain cells (tracheoblasts) separate from the tracheal epithelium and grow out in a stellate form towards the tissues. It is in these cells that the tracheoles develop as fine intracellular tubes; where the tracheoles invade the cells of different organs they are apparently formed by branches of the tracheoblasts directly penetrating the cytoplasm.

The Salivary Glands – These appear as a pair of ectodermal ingrowths of the labial segment (Fig. 169). As they increase in depth their apertures approximate, and become drawn into the preoral food cavity where they finally open by a median pore into the salivarium.

The Body-wall – The body-wall is directly derived from the superficial ectoderm, as are also the invaginations that give rise to the apodemes, tentorium and other endoskeletal structures. The sensory neurons and integumentary components of the sense organs are also formed from the general ectodermal surface of the embryo and the corpora allata arise as invaginations of the cephalic ectoderm.

The Body-cavity and Dorsal Vessel – The permanent body-cavity begins as a space – the *epineural sinus* – which is mainly produced by the separation of the yolk from the embryo over the region of the ventral nerve-cord (Fig. 176). The process of separation extends laterally, and in some insects the walls of the coelom sacs are stated to break through in such a manner that their cavities become confluent, both with one another and with the epineural sinus; in other cases, however, the coelomic cavities are known not to unite with the epineural sinus. The developing haemocoele extends upwards along with the mesoderm, on either side, until the formation of the body-cavity is completed. The upward migration of the mesoderm carries the cardioblasts with it; the latter subsequently meet along the dorsal line of

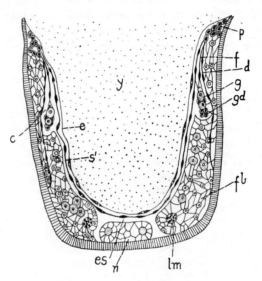

FIG. 177   Transverse section of the abdomen of the embryo of *Blattella germanica* where the germ band is beginning to grow around the yolk

For lettering, see Fig. 178.

the embryo, and arrange themselves as the tubular rudiment of the heart. A single layer of cells unites the cardioblasts to the somatic mesoderm on either side and eventually gives rise to the dorsal diaphragm (Figs. 175, 177). The aorta is formed by the union in the mid-dorsal line of the two coelom sacs of the intercalary segment (*Donacia*), or of the antennal segment (*Forficula, Apis*, etc.); by its backward extension the developing aorta comes to unite with the heart.

   The Reproductive System – The development of the reproductive system is somewhat complicated since it involves the differentiation of the primordial germ cells, the remainder of the gonad, and the efferent ducts, which are themselves usually partly mesodermal and partly ectodermal. In some insects, such as the Orthopteroid groups, the primordial germ cells first become recognizable during gastrulation. In others, like the Hemipteroids

(Goss, 1953) and many Endopterygotes, they are first seen in the blastoderm, while in the Diptera (Davis, 1967), parasitic Hymenoptera (Bronskill, 1959, 1964) and some Coleoptera (Mulnard, 1947; Jung, 1966) the primordial germ cells are segregated precociously as a group of rounded, posterior pole cells (but cf. Günther, 1971). These are preceded in the egg by a specially differentiated region, the so-called oosome, composed of 'polar granules' which contain RNA and protein (Mahowald, 1972). In any event, however, the primordial germ cells eventually migrate to the splanchnic

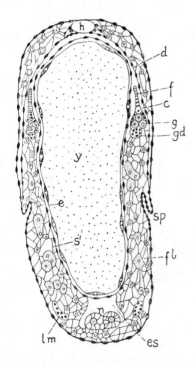

FIG. 178
Transverse section of the abdomen of the embryo of *Blattella germanica*, after the yolk has become enclosed by the germ band

c, coelom; *d*, dorsal diaphragm; *e*, endoderm; *es*, epineural sinus; *f*, filament plate; *fb*, fat-body; *g*, genital cells; *gd*, rudiment of genital duct; *h*, heart; *lm*, ventral longitudinal muscles; *n*, rudiment of nerve-cord; *ng*, neural groove; *p*, cardioblasts; *s*, somatic mesoderm; *s¹*, splanchnic do.; *sp*, spiracle. This and Figs. 176, 177 *after* Heymons (with different lettering).

mesoderm, coming to lie in the walls of some of the coelom sacs – in the case of *Blattella* in those of the 2nd to 7th abdominal segments. The germ cells become surrounded by mesoderm, which forms the *genital ridges*, and these fuse into a cell-strand lying on each side of the dorsal wall of the coelom. The primitive germ cells give rise postembryonically to the gametes, while the enveloping mesoderm produces all other parts of the gonads and their primitive ducts. At an early stage a sheet of cells, the *filament plate*, is differentiated in female embryos, and connects the apex of the genital rudiment with the heart rudiment of the same side of the body. With the migration of the heart rudiments towards the mid-dorsal line, the genital rudiments follow. Their primitive metamerism becomes lost, and it is only in later embryonic stages that sexual differences can be recognized. In the female the filament plate divides into several strands or terminal threads, and these are connected with the divisions of the ovary which represent the ovarioles. The undivided basal portion of the genital rudiment gives rise to the efferent duct of its side. The paired mesodermal efferent ducts run posteriorly and each ends in a closed, dilated ampulla (Fig. 179). The genital ampullae of the male are found in the 10th abdominal segment, those of the female in the 7th segment; in both cases they represent the coelomic sacs of

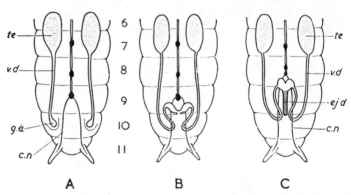

FIG. 179 A, B. Stages in development of male reproductive ducts. C. Adult condition (*after* Snodgrass, 1941)

c.n, cercal nerve; *ej.d*, ejaculatory duct; *g.a*, genital ampulla; *te*, developing testis; *v.d*, vas deferens. 6–11 Abdominal segmentation.

the relevant segment. The development of the median, terminal, ectodermal part of the reproductive system is largely a postembryonic process and is discussed on p. 287.

Sequence of the Developmental Stages – Data concerning the rate of development, and the time taken to arrive at the principal stages, are given in great detail by several authors and tabulated for six species by Anderson (1972). Details for *Hydrophilus* are quoted here after Heider (1889) who divides the developmental period into three phases, occupying altogether 11 days.

1ST PHASE

1st Day. Blastoderm completely formed

2nd Day. Metamerization of the germ band; formation of amniotic folds, procephalic lobes, and inner layer.

3rd Day. Appearance of neural groove and antennae; closure of amniotic folds.

2ND PHASE

4th Day. Appearance of buccal and trunk appendages together with the stomodaeum, which is followed by the proctodaeum; appearance of tracheal invaginations.

5th Day. Elongation of appendages; mouths of tracheal invaginations reduced to rounded orifices representing the spiracles.

7th Day. Elongation of neural groove; embryonic membranes rupture, exposing the embryo.

During this phase the mid gut rudiments form the mid intestine, the mesodermal somites and body-cavity appear, and yolk cleavage occurs.

3RD PHASE

End of 7th Day. Appearance of secondary dorsal organ.

8th Day. Dorsal organ completely formed.

9th Day. Pigmentation of the eyes.

10th Day. Eyes more pigmented; main tracheal trunks visible.

11th Day. Embryo becomes strongly pigmented and exhibits movements beneath the chorion.

12th Day. Eclosion of the larva.

This phase is one of histological differentiation and no new permanent organs are developed.

Physiology of Embryonic Development – The physiological processes which determine or co-ordinate some phases of embryonic development have been elucidated in experimental studies which show that three centres control early development in a number of cases. Cleavage and nuclear migration begin at the *cleavage centre* which is situated in that part of the egg where the future head-rudiment normally develops. The mechanisms which ensure approximately synchronous division of cleavage nuclei and their peripheral migration are not known but the arrival of cleavage nuclei at the posterior pole of the egg causes the release from an *activation centre* there of a substance which diffuses forwards and determines the formation of a germ band. Experimental elimination of this part of the egg by excision, ligaturing or ultra-violet radiation prevents the development of a germ band though an extra-embryonic blastoderm can develop and yolk cleavage occurs. Once the centre has had time to exert its effect the posterior pole of the egg can be constricted off by a ligature without effect on the formation of the germ band. Finally, there is a *differentiation centre* in the cortex of the egg near the future thoracic region of the embryo. This is stimulated by the products of the activation centre and is the region from which visible differentiation of the blastoderm proceeds. The differentiation centre apparently behaves as a localized zone where the yolk undergoes retraction from the chorion, the blastoderm nuclei assembling in the resulting space and forming the germ band.

The earliest cleavage nuclei are isopotent, that is, any one of them can, under experimental conditions, give rise to tissues other than that into which it would normally have developed. The elimination of early cleavage nuclei by ultra-violet radiation therefore does not prevent the remaining nuclei from reorganizing to form a normal embryo. The early egg is thus capable of considerable regulation. After the differentiation centre has completed its function, however, the fate of the various parts is determined and the removal of part of the developing egg then causes corresponding deficiencies in the postembryonic insect. After determination the egg can be described as 'mosaic' and maps of the prospective organ-forming areas of several species have been constructed. The stage at which determination is complete varies with the species and some regions are determined before others. The eggs of *Platycnemis* (Seidel, 1926–35), *Tachycines* (Krause, 1934–39), and *Chironomus* (Yajima, 1960, 1964) are capable of regulation up to the beginning of germ band formation, whereas those of *Drosophila* and other Cyclorrhaphan Diptera are determined by the time they are laid (Geigy, 1931; Hathaway and Selman, 1961). In other species the eggs become determined at times between these extremes – e.g. *Schistocerca* (Moloo, 1971), *Apis* (Schnetter, 1934*a*, *b*) and *Tenebrio* (Ewest, 1937). In *Drosophila* (Geigy, 1931; Nöthiger and Strub, 1972) and in *Tineola* (Lüscher, 1944) the egg is first determined in respect of larval characters and then, shortly afterwards, in respect of imaginal features.

The causal analysis of later embryonic development and the phenomena of embryonic induction have been less fully studied in insects but Bock (1939; 1941; also Seidel, Bock and Krause, 1940) has shown in *Chrysopa* that the ectoderm forms a self-differentiating system and determines the differentiation of the underlying mesoderm. Further evidence for similar relationships comes from the analysis of developmental processes in *Drosophila* mutants, which have also shown that the differentiation of the gonad sheath requires the presence of the germ cells (see literature in Anderson, 1966). Some of the later morphogenetic processes depend on the establishment of normal muscle attachments to the developing epidermis, indicating the part played by muscular contraction in determining changes of form (Wright, 1960).

In many insects the embryo develops without interruption from oviposition to the hatching of the egg. In others, however, development may be interrupted or considerably retarded even though external conditions appear favourable. This form of dormancy is referred to as diapause (Lees, 1955) and is ecologically comparable to the postembryonic diapause described on p. 384. The embryonic stage at which diapause occurs may vary from one species to another. In some insects, such as *Leptohylemyia coarctata* it takes place at the end of development (Way, 1959) while in the Geometrid *Chesias legatella* it occurs at a much earlier stage, when the germ band is becoming immersed in the yolk (Wall, 1973). The egg of the Phasmid *Didymuria violescens* may undergo two diapauses – one in the first winter before germ band formation and another during the second winter when the embryo is completely formed (Bedford, 1970). Of two related species, one may have diapausing eggs and the other not (e.g. Böhle, 1969; Wall, 1974). Even a single species may have genetic or geographical forms that differ in their diapause behaviour. When the egg has entered diapause it undergoes a period of physiological diapause development and only when this is completed does it resume embryonic differentiation. Diapause development commonly requires a period of lower temperature such as occurs in the winter. In *Leptohylemyia*, for example, it takes place under natural conditions between $-6°C$ and $12°C$, with an optimum at $3°C$, whereas pre- and post-diapause development can occur between $0°C$ and $30°C$. Some indication of the complex of factors that may influence egg diapause is seen in *Bombyx mori*, where maternal effects are important. In some races the eggs of every generation enter diapause (univoltine races) while in others there may be one or two intervening generations without diapause (bivoltine or multivoltine forms). Eggs exposed to low temperatures and short photoperiods tend to produce females that lay non-diapausing eggs and conversely. These latter eggs are, in fact, influenced while still in the ovary by a diapause-inducing hormone produced by cells in the suboesophageal ganglion of the female; and the activity of these cells is, in turn, subjected to inhibitory control by the brain (Fukuda, 1963).

Literature on Embryonic Development

AMY, R. L. (1961), The embryology of *Habrobracon juglandis* (Ashmead), *J. Morph.*, **109**, 199–217.

ANDERSON, D. T. (1962), The embryology of *Dacus tryoni* (Frogg.) (Diptera, Trypetidae (=Tephritidae)), the Queensland Fruit-fly, *J. Embryol. exp. Morph.*, **10**, 248–292.

—— (1966), The comparative embryology of the Diptera, *A. Rev. Ent.*, **11**, 23–46.

—— (1972), The development of hemimetabolous insects. The development of holometabolous insects, In: Counce, S. J. and Waddington, C. H. (eds) (1972), *Developmental Systems: Insects*, **1**, 95–163; 165–242.

—— (1973), *Embryology and Phylogeny in Annelids and Arthropods*, Pergamon Press, London, 495 pp.

ANDERSON, D. T. AND WOOD, E. C. (1968), The morphological basis of embryonic movements in the light brown apple moth, *Epiphyas postvittana* (Walk.) (Lepidoptera: Tortricidae), *Aust. J. Zool.*, **16**, 763–793.

ANDO, H. (1962), *The Comparative Embryology of Odonata with Special Reference to the Relic Dragonfly* Epiophlebia superstes *Selys*, Jap. Soc. Promotion Sci., Tokyo, 205 pp.

BEDFORD, G. O. (1970), The development of the egg of *Didymuria violescens* (Phasmatodea: Phasmatidae: Podacanthinae) – embryology and determination of the stage at which first diapause occurs, *Aust. J. Zool.*, **18**, 155–169.

BERLESE, A. (1913), Intorno alle metamorfosi degli insetti, *Redia*, **9**, 121–138.

BOCK, E. (1939), Bildung und Differenzierung der Keimblätter bei *Chrysopa perla* (L.), *Z. Morph. Ökol. Tiere*, **35**, 615–702.

—— (1941), Wechselbeziehung zwischen den Keimblättern bei der Organbildung von *Chrysopa perla* (L.). I, *Arch. EntwMech. Org.*, **141**, 159–247.

BÖHLE, H. W. (1969), Untersuchungen über die Embryonalentwicklung und die embryonale Diapause bei *Baetis vernus* Curtis und *Baetis rhodani* (Pictet) (Baetidae, Ephemeroptera), *Zool. Jb. (Anat.)*, **86**, 493–575.

BRONSKILL, J. F. (1959), Embryology of *Pimpla turionellae* (L.) (Hymenoptera: Ichneumonidae), *Can. J. Zool.*, **37**, 655–688.

—— (1964), Embryogenesis of *Mesoleius tenthredinis* Morl. (Hymenoptera: Ichneumonidae), *Can. J. Zool.*, **42**, 439–453.

BUTT, F. H. (1949), Embryology of the Milkweed Bug, *Oncopeltus fasciatus* (Hemiptera), *Mem. Cornell Univ. agric. Exp. Sta.*, **283**, 43 pp.

BUCHNER, R. (1965), *Endosymbiosis of Animals with Plant Micro-organisms*, Interscience, New York, 909 pp.

CAVALLIN, M. (1970), Développement embryonnaire de l'appareil génital chez le phasme *Carausius morosus* Br., *Bull. biol. Fr. Belg.*, **104**, 343–366.

CHRISTENSEN, P. J. H. (1943), Embryologische und zytologische Studien über die erste und frühe Eientwicklung bei *Orgyia antiqua* Linné (Fam. Lymantriidae, Lepidoptera), *Vidensk. Meddr dansk naturh. Foren.*, **106**, 1–223.

COUNCE, S. J. (1961), The analysis of insect embryogenesis, *A. Rev. Ent.*, **6**, 295–312.

—— (1973), The causal analysis of insect embryogenesis, In: Counce, S. J. and Waddington, C. H. (1973) (eds), *Developmental Systems: Insects*, **2**, 1–156.

DAVIS, C. W. (1967–70), A comparative study of larval embryogenesis in the mosquito *Culex fatigans* Wiedemann (Diptera: Culicidae) and the sheep fly *Lucilia sericata* Meigen (Diptera: Calliphoridae). I, II, *Aust. J. Zool.*, **15**, 547–579; **18**, 125–154.

EASTHAM, L. E. S. (1930), The formation of germ layers in insects, *Biol. Rev.*, **5**, 1–29.

EWEST, A. (1937), Struktur und erste Differenzierung im Ei des Mehlkäfers *Tenebrio molitor*, *Arch. EntwMech. Org.*, **135**, 689–752.

FAROOQUI, M. M. (1963), The embryology of the mustard sawfly *Athalia proxima* Klug. (Tenthredinidae, Hymenoptera), *Aligarh Musl. Univ. Publ. (Zool.)*, **6**, 1–68.

FISH, W. A. (1947–52), Embryology of *Lucilia sericata* (Meigen) (Diptera: Calliphoridae), *Ann. ent. Soc. Am.*, **40**, 15–28; **40**, 677–687; **42**, 121–133; **45**, 1–22.

FOURNIER, B. (1967), Echelle résumée des stades du développement embryonnaire du phasme *Carausius morosus* Br., *Actes Soc. Linn. Bordeaux*, **104**, 1–30.

FUKUDA, S. (1963), Déterminisme hormonal de la diapause chez le ver à soie, *Bull. Soc. zool. Fr.*, **88**, 151–179.

GEIGY, R. (1931), Erzeugung rein imaginaler Defekte durch ultra-violette Bestrahlung bei *Drosophila melanogaster*, *Arch. EntwMech. Org.*, **125**, 406–447.

GÖRG, I. (1959), Untersuchungen am Keim von *Hierodula* (*Rhombodera*) *crassa* Giglio-Tos, ein Beitrag zur Embryologie der Mantiden (Mantodea), *Dt. ent. Z.*, **6** (N.F.), 389–450.

GOSS, R. J. (1952), The early embryology of the book louse, *Liposcelis divergens* Badonnel (Psocoptera: Liposcelidae), *J. Morph.*, **91**, 135–167.

—— (1953), The advanced embryology of the book louse, *Liposcelis divergens* Badonnel (Psocoptera: Liposcelidae), *J. Morph.*, **92**, 157–191.

GÜNTHER, J. (1971), Entwicklungsfähigkeit, Geschlechtsverhältnis und Fertilität von *Pimpla tuironellae* L. (Hymenoptera, Ichneumonidae) nach Röntgenbestrahlung oder Abschnürung des Eihinterpols, *Zool. Jb. (Anat.)*, **88**, 1–46.

HAGAN, H. R. (1951), *Embryology of the Viviparous Insects*, The Ronald Press Co., New York, 472 pp.

HALLEZ, P. (1886), Loi de l'orientation de l'embryon chez les insectes, *C.r. Acad. Sci., Paris*, **103**, 606–608.

HATHAWAY, D. S. AND SELMAN, G. G. (1961), Certain aspects of cell lineage and morphogenesis studied in embryos of *Drosophila melanogaster* with an ultra-violet micro-beam, *J. Embryol. exp. Morph.*, **9**, 310–325.

HEIDER, K. (1889), *Die Embryonalentwicklung von* Hydrophilus piceus *L.*, Fischer, Jena, 98 pp.

HENSON, H. (1946), The theoretical aspects of insect metamorphosis, *Biol. Rev.*, **21**, 1–14.

HEYMONS, R. (1895), *Die Embryonalentwicklung von Dermapteren und Orthopteren unter besonderer Berücksichtigung der Keimblätterbildung*, Fischer, Jena, 136 pp.

—— (1896), Grundzüge der Entwicklung und des Körperbaues von Odonaten und Ephemeriden, *Anh. Abl. Kgl. Akad. Wiss. Berlin*, **1896**, 1–66.

IBRAHIM, M. M. (1958), Grundzüge der Organbildung im Embryo von *Tachycines* (Insecta, Saltatoria), *Zool. Jb. (Anat.)*, **76**, 541–594.

IVANOVA-KASAS, O. M. (1959), Die embryonale Entwicklung der Blattwespe *Pontania capreae* L. (Hymenoptera, Tenthredinidae), *Zool. Jb. (Anat.)*, **77**, 193–228.

—— (1961), *Studies on the Comparative Embryology of Hymenoptera*, Univ. Leningrad, Leningrad, 266 pp.

JOHANNSEN, O. A. AND BUTT, F. H. (1941), *Embryology of Insects and Myriapods*, McGraw-Hill, New York, 462 pp.

JONES, B. M. (1956), Endocrine activity during insect embryogenesis. Control of events in development following the embryonic moult (*Locusta migratoria* and *Locustana pardalina*, Orthoptera), *J. exp. Biol.*, 33, 685–696.

JUNG, E. (1966), Untersuchungen am Ei des Speisebohnenkäfers *Bruchidius obtectus* Say (Coleoptera) I, II, *Z. Morph. Ökol. Tiere*, 56, 444–480; *Arch. EntwMech. Org.*, 157, 320–392.

JURA, C. (1965), Embryonic development of *Tetrodontophora bielanensis* (Waga) (Collembola) from oviposition till germ band formation, *Acta Biol. Cracow, Zool.*, 8, 141–157.

—— (1966), Origin of the endoderm and embryogenesis of the alimentary system in *Tetrodontophora bielanensis* (Waga) (Collembola), *Acta Biol. Cracow, Zool.*, 9, 95–102.

—— (1967), The significance and function of the primary dorsal organ in embryonic development of *Tetrodontophora bielanensis* (Waga) (Collembola), *Acta Biol. Cracow, Zool.*, 10, 301–311.

—— (1972), Development of apterygote insects, In: Counce, S. J. and Waddington, C. H., (eds), *Developmental Systems – Insects*, 1, 49–94.

KESSEL, E. L. (1939), The embryology of fleas, *Smithson. misc. Collns*, 98 (3), 78 pp.

KRAUSE, G. (1934), Analyse erste Differenzierungsprozesse im Keim der Gewächshausheuschrecke durch künstlich erzeugte Doppel-, Zwillings- und Mehrfachbildungen, *Arch. EntwMech. Org.*, 132, 115–205.

—— (1938), Einzelbeobachtungen und typische Gesamtbilder der Entwicklung von Blastoderm und Keimanlage im Ei der Gewächshausheuschrecke *Tachycines asynamorus* Adelung, *Z. Morph. Ökol. Tiere*, 34, 1–78.

—— (1938a), Die Ausbildung der Körpergrundgestalt im Ei der Gewächshausheuschrecke *Tachycines asynamorus*, *Z. Morph. Ökol. Tiere*, 34, 499–564.

—— (1939a), Die Eitypen der Insekten, *Biol. Zbl.*, 59, 495–536.

—— (1939b), Die Regulationsfähigkeit der Keimanlage von *Tachycines* (Orthoptera) im Extraovatversuch, *Arch. EntwMech. Org.*, 139, 639–723.

—— (1957), Neue Beiträge zur Entwicklungsphysiologie des Insektenkeimes, *Verh. dt. zool. Ges. (Graz)*, 1957, 396–424.

—— (1958), Induktionssysteme in der Embryonalentwicklung von Insekten, *Ergeb. Biol.*, 20, 159–198.

KÜHN, A. (1971), *Lectures on Developmental Physiology*, Springer, Berlin and New York, 2nd edn, 535 pp.

LARINK, O. (1969), Zur Entwicklungsgeschichte von *Petrobius brevistylis* (Thysanura, Insecta), *Helgoländer wiss. Meeresunters.*, 19, 111–155.

—— (1970), Die Kopfentwicklung von *Lepisma saccharina* L. (Insecta, Thysanura), *Z. Morph. Ökol. Tiere*, 67, 1–15.

LEES, A. D. (1955), *The Physiology of Diapause in Arthropods*, Cambridge Univ. Press, Cambridge, 150 pp.

LEUZINGER, H., WIESMANN, R. AND LEHMANN, F. E. (1926), *Zur Kenntnis der Anatomie und Entwicklungsgeschichte der Stabheuschrecke* Carausius morosus Br, Fischer, Jena, 414 pp.

LOUVET, J. P. (1964), La ségrégation du mésoderme chez l'embryon du Phasme *Carausius morosus* Br., *Bull. Soc. zool. Fr.*, 89, 688–701.

LÜSCHER, M. (1944), Experimentelle Untersuchungen über die larvale und die imaginale Determination im Ei der Kleidermotte (*Tineola biselliella* Humm.), *Revue suisse Zool.*, 51, 531–627.

MAHOWALD, A. P. (1972), Oogenesis, In: Counce, S. J. and Waddington, C. H. (eds), *Developmental Systems: Insects*, 1, 1–47.

MAHR, E. (1960), Normale Entwicklung, Pseudofurchung und die Bedeutung des Furchungszentrum im Ei des Heimchens (*Gryllus domesticus*), *Z. Morph. Ökol. Tiere*, 49, 263–311.

MALZACHER, P. (1968), Die Embryogenese des Gehirns paurometaboler Insekten: Untersuchungen an *Carausius morosus* und *Periplaneta americana*, *Z. Morph. Ökol. Tiere*, 62, 103–161.

MARTIN, F. (1914), Zur Entwicklungsgeschichte des polyembryonalen Chalcidiers *Ageniaspis* (*Encyrtus*) *fuscicollis*, *Z. wiss. Zool.*, 110, 419–479.

MELLANBY, H. (1935), The early embryonic development of *Rhodnius prolixus*, *Q. Jl microsc. Sci.*, 78, 71–90.

—— (1936), The later development of *Rhodnius prolixus*, *Q. Jl microsc. Sci.*, 79, 1–42.

MILLER, A. (1939), The egg and early development of the stonefly, *Pteronarcys proteus* Newman (Plecoptera), *J. Morph.*, 64, 555–609.

—— (1940), Embryonic membranes, yolk cells and morphogenesis of the stonefly *Pteronarcys proteus* Newman (Plecoptera: Pteronarcidae), *Ann. ent. Soc. Am.*, 33, 437–477.

MOLOO, S. K. (1971), The degree of determination of the early embryo of *Schistocerca gregaria* (Forskål) (Orthoptera: Acrididae), *J. Embryol. exp. Morph.*, 25 (3), 277–299.

MULNARD, J. (1947), Le développement embryonnaire d'*Acanthoscelides obtectus* Say (Col.), *Archs Biol., Liège*, 58, 289–314.

NELSON, J. A. (1915), *The Embryology of the Honey Bee*, Princeton Univ. Press, Princeton, 282 pp.

NÖTHIGER, R. AND STRUB, S. (1972), Imaginal defects after UV-microbeam irradiation of early cleavage stages of *Drosophila melanogaster*, *Revue suisse Zool.*, 79 (Suppl.), 267–279.

PFLUGFELDER, O. (1958), *Entwicklungsphysiologie der Insekten*, Akademische Verlagsgesellschaft, Leipzig, 2. Auflage, 490 pp.

PIOTROWSKI, F. (1953), The embryological development of the body louse – *Pediculus vestimenti* Nitzsch – Part 1, *Acta parasit. pol.*, 1, 61–84.

POLIVANOVA, E. N. (1965), Excretory system and excretion processes in the embryonic development of Hemiptera Pentatomoidea, *Zh. obshch. Biol.*, 26, 700–710.

POULSON, D. F. (1950), Histogenesis, organogenesis and differentiation in the embryo of *Drosophila melanogaster* Meigen, In: Demerec, M. (ed.), *The Biology of* Drosophila. New York, pp. 168–274.

REED, E. M. AND DAY, M. R. (1966), Embryonic movements during development of the Light Brown Apple Moth, *Aust. J. Zool.*, 14, 253–263.

REMPEL, J. G. AND CHURCH, N. S. (1965–1971), The embryology of *Lytta viridana* Le Conte (Coleoptera: Meloidae). I, V–VII, *Can. J. Zool.*, 43, 915–925; 47, 1157–1171; 49, 1563–1570; 49, 1571–1581.

ROONWAL, M. L. (1936–37), Studies on the embryology of the African Migratory

Locust, *Locusta migratoria migratorioides*. I, II, *Phil. Trans. R. Soc. Ser. B*, **226**, 391–421; **227**, 175–244.

SCHNETTER, M. (1934), Physiologische Untersuchungen über das Differenzierungszentrum in der Embryonalentwicklung der Honigbiene, *Arch. EntwMech. Org.*, **131**, 285–323.

—— (1934a), Morphologische Untersuchungen über das Differenzierungszentrum in der Embryonalentwicklung der Honigbiene, *Z. Morph. Ökol. Tiere*, **29**, 114–195.

SCHOLL, G. (1969), Die Embryonalentwicklung des Kopfes und Prothorax von *Carausius morosus* Br. (Insecta, Phasmida), *Z. Morph. Ökol. Tiere*, **65**, 1–142.

SEIDEL, F. (1924), Die Geschlechtsorgane in der embryonalen Entwicklung von *Pyrrhocoris apterus*, *Z. Morph. Ökol. Tiere*, **1**, 429–506.

—— (1926–29a), Die Determination der Keimanlage bei Insekten. I–III, *Biol. Zbl.*, **46**, 321–343; **48**, 230–251; **49**, 577–607.

—— (1929b), Untersuchungen über das Bildungsprinzip der Keimanlage im Ei der Libelle *Platycnemis pennipes*. I–V, *Arch. EntwMech. Org.*, **119**, 322–440.

—— (1932), Die Potenz der Furchungskerne im Libellenei und ihre Rolle bei der Aktivierung des Bildungszentrums, *Arch. EntwMech. Org.*, **126**, 213–276.

—— (1934), Das Differenzierungszentrum in Libellenkeim. I, *Arch. EntwMech. Org.*, **131**, 135–187.

—— (1935), Der Anlagenplan im Libellenei, zugleich eine Untersuchung über die allgemeinen Bedingungen für defekte Entwicklung und Regulation bei dotterreichen Eiern, *Arch. EntwMech. Or.*, **132**, 671–751.

—— (1936), Entwicklungsphysiologie des Insekten-Keims, *Verh. dt. zool. Ges.*, **38**, 291–336.

—— (1961), Entwicklungsphysiologische Zentren im Eisystem der Insekten, *Verh. dt. zool. Ges.*, **1960**, 121–142.

SEIDEL, F., BOCK, E. AND KRAUSE, G. (1940), Die Organisation des Insekteneies (Reaktionsablauf, Induktionsvorgänge, Eitypen), *Naturwissenschaften*, **28**, 433–446.

SHAFIQ, S. A. (1954), A study of the embryonic development of the gooseberry sawfly *Pteronidea ribesii*, *Q. Jl microsc. Sci.*, **95**, 93–114.

SHAROV, A. G. (1953), Development of bristle tails (Thysanura, Apterygota) in connection with the problem of insect phylogeny, *Trud. Inst. morf. zhivot.*, **8**, 63–127.

SLIFER, E. H. (1937), The origin and fate of the membranes surrounding the grasshopper egg; together with some experiments on the source of the hatching enzyme, *Q. Jl microsc. Sci.*, **79**, 493–506.

SONNENBLICK, B. P. (1950), The early embryology of *Drosophila melanogaster*, In: Demerec, M. (ed.), *The Biology of Drosophila*, Wiley, New York, pp. 62–167.

STRIEBEL, H. (1960), Zur Embryonalentwicklung der Termiten, *Acta trop.*, **17**, 193–260.

TIEGS, O. W. (1942a), The dorsal organ of Collembolan embryos, *Q. Jl microsc. Sci.*, **83**, 153–170.

—— (1942b), The dorsal organ of the embryo of *Campodea*, *Q. Jl microsc. Sci.*, **83**, 35–47.

TIEGS, O. W. AND MURRAY, F. V. (1938), The embryonic development of *Calandra oryzae*, *Q. Jl microsc. Sci.*, **80**, 159–284.

ULLMANN, S. L. (1964), The origin and structure of the mesoderm and the formation of the coelomic sacs in *Tenebrio molitor* L. (Insecta, Coleoptera), *Phil. Trans. R. Soc. Ser. B*, **248**, 245–277.

—— (1967), The development of the nervous system and other ectodermal derivatives in *Tenebrio molitor* L. (Insecta, Coleoptera), *Phil. Trans. R. Soc. Ser. B*, **252**, 1–25.

WALL, C. (1973), Embryonic development in two species of *Chesias* (Lepidoptera: Geometridae), *J. Zool.*, **169**, 65–84.

—— (1974), Effect of temperature on embryonic development and diapause in *Chesias legatella* (Lepidoptera: Geometridae), *J. Zool.*, **172**, 147–168.

WAY, M. J. (1959), The effect of temperature, particularly during diapause, on the development of the egg of *Leptohylemyia coarctata* Fallén (Diptera: Muscidae), *Trans. R. ent. Soc. Lond.*, **111**, 351–364.

WELLHOUSE, W. T. (1953), The embryology of *Thermobia domestica* Packard (Thysanura), *Iowa State Coll. J. Sci.*, **28**, 416–417.

WHEELER, W. M. (1893), A contribution to insect embryology, *J. Morph.*, **8**, 1–160.

WIESMANN, R. (1926), In: Leuzinger, H., Wiesmann, R. and Lehmann, F. E. (eds), *Zur Kenntnis der Anatomie und Entwicklungsgeschichte der Stabheuschrecke* Carausius morosus *Br.*, Fischer, Jena, 414 pp.

WOODLAND, J. T. (1957), A contribution to our knowledge of Lepismatid development, *J. Morph.*, **101**, 523–577.

WRIGHT, T. R. F. (1960), The phaenogenetics of the embryonic mutant, *lethal myospheroid*, in *Drosophila melanogaster*, *J. exp. Zool.*, **143**, 77–99.

YAJIMA, H. (1960, 1964), Studies on embryonic determination of the Harlequin Fly, *Chironomus dorsalis*. I, II, *J. Embryol. exp. Morph.*, **8**, 198–215; **12**, 89–100.

Chapter 19

POSTEMBRYONIC DEVELOPMENT

Metamorphosis

Eclosion from the Egg – The process of eclosion or hatching from the egg varies greatly in the different groups but usually begins with the swallowing of amniotic liquid so that air enters the egg (Sikes and Wigglesworth, 1931). The chorion and other embryonic membranes are then ruptured, the former either along a preformed line of weakness, such as surrounds the cap of some eggs, or in a more irregular manner. Lepidopteran larvae eat their way through the membranes but other insects exert pressure through muscle contractions. The forces involved may be aided by the swallowing of air or the presence of special structures such as the blood-filled eversible *cervical ampullae* of some Orthoptera or the various types of *hatching spines* (egg-bursters). When present, the egg-bursters (van Emden, 1946) are solid tooth-like or spine-like cuticular outgrowths. They may be borne on the frontal region of the embryonic cuticle which is shed at or soon after eclosion (most Exopterygotes, Neuroptera, Trichoptera). Alternatively, egg-bursters occur on the cuticle of the normal first-stage larva and are lost only when this moults (Lepismatidae, Nematoceran Diptera, Siphonaptera, Adephagan Coleoptera). Or again, the cuticle of the first-stage larva bears variously situated dorsal structures on the thorax or abdomen or both (many Polyphagan Coleoptera). In general, the egg-burster ruptures the chorion and embryonic membranes but in the Mallophaga and Siphunculata it is used only to break the inner membranes, the chorion splitting around a cap (Weber, 1939). The chorion of the uterine egg of *Glossina* is broken by the larva but then stripped off by a maternal organ, the choriothete (Jackson, 1948). Some insects hatch with the tracheal system still full of liquid, but in others the rise of osmotic pressure in the tissue fluids caused by muscular activity results in a withdrawal of the liquid and the appearance of air in the tracheae before hatching.

Developmental Stages – The postembryonic life of an insect is divided by moulting of the cuticle into successive developmental stages or *instars*, which may differ more or less extensively from one another. When an insect

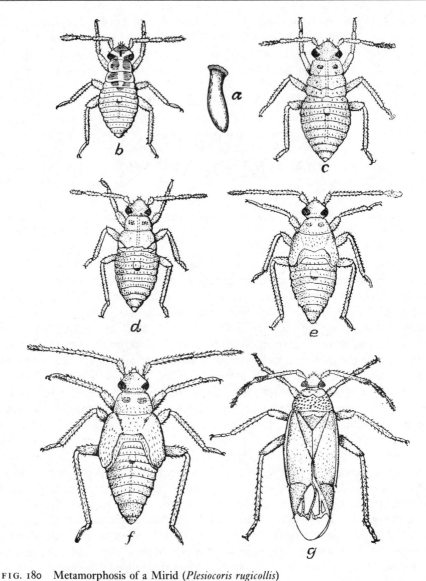

FIG. 180   Metamorphosis of a Mirid (*Plesiocoris rugicollis*)

a, egg; b–f, larval instars (wing-rudiments minute in d, larger in e and f) × 20; g, imago ×
From Carpenter, *after* Petherbridge and Husain.

emerges from the egg it is said to be in its first instar. The exact point at
which each successive instar becomes recognizable has been debated and
depends on a detailed analysis of the sequence of events that culminate in the
shedding of the cuticle. This is discussed further on p. 363 but it is possible
to distinguish between the separation of the old cuticle from the epidermis –
a process called *apolysis* by Jenkin and Hinton (1966) – and its final rupture
and shedding, which constitutes *ecdysis*. Each instar can then be defined as

beginning with apolysis, and during the interval between apolysis and ecdysis (which may be quite prolonged) the new stage lying within the old, unshed cuticle is referred to as a pharate instar (Hinton, 1958, 1971). In the majority of insects the final instar is the fully mature form, capable of reproduction and known as the adult or *imago*. In most Exopterygotes, the Neuroptera and Trichoptera, the insect that emerges from the egg is enclosed in the so-called embryonic cuticle which is either shed during eclosion so that it remains behind in the egg-shell or is cast shortly after hatching is complete, the insect bearing it being variously known as the pronymph (Odonata), vermiform larva (Acrididae) or primary larva (Cicadidae). There seems little doubt that this stage represents a greatly abbreviated first instar (van Emden, 1946; Bullière, 1973) though it is usually excluded from the system of numbering the instars in life-cycle studies. In *Carausius* three cuticles are secreted by the embryo, which sheds the first two when it hatches (Louvet, 1974).

Metamorphosis – One of the most characteristic features of insects is

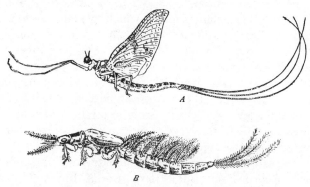

FIG. 181   Metamorphosis of *Ephemera*
A, male imago. B, larva. From Comstock *after* Needham.

the fact that they almost always hatch in a condition morphologically different from that of the imago. In order to reach the latter instar they consequently have to pass through changes of form which are collectively termed *metamorphosis*; these are usually most pronounced towards the end of postembryonic development and they are accompanied by physiological and biochemical changes (Chen, 1971; L'Helias, 1970).

Some insects emerge from the egg in a form which differs from the imago only in the undeveloped state of the reproductive organs and external genitalia and in morphologically unimportant details of shape or chaetotaxy and the segmentation of antennae and cerci. Such insects include the Apterygotes and secondarily apterous Exopterygotes like the Mallophaga, Siphunculata and female Embioptera. They are often regarded as having no metamorphosis, but the changes mentioned above are generally sufficient to constitute a slight metamorphosis. The majority of insects, however, pass through a rather more profound metamorphosis, on the basis of which they may be divided into two main groups, the Hemimetabola and the Holometabola (Snodgrass, 1954). More elaborate classifications of metamorphosis entail the subdivision of these major groups (Weber, 1933; Fischer, 1968).

The hemimetabolous insects consist of most of the Exopterygotes and pass through a simple metamorphosis, often described as *direct* or *incomplete*. The immature stages differ from the adult mainly in that the wings and genitalia are present only in an incompletely developed condition. Wing rudiments are not usually discernible in the first instar but they later become visible as external wing-pads, which gradually increase in size with each moult. The mouthparts exhibit the same general form as in the adult, compound eyes are almost always present, and the habits of the young and the adults are often alike. There is no pupal instar and though the young stages are sometimes referred to as nymphs, thus emphasizing the differences between them and the immature holometabolous insects, the modern tendency is to call them *larvae* (Hinton, 1948; Wigglesworth, 1954; and cf. Davies, 1958). The degree of metamorphosis which prevails in hemimetabolous insects varies among the different orders. Where the young resemble the adults in general form and mode of life (Fig. 180), postembryonic development is gradual and involves mainly the acquisition of wings and genitalia. In the Ephemeroptera (Fig. 181), Odonata and Plecoptera, however, the aquatic larvae differ from the terrestrial adults in possessing adaptive provisional organs concerned chiefly with respiration, feeding and locomotion. The tracheal gills of the larva are lost or, as in the Plecoptera, persist only as non-functional vestiges in the adult, which acquires open spiracles. The highly specialized labium of the Odonate larva undergoes extensive remodelling to form the adult structure (Munscheid, 1933). The Ephemeroptera are unique in that the last immature stage, known as the subimago, possesses functional wings; this condition may represent an evolutionary vestige of the multiple adult instars thought to have occurred in the ancestral Pterygotes (p. 485).

The holometabolous insects include the Endopterygota and pass through an *indirect* or *complete* metamorphosis. The life-cycle (Fig. 182) incorporates a sequence of larval instars, successive larvae usually resembling each other closely but differing considerably from the adult. Their mouthparts and feeding habits are usually unlike those of the adult, their eyes very rarely consist of more than a few simple ocelli, and their wings are represented at most by internal rudiments that may be sunk in epidermal sacs beneath the general body surface (p. 58). The last larval instar is followed by a more or less quiescent pupal instar, within whose cuticle most of the external and internal transformation to the adult occurs. Among the Exopterygotes, the Aleyrodoidea, male Coccoidea and Thysanoptera have each evolved independently a metamorphosis not unlike that of the holometabolous Endopterygota. Their larvae are without external wing rudiments and differ more or less in form and habits from the adults. In the male Coccoidea and the Thysanoptera there are two or three pupa-like instars while the Aleyrodoidea transform directly from the fourth-instar larva to the adult without a separate pupal stage.

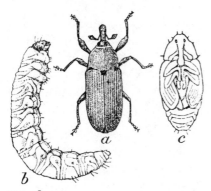

FIG. 182
Metamorphosis of a weevil (*Trichobaris trinotata*)

a, imago; *b*, larva; *c*, pupa. *After* Chittenden, *U.S. Dept. Agric. ent. Bull.*, 33 (n.s.).

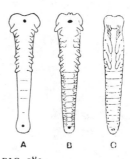

FIG. 183
Embryonic phases

A, protopod. B, polypod. C, oligopod. *After* Berlese, *Redia*, 9.

Endopterygote Larvae – The larvae of Endopterygote insects assume an immense variety of forms, many of which are clearly adaptive and are discussed in detail in the systematic sections of this work (see also Peterson, 1948–51; Paulian, 1950; and Hayes, 1941). Several anatomical classifications of larvae are possible but the larger orders contain several different types of larvae so that transitional forms are not uncommon. In addition, special descriptive terms are widely used for the larvae of particular groups such as the 14 types of first-instar larvae recognized in the parasitic Hymenoptera by Clausen (1940), some of which are mentioned below and on p. 1197. Berlese (1913) based a classification of insect larvae on the resemblances which they show to the three phases – protopod, polypod and oligopod – which sometimes succeed each other in embryonic development (Fig. 183; see also p. 333). He considered that, in general, the development of Endopterygote insects is temporarily arrested in one or other of these phases when eclosion from the egg takes place so that the larva is, so to speak, a protracted, free-living embryo. This is an unacceptable theory (Hinton, 1955) but Berlese's terms can be used descriptively to help in classifying the larvae. In the classification which follows, a fourth, apodous type of larva is also recognized and the oligopod and apodous types are respectively subdivided on the basis of their general facies or degree of cephalic development (see also Kruel, 1962).

1. *Protopod Larvae.* This type is found in the primary larvae of certain parasitic Hymenoptera. The eggs of such species contain little yolk and the larvae emerge in an early embryonic phase. Their survival is possible because they occur in the eggs or bodies of other insects where they develop

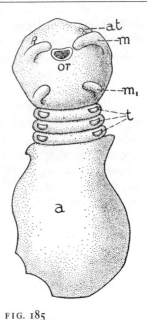

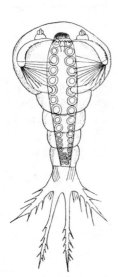

FIG. 184
First larval instar
(protopod) of
Platygaster
After Ganin.

FIG. 185
Protopod 1st instar larva of
Platygaster herrickii

a, abdomen; *at*, antenna; *m*,
mandible; m_1, maxilla; *or*, oral
aperture; *t*, thoracic appendages.
After Kulagin.

immersed in a highly nutritive medium. Protopod larvae are characteristic of
the Platygasteridae (Fig. 185) where, in *P. herrickii*, the larva is little more
than a prematurely hatched embryo, devoid of segmentation in the abdomen
and with rudimentary cephalic and thoracic appendages. The nervous and
respiratory systems are undeveloped, and the digestive organs are still largely
embryonic. In the cyclopoid primary larva of other species of *Platygaster* (Fig.
184) and of *Synopeas* there is a greatly developed cephalothorax, powerful jaws
and elaborate caudal outgrowths. The so-called eucoiliform larva of Figitidae is
also a specialized protopod type with greatly developed thoracic appendages.
Other examples include the first instars of Dryinidae and Scelionidae.

2. *Polypod Larvae*. Typical examples of this type (Fig. 186) are the so-
called *eruciform larvae* of most Lepidoptera, sawflies and scorpion flies.

FIG. 186
Eruciform (polypod) larva of *Pieris
brassicae*

Their essential features are the well-defined segmentation, the presence of abdominal limbs or prolegs (whose serial homology with the thoracic ones is, however, uncertain) and a peripneustic tracheal system. The antennae and thoracic legs are present but little developed, and such larvae are relatively inactive, living in proximity to their food. The existence of a somewhat reduced polypod instar among endoparasitic Hymenopterous larvae is recorded in the Figitidae where there are ten pairs of trunk appendages (James, 1928); in *Ibalia* where there are twelve pairs (Chrystal, 1930) and in the Proctotrupid *Phaenoserphus* where there are eight pairs of such organs (Eastham, 1929). In the Figitidae the polypod instar follows the protopod stage but, in the other examples, it is the first-stage larva. Owing to their endoparasitic mode of life the spiracles are undeveloped in these examples.

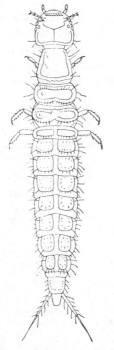

FIG. 187
Campodeiform
(oligopod) larva of a
Staphylinid (*Philonthus nitidus*)
After Schiödte.

3. *Oligopod Larvae.* These are characterized by the presence of more or less well-developed thoracic legs and the absence of abdominal appendages except sometimes for a pair of urogmphi or similar caudal processes. The head-capsule and its appendages are usually well developed but otherwise there is appreciable variation in their general appearance though two common types can be distinguished: (*a*) *Campodeiform larvae* (Figs. 187, 188A) are so called from their resemblance to *Campodea* and typically possess a long, more or less fusiform, somewhat depressed body which is often well sclerotized, a markedly prognathous head, long thoracic legs and usually a pair of terminal abdominal processes. They are generally active predators with well developed sensory equipment and are commonly regarded as the

most primitive Endopterygote larvae, differing mainly from those of exopterygote insects in the absence of compound eyes, dorsal ocelli and wing
pads. Varying degrees of reduction occur and in some hypermetamorphic
forms a campodeiform first instar is succeeded by more degenerate oligopod
stages (e.g. Meloidae, Rhipiphoridae, *Aleochara*, etc., Fig. 188). Campodeiform larvae occur in the Neuroptera, some Coleoptera (especially
Adephaga), Strepsiptera and Trichoptera. (*b*) *Scarabaeiform larvae* (Fig. 188,
C) are stout subcylindrical, C-shaped larvae with shorter thoracic legs, a
soft, fleshy body and no caudal processes. They are typical of the
Scarabaeoidea but also occur, sometimes in a more reduced condition, in
other Coleopteran families (e.g. Ptinidae, Anobiidae) and are evidently
derived from the campodeiform type, leading a less active life in the
presence of ample supplies of food.

    4. *Apodous Larvae.* In this type the trunk appendages are completely
absent and most cases are probably derived from the oligopod type. Among
Coleoptera the apodous condition occurs in several families and in the
Bruchidae it is preceded by a degenerate oligopod instar, while in the
Curculionidae the last rudiments of thoracic limbs are retained as sensory
protuberances in *Phytonomus* (Pérez, 1911). In the Hymenoptera the
apodous condition occurs in almost all Apocritan larvae – traces of an
oligopod stage have been detected in *Polysphincta* and certain other
Ichneumonoidea and a few polypod types are mentioned above. In the
Symphyta a long series of transitional forms occur from typical polypod
types to oligopod forms, finally culminating in the apodous larva of *Orussus*.
In Diptera the larvae are usually apodous but bear three pairs of sensory
papillae occupying the positions of ancestral thoracic limbs, of which they are
perhaps the vestiges (Keilin, 1915; Peters, 1961). Apodous Dipteran larvae
are, therefore, probably to be regarded as highly specialized derivatives of
the oligopod type. Polypod larvae are rare in the Diptera, but secondarily
evolved prolegs occur in *Atherix* and a polypod condition is also recalled by
the creeping welts of other species (Hinton, 1955). In general, apodous
larvae may be divided into three types depending on the degree of development of the head. The *eucephalous* larvae (most Nematocera; Buprestidae,
Cerambycidae; Aculeate Hymenoptera) have a more or less well sclerotized
head-capsule with relatively little reduction of the cephalic appendages. In
the *hemicephalous* larvae (Tipulidae, most Brachycera) there is appreciable
reduction of the head-capsule and its appendages accompanied by marked
retraction of the head into the thorax while the acephalous larvae of
Cyclorrhaphan Diptera have no obvious head-capsule though some cephalic
structures contribute to the largely concealed buccopharyngeal apparatus
and the mouth-hooks which it bears (p. 972).

    Hypermetamorphosis – When a developing insect passes through two
or more markedly different larval instars it is said to undergo *hypermetamorphosis* (Fig. 188). This phenomenon is usually accompanied by a marked
change in habits. In most cases of hypermetamorphosis the first larval instar

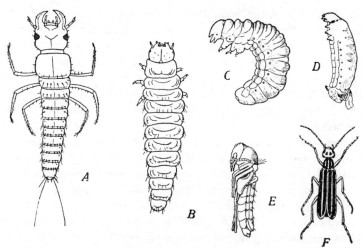

FIG. 188   Hypermetamorphosis of *Epicauta*

A, triungulin; B, caraboid second instar; C, ultimate form of second instar; D, coarctate larva; E, pupa; F, imago. All refer to *E. vittata* except E, which is *cinerea* (F nat. size, others enlarged). From Folsom's *Entomology. After* Riley.

is campodeiform. During this stage it seeks out its future food and, having discovered it, undergoes subsequent morphological transformations which adapt it to the changed mode of life. Since various examples of hyper-metamorphosis are described under the several families concerned, only the principal instances are enumerated below.

NEUROPTERA. Mantispidae.

COLEOPTERA. Carabidae (*Lebia scapularis*). Staphylinidae (*Aleochara* spp.). Meloidae. Rhipiphoridae.

STREPSIPTERA. All species.

DIPTERA. Bombyliidae. Acroceridae. Nemestrinidae. Some Tachinidae.

HYMENOPTERA. In all the chief groups of Parasitica.

LEPIDOPTERA. Gracilariidae.

Lesser degrees of morphological and biological change, hardly pronounced enough to constitute hypermetamorphosis, are not uncommon in many other groups.

Moulting of the Cuticle – The more or less rigid integument cannot easily accommodate itself to the increasing size of a growing insect and must therefore be shed and renewed periodically. At each moult there is cast off not only the general cuticle that invests the body and its appendages exter-nally, but also the various endoskeletal structures and the intima or lining of most of the tracheal system, fore and hind gut, ectodermal glands and efferent reproductive ducts. All these, together with hairs, scales and cuticular sensilla, are renewed by the underlying epidermal cells.

The histological and other changes that accompany moulting are sum-marized by Wigglesworth (1972), Locke (1974) and in works cited on pp.

385–394. The insect which is about to moult becomes less active, its epidermal cells enlarge, and apolysis takes place: the cuticle becomes detached through retraction of the epidermis and, apparently, withdrawal of the cytoplasmic filaments in the pore canals. An 'ecdysial membrane', formed from the innermost layer of procuticle, may also become visible at about this time (Taylor and Richards, A. G., 1965). Mitoses occur and the first new epicuticular material, the cuticulin layer (p. 12), is laid down by the epidermal cells and supplemented by the secretion of paraffins from the oenocytes (e.g. Locke, 1969b; Diehl, 1973). The epicuticle becomes folded under the control of the epidermal cells (Wigglesworth, 1973a) and it continues to be deposited while development of the new procuticle is taking place. This latter process occurs through secretion around cytoplasmic filaments that later fill the pore canals and by direct transformation of epidermal cell material (e.g. Condoulis and Locke, 1966). It may also involve the transport of unaltered haemolymph proteins through the epidermis and their incorporation into the cuticle (Koeppe and Gilbert, 1973). Soon after the secretion of new cuticle has begun, the space between old and new cuticles becomes filled with the moulting fluid, also secreted by the epidermis (Passonneau and Williams, 1953; Katzenellenbogen and Kafatos, 1970, 1971). The moulting fluid contains proteases and chitinases and so dissolves the old endocuticle. Eventually the fluid and the products of endocuticle digestion are resorbed by the epidermis. The old cuticle, reduced to its undigested exo- and epicuticle, now splits along preformed lines of weakness over the head and thorax (Henriksen, 1932; Snodgrass, 1947) as the insect expands the anterior part of its body by swallowing air or water and contracting the abdominal muscles (Cottrell, 1964). Ecdysis, the shedding of the old cuticle, is then completed by the insect extricating itself from the remains of its former integument, aided by cuticular spines or processes (e.g. Bernays, 1972a). Ecdysis is followed by a further increase in cuticular thickness and sclerotization and by the development of full pigmentation. Many of the changes of form which become evident at a moult are due to differential growth of the epidermis but tonic muscular contractions probably play some role, especially in the formation of apodemes.

In some insects the old cuticle is not ruptured but remains as a protective covering for the new instar, as in the puparium of Cyclorrhaphan Diptera, the larval sac of Dryinids and a few other cases. In other instances the successive shrivelled exuviae are retained at the hind end of the body (e.g. the larvae of some parasitic Hymenoptera or Cassidine beetles) or are incorporated into the 'scale' of some Coccoidea. For information on the hormonal control of moulting, see p. 383.

Different insect species differ greatly in the number of moults which they undergo in the course of their life, and details are given in the appropriate systematic sections. Though little phylogenetic significance is to be attached to the number of moults it is broadly true that specialized forms have fewer instars than do more primitive ones. Among the Apterygota, *Ctenolepisma* moults 14 times before reaching a mature state and then continues moulting at intervals without change of form until it dies. Similar moults occur after sexual maturity in *Campodea* and the Collembola but in no Pterygote insects do the adults ever moult. The larvae of Ephemeroptera moult more than 20 times whereas most Plannipennian Neuroptera undergo 4 ecdyses in development (or 5 if one includes the throwing off of the embryonic cuticle at eclosion). Most other insects fall between these extremes. In some groups the number of moults is remarkably constant (e.g. 5 in the Heteroptera, 6 in the Psocoptera, both excluding the embryonic moult) while the Lepidoptera, for example, have from 2 to 9 larval instars according to species, with not infrequent variations between individuals of the same species. Under adverse nutritional conditions the number of larval instars in some Coleoptera and Lepidoptera can be greatly increased, though there is little growth in size and sometimes even a decrease, as in the retrogressive moults of *Trogoderma* (Beck, 1973). In several species the sexes differ in the number of moults, the female commonly having more instars.

Definition of Instars – As mentioned above, Hinton (1958*a*, 1971) argues that the instar should be defined to begin with the separation of the epidermis from the cuticle (apolysis). The pharate instar then frees itself from the old cuticle (ecdysis) after a variable length of time, during the first part of which its epidermis lacks the new cuticle. This is the short *exuvial* phase and it is followed by the secretion of the new cuticle so that the instar enters a *cuticular* phase that lasts for the remainder of the stage. In most larval moults the pharate condition occupies a very short part of the instar, but in some cases it is more prolonged and this can lead to a misinterpretation of the instars unless the times of apolysis and ecdysis are carefully distinguished. Thus, the pharate pupa and pharate adult are usually relatively long phases and in a Tipulid cited by Hinton, for example, the larval-pupal ecdysis occurs at the same time as the pupal-adult apolysis. The instar that emerges from the last larval cuticle in this case is therefore a pharate adult and the non-pharate pupa has been eliminated from the life-cycle, as it has also in the Cyclorrhapha where both larval and pupal cuticles are shed simultaneously (but see p. 367). Again, in the Gracillariid moth *Mamara* and in some Meloid beetles the fourth and fifth instars respectively are entirely pharate. It remains true, of course, that the pupal-adult ecdysis is accompanied by far greater changes in the behaviour and external appearance of the insect than is the corresponding apolysis. There are also variations in the time of apolysis in different parts of the body and often appreciable changes in the external form of the insect during the pharate pupa and pharate adult. In these cases, and where the pharate phase is very short, the older Linnaean definition of the instar, starting with ecdysis, may still retain some convenience (Wigglesworth, 1973*b*, but cf. Hinton, 1973).

Growth – The larval or nymphal period is pre-eminently one of growth and in this respect the insect larva differs from that of almost all other invertebrates, in which most of the growth follows metamorphosis. The rapidity with which growth takes place, and the great increase in size which accompanies it, are particularly evident in many holometabolous insects. A comparison of the weight of a mature larva with that at the time of eclosion from the egg has been made in several species. Thus Trouvelot found that the silkworm *Telea polyphemus* when fully grown is 4140 times its original weight. In the larva of the bee *Anthophora retusa* the corresponding increase is 1020 times (Newport); in the larva of *Cossus cossus* which lives for three years, it is 72 000 times (Lyonnet); in *Sphinx ligustri*, 9976 times (Newport); in the silkworm *Bombyx mori* the increase varies according to racial and other factors between about 9100 times and 10 500 times. Growth in weight is not, however, continuous but is interrupted at each ecdysis, though the habit of swallowing water in aquatic forms in order to split the old cuticle may obscure the interruption on a growth curve.

Growth of the tissues occurs partly through cell-division and partly by an increase in the size of individual cells, the latter process predominating in the brain and gut of *Popillia japonica* larvae, for example (Abercrombie, 1936). In *Aedes* larvae, Trager (1937) finds that some tissues, which are destroyed at pupation, grow by an increase in cell size whereas others, which persist into the adult stage, grow by cell-multiplication, and in *Calliphora*, where destruction of larval tissues at metamorphosis is very extensive, larvae develop only by growth of the cells without division.

Dyar (1890) showed from observations on 28 species of Lepidopteran larvae that the width of the head-capsule increases in a regular geometrical progression in successive instars by a ratio of about 1·4 and the same is true for linear measurements of many cuticular structures in widely different insects (Teissier, 1936). Dyar's 'law' has consequently often been used successfully to determine the number of instars in a life-cycle, but there are many exceptions to it, the progression factor often changing in successive instars (Gaines and Campbell, 1935; Ludwig and Abercrombie, 1940). Richards, O. W. (1949) found that in his examples the progression is regular only if account is taken of the fact that the different instars differ in duration, growth apparently proceeding in a uniform manner so that the longer the instar the greater the amount of growth, but other cases do not follow this rule.

The various parts of the body tend to grow at rates which differ from each other and from the growth-rate of the whole body. In many cases, the phenomenon of allometric growth occurs (Huxley, 1932) in which the size of a part is a power function of the whole. This type of differential growth serves to explain such features as the relatively larger mandibles in male Lucanidae of greater absolute size, but it is widespread in many less striking features and seems to account for much of the variation that makes up the general pattern of an insect's growth (Brown and Davies, 1972; Davies and Brown, 1972). It also follows that taxonomic characters dependent on simple statements of the relative size of a structure may be unreliable if allometric growth occurs in the parts concerned (e.g. Boratynski, 1952).

The Pupa – The term *pupa* denotes the relatively inactive instar that intervenes between the larva and adult in Endopterygote insects, though it may be extended to include comparable stages in the life-cycle of aberrant holometabolous Exopterygotes like the Thysanoptera and male Coccoidea. During the pupal stage the insect is usually incapable of feeding and is quiescent. It is to be regarded as a transitional instar during which the larval body and its internal organs are remodelled to the extent necessary to produce the future imago though some of these processes begin in the larva, continue into the pharate adult and may not be complete until after the pupal-adult ecdysis. As often used, the term refers to the phase between the

FIG. 189
Exarate or free pupa
of a Hymenopteran
(Ichneumonidae):
lateral view

larval-pupal and pupal-adult ecdyses, but it must now be recognized that for a varying period before adult emergence the 'pupa' so defined is actually a pharate adult (p. 363) – i.e. the more or less fully formed adult lies separate within the old pupal cuticle and the limited degree of locomotion which such 'pupae' undertake is usually the result of movements by the pharate adult.

Among some Neuroptera including *Raphidia*, *Hemerobius* and *Chrysopa* the pharate adults are relatively active and are able to crawl about. Those of certain Trichoptera exhibit adaptive modifications which enable them to swim to the surface of the water to allow the adults to escape. In the Culicidae and certain Chironomidae the pupae are active throughout the instar, and can swim by vigorous caudal movements. Movements of a less pronounced character are made by many pupae which occur in the soil, in wood or in stems.

The following types of Endopterygote pupae are distinguished by Hinton (1946), the primary division reflecting the way in which the pharate adult escapes from the cocoon or pupal cell before emergence of the adult occurs (see also p. 368).

1. *Decticous Pupae* (Fig. 540). These have relatively powerful, sclerotized, articulated mandibles which are actuated by the musculature of the pharate adult when they are used to escape from the cocoon or cell. This is the primitive type of pupa and is always exarate – i.e. the appendages are not adherent to the rest of the body and can be used in locomotion. It occurs in the more primitive Endopterygotes – Neuroptera, Mecoptera, most Trichoptera, in the Lepidopteran families Micropterigidae and Eriocraniidae and in the Xyelidae (Hymenoptera).

2. *Adecticous Pupae*. Pupae of this type have non-articulated mandibles which are often reduced and are not used in escaping from the cocoon or cell when this is present. Two main forms of adecticous pupae are recognizable, though intermediates are known.

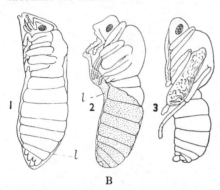

FIG. 190

External changes during pupal instar of *Formica rufa*

1, pharate pupa; 2, emergence of pupa; 3, pupa; *l*, larval skin. *After* C. Pérez, 1902.

(*a*) *Exarate adecticous pupae*. As in the exarate pupae of the other main group, the appendages are free of any secondary attachment to the body (Fig. 189). They are found in the Siphonaptera and Strepsiptera, most Coleoptera and Hymenoptera, and the Cyclorrhaphan and a few Brachyceran Diptera.

(*b*) *Obtect adecticous pupae*. Here the appendages of the pupa are firmly pressed against its body and are soldered down to it by a secretion produced at the last larval moult; the exposed surfaces of the appendages are more heavily cuticularized than those adjacent to the body. Such pupae are best known from all higher Lepidoptera (Fig. 191) but also occur in some Coleoptera, the Nematoceran and most Brachyceran Diptera and in many Chalcidoids.

The so-called coarctate pupa of Cyclorrhaphan Diptera (Fig. 192) is clearly an adecticous, exarate pupa enclosed in a puparium which is formed from a preceding larval cuticle and which is functionally comparable to the cocoon or pupal cell of other forms.

The Prepupa – This term is used in more than one sense but it refers in most cases to the pharate pupa. Near the end of larval life the insect prepares for transformation into the pupa, usually constructing a cocoon, cell or other form of protection. Meanwhile the larval cuticle becomes separated from its epidermis and the developing pupa, with everted wings and appendages, lies within the unshed larval cuticle. The changes in form which accompany the development of the pupa often cause the deformation of this larval cuticle so that the prepupal phase differs somewhat in external appearance from the earlier larva. The body is commonly more depressed and the abdomen relatively shorter. The insect at this time is usually quiescent and does not

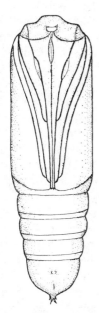

FIG. 191
Obtect pupa of a
Lepidopteron
(Noctuidae)
ventral view

FIG. 192
Puparium of a
Dipteron
(Muscidae):
dorsal view

feed, though the pharate pupa of *Simulium* feeds and spins its cocoon (Hinton, 1958*b*). The prepupal phase does not, however, constitute a distinct instar and when the old larval skin is ultimately cast off it reveals the pupa (Fig. 190).

In the Cyclorrhaphan Diptera conditions are somewhat different and it is convenient here to indicate an appropriate terminology for the successive phases of metamorphosis in these insects (Fraenkel and Bhaskaran, 1973). The third-instar larva eventually stops feeding and from then until the onset of puparium formation (pupariation) it may be referred to as the post-feeding larva. Between pupariation and the larval-pupal apolysis the third-instar larva is called a prepupa; it is not a separate instar and it gives rise, after the larval-pupal apolysis, to a cryptocephalic pupa. The latter then undergoes marked changes of form, through evagination of the head though without moulting, to form a phanerocephalic pupa, after which the pupal-adult apolysis leads to the pharate adult. This, in turn, finally emerges simultaneously from the pupal and persistent third-instar cuticles. Each of the two apolyses occur over a period of some hours, the larval-pupal apolysis beginning anteriorly while the pupal-adult one starts posteriorly. Both cryptocephalic and phanerocephalic phases of the pupal instar are technically pharate, though the phanerocephalic phase is in many ways more like a nonpharate pupa enclosed in a cocoon.

Exceptionally among Endopterygotes there is an anatomically very distinct instar which precedes the pupa, e.g. of the hypermetamorphic Carabid *Lebia scapularis*, and this is known as the prepupa (Silvestri, 1904). Finally, in male Coccoidea and in Thysanoptera the first pupa-like instar in which external wing-pads appear is sometimes known as the prepupa or propupa.

Pupal Protection – During transformation into the pupa and throughout the pupal and pharate adult instars insects are particularly vulnerable. Since at these periods they are provided with very limited powers of movement and defence, special methods of protection are necessary. Most pupae are concealed in one way or another from their enemies, and also from such adverse influences as excess of moisture, sudden marked variations of temperature, shock and other mechanical disturbances. Provision against such influences is usually made by the larva in its last instar. Many Lepidopteran and Coleopteran larvae burrow beneath the ground and there construct earthen cells in which to pupate. Most insects, however, construct cocoons which are special envelopes formed either of silk alone, or of extraneous material bound together by silken threads. Thus many wood-boring larvae utilize wood-chips; larvae which transform in the ground select earth-particles; many Arctiid larvae use their body-hairs, and some Trichoptera use pebbles, vegetable fragments, etc., their larval cases functioning as cocoons. In these instances the materials are held together by a warp of silk and worked up to form cocoons. Many other insects, including some Neuroptera and Trichoptera, many Lepidoptera and Hymenoptera and the Siphonaptera, use only silk in forming their cocoons. Great variations exist in the colour and nature of the silk and in the texture and form of the completed cocoons. Among Lepidoptera the densest and most perfect types of cocoon are found in the Saturniidae, while the other extreme is met with in the Papilionoidea, where the pupa may be suspended by its caudal extremity, which is hooked on to a small pad of silk representing the last vestige of a cocoon. Among the Tenthredinidae cocoons of a parchment-like or shell-like consistency are frequent; in some cases the outer cocoon encloses an inner one of more delicate texture. In the Diptera Cyclorrhapha a cocoon is almost always wanting, and the hard puparium forms the sole protection to the pupa.

Some specialized Lepidoptera, Coleoptera, Diptera and Hymenoptera have pupae which are not protected by a cell, cocoon or puparium though they may be concealed in the integument of the host (Chalcidoidea) or are strongly sclerotized, obtect structures which are sometimes protectively coloured.

Emergence from the Cocoon – Escape from the cocoon or cell is effected in a variety of ways, many of which are described by Hinton (1946; 1949). Most of the more primitive Endopterygotes escape as the pharate adult, the mandibles of their decticous pupae, aided sometimes by backwardly directed spines on the body, enabling them to cut open the cocoon or cell and burrow through the soil, rotten wood, etc., in which it lay. The pupal cuticle is then shed some distance from the cocoon. Insects with adecticous pupae have evolved many different methods of escape, especially in the Lepidoptera (p. 1100) and the Diptera. In the latter order, the primitive method of escape is probably one in which the pupa, aided sometimes by body spines and sharp cephalic processes or cocoon-cutters, makes its way to the surface of the pupation medium from which its body projects while the adult escapes. Other special methods occur in forms with aquatic pupae and in the Cyclorrhapha where the newly emerged adult has an eversible cephalic sac, the ptilinum, to force open the puparium and escape from the surrounding medium (p. 953). In the Siphonaptera, Coleoptera, male Strepsiptera and Hymenoptera the adult emerges within the cocoon or cell and escapes by using its mandibles (including the special deciduous mandibular appendages of some Curculionids) or, as in most fleas, by means of imaginal cephalic cocoon-cutters.

Emergence of the Imago – As the time for the emergence of the imago approaches the pupa darkens noticeably in colour. In some Papilionoidea the colours of the imago are distinctly visible through the transparent pupal cuticle a short period before its emergence. When this is due the contained insect ruptures the pupal cuticle by convulsive movements of its legs and body. A longitudinal fracture occurs down the back of the thorax and there are often other fractures in the region of the legs and elsewhere. The insect withdraws its appendages from within those of the pupa or last nymphal instar and emerges completely formed except for the wings. It may crawl up the nearest available support and rest there in such a position that the folded miniature wings are inclined downwards (Fig. 193). By the influx of blood from the body, and pressure exerted upon that fluid by muscular action, the complete expansion of the wings is rapidly achieved (Moreau, 1974). During this preliminary phase drops of liquid (the *meconium*) are discharged from the anus; they represent the waste products of pupal metabolism, and in Lepidoptera are coloured with residual pigments which have not been utilized. A short period elapses after eclosion before the insect is able to make its trial flight. The time of emergence varies greatly in different species and may show diurnal periodicity (e.g. Moriarty, 1959). In some Lepidoptera, for example, it occurs in early morning and in others towards evening; at such times they may sometimes be observed resting upon tree trunks, etc., awaiting their normal hours of activity. Among certain aquatic insects the imago is able to take to the wing almost immediately after eclosion.

Origin and Significance of the Pupal Instar – Several more or less distinct interpretations of metamorphosis have been proposed, accounting in different ways for the origin of the Endopterygote pupa. While it is clear that the pupa has considerable selective value in making possible a wide divergence in form and habits between the adult and larva and thus permitting the latter to exploit many habitats which have proved less accessible to Exopterygote nymphs, there is as yet no general agreement on the evolutionary relationship between the typical Exopterygote and Endopterygote modes of development.

Berlese (1913) supposed that the Exopterygote nymph represents a post-oligopod stage of development whereas the larvae of the Endopterygotes evolved through eclosion occurring at a developmental stage corresponding

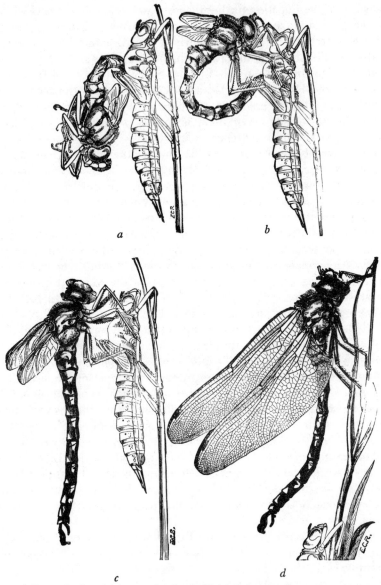

FIG. 193   Stages in the emergence of a dragonfly (*Aeshna cyanea*)

a–c, emergence from the nymphal cuticle; d, fully-formed imago. *After* Latter, *Nat. Hist. of Common Animals.*

to one of the preceding embryonic phases (p. 333). On this theory the larva is a sort of free-living embryo while the pupa represents the sequence of nymphal stages compressed, as it were, into a single instar. A fundamental distinction is thus made between larva and nymph, supported, it is claimed, by such characters as those mentioned on p. 356. Berlese's theory has been adversely criticized by Hinton (1955) and others and it no longer finds many adherents. There are considerable differences in the rates at which different organs develop in different groups of insects, so that no simple relationship holds between the levels of organization of the different larval forms and the similarly named embryonic stages. Nor is it certain, as Berlese's theory demands, that the prolegs of polypod larvae are always homologous with the segmental appendages of polypod embryos. In a quite different interpretation Poyarkoff (1914), supported at first by Hinton (1948), claimed that while the nymphal and larval stages are equivalent, the pupa represents the first of two imaginal instars whose principal function it is to provide a cuticular 'mould' that is mechanically necessary to ensure the adequate development of the skeletal musculature peculiar to the adult. The muscles then become attached to the adult cuticle when the latter is laid down beneath the similarly shaped pupal cuticle. The absence of such an 'imaginal moult' in the normal Exopterygotes is explained on the grounds that the differences between the musculature of their adults and immature stages are few and slight and the necessary changes are accomplished without the need for a moult to secure the simultaneous attachment of a large number of newly-formed muscles to the adult cuticle. This theory also now seems improbable as the pupal epidermis may undergo considerable changes of form while the adult musculature is developing (Hinton, 1963; Daly, 1964; and, for the Thysanoptera, Davies, 1969). Yet another viewpoint, due originally to Pérez (1910) and Handlirsch (1928), recalls that even in the Exopterygotes there are commonly rather distinct changes during post-embryonic development and supposes that by their accentuation and concentration towards the later part of development the larval stages arose as specialized nymphs and that the pupa is a modified last-instar nymph whose quiescence accompanies the more extensive and abrupt changes now necessary for transformation into the adult (see also Davies, 1958). Such a view denies that there is any fundamental distinction between nymph and larva and regards those differences which exist as merely due to coenogenetic modification and to changes in the rate at which adult characters develop in the two types of metamorphosis. This theory is consistent with the physiological evidence, which indicates that the metamorphosis of both Exo- and Endopterygotes is controlled by qualitatively similar hormonal changes (Wigglesworth, 1954, and see p. 383). The various theories of the origin of the pupa and the relations between the developmental stages of the major groups of insects are summarized in Fig. 194.

The Development of the Imago

The culminating feature of metamorphosis is the formation of the imago which, as indicated on p. 356, differs between the hemi- and holometabolous insects. In hemimetabolous forms it is accomplished through a gradual series of external and internal changes and alterations of form, which may be

Embryo		Nymphs			Adults	Thysanura
Embryo		Nymphs			sub-imago	Adult	Ephemeroptera
Embryo		Young Nymphs		Older Nymphs	Adult		other Exopterygota
Embryo	Larvae	Pupa			Adult		Berlese
Embryo		Larvae			Pupa	Adult	Poyarkoff & Hinton
Embryo		Larvae		Pupa	Adult		Pérez, Handlirsch
Embryo		Larvae		Pupa	Adult		Davies

FIG. 194   Diagram to illustrate homologies between various developmental stages of different groups of insects, as suggested in the theories proposed by different authorities. The lower four lines of the diagram refer to the Endopterygote life-cycle. For further discussion, see text.

traced back to simple growth during the immature stages, though the changes at the last moult are generally greater than those at previous ones. In holometabolous insects the transformation from the larva to the adult is accomplished largely during the quiescent pupal instar and a more detailed consideration of the complex changes which occur there is required. The extent of the transformation varies according to the species and the organs or tissues concerned as may be seen on referring to such general accounts as

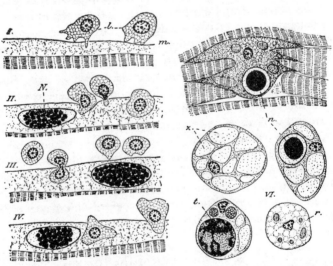

FIG. 195   Phagocytosis of muscle in *Calliphora*
I–IV, stages in the immigration of phagocytes into a muscle. *V*, a phagocyte within a muscle the nucleus of the latter has been ingested by the phagocyte. *VI*, diverse phases of the granular spheres. *l*, phagocytes; *m*, sarcolemma; *N*, muscle nucleus; *n*, do. within a phagocyte; *z*, sarcolytes or muscle fragments; *t*, other tissue engulfed by phagocyte; *r*, phagocyte which has almost digested its inclusions. *After* C. Pérez.

those by Snodgrass (1954, 1961), Nüesch (1965), Fristrom (1970) and the more detailed references cited below and on p. 385 ff. The simplest transition is effected when the cells of a particular larval tissue are carried over to form the corresponding tissue of the adult, the process being accomplished with relatively minor alterations due to differential growth and cellular differentiation. In other cases the larval tissue is largely or completely destroyed, a process known as *histolysis*, and the adult tissue which replaces it is developed by *histogenesis* from certain masses of formative cells known variously as *imaginal buds, imaginal disks* or *histoblasts*. The details of the histolytic processes differ in different insects but the first event is the death of the tissue concerned, probably as a result of lysosomal breakdown (Lockshin and Williams, 1964–65). Even when the cell is not entirely destroyed, larval organelles may be isolated through enclosure in special cell-membranes and then aggregate into autophagic vacuoles (Locke, 1966). In many cases, phagocytic haemocytes may deal with the larger fragments or, again, the recently dead but intact tissues may be actively attacked by numerous phagocytes (Figs. 195, 203 and see Crossley, 1968). The cells of the centres of imaginal differentiation which carry out histogenesis are either indistinguishable or relatively inconspicuous during the earlier larval stages though they are constant in position for a given species and in the higher Diptera, for example, they are already evident in the later embryo. They are best regarded as zones of persistent embryonic tissue whose capacity for growth and differentiation is largely suppressed during the larval stages and is fully realized only when the hormonal balance which controls metamorphosis is altered in the later part of larval life (see p. 383; also Henson, 1946). In the more specialized forms, such as *Drosophila*, the structure of the fully-developed imaginal disks of the wings and legs allows them to alter rapidly in shape in giving rise to the adult organs at metamorphosis. These changes in form can also be induced experimentally by treatment with dilute trypsin preparations, suggesting that they depend in part on changes in cellular adhesivity (Poodry and Schneiderman, 1971).

The different Endopterygote orders differ considerably in regard to which larval tissues pass over to the adult without great change and which are destroyed and replaced. There are also variations in the mode of replacement and the following account summarizes the changes which can occur.

The Epidermis – The development of the imaginal epidermis is a particularly important process since this tissue imparts to the pupa or adult the characteristic external form which distinguishes it from the larva. Even in the hemimetabolous insects the epidermis plays a major role in determining the changing form of the insect (Wigglesworth, 1959, 1973a), control extending to minute details of surface pattern and the distribution of setae (Lawrence, 1970, 1973; Lawrence and Hayward, 1971; Lawrence *et al.*, 1972). In the holometabolous groups the epidermis may undergo complicated cytological changes in the differentiation of characteristically adult structures like the scales of Lepidoptera (e.g. Henke and Pohley, 1952;

Esser, 1961; Greenstein, 1972). It also helps to determine the arrangement of tracheae and muscles (Hinton, 1961) and plays a leading part in those developmental processes that result in the formation of the segmental appendages and wings, which become everted in the pupal stage, together with the elaborate alterations by which the highly differentiated imaginal head is formed in species whose larvae have a less well-developed cephalic region.

In the Coleoptera and the sawfly *Pontania* much of the larval epidermis passes without destruction into the adult (Poyarkoff, 1910; Murray and Tiegs, 1935; Patay, 1939; Ivanova-Kazas and Ivanova, 1964). Something like this also occurs in the metamorphosis of the Lepidopteran leg (Kim, 1959; Kuske, 1963) where the pupal leg is formed directly from that of the larva. There is no discrete imaginal disk but the larval epidermis represents a mosaic-like pattern of adult *Anlagen* with four centres of mitotic activity.

In higher Hymenoptera and Diptera, however, there is a gradual destruction of the larval epidermis with accompanying replacement by the growth of cells from imaginal disks. The new epidermis of the head and thorax arises from part of the imaginal discs of the cephalic appendages, wings and legs, while that of the abdomen is formed from separate centres of regeneration – in *Drosophila*, for example, there are six such centres on each spiracle-bearing abdominal segment (Robertson, 1936). In lower Endopterygotes where the appendages of the larval head and thorax do not differ greatly from those of the pupa or adult, the latter develop within the larval appendages, often being folded until they are freed at the moult into the pupa. In other cases, the imaginal buds from which they differentiate are not always so closely associated with the larval structures (Fig. 32) though their position may be indicated externally by sensilla which lie on or replace the corresponding larval appendages (e.g. Eassa, 1953). They usually appear as evaginations from previously developed epidermal pockets. The entrance to the pocket narrows or closes up, and the space surrounding the imaginal bud is the *peripodial cavity*, whose wall is known as the *peripodial membrane* and is continuous with the general epidermis (e.g. Pohley, 1959 and see p. 58). As the imaginal buds develop, the peripodial sacs enlarge accordingly and the peripodial membrane becomes attentuated. The mouths of the cavities eventually open, the buds begin to protrude and in the pupa they are completely everted and appear outside the body.

The imaginal buds are exhibited in a relatively simple condition in the larva of *Anopheles*. Those of the head appendages appear at the bases of the larval organs which they replace, and the imaginal head is formed within that of the larva. The largest buds are those of the antennae and the future labrum, maxillary palps and labium. In the thorax two pairs of imaginal buds are present on each segment – a dorsal pair and a ventral pair. The dorsal buds give rise to the pupal respiratory horns, the wings and the halteres in their respective segments. Each pair of ventral buds forms the legs of its segment. In the abdomen there is a conspicuous pair of posterior dorsal buds which form the pupal caudal lamellae.

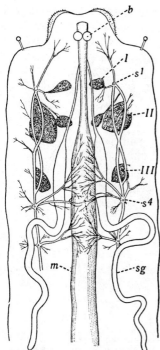

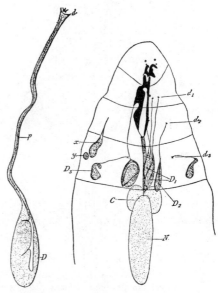

FIG. 196

Fully-grown larva of *Pieris* dissected from above showing imaginal buds

b, brain; *m*, mid gut; *s¹*, prothoracic spiracle; *s⁴*, 1st abdominal spiracle; *sg*, silk gland; *I*, prothoracic bud; *II* and *III*, buds of fore and hind wings. From Folsom's *Entomology*. *After* Gouin.

FIG. 197

Imaginal buds of the larva of *Erioischia brassicae* showing the filamentous pedicels connecting them with the epidermis

D_1–D_3, leg-buds; d_1–d_3, cutaneous sensory organs (vestiges of larval legs); *x*, wing-bud; *y*, bud of haltere; *C*, brain; *N*, ventral nerve-centre. × 36. On the left – imaginal bud *D* of the fore leg with its pedicel *p*. × 75. *After* Keilin, 1915.

The classical researches of Weismann (1864–66), Pérez (1910) and others have shown that the imaginal buds attain increasing complexity from the lower to the higher Diptera. In the Cyclorrhapha they are deeply sunk into the body, and it is often difficult to trace their connections with the epidermis as the peripodial membrane is reduced to a greatly attenuated cord (Fig. 197). In *Melophagus* the buds, though superficial in position, become disconnected from the epidermis.

The most complex feature is exhibited in the imaginal buds of the head, the development of which is associated with the position of the cerebral ganglion in the larva, and the fact that the larval head is no longer able to accommodate the developing head of the imago. In *Chironomus* Miall and Hammond (1892) have shown that the cerebral ganglion lies in the larval prothorax, within which the imaginal head is formed. In a larva about a centimetre long, the epidermis becomes infolded along two nearly longitudinal lines, corresponding to the margins of the

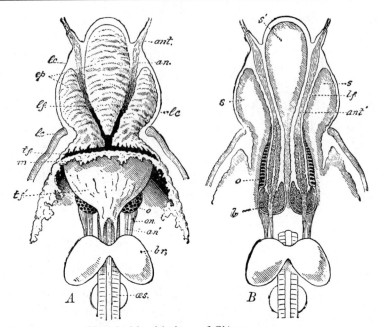

FIG. 198   Formation of imaginal head in larva of *Chironomus*

A, The new epidermis thrown into folds which have been cut away in places. B, The same parts in horizontal section. *lc*, larval cuticle; *t.f*, transverse fold; *t.f'*, upper wall of ditto; *ep*, epidermis; *m*, cut edge of new epidermis; *ant*, larval antenna; *a.n*, nerve to ditto; *ant'*, antenna of fly; *l.f*, longitudinal fold; *o*, eye of fly; *on*, optic nerve; *a.n'*, root of antennal nerve; *br*, brain; *oes*, oesophagus; *b*, enlarged second joint of antenna of fly; *s, s, s'*, blood-spaces. *After* Miall & Hammond.

larval clypeus. The imaginal buds of the compound eyes and antennae arise from the inner extremities of these cephalic folds and are thus far removed from the surface. The folds gradually extend backwards into the prothorax and this is accompanied by the formation of a transverse fold which runs back from the junction of the larval head and prothorax (Fig. 198). During the change to the pupa the parts of the head, thus formed in the larva, assume their final exterior position by a process of eversion, with the result that the now evaginated folds form the wall of the imaginal head and carry the eyes and antennae with them. Similar processes occur in *Psychoda* (Feuerborn, 1927).

In the Cyclorrhaphan Diptera (Snodgrass, 1953; Schoeller, 1964; Anderson, 1963), the head becomes invaginated during the later embryonic period into the region which follows it, and its outwardly visible portion is reduced in the larva to a small apical papilla. The invaginated part of the head forms the so-called larval 'pharynx' and the true mouth opens into the posterior end of this pouch. A pair of cephalic buds extend as diverticula from the so-called pharynx to the cerebral ganglion (which is located in the metathorax), and the imaginal eyes and antennae develop from the inner wall of each sac. During the pupal stage the cerebral ganglion and cephalic buds move forwards until the former come to lie in the prothorax. At the same time the openings of the buds into the 'pharynx' widen, and ultimately both the pharynx and its diverticula become confluent, forming a single sac or *cephalic vesicle* (Fig. 199C). The latter is finally everted through the mouth of the ɔharynx, and becomes turned inside out to form the completed imaginal head very much as in *Chironomus* (Fig. 199D). In the embryo of *Melophagus* Pratt (1900) states that the cephalic buds, which ultimately form the adult head, develop as paired dorsal and unpaired ventral thickenings which later on become invaginated. The dorsal pair corresponds to the cephalic buds of *Calliphora*: they are destined to form the dorsal and lateral portion of the imaginal head together with the compound eyes. The ventral cephalic bud has no counterpart in *Calliphora*: it forms the floor of

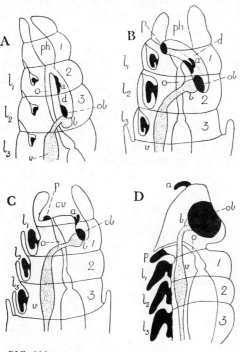

FIG. 199
Development of imaginal buds in the Muscidae

A, larva; B–D, pupa. 1–3, thoracic segments; l_1–l_3, leg-buds; *ph*, 'pharynx'; *o*, oesophagus; *b*, brain; *cv*, cephalic vesicle; *v*, ventral nerve-centre; *d*, diverticulum of pharynx; *m*, mouth; *a*, antennal bud; *ob*, optic bud; *p*, proboscis rudiment. Based on Korschelt & Heider *after* Kowalevsky & van Rees.

the imaginal head together with the proboscis. Involution of the embryonic head takes place as in *Calliphora* and the cephalic buds become drawn into the secondary pharynx' thus developed. Owing to the early fusion of the dorsal buds the cavities of the latter open into the pharynx by a common connection but they retain their paired formation posteriorly. Finally the dorsal and ventral diverticula combine to form the cephalic vesicle, which subsequently becomes evaginated.

The imaginal disks of some insects can be grown *in vitro* or transplanted into host larvae and are being used increasingly to investigate some of the more fundamental processes of development (Ursprung and Nöthiger, 1972; Gehring and Nöthiger,

1973). Experimental analysis, mainly of *Drosophila*, shows that different parts of an adult structure may arise from localized regions of the larval disk. Maps of the disks can therefore be drawn to show the positions of the presumptive areas (Hadorn *et al.*, 1949; Hadorn, 1953; Ursprung, 1959; Murphy, 1971). Each of the various fields in the disk seem to grow and differentiate more or less autonomously, but within a given field there may be considerable regulation; a portion of a field can therefore give rise experimentally to normally formed adult structures.

Alimentary Canal – The epithelium of stomodaeum and proctodaeum may pass from larva to adult with little change, as in the Coleoptera. In *Bombyx* (Verson, 1905) and *Malacosoma* (Deegener, 1908), this simple transition is supplemented by extension through the activity of cells which form the so-called imaginal rings at the inner ends of the stomodaeum and proctodaeum (cf. Fig. 200, *i*; Fig. 201, *a* and *c*). Finally, in *Mormoniella* (Tiegs, 1922) and the Cyclorrhaphan Diptera (Pérez, 1910; Robertson, 1936) the epithelium of the larval fore and hind gut is destroyed and replaced by cells proliferating from the imaginal rings and sometimes also from other centres. In all Endopterygotes the epithelium of the larval mesenteron is largely or entirely replaced. The remains of the old tissue are cast off into the lumen

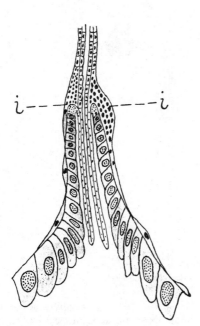

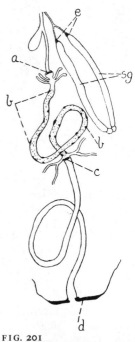

FIG. 200
Median longitudinal section at the junction of fore and mid intestine of *Formica rufa*

i, annular imaginal bud of fore gut. *After* C. Pérez.

FIG. 201
Alimentary canal and salivary glands (*sg*) of a Muscid larva showing imaginal buds

a, of fore gut; *b*, of mid gut; *c*, of hind gut; *d*, of rectum; *e*, of salivary glands.   *After* Kowalevsky.

and a new pupal epithelium is formed either from the persistent replacement cells of the larval mid gut (Trichoptera, Lepidoptera, Diptera, Hymenoptera, many Coleoptera) or from histoblasts at the posterior end of the mesenteron, as in some Coleoptera (Mansour, 1927; 1934). In some insects there is a further destruction of the pupal epithelium and replacement to form the definitive adult mesenteron (e.g. *Cybister* – Deegener, 1904; *Leptinotarsa* – Patay, 1939). The bacterial flora of the gut survives metamorphosis in the butterfly *Danaus plexippus* (Kingsley, 1972).

Malpighian Tubules – In the Diptera (Pérez, 1910), *Heterogenea* (Samson, 1908) and *Leptinotarsa* (Patay, 1939) the larval Malpighian tubules are directly transformed into those of the adult without histolysis but in other species there is partial or complete destruction followed respectively by regeneration from replacement cells or the differentiation of new tubules from the extreme anterior end of the proctodaeum. The perirectal tubules of the *Tenebrio* larva survive metamorphosis but experience partial cellular breakdown and redifferentiation (Byers, 1971).

Salivary Glands and Silk Glands – These glands degenerate and are usually destroyed by phagocytosis. The imaginal glands are built up by a pair of annular buds situated at the junction of each gland with its duct (Fig. 201). In *Galerucella* Poyarkoff states that the new glands are formed as invaginations at the bases of the maxillae.

The Fat-body – The extent to which the larval fat-body is destroyed is correlated with the extent of changes in the other tissues since its disintegration provides the materials necessary for histogenesis. In most Coleoptera there is little change in the larval fat-body whereas in *Calliphora* (Fig. 202) it is entirely replaced by imaginal tissue derived from mesenchyme on the inner surface of the epidermis, the process being completed only after the emergence of the adult (Pérez, 1910).

Oenocytes – Apparently a new imaginal complement is usually formed by localized proliferation from the epidermis, the larval oenocytes being histolysed. In *Calpodes* the so-called 'permanent oenocytes' undergo extensive redifferentiation at the larval-pupal moult while another group, the subdermal oenocytes, are essentially adult cells that develop from the epidermis before pupation (Locke, 1969a). See also Schmidt (1961).

Tracheal System – The tracheal matrix cells are generally carried over from the larva without histolysis though modifications result from the proliferation of new branches and the opening of trunks to newly functional spiracles. Extensive replacement occurs in *Mormoniella* from special cells at the base of the larval spiracular trunks (Tiegs, 1922) and in *Calliphora* some replacement is brought about by histoblasts scattered in the walls of the larval tracheae (Pérez, 1910).

Dorsal Vessel – The dorsal vessel of the larva normally becomes that of the adult with very little change, the heart continuing to beat throughout pupation. In *Mormoniella* (Tiegs, 1922) there is degeneration of the larval cells and replacement from scattered histoblasts.

Central Nervous System – The metamorphosis of this system has attracted much recent study (Edwards, 1969; Young, 1973). The larval ganglia commonly become displaced by shortening of the connectives to give the more highly concentrated ventral chain of the adult. In the Lepidoptera this results from the withdrawal and migration of the glial cells from around the axons, which then shorten by forming coils that eventually disappear through a reduction in the axon lengths (Pipa, 1963, 1967; Pipa and Woolever, 1965; Robertson and Pipa, 1973). As well as this condensation

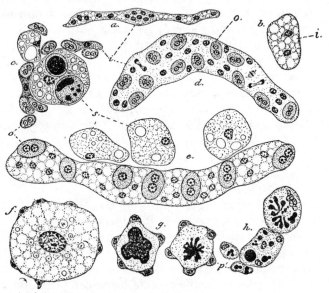

FIG. 202   Origin of imaginal fat-body in *Calliphora*

a, b, in the head and thorax; c–e, in the abdomen in close relation with the imaginal oenocytes; s, granular spheres; o, oenocytes; i, initial cells of imaginal fat-body; f–h, destruction of remaining fat-body in the imago by phagocytes p. *After* C. Pérez.

there is breakdown and phagocytosis of the neural lamella of the larva, followed by secretion of an adult lamella from the perineurial cells (Ali, 1973; McLaughlin, 1974). New larval neurons may be formed throughout the earlier stages; some of these persist into the adult but others are lost and their place is taken by newly formed adult nerve-cells (Taylor and Truman, 1974). There are also extensive alterations in the glomerular bodies, axonal tracts and localized aggregations of neurons, especially in the brain. These have been followed in considerable histological detail in several species (e.g. Lucht-Bertram, 1961; Gieryng, 1965; Nordlander and Edwards, 1969–70; Ali, 1974), and are particularly noticeable in the optic lobes. Striking developmental changes also enable the large adult compound eyes to replace the small ocelli of the larva. In more specialized forms the eye disk is segregated early, but in the Culicidae it extends during larval life by incorporating an adjacent zone of epidermis (Haas, 1956; White, 1961), a process that recalls the recruitment of epidermal cells into the developing compound eye of Exopterygote insects (Mouze, 1975; Green and Lawrence, 1975).

Muscular System – There are considerable variations in the fate of the skeletal muscles at metamorphosis and even in the Exopterygotes the flight musculature undergoes appreciable growth and differentiation (Tiegs, 1955). In relatively simple cases like the Acrididae (Wiesend, 1957; Bernays, 1972*b*) almost all the muscles are formed in the embryo, though the flight muscles are at first weak and only attain their full development in the adult. In *Oncopeltus* and *Cenocorixa*, however, the dorsal longitudinal thoracic muscles of the first instar soon disappear and are replaced by new muscles that arise by myoblast aggregation in the second instar and become functional flight muscles in the adult (Scudder, 1971; Scudder and Hewson, 1971). Hinton (1959) distinguishes five modes of growth in the indirect flight muscles of insects. The pre-existing fibres may divide without the incorporation of free myoblasts, or the muscles increase in cell number by incorporating free myoblasts without themselves dividing, or both processes may occur. In many cases larval muscles degenerate and the adult muscles that replace them develop either from myoblasts which were present within the muscles or from free myoblasts. These may adhere to or penetrate degenerating larval muscles (Fig. 203) or they may become organized independently

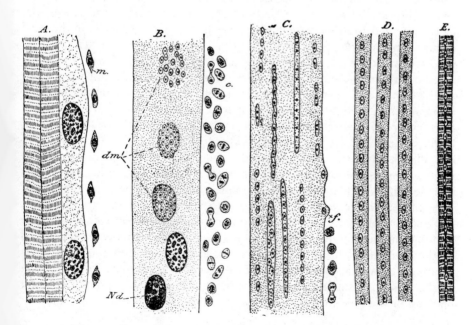

FIG. 203   Metamorphosis of a muscle of *Calliphora* (semi-diagrammatic)

A, larval stage; *m*, myoblast. B, commencement of metamorphosis; *c*, mitosis of myoblasts; *dm*, multiple division of larval nuclei; *Nd*, degenerating nucleus. C, direct division, in chains, of imaginal nuclei; *f*, fusion of myoblasts with differentiated muscles. D, cleavage into separate fibres. E, final stage of imaginal muscle fibre. *After* C. Pérez.

into entirely new adult muscles. The histological and ultrastructural changes that accompany some of these processes are reported in detail in papers cited on pp. 385–394. The leg muscles of *Drosophila* appear to arise from the so-called adepithelial cells of the imaginal disks (Poodry and Schneiderman, 1970). These are ultrastructurally indistinguishable from the disk epithelium though the presence of muscle stem-cells in the disks has been demonstrated by electrophoretic techniques (Ursprung *et al.*, 1972). The abdominal muscles of *Calliphora* disintegrate and are replaced in the pupa from myoblasts of epidermal origin (Crossley, 1965). The adult intersegmental muscles of the same species have a dual origin: they contain large residual polyploid nuclei from the larval muscles as well as smaller ones derived from fused myoblasts (Crossley, 1972).

Postmetamorphic Development

Though the adult Pterygote insect does not moult it generally undergoes some further development after emergence. Coloration, thickening and hardening of the cuticle take a little time to complete and in the majority of insects there is a period of some days during which the gonads and accessory reproductive glands attain their full size and become functional. The protein content of *Glossina* flight muscles increases two- to threefold in the young adult (Bursell, 1973) and in Corixidae the immature flight muscles of the newly emerged adult increase in size in the flying morphs but not in the non-flying forms (Young, 1965; Scudder, 1971). In a few insects sexual maturity is accompanied by more striking changes. In the Isoptera (p. 617) the wings are shed, the mandibular and thoracic muscles degenerate and the ovaries undergo marked hypertrophy. Shedding of the wings also occurs in some Hippoboscidae and Streblidae and in the Zoraptera and Formicidae and degeneration of the thoracic muscles, with consequent loss of the power of flight, in some Scolytidae, Pyrrhocoridae, Aphididae and Culicidae (see p. 96). Physogastry, the distension of the abdomen by swelling of the ovaries or gut, is well shown in the flea *Tunga penetrans*, some Coccoidea, queen ants and termites and many myrmecophile and termitophile Diptera and Coleoptera. The female of *Ascodipteron* (Streblidae) changes greatly in the adult stage (Jobling, 1939). After burying itself in the wound which its large mouthparts cause in its host (a bat), the legs and wings are cast off and a fold of the abdominal integument grows forward to surround the body, thus forming a flask-shaped structure. All that remains protruding from the body of the host is a small prominence bearing the spiracles and the conjoint opening of the rectum and vagina.

Physiology of Postembryonic Development

Two topics are discussed here, the hormonal control of growth and metamorphosis and the phenomenon of diapause.

Hormonal Control of Metamorphosis – As described on p. 272 ff. the anterior part of the body of insects contains a number of endocrine glands whose interacting secretions control postembryonic growth and differentiation in a way which is fundamentally very similar in both hemi- and holometabolous insects (Wyatt, 1972; Doane, 1973; Gilbert and King, 1973; and other references cited on p. 385 ff.). Well-defined neurosecretory cells in the brain produce a polypeptide hormone (brain hormone or prothoracicotropic hormone) which passes to the corpora cardiaca where it is stored, and perhaps modified, before being released into the blood. Here it is transported to the prothoracic glands, causing them to secrete α-ecdysone, probably by conversion of cholesterol and/or β-cholesterol. The α-ecdysone is then rapidly converted to β-ecdysone by various insect tissues. β-ecdysone is the primary moulting hormone which activates the epidermal cells, initiating apolysis and causing them to secrete the new cuticle. While the juvenile hormone secreted by the corpora allata is also present in sufficient quantity, however, the latent imaginal characters (in respect of which the egg had already been determined) remain largely suppressed and development progresses through the normal series of larval or nymphal forms. In the last-instar nymph or larva the production of juvenile hormone by the corpora allata falls temporarily and development is directed towards the formation of the adult. Juvenile hormone is therefore a modifying agent which favours the development of larval structures and opposes adult differentiation. Experimental proof of this mechanism has been obtained in several ways. Thus, elimination of the brain prevents the growth and moulting of immature stages while extirpation of the corpora allata at an early developmental stage induces the premature appearance of adult characteristics. Again, the implantation of corpora allata from young nymphs of *Rhodnius* into mature nymphs causes one or two extra moults with a marked suppression of adult features. Furthermore the imaginal disks of mature larvae show little change when cultured in larval blood but unfold and differentiate when transferred to pupal blood or when treated with ecdysone *in vivo* or *in vitro* (Postlethwait and Schneiderman, 1970; Agui and Fukaya, 1973).

There are some observations which do not fit readily into the classical scheme of endocrine control outlined above. These are indicated by Willis (1974) and Gilbert (1974), the latter also discussing in detail some of the biochemical processes by which the hormonal effects may be produced. Brain hormone, for instance, may regulate the thoracic glands by activating a membrane-bound adenyl cyclase system there, this in turn affecting steroid conversions through a mechanism involving cyclic AMP. The effect of ecdysones on the puffing pattern of polytene chromosomes is now well known (Clever and Karlson, 1960; and see p. 276); it suggests that moulting hormone operates by inducing the synthesis of specific forms of messenger RNA, which themselves subsequently participate in protein syntheses (see Beerman, 1972; Ilan and Ilan, 1973; Williams and Kafatos, 1971). Two additional neurosecretory hormones have also been found to control certain developmental processes. Bursicon is concerned in tanning the cuticle of newly-emerged insects (p. 274) and an eclosion hormone is active at the pupal-adult ecdysis of Endopterygotes, where it was found in several Lepidoptera by Truman (1971–73). Eclosion hormone activity is confined to the brain and corpora cardiaca of the pharate adult and initiates a sequence of events, including abdominal movements, ecdysial behaviour, wing-spreading, and also the secretion of bursicon.

   Diapause – In a favourable environment the development of many insects will proceed without interruption and even if it is stopped tempor-arily by adverse conditions (e.g. low temperature) it is immediately resumed when these improve. On the other hand many insects can, under certain conditions, pass into a state of diapause during which development is inhibited even though external circumstances appear suitable (Andrewartha, 1952; Lees, 1955; Harvey, 1962; Tauber and Tauber, 1976). Diapause may consist merely of a retardation in growth for a few weeks or a cessation of development for several years but it is eventually 'broken' and the life-cycle then continues normally. Diapause occurs at a definite stage of the life-history which may be the egg, one of the other immature stages or the adult according to species. Some species enter diapause in every generation while others do so every other generation or less regularly; in some Lepidoptera different strains of the same species differ in this respect (e.g. *Bombyx mori*). Many external factors appear to be capable of inducing diapause, differing in different species and sometimes interacting. Unfavourable nutrition, desic-cation and variations in illumination and temperature are variously known to be responsible. Of these the most important is the length of day (Danilevskii, 1965; Beck, 1968). In some insects, such as the Lepidoptera *Diataraxia oleracea* and *Acronycta rumicis*, pupal development proceeds without inter-ruption when there is more than about 16 hours of light per day, while at shorter periods of illumination the pupae enter diapause – the so-called 'long day response'. Conversely, in *Bombyx mori* the eggs or young larvae exposed to a short day (13 hours or less of light) yield moths which lay diapausing eggs. Other forms of response to light are known and Beck (1974) has devised theoretical models to account for many forms of photoperiodically induced diapause, though they have as yet no physiological basis. As the environmental changes often exert their effects some time before the onset of diapause and can even act through the female parent, the phenomenon often appears to originate with some semblance of spontaneity. During diapause there is a considerable depression of metabolism, the diapausing eggs of *Melanoplus*, for example, having only about a quarter of the normal oxygen requirement. This is probably connected with a disturbance of the cyto-chrome system since diapausing eggs of *Melanoplus* are relatively insensitive to hydrogen cyanide and pupae of *Hyalophora* contain virtually no cytochrome *c* while in diapause. Diapause may be broken naturally or experimentally by a number of factors: exposure to low temperature is a common agent but mechanical injury and changes in illumination or water relations may also be effective, depending on the species. The immediate cause of postembryonic diapause is probably the absence of a growth hormone and Williams (see p. 276) has shown that the diapause of *Hyalophora* is broken by chilling, the effect of which is to stimulate the brain to produce a factor which activates the prothoracic glands whose secretion leads to renewed development. In *Bombyx mori* the neurosecretory cells of the suboesophageal ganglion of the mother induce diapause in the eggs she lays (Fukuda and Takeuchi, 1967; and see p. 346). Again, in the moth *Phalaenoides glycinae* a substance produced by the suboesophageal ganglion of the larva induces a facultative pupal diapause (Andrewartha *et al.*, 1974). Ecologically, diapause permits survival under adverse environmental circumstances and allows synchronized resumption of development when conditions improve. It has no doubt been evolved many times and probably depends on rather diverse physiological mechanisms. Until these have been more fully elucidated, simple classifications of diapause based primarily on the environmental factors that control it (Müller, 1970; Thiele, 1973) seem of limited value.

Literature on Postembryonic Development

ABERCROMBIE, W. F. (1936), Studies on cell-number and the progression factor in the growth of Japanese Beetle larvae (*Popillia japonica* Newman), *J. Morph.*, **59**, 91–112.

AGUI, N. AND FUKAYA, M. (1973), Effects of moulting hormones and prothoracic glands on the development of wing discs of the cabbage armyworm (*Mamestra brassicae* L.) *in vitro* (Lepidoptera: Noctuidae), *Appl. Ent. Zool.*, **8**, 73–82.

ALI, F. A. (1973), Post-embryonic changes in the central nervous system and perilemma of *Pieris brassicae* (L.) (Lepidoptera: Pieridae), *Trans. R. ent. Soc. Lond.*, **124**, 463–498.

—— (1974), Structure and metamorphosis of the brain and suboesophageal ganglion of *Pieris brassicae* (L.) (Lepidoptera: Pieridae), *Trans. R. ent. Soc. Lond.*, **125**, 363–412.

ANDERSON, D. T. (1963), The larval development of *Dacus tryoni* (Frogg.). I, II, *Aust. J. Zool.*, **11**, 202–218; **12**, 1–8.

ANDREWARTHA, H. G. (1952), Diapause in relation to the ecology of insects, *Biol. Rev.*, **27**, 50–107.

ANDREWARTHA, H. G., MIETHKE, P. M. AND WELLS, A. (1974), Induction of diapause in the pupa of *Phalaenoides glycinae* by a hormone from the suboesophageal ganglion, *J. Insect Physiol.*, **20**, 679–701.

BECK, S. D. (1968), *Insect Photoperiodism*, Academic Press, New York, 288 pp.

—— (1973), Growth and retrogression in larvae of *Trogoderma glabrum* (Coleoptera: Dermestidae). 4. Developmental characteristics and adaptive functions, *Ann. ent. Soc. Am.*, **66**, 895–900.

—— (1974), Photoperiodic determination of insect development and diapause. I, II, *J. comp. Physiol.*, **90**, 275–295; 297–310.

BEERMANN, W. (ed.) (1972), *Developmental Studies on Giant Chromosomes*, Springer, Berlin, 227 pp.

BEINBRECH, G. (1968), Elektronenmikroskopische Untersuchungen über die Differenzierung von Insektenmuskeln während der Metamorphose, Z. Zellforsch., 90, 463–490.

BERLESE, A. (1913), Intorno alle metamorfosi degli insetti, Redia, 9, 121–138.

BERNAYS, E. A. (1972a), The intermediate moult (first ecdysis) of Schistocerca gregaria (Forskål) (Insecta, Orthoptera), Z. Morph. Tiere, 71, 160–179.

—— (1972b), The muscles of newly hatched Schistocerca gregaria larvae and their possible functions in hatching, digging and ecdysial movements, J. Zool., Lond., 166, 141–158.

BIENZ-ISLER, G. (1968), Elektronenmikroskopische Untersuchungen über die Entwicklung der dorsolongitudinalen Flugmuskeln von Antheraea pernyi Guér. (Lepidoptera). 2. Teil, Acta anat., 70, 524–553.

BORATYNSKI, K. L. (1952), Matsucoccus pini (Green, 1925) (Homoptera, Coccoidea: Margarodidae): Bionomics and external anatomy with reference to the variability of some taxonomic characters, Trans. R. ent. Soc. Lond., 103, 285–326.

BROWN, V. AND DAVIES, R. G. (1972), Allometric growth in two species of Ectobius (Dictyoptera: Blattidae), J. Zool., Lond., 166, 97–132.

BULLIÈRE, F. (1973), Cycles embryonnaires et sécrétion de la cuticule chez l'embryon de blatte, Blabera craniifer, J. Insect Physiol., 19, 1465–1479.

BURSELL, E. (1973), Development of mitochondrial and contractile components of the flight muscle in adult tsetse flies, Glossina morsitans, J. Insect Physiol., 19, 1079–1086.

BYERS, J. R. (1971), Metamorphosis of the perirectal Malpighian tubules in the mealworm Tenebrio molitor L. (Coleoptera, Tenebrionidae). I. Histology and histochemistry, Can. J. Zool., 49, 823–830.

CAMATINI, M. AND SAITA, A. (1968), Studio ultrastrutturale dell'istogenesi dei muscoli del volo di Bombyx mori L., Rc. Ist. Lomb. sci. Lett. (B), 102, 227–242.

CHEN, P. S. (1971), Biochemical Aspects of Insect Development, Karger, New York, 230 pp.

CHRYSTAL, R. N. (1931), Studies of the Sirex parasites, Oxford Forestry Mem., 11, 63 pp.

CLAUSEN, C. P. (1940), Entomophagous Insects, McGraw-Hill, New York, 688 pp.

CLEVER, U. AND KARLSON, P. (1960), Induktion von Puff-Veränderungen in den Speicheldrüsenchromosomen von Chironomus tentans durch Ecdyson, Expl Cell Res., 20, 623–626.

CONDOULIS, W. V. AND LOCKE, M. (1966), The deposition of endocuticle in an insect, Calpodes ethlius Stoll (Lepidoptera, Hesperiidae), J. Insect Physiol., 12, 311–323.

COTTRELL, C. B. (1964), Insect ecdysis with particular emphasis on cuticular hardening and darkening, Adv. Insect Physiol., 2, 175–218.

CROSSLEY, A. C. S. (1965), Transformations in the abdominal muscles of the blue blow-fly, Calliphora erythrocephala (Meig.) during metamorphosis, J. Embryol. exp. Morph., 14, 89–110.

—— (1968), The fine structure and mechanism of breakdown of larval intersegmental muscles in the blowfly Calliphora erythrocephala, J. Insect Physiol., 14, 1389–1407.

—— (1972), Ultrastructural changes during transition of larval to adult intersegmental muscle at metamorphosis in the blowfly *Calliphora erythrocephala*. I. II, *J. Embryol. exp. Morph.*, **27**, 43–74; **27**, 75–101.

DALY, H. V. (1964), Skeleto-muscular morphogenesis of the thorax and wings of the honey bee *Apis mellifera* (Hymenoptera: Apidae), *Univ. Calif. Publs Ent.*,**39**, 1–77.

DANILEVSKII, A. S. (1965), *Photoperiodism and Seasonal Development of Insects*, Oliver & Boyd, Edinburgh and London, 282 pp.

DAVIES, R. G. (1958), The terminology of the juvenile phases of insects, *Trans. Soc. Br. Ent.*, **13**, 25–36.

—— (1969), The skeletal musculature and its metamorphosis in *Limothrips cerealium* Haliday (Thysanoptera: Thripidae), *Trans. R. ent. Soc. Lond.*, **121**, 167–233.

DAVIES, R. G. AND BROWN, V. K. (1972), A multivariate analysis of postembryonic growth in two species of *Ectobius* (Dictyoptera: Blattidae), *J. Zool., Lond.*, **168**, 51–79.

DEEGENER, P. (1904, 1908), Die Entwicklung des Darmcanals der Insekten während der Metamorphose. I, II, *Zool. Jb. (Anat.)*, **20**, 499–676; **26**, 45–182.

DIEHL, P. A. (1973), Paraffin synthesis in the oenocytes of the desert locust, *Nature*, **243**, 468–470.

DOANE, W. W. (1973), Role of hormones in insect development, In: Counce, S. J. and Waddington, C. H. (eds), *Developmental Systems: Insects*, **2**, 291–497.

DYAR, H. G. (1890), The number of moults of Lepidopterous larvae, *Psyche*, **5**, 420–422.

EASSA, Y. E. E. (1953), The development of imaginal buds in the head of *Pieris brassicae* Linn. (Lepidoptera), *Trans. R. ent. Soc. Lond.*, **104**, 39–50.

EASTHAM, L. E. S. (1929), The post-embryonic development of *Phaenoserphus viator* Hal. (Proctotrypoidea), a parasite of the larva of *Pterostichus niger* (Carabidae), with notes on the anatomy of the larva, *Parasitology*, **21**, 1–21.

EDWARDS, J. S. (1969), Postembryonic development and regeneration of the insect nervous system, *Adv. Insect Physiol.*, **6**, 97–137.

EIGENMANN, R. (1965), Untersuchungen über die Entwicklung der dorsolongitudinalen Flugmuskeln von *Antheraea pernyi* (Lepidoptera), *Revue suisse Zool.*, **72**, 789–840.

EMDEN, F. I. VAN (1946), Egg-bursters in some families of polyphagous beetles and some general remarks on egg-bursters, *Proc. R. ent. Soc. Lond.* (A), **21**, 89–97.

ESSER, H. (1961), Untersuchungen zur Entwicklung des Puppenflügels von *Ephestia kühniella*, *Arch. EntwMech. Org.*, **153**, 176–212.

FEUERBORN, H. J. (1927), Die Metamorphose von *Psychoda alternata* Say, *Zool. Anz.*, **70**, 315–328.

FISCHER, M. (1968), Die Verwandlung der Insekten, In: Helmcke, J.-G., Starck, D. and Wermuth, H. (eds), *Handbuch der Zoologie*, **4** (2) Lief. 8, 68 pp.

FRAENKEL, G. AND BHASKARAN, G. (1973), Pupariation and pupation in cyclorrhaphan flies (Diptera): terminology and interpretation, *Ann. ent. Soc. Am.*, **66**, 418–422.

FRISTROM, J. W. (1970), The developmental biology of *Drosophila*, *A. Rev. Genet.*, **4**, 325–346.

FUKUDA, S. AND TAKEUCHI, S. (1967), Diapause factor-producing cells in the suboesophageal ganglion of the silkworm, *Bombyx mori* L, *Proc. Japan Acad.*, **43**, 51–56.

GAINES, J. C. AND CAMPBELL, F. L. (1935), Dyar's rule as related to the number of instars of the corn ear worm, *Heliothis obsoleta* (Fab.), collected in the field, *Ann. ent. Soc. Am.*, 28, 445–461.

GEHRING, W. J. AND NÖTHIGER, R. (1973), The imaginal discs of *Drosophila*, In : Counce, S. J. and Waddington, C. H. (eds), *Developmental Systems: Insects*, 2, 211–290.

GIERYNG, R. (1965), Veränderungen der histologischen Struktur des Gehirns von *Calliphora vomitoria* (L.) (Diptera) während der postembryonalen Entwicklung, *Z. wiss. Zool.*, 171, 80–96.

GILBERT, L. I. (1974), Endocrine action during insect growth, *Recent Prog. Horm. Res.*, 30, 347–390.

GILBERT, L. I. AND KING, D. S. (1973), Physiology of growth and development: endocrine aspects, In: Rockstein, M. (ed.), *The Physiology of Insecta*, 1, 249–370.

GREEN, S. M. AND LAWRENCE, P. A. (1975), Recruitment of epidermal cells by the developing eye of *Oncopeltus* (Hemiptera), *Arch. EntwMech. Org.*, 177, 61–65.

GREENSTEIN, M. E. (1972), The ultrastructure of developing wings in the giant silkmoth, *Hyalophora cecropia*. I, II, *J. Morph.*, 136, 1–22; 136, 23–52.

GREGORY, D. W., LENNIE, R. W. AND BIRT, L. M. (1967), An electron microscope study of flight muscle development in the blowfly *Lucilia cuprina*, *Jl R. microsc. Soc.*, 88, 151–175.

HAAS, G. (1956), Entwicklung des Komplexauges bei *Culex pipiens* und *Aedes aegypti*, *Z. Morph. Ökol. Tiere*, 45, 198–216.

HADORN, E. (1953), Regulation and differentiation within field districts in imaginal discs of *Drosophila*, *J. Embryol. exp. Morph.*, 1, 213–216.

HADORN, E., BERTANI, G. AND GALLERA, J. (1949), Regulationsfähigkeit und Feldorganisation der männlichen Genitalimaginalscheibe von *Drosophila melanogaster*, *Arch. EntwMech. Org.*, 144, 31–70.

HANDLIRSCH, A. (1928), Die postembryonale Entwicklung, In: Schröder, C. (ed.), *Handbuch der Entomologie*, 1, 1117–1185.

HARVEY, W. R. (1962), Metabolic aspects of insect diapause, *A. Rev. Ent.*, 7, 57–80.

HAYES, W. P. (1941), Some recent works on the classification of immature insects, *J. Kansas ent. Soc.*, 14, 3–11.

HENKE, K. AND POHLEY, H.-J. (1952), Differentielle Zellteilungen und Polyploidie bei der Schuppenbildung der Mehlmotte *Ephestia kühniella* Z, *Z. Naturf.*, 7b, 65–79.

HENRIKSEN, K. (1932), The manner of moulting in Arthropoda, *Notul. ent.*, 11, 103–127.

HENSON, H. (1946), The theoretical aspect of insect metamorphosis, *Biol. Rev.*, 21, 1–14.

HINTON, H. E. (1946), A new classification of insect pupae, *Proc. zool. Soc. Lond.*, 116, 282–328.

—— (1948), On the origin and function of the pupal stage, *Trans. R. ent. Soc. Lond.*, 99, 395–409.

—— (1949), On the function, origin and classification of pupae, *Trans. S. London ent. nat. Hist. Soc.*, 1947–48, 111–154.

—— (1955), On the structure, function and distribution of the prolegs of the Panorpoidea, with a criticism of the Berlese-Imms theory, *Trans. R. ent. Soc. Lond.*, 106, 455–545.

—— (1958a), Concealed phases in the metamorphosis of insects, *Science Progress*, 182, 260–275.

—— (1958b), The pupa of the fly *Simulium* feeds and spins its own cocoon, *Entomologist's mon. Mag.*, 94, 14–16.

—— (1959), How the indirect flight muscles of insects grow, *Sci. Prog., Lond.*, 47, 321–333.

—— (1961), The role of the epidermis in the disposition of tracheae and muscles, *Sci. Prog., Lond.*, 49, 329–339.

—— (1963), The origin and function of the pupal stage, *Proc. R. ent. Soc. Lond.* (A), 38, 77–85.

—— (1971), Some neglected phases in metamorphosis, *Proc. R. ent. Soc. Lond.* (C), 35, 55–64.

—— (1973), Neglected phases in metamorphosis: a reply to V. B. Wigglesworth, *J. Ent.* (A), 48, 57–68.

HUXLEY, J. S. (1932), *Problems of Relative Growth*, Methuen, London, 276 pp.

ILAN, J. AND ILAN, J. (1973), Protein synthesis and insect morphogenesis, *A. Rev. Ent.*, 18, 167–182.

IVANOVA-KAZAS, O. M. AND IVANOVA, N. A. (1964), Metamorphosis of *Pontania capreae* L. (Hymenoptera, Tenthredinidae). I. Hypodermis, *Ent. Rev.*, 43, 157 166.

JACKSON, C. H. N. (1948), The eclosion of tsetse (*Glossina*) larvae (Diptera), *Proc. R. ent. Soc. Lond.* (A), 23, 36–38.

JAMES, H. C. (1928), On the life histories and economic status of certain Cynipid parasites of Dipterous larvae with descriptions of some new larval forms, *Ann. appl. Biol.*, 15, 287–316.

JENKIN, P. M. AND HINTON, H. E. (1966), Apolysis in arthropod moulting cycles, *Nature*, 211, 871.

JOBLING, B. (1939), On the African Streblidae (Diptera, Acalypterae) including the morphology of the genus *Ascodipteron* Adens. and a description of a new species, *Parasitology*, 31, 147–165.

KATZENELLENBOGEN, B. S. AND KAFATOS, F. C. (1970), Some properties of silkmoth moulting gel and moulting fluid, *J. Insect Physiol.*, 16, 2241–2256.

—— (1971), Proteinases of silkmoth moulting fluid: physical and catalytic properties, *J. Insect Physiol.*, 17, 775–800.

KEILIN, D. (1915), Recherches sur les larves de Diptères cyclorrhaphes, *Bull. sci. Fr. Belg.*, 49, 15–198.

KIM, C. W. (1959), The differentiation centre inducing the development from larval to adult leg in *Pieris brassicae* (Lepidoptera), *J. Embryol. exp. Morph.*, 7, 572–582.

KINGSLEY, V. V. (1972), Persistence of intestinal bacteria in the developmental stages of the monarch butterfly (*Danaus plexippus*), *J. Invert. Path.*, 20, 51–58.

KOEPPE, J. K. AND GILBERT, L. I. (1973), Immunochemical evidence for the transport of haemolymph protein into the cuticle of *Manduca sexta*. *J. Insect Physiol.*, 19, 615–624.

KRUEL, W. (1962), Ein didaktisches System der Insektenlarven auf entwicklungsgeschichtlicher Grundlage nach Massgabe essentieller und adaptiver Gruppenmerkmale, *Wanderversamml. dt. Ent.*, 9, (45), 89–100.

KUSKE, G. (1963), Untersuchungen zur Metamorphose der Schmetterlingsbeine, *Arch. EntwMech. Org.*, 154, 354–377.

LAWRENCE, P. A. (1970), Polarity and patterns in the postembryonic development of insects, *Adv. Insect Physiol.*, **7**, 197–266.

—— (1973), The development of spatial patterns in the integument of insects, In: Counce, S. J. and Waddington, C. H. (eds), *Developmental Systems: Insects*, **2**, 157–209.

LAWRENCE, P. A., CRICK, F. H. C. AND MUNRO, M. (1972), A gradient of positional information in an insect, *Rhodnius prolixus*, *J. Cell Sci.*, **11**, 815–853.

LAWRENCE, P. A. AND HAYWARD, P. (1971), The development of a simple pattern: spaced hairs in *Oncopeltus fasciatus*, *J. Cell Sci.*, **8**, 513–524.

LEES, A. D. (1955), *The Physiology of Diapause in Arthropods*, Cambridge, 150 pp.

L'HELIAS, C. (1970), Chemical aspects of growth and development in insects, In: Florkin, M. and Scheer, B. T. (eds), *Chemical Zoology*, **5**, 343–393.

LOCKE, M. (1966), Isolation membranes in insect cells at metamorphosis, *J. Cell Biol.*, **31**, 132 A.

—— (1969a), The ultrastructure of the oenocytes in the molt/intermolt cycle of an insect, *Tissue and Cell*, **1**, 103–154.

—— (1969b), The structure of an epidermal cell during the development of the protein epicuticle and the uptake of molting fluid in an insect, *J. Morph.*, **127**, 7–39.

—— (1974), The structure and formation of the integument in insects, In: Rockstein, M. (ed.), *The Physiology of Insecta*, 2nd edn, Academic Press, New York, **6**, 124–213.

LOCKSHIN, R. A. AND WILLIAMS, C. M. (1964–65), Programmed cell death. I–V, *J. Insect Physiol.*, **10**, 643–649; **11**, 123–133; 601–610; 803–809; 831–844.

LOUVET, J. P. (1974), Observation en microscopie électronique des cuticules édifiées par l'embryon, et discussion du concept de 'mue embryonnaire' dans le cas du phasme *Carausius morosus* Br. (Insecta, Phasmida), *Z. Morph. Ökol. Tiere*, **78** (1–2), 159–179.

LUCHT-BERTRAM, E. (1961), Das postembryonale Wachstum von Hirnteilen bei *Apis mellifica* L. und *Myrmeleon europaeus* L., *Z. Morph. Ökol. Tiere*, **50**, 543–575.

LUDWIG, D. AND ABERCROMBIE, W. F. (1940), The growth of the head capsule of the Japanese beetle larva, *Ann. ent. Soc. Am.*, **33**, 385–390.

MANSOUR, K. (1927), The development of the larval and adult mid-gut of *Calandra oryzae* (Linn.): the rice weevil, *Q. Jl microsc. Sci.*, **71**, 313–352.

—— (1934), The development of the adult mid-gut of coleopterous insects and its bearing on systematics and embryology, *Bull. Fac. Sci. Egyptian Univ.*, **2**, 34 pp.

MCLAUGHLIN, B. J. (1974), Fine-structural changes in a Lepidopteran nervous system during metamorphosis, *J. Cell Sci.*, **14**, 369–387.

MIALL, L. C. AND HAMMOND, A. R. (1892), The development of the head of the imago of *Chironomus*, *Trans. Linn. Soc. Lond.*, **5**, 265–279.

MOREAU, R. (1974), Variations de la pression interne au cours de l'émergence et de l'expansion des ailes chez *Bombyx mori* et *Pieris brassicae*, *J. Insect Physiol.*, **20**, 1475–1480.

MORIARTY, F. (1959), The 24-hr rhythm of emergence of *Ephestia kühniella* Zell. from the pupa, *J. Insect Physiol.*, **3**, 357–366.

MOUZE, M. (1975), Croissance et regeneration de l'œil de la larve d'*Aeshna cyanea* Müll. (Odonate, Anisoptère), *Arch. EntwMech. Org.*, **176**, 267–283.

MÜLLER, H. J. (1970), Formen der Dormanz bei Insekten, *Nova Acta Leopoldina*, **35**, 1–27.

MUNSCHEID, L. (1933), Die Metamorphose des Labiums der Odonaten, *Z. wiss. Zool.*, **143**, 201–240.

MURPHY, C. (1971), Localization of primordia within the dorsal mesothoracic disc of *Drosophila*, *J. exp. Zool.*, **179**, 51–62.

MURRAY, F. V. AND TIEGS, O. W. (1935), The metamorphosis of *Calandra oryzae*, *Q. Jl microsc. Sci.*, **77**, 405–495.

NORDLANDER, R. H. AND EDWARDS, J. S. (1969–70), Postembryonic brain development in the monarch butterfly, *Danaus plexippus plexippus* L. I–III, *Arch. EntwMech. Org.*, **162**, 197–217; **163**, 197–220; **164**, 247–260.

NÜESCH, H. (1965), Die Imaginalentwicklung von *Antheraea polyphemus* Cr. (Lepidoptera), *Zool. Jb.* (*Anat.*), **82**, 393–418.

PASSONNEAU, J. V. AND WILLIAMS, C. M. (1953), The moulting fluid of the Cecropia silkworm, *J. exp. Biol.*, **30**, 545–560.

PATAY, R. (1939), Contribution à l'étude d'un Coléoptère (*Leptinotarsa decemlineata* (Say)). Evolution des organes au cours du développement, *Bull. Soc. sci. Bretagne*, **16**, 1–145.

PAULIAN, R. (1950), La vie larvale des insectes, *Mém. Mus. natn. Hist. nat., Paris*, **30**, 1–206.

PÉREZ, C. (1910), Recherches histologiques sur la métamorphose des Muscides (*Calliphora erythrocephala* Mg.), *Archs Zool. exp. gén.* (5), **4**, 1–274.

—— (1911), Disques imaginaux des pattes chez le *Phytonomus adspersus* Fabr, *C.r. Soc. biol., Paris*, **71**, 498–501.

PERISTIANIS, G. C. AND GREGORY, D. W. (1971), Early stages of flight-muscle development in the blowfly *Lucilia cuprina*: a light- and electron-microscopic study, *J. Insect Physiol.*, **17**, 1005–1022.

PETERS, W. (1961), Die sogenannten Fussstummelsinnesorgane der Larven von *Calliphora erythrocephala* Mg. (Diptera), *Zool. Jb.* (*Anat.*), **79**, 339–346.

PETERSON, A. (1948–51), *Larvae of Insects. An Introduction to Nearctic Species*. I, II, The author, Columbus, 315 pp.; 416 pp.

PIPA, R. L. (1963), Studies on the hexapod nervous system. IV. Ventral nerve cord shortening: a metamorphic process in *Galleria mellonella* (L.) (Lepidoptera, Pyralidae), *Biol. Bull.*, **124**, 293–302.

—— (1967), Insect neurometamorphosis. III. Nerve cord shortening in a moth *Galleria mellonella* (L.) may be accomplished by humeral potentiation of neuroglial mobility, *J. exp. Zool.*, **164**, 47–60.

PIPA, R. L. AND WOOLEVER, P. S. (1965), Insect neurometamorphosis, I, II, *Z. Zellforsch.*, **63**, 405–417; **68**, 80–101.

POHLEY, H.-J. (1959), Über das Wachstum des Mehlmottenflügels unter normalen und experimentellen Bedingungen, *Biol. Zbl.*, **78**, 232–250.

POODRY, C. A. AND SCHNEIDERMAN, H. A. (1970), The ultrastructure of the developing leg of *Drosophila melanogaster*, *Arch. EntwMech. Org.*, **166**, 1–44.

—— (1971), Intercellular adhesivity and pupal morphogenesis in *Drosophila melanogaster*, *Arch. EntwMech. Org.*, **168**, 1–9.

POSTLETHWAIT, J. H. AND SCHNEIDERMAN, H. A. (1970), Induction of metamorphosis by ecdysone analogues: *Drosophila* imaginal discs cultured in vivo, *Biol. Bull.*, **138**, 47–55.

POYARKOFF, E. (1910), Recherches histologiques sur la métamorphose d'un Coléoptère (la Galéruque de l'orme), *Arch. Anat. microsc.*, **12**, 333–474.

—— (1914), Essai d'une theorie de la nymphe des insectes holométaboles, *Arch. zool. Paris*, **54**, 221–265.

PRATT, H. S. (1900), The embryonic history of imaginal discs in *Melophagus ovinus* L., together with an account of the earlier stages in the development of the insect, *Proc. Boston Soc. nat. Hist.*, **29**, 241–272.

RICHARDS, O. W. (1949), The relation between measurements of the successive instars of insects, *Proc. R. ent. Soc. Lond.* (A), **24**, 8–10.

ROBERTSON, C. W. (1936), The metamorphosis of *Drosophila melanogaster*, including an accurately timed account of the principal morphological changes, *J. Morph.*, **59**, 351–400.

ROBERTSON, J. AND PIPA, R. (1973), Metamorphic shortening of interganglionic connectives of *Galleria mellonella* (Lepidoptera): in vitro stimulation by ecdysone analogues, *J. Insect Physiol.*, **19**, 673–679.

SAMSON, K. (1908), Über das Verhalten der Vasa Malpighii und die exkretorische Funktion der Fettzellen während der Metamorphose von *Heterogenea limacodes* Hufn., *Zool. Jb.* (*Anat.*), **26**, 403–422.

SCHMIDT, G. H. (1961), Sekretionsphasen und cytologische Beobachtungen zur Funktion der Oenocyten während der Puppenphase verschiedener Kasten und Geschlechter von *Formica polyctena* Foerst (Ins. Hym. Form.), *Z. Zellforsch.*, **55**, 707–723.

SCHOELLER, J. (1964), Recherches descriptives et expérimentales sur la céphalogenèse de *Calliphora erythrocephala* (Meigen), au cours des développements embryonnaire et postembryonnaire, *Archs Zool. exp. gén.*, **103**, 1–216.

SCHWAGER-HÜBNER, M. (1970), Untersuchungen über die Entwicklung des thorakalen Nerven-Muskel-Systems bei *Apis mellifera* L, *Revue suisse Zool.*, **77**, 807–849.

SCUDDER, G. G. E. (1971), The post-embryonic development of the indirect flight-muscles in *Cenocorixa bifida* (Hung.) (Hem., Corixidae), *Can. J. Zool.*, **49**, 1387–1398.

SCUDDER, G. G. E. AND HEWSON, R. J. (1971), The postembryonic development of the indirect flight muscles in *Oncopeltus fasciatus* (Dallas) (Hemiptera: Lygaeidae), *Can. J. Zool.*, **49**, 1377–1386.

SIKES, E. K. AND WIGGLESWORTH, V. B. (1931), The hatching of insects from the egg and the appearance of air in the tracheal system, *Q. Jl microsc. Sci.*, **74**, 165–192.

SILVESTRI, F. (1904), Contribuzione alla conoscenza della metamorfosi e dei costumi della *Lebia scapularis* Fourc. con descrizione dell'apparato sericiparo della larva, *Redia*, **2**, 63–83.

SNODGRASS, R. E. (1947), The insect cranium and the 'epicranial suture', *Smithson. misc. Collns*, **107** (7), 52 pp.

—— (1953), The metamorphosis of a fly's head, *Smithson. misc. Collns*, **122** (3), 25 pp.

—— (1954), Insect metamorphosis, *Smithson. misc. Collns*, **122** (9), 124 pp.

—— (1961), The caterpillar and the butterfly, *Smithson. misc. Collns*, **143** (6), 1–51.

TAUBER, M. J. AND TAUBER, C. A. (1976), Insect seasonality: diapause maintenance, termination and postdiapause development, *A. Rev. Ent.*, **21**, 81–107.

TAYLOR, H. M. AND TRUMAN, J. W. (1974), Metamorphosis of the abdominal ganglia of the tobacco hornworm, *Manduca sexta*. Changes in populations of identified motor neurons, *J. comp. Physiol.*, **90**, 367–388.

TAYLOR, R. L. AND RICHARDS, A. G. (1965), Integumentary changes during moulting of arthropods with special reference to the subcuticle and ecdysial membrane, *J. Morph.*, **116**, 1–22.

TEISSIER, G. (1936), La loi de Dyar et la croissance des Arthropods, *Livre jubil. Bouvier*, Paris, 335–342.

THIELE, H. U. (1973), Remarks about Mansingh's and Müller's classification of dormancies in insects, *Can. Ent.*, **105**, 925–928.

TIEGS, O. W. (1922), Researches on the insect metamorphosis, I, II, *Trans. Proc. R. Soc. S. Aust.*, **46**, 319–527.

—— (1955), The flight muscles of insects – their anatomy and histology, with some observations on the structure of striated muscle in general, *Phil. Trans. R. Soc. Ser. B*, **238**, 221–348.

TRAGER, W. (1937), Cell size in relation to growth and metamorphosis of the mosquito *Aedes aegypti*, *J. exp. Zool.*, **76**, 467–489.

TRUMAN, J. W. (1971–73), Physiology of insect ecdysis. I–III, *J. exp. Biol.*, **54**, 805–814; *Biol. Bull. mar. biol. Lab., Woods Hole*, **144**, 200–211; *J. exp. Biol.*, **58**, 821–829.

URSPRUNG, H. (1959), Fragmentierungs- und Bestrahlungsversuche zur Bestimmung vom Determinationszustand und Anlageplan der Genitalscheiben von *Drosophila melanogaster*, *Arch. EntwMech. Org.*, **151**, 504–558.

URSPRUNG, H., CONSCIENCE-EGLI, M., FOX, D. J. AND WALLIMAN, T. (1972), Origin of leg musculature during *Drosophila* metamorphosis, *Proc. natn. Acad. Sci. U.S.A.*, **69**, 2812–2813.

URSPRUNG, H. AND NÖTHIGER, R. (eds) (1972), *The Biology of Imaginal Disks*, Springer, Berlin, 172 pp.

VERSON, E. (1905), Zur Entwicklung des Verdauungscanal bei *Bombyx mori*, *Z. wiss. Zool.*, **82**, 523–600.

WEBER, H. (1933), *Lehrbuch der Entomologie*, Fischer, Jena, 726 pp.

—— (1939), Zur Eiablage und Entwicklung der Elefantenlaus *Haematomyzus elephantis* Piaget, *Biol. Zbl.*, **59**, 98–109; 397–409.

WEISMANN, A. (1864), Die nachembryonale Entwicklung der Musciden nach Beobachtungen an *Musca vomitoria* und *Sarcophaga carnaria*, *Z. wiss. Zool.*, **14**, 187–336.

—— (1866), Die Metamorphose der *Corethra plumicornis*, *Z. wiss. Zool.*, **16**, 45–127.

WHITE, R. H. (1961), Analysis of the development of the compound eye in the mosquito, *Aedes aegypti*, *J. exp. Zool.*, **148**, 223–237.

WIESEND, P. (1957), Die postembryonale Entwicklung der Thoraxmuskulatur bei einigen Feldheuschrecken mit besonderer Berücksichtigung der Flugmuskeln, *Z. Morph. Ökol. Tiere*, **46**, 529–570.

WIGGLESWORTH, V. B. (1954), *The Physiology of Insect Metamorphosis*, Cambridge Univ. Press, Cambridge, 152 pp.

—— (1959), *The Control of Growth and Form; A Study of the Epidermal Cell in an Insect*, Cornell Univ. Press, Ithaca, New York, 136 pp.

—— (1972), *The Principles of Insect Physiology*, Chapman & Hall, London, 7th edn, 827 pp.

—— (1973a), The role of the epidermal cells in moulding the surface pattern of the cuticle in *Rhodnius* (Hemiptera), *J. Cell Sci.*, **12**, 683–705.

WIGGLESWORTH, V. B. (1973*b*), The significance of 'apolysis' in the moulting of insects, *J. Ent.* (A), **47**, 141–149.

WILLIAMS, C. M. AND KAFATOS, F. C. (1971), Theoretical aspects of the action of juvenile hormone, *Bull. Soc. ent. Suisse*, **44**, 151–162.

WILLIS, J. J. (1974), Morphogenetic action of insect hormones, *A. Rev. Ent.*, **19**, 97–115.

WYATT, G. R. (1972), Insect hormones, In: Litwack, G. (ed.), *Biochemical Actions of Hormones*, New York, **2**, 385–490.

YOUNG, D. (ed.) (1973), *Developmental Neurobiology of Arthropods*, Cambridge Univ. Press, Cambridge, 263 pp.

YOUNG, E. C. (1965), Teneral development in British Corixidae, *Proc. R. ent. Soc. Lond.* (A), **40**, 159–168.

INDEX

Names of authors are in small capitals. Generic names are in italics. Synonyms are indicated by a cross-reference (e.g., *Calotermes*, see *Kalotermes*). Page numbers in heavy type denote illustrations.

abdomen, 26, 73 ff.; appendages of, **74** ff., 333 ff.; of embryo, 333; endoskeleton, 85; musculature, 93, **94**; segmentation, 73
abdominal gill, 226
ABEL, 262
ABERCROMBIE, 364, 385, 390
ABRAHAM, 103, 115
Abricta, sound production, 184
absorption, in gut, 201, 203
accessory antennal nerve, 106
accessory glands, female, 298; male, 292 ff.
accessory hearts, 237 ff.
acephalous larva, 360
acetylcholine, 112
acetylcholine esterase, 112
Acherontia, larval mandibular gland, **267**; sound production, 185
Acheta, tympanum, 133
Acilius, ocellus, 151
acinus, 264
acone eye, 155
acoustic interneuron, 113
ACREE, 144, 162
Acrididae, flight muscles, **93**; food-plant selection, 139; hind leg, **180**; tympanum, 131, 132
acron, 37
Acronycta, diapause, 384
acrotergite, 43
acrotrophic ovariole, **296**
actin, 87
action potential, 112, 124
activation centre, 345
ACTON, 93, 96
ADAMS, J. R., 136, 138, 139, 162
ADAMS, W. B., 162
adecticous pupa, 366, 368
adenosine triphosphate, 96
adenotrophic viviparity, 302
adenyl cyclase, 383
adepithelial cells, 382
adipohaemocyte, 241

adipokinetic hormone, 274
ADIYODI, K. G., 288, 309
ADIYODI, R. G., 288, 309
adult, development of, 371; emergence from cocoon, 368
adventitia, of heart, 236
aedeagus, 78, 291
Aedes, fat-body, 254; growth, 364; hearing, 137; host location, 144; humidity receptors, 147; intersexes, 289; larval ocellus, 151; pigmentation, 17
aerial plankton, 66
aerodynamics, of flight, 64
aeropyle, 300
aeroscopic plate, 228
Aeshna, emergence, **370**; flight-speed, 65; larval heart, 237; larval velocity receptors, 129
afferent neuron, 101
AFZELIUS, 218, 229
Ageniaspis, cleavage, 324; embryo chain, **307**; polyembryony, 305, **306**
AGGARWAL, 275, 278, 282, 313
aggregation, 143
aggregation pheromone, 142
Agrion, embryonic development, 330; gills, 227
Agrius, accessory heart, 238; circulatory system, **238**
AGUI, 383, 385
AHEARN, 224, 229
AIDLEY, 94, 96, 184, 187
air-flow receptor, 125, 129
air sac, 131, 221ff.; functions of, 222
air store, 225
AKAI, 241, 244
alarm pheromone, 142, 270
alary muscles, 234
alary polymorphism, 93
ALDRICH, 99
Aleochara, hypermetamorphosis, 360

ALEXANDER, N. J., 200, 204
ALEXANDER, R. D., 67, 179, 181, 187, 301, 309
ALEXANDER, R. M., 50, 67
ALI, 380, 385
ALICATA, 89, 97
aliform muscles, 234
alimentary canal, 192ff., **194**, **198**; embryology of, 339 ff.; metamorphosis, 378; water movement in, **203**, see also absorption.
allantoic acid, 251
allantoin, 251
allergy, 272
Alloeorhynchus, haemocoelic fecundity, 302
allometry, 364
Alloperla, thoracic endoskeleton, 83
ALLOWAY, 114, 115
alpha-lobe, of brain, 103
ALTMANN, 200, 204
ALTNER, 255, 259
Amicroplus, polyembryony, 305
amino-acids, in diet, 200; of blood, 239; excretion of, 251
aminopeptidases, 201
amnion, 329
amnioserosal sac, 335
amniotic cavity, **329**
amniotic fluid, 332, 353
amniotic folds, 329
amphipneustic respiratory system, 210
amphitoky, 303
amplitude modulation, 180, 182
AMY, 328, 338, 347
amylase, 268
anal angle, of wing, 51
anal gill, 228
anal vein, 60
anatrepsis, 331
Anax, nerve roots, 107
ANDERSEN, L. W., 147, 162
ANDERSEN, S. O., 14, 20